Forward Recoil Spectrometry

Applications to Hydrogen Determination in Solids

Forward Recoil Spectrometry

Applications to Hydrogen Determination in Solids

Jorge Tirira

SECTOR
Courtabœuf, France

Yves Serruys

DTA/CEREM/DECM/SRMP
CEA Saclay, France

and
Patrick Trocellier

CEA–CNRS Laboratoire Pierre Sue
DSM/DRECAM
CEA Saclay, France

Plenum Press • New York and London

Library of Congress Cataloging-in-Publication Data

Tirira, Jorge.
    Forward recoil spectrometry : applications to hydrogen
  determination in solids / Jorge Tirira and Yves Serruys and Patrick
  Trocellier.
        p.   cm.
    Includes bibliographical references and index.
    ISBN-13:978-1-4613-8012-2   e-ISBN-13:978-1-4613-0353-4
    DOI:10.1007/978-1-4613-0353-4
    1. Backscattering--Measurement.   2. Solids--Spectra--Measurement.
  3. Materials--Hydrogen content--Measurement.   I. Serruys, Yves.
  II. Trocellier, Patrick. III. Title.
  QC794.6.S3T57 1996
  543'.0873--dc20   96-32552
                                                              CIP

ISBN-13:978-1-4613-8012-2

© 1996 Plenum Press, New York
Softcover reprint of the hardcover 1st edition 1996

A Division of Plenum Publishing Corporation
233 Spring Street, New York, N. Y. 10013

To unemployed scientists

Foreword

The practical properties of many materials are dominated by surface and near-surface composition and structure. An understanding of how the surface region affects material properties starts with an understanding of the elemental composition of that region. Since the most common contaminants are light elements (for example, oxygen, nitrogen, carbon, and hydrogen), there is a clear need for an analytic probe that simultaneously and quantitatively records elemental profiles of all light elements. Energy recoil detection using high-energy heavy ions is unique in its ability to provide quantitative profiles of light and medium mass elements. As such this method holds great promise for the study of a variety of problems in a wide range of fields.

While energy recoil detection is one of the newest and most promising ion beam analytic techniques, it is also the oldest in terms of when it was first described. Before discussing recent developments in this field, perhaps it is worth reviewing the early days of this century when the first energy recoil detection experiments were reported.

In a paper published in two parts in 1914 and 1915, E. Marsden, a researcher from New Zealand working in Rutherford's laboratory in Manchester, described a series of experiments on *"Passage of Particles Through Hydrogen."* These papers are interesting not only because they describe the first observation of protons recoiling from targets bombarded with α particles, but they also introduce many of the techniques rediscovered in recent years to identify recoiling particles and lower experimental backgrounds. Marsden was interested in testing the model of the atom recently proposed by Rutherford. He states:

> In the case of the encounter of an α particle of velocity V with the nucleus of the hydrogen atoms, if the latter is projected in a direction making an angle θ^1 with the original direction of the α particle, then the velocity of the "H" particle is given by... u = 1.6 V cos θ ... in an end-on collision, i.e. θ = 0, the "H" particle will have about four times the range of the α particle producing it.

From these few words, it is clear that Marsden in 1914 understood two-body kinematics and how to use energy loss to identify recoiling particles and exclude bombarding ions. Although we now know that it does not apply in his particular case, Marsden also knew the angle-dependent recoil yield based on Rutherford's cross section. Perhaps more surprising, Marsden used magnetic filters to reduce the background at his scintillation detector screen: *"By placing the apparatus in a strong magnetic field, the β-rays were prevented from striking the screen."* While Marsden's goal was to test the fundamental theories of his day, this led him to materials analysis. Quoting again from the second of these papers:

> An estimate was therefore made of the amount of hydrogen which would be required to produce the number of "H" particles observed. A film of wax, of thickness about 10 μ, . . . was placed round the α-ray tube and comparative measurements were made of the number of "H" particles with and without the wax.

By using this wax film as a standard, he was surprised at the amount of hydrogen his measurement showed to be present in the thin glass that covered the α source.

All of the preceding is strikingly similar to methods used today in modern energy recoil detection analysis. Marsden's observation of a high hydrogen content in a nominally hydrogen-free material was also rediscovered by many modern practitioners of ion beam analysis. The fact that so much could be done in the days before electronics, accelerators, energy-sensitive detectors instills us with both humility and inspiration. Humility that all this was done so long ago and inspiration that we should be able to do much more with modern equipment and just a fraction of Marden's wits.

After publication of these papers by Marsden, the concept of using particle beams for materials analysis was lost in the excitement associated with the development of our modern understanding of atomic structure and quantum mechanics in the 1920s and 1930s. In the 1950s Sylvan Rubin began developing backscattering as an analytical probe, which under the later leadership of Jim Mayer, Jim Ziegler, Georges Amsel, and others led to the remarkable expansion of ion beam techniques, most importantly backscattering, in a wide range of different disciplines.

Modern energy recoil detection analysis did not appear in the published literature until B. Cohen and coworkers published a paper in 1972 on use of high-energy (17-MeV) protons in recoil geometry as a probe for hydrogen in thin foils. This analytic procedure may have been inspired by a practice among nuclear physicists of accurately determining laboratory scattering angles by measuring the energy of elastically recoiled protons observed to come from any thin film target. Cohen was able to demonstrate part per million sensitivity for hydrogen in thin foils. In spite of its potential, Cohen's method did not find wide use in practical analysis.

In 1979 B. Doyle and P. Peercy pointed out that hydrogen can easily be measured by recoil using He beams similar to those commonly used in Rutherford backscattering. This approach was the first recoil technique widely adopted at laboratories around the world.

In all of the preceding, forward recoil is used as a probe for hydrogen in a target. While analysis for hydrogen is important, energy recoil detection analysis's future is tied to its ability to provide simultaneous quantitative concentration profiles for all light- and medium-mass elements in any target. It is clear that energy recoil detection has uniquely demonstrated its ability to satisfy this important analytic need.

The first to demonstrate energy recoil detection as a technique for simultaneously measuring profiles of all light elements in a target were J. L'Ecuyer and his coworkers at the University of Montreal when in 1976, they used a 30-MeV ^{35}Cl beam to detect elements from hydrogen through oxygen in thin targets. In their paper the identity of the recoiling elements was inferred from two-body kinematics. Thereafter R. Groleau and coworkers, and J. P. Thomas and coworkers demonstrated the use of time of flight to identify recoiling elements and record elemental profiles for all light elements simultaneously even in cases where kinematics alone could not identify the elements.

In recoil analysis there is always a need to detect recoiling nuclei in the intense flux of elastically scattered beam particles. The most commonly used method then (and now) is to exclude the elastic particles by placing an absorber in front of the detector. This absorber must be thick enough to exclude the elastic particles while including recoils. This method has the disadvantage of reducing depth resolution due to energy straggle in the absorber. G. Ross was quick to realize the advantage of using a purely electromagnetic filter that could separate recoils from elastics without introducing increased straggle in the recoils.

These important papers were the basis of modern energy recoil detection analysis. The present book records for the first time in a single coherent work all the detailed theory and practice required to use energy recoil detection to analyze hydrogen and other light- and medium-mass elements. The authors have paid special attention to the important question of when the recoil cross section is Rutherford, as well as to such practical issues, as data-processing methods and ion-beam-induced damage in solids. They describe the rich array of spectroscopic methods used in recoil analysis, from the conventional surface barrier detector to coincidence methods, to time of flight, *ExB*, and energy-loss detector telescopes. Care has been taken to provide many illustrations of the analytic capabilities of energy recoil detection. As such this work is likely to become a major contributor to the continued development of this approach in materials analysis. It is a resource for both researchers and students who need to evaluate this method against the needs of their particular analytic problems.

Because determining hydrogen in solids has been one of the major driving forces in the development of energy recoil detection, the authors include a chapter on applications of recoil analysis to hydrogen determination in solids and a comparative chapter on nuclear reaction analysis for hydrogen in materials.

The authors, J. Tirira, Y. Serruys, and P. Trocellier are all major contributors to ion beam analysis who write with authority on this subject. To this group, where appropriate, other expert authors have been added. These contributors include N. Dytlewski, G. Ross, and H. Hofsäss. This approach has resulted in a book that covers even the most specialized subtopic with both authority and uniformity of presentation.

Publication of a major book that intends to relay to a wide audience only the status of a developing field sometimes has unanticipated and important effects on that field.

A classic example is the publication in 1978 of *Backscattering Spectrometry* by W. K. Chu, J. W. Mayer, and M. A. Nicolet. That book became both the bible and the handbook for a whole generation of new practitioners of Rutherford backscattering and as such greatly speeded up the development of this technique. This present book on *Energy Recoil Detection* may well have a similar impact.

William A. Lanford

State University of New York
Albany

1. Note that in the following chapters, θ is used for scattering angle and φ for recoil.

BIBLIOGRAPHY

Chu, W.-K., Mayer, J. W., and Nicolet, M.-A., Backscattering Spectrometry (Academic Press, New York 1978).

Cohen, B. L., Fink, C. L., and Degnan, H., Nondestructive analysis for trace amounts of hydrogen, *J. Appl. Phys.* **43**, 19–25 (1972).

Doyle, B. L., and Peercy, P. S., Technique for profiling H with 2.5-MeV Van De Graaff accelerators, *Appl. Phys. Lett.* **34**, 811–13 (1979).

Groleau, R., Gujrathi, S. C., and Martin, J. P., Time-of-flight system for profiling recoiled light elements, *Nucl. Instrum. Methods Phys. Res.* **218**, 11–15 (1983).

L'Ecuyer, J., Brassard, C., Cardinal, C., Chabbal, J., Deschenes, L., Labrie, J. P., Terreault, B., Martel, J. G., and St. Jacques, R., An accurate and sensitive method for the determination of the depth distribution of light elements in heavy materials, *J. Appl. Phys.* **47**, 381–82 (1976).

Marsden, E., The passage of particles through hydrogen, *Philosophical Mag.* **27**, 824–30 (1914).

Marsden, E., and Lantsberry, W. C., The passage of particles through hydrogen-II, *Philosophical Mag.* **30**, 240–43 (1915).

Ross, G. G., Terreault, B., Gobeil, G., Abel, G., Boucher, C., and Veilleux, G., Inexpensive, quantitative hydrogen depth profiling for surface probes, *J. Nucl. Mat.* **128/129**, 730–33 (1984).

Thomas, J. P., Fallavier, M., Ramdane, D., Chevarier, N., and Chevarier, A., High resolution depth profiling of light elements in high atomic mass materials, *Nucl. Instrum. Methods Phys. Res.* **218**, 125–28 (1983).

Preface

Why an entire book dedicated to elastic recoil detection analysis? Why did *we* write it? These questions are not amenable to scientific discussion. Yet there are some objective reasons as an answer to the first question. Time has come in 1995 to make a detailed review of the state of knowledge about elastic recoil process and its applications nearly 20 years after the first published papers in this field. Unlike Rutherford-backscattering spectrometry and proton-induced X-ray emission, a handbook and practical guide for elastic recoil practitioners do not exist.

Was there any reason for *us* to fill up this vacancy? No determining one whatsoever except that someone had to do it. Yet it was somewhat frightening to attempt what more qualified scientists might do much better. Remembering the exhortation of the poet:

> Freue dich und treibe neue Triebe
> Gib dich hin und fürcht das Leben nicht,[1]

we decided to try and write this book.

Naturally difficulties immediately arose, and two of them appeared as unavoidable in such a task. First we had to arbitrate between contradictory arguments, and sometimes we felt that no peremptory argument was acceptable but rather that we had to introduce lights and shades depending on the point of view. *"Ne dites pas: 'J'ai trouvé la vérité,' mais plutôt 'J'ai trouvé une vérité,'"*[2] Second a mere review of present knowledge would have been disappointing. Trying to give some insight into the limits of present theories or into still controversial questions, we encountered the difficulty of giving useful advice while standing on moving sands. *"Aucun homme ne peut rien vous révéler, sinon ce qui repose à demi endormi dans l'aube de votre connaissance."*[3]

Our ambition—perhaps an "uncommon want"[4]—was to make this book useful for students learning the physics of ion beam interactions with condensed matter as well as for young scientists needing to carry out elastic recoil spectrometry measurements and for confirmed scientists wishing to know about more recent developments of the technique. We hope such a purpose does not require too much patience from every reader.

Above all writing this book has been a rich experience. Three French authors and three collaborators abroad have learned a lot from each other and come face-to-face with the limits of their knowledge and their pedagogic capabilities. *"Docere vos coepi quod ipse non didici. Itaque factum est ut prius docere inciperem, quam discere. Discendum igitur mihi simul et docendum est."*[5] This exceptional opportunity for scientific exchange has been an unquestionable positive result of our undertaking.

Now this book stands before the trial of readers. "Be thoroughly attentive to the people's judgement, the even-handed judgement."[6] Let this somewhat idealistic citation be an incentive to readers to provide authors and contributors with comments and suggestions to improve the quality of this book and—let us once again be optimistic—of further revised versions.

The authors

Saclay

1. "Rejoice and shoot new buds, give yourself up to action and fear not life." H. Hesse, *Sprache des Frühlings* (1952) In: *Herman Hesse* (edited by A. Piot) (Compagnons du Devoir du Tour de France, Paris).
2. "Do not say 'I have found the truth,' but rather 'I have found a truth.'" K. Gibran, *Le Prophète* (Casterman, Paris, 1956).
3. "No man can reveal anything to you, be it not what already lies half-asleep in the dawn of your knowledge." K. Gibran, *Le Prophète* (Casterman, Paris, 1956).
4. Lord Byron. *Don Juan.* (1819). Canto the first, v. 1.
5. "I have begun to teach you what I did not learn. So it occurred that I had to begin to teach before having learnt. Thus I have simultaneously to learn and to teach." Saint Ambrose of Milan, (about 374–397), *De officiis* I, 1–4.
6. M. P. Mussorgsky, *Boris Godunov*, (1825). Act IV, sc. 2.

Contents

Chapter 5. Conventional Recoil Spectrometry
    With Hans Hofsäss

Chapter 6. Time of Flight ERDA
    With Nick Dytlewski

Chapter 7. Depth Profiling by Means of the ERDA *ExB* Technique
        With Guy Ross

Chapter 8. Recoil Spectrometry with a ΔE-E Telescope

1

Introduction

1.1. GENERAL DESCRIPTION

When a beam of positive ions, produced by a single-ended Van de Graaff accelerator, a tandem, or a cyclotron (typically in the energy range from 0.1–5 MeV/amu) hits the near-surface region of a solid target, most incident particles are elastically scattered by the target atoms. Only a small number of incident ions are able to induce nuclear reactions on isotopes of light elements contained in the target. Energy transferred from an incident ion to a target nucleus during their elastic collision can be large enough so that the target nucleus recoils from the target surface. This elastic recoil process is simply described using kinematic equations from the physics of elastic collision. *Elastic recoil detection analysis* (ERDA[*]) consists in detecting recoiling nuclei to acquire information about the target composition.

The principle of an elastic recoil experiment is quite simple. A beam of monoenergetic ions is directed perpendicularly to the surface of a target. When the target is a thin foil, almost all incident particles and recoils emerge at the back surface of the target. If the target is a thick sample, only particles with large enough energy can reach the back surface. In both cases recoils emerging at a given angle can be detected. This experiment, called *transmission ERDA*, is illustrated in Figure 1.1. Another elastic recoil experiment, *reflexion ERDA*, consists of an ion beam impinging at grazing incidence onto the target and detecting recoils emerging at the front surface, generally also at a grazing angle. This arrangement is illustrated in Figure 1.2. The same experimental arrangements can also be used to determine physical parameters of elastic scattering and energy loss.

---

[*]There are many names and acronyms for the same technique: ERD (elastic recoil detection), ERDA (elastic recoil detection analysis), ERS (elastic recoil spectrometry), FRD (forward-recoil detection), FRS (forward-recoil spectrometry) are the most widespread. A special session of the IBA-12 conference found a general agreement in favor of ERDA, with eventual complements as HI-ERDA for heavy ion-elastic recoil detection analysis, TOF-ERDA for time-of-flight-ERDA, and so on. We have adopted this convention throughout this book.

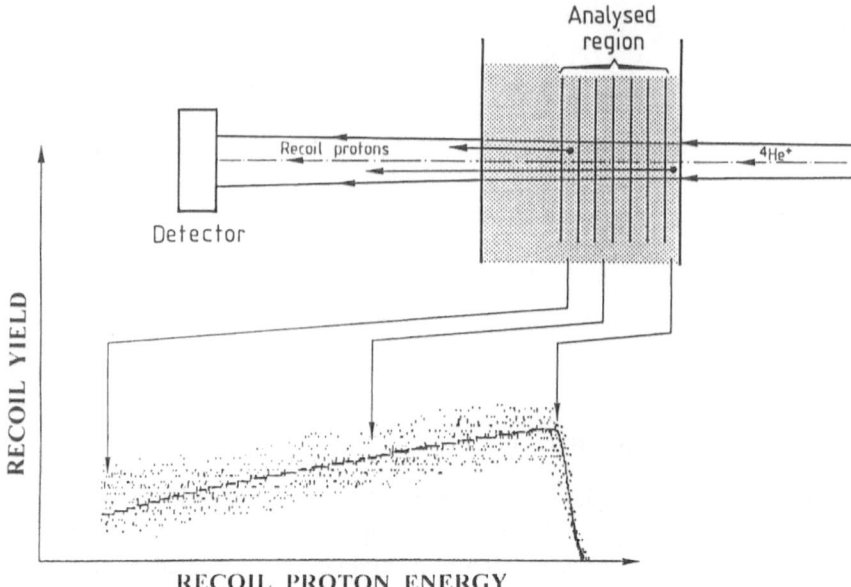

Figure 1.1. Schematic of a typical transmission ERDA experiment (hydrogen profiling with an helium beam) with the resulting spectrum in the particular case of detection at 0° angle and a target thick enough to stop completely incident particles.

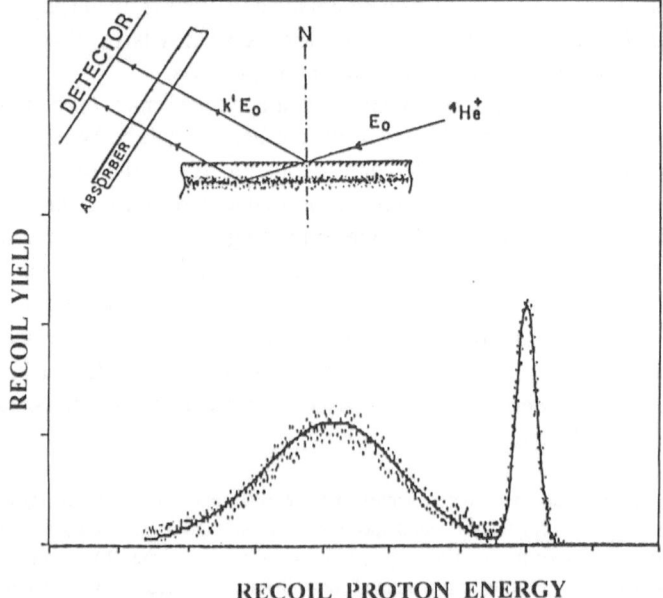

Figure 1.2. Typical arrangement for reflexion ERDA (hydrogen profiling with an helium beam) and resulting spectrum from an implanted target. (The thin peak is due to hydrogen adsorbed on the surface.)

Different detection systems are used to discriminate scattered projectiles and various recoil species. The most usual solution for profiling light elements with heavier incident ions consists in stopping these latter in an absorber foil placed in front of the detector. More sophisticated setups use electric or magnetic mass spectrometers; time of flight detection; electromagnetic filters (*ExB* filter); combined detectors, known as telescopes; or coincidence detection for discriminating different species without ambiguity.

However it is not sufficient to expose a sample to an ion beam under such conditions that recoiling nuclei escape through the back or front face and to place a detector in their path. The situation is more complex, and some other basic phenomena occur simultaneously with the elastic collision process. All of them must be examined in detail. For the moment let us just enumerate them and explain briefly their nature.

First of all we are not interested only in the nature of recoiling nuclei; that can be determined through *energy transfer* during collision characterizing their atomic mass. A quantitative analysis requires determining the number of nuclei of a given species present in the target, which is related to the number of detected nuclei—the *recoil yield*—through the *scattering cross section* that characterizes the probability that scattering events occur. Scattering cross sections involved in elastic recoil analysis are generally nonclassical and must be thoroughly studied. In fact the number of scattering events in a finite time period obeys a statistical distribution. The scattering cross section describes the mean value of this distribution, but statistical fluctuations of the recoil yield are also taken into account, and these are known as *counting fluctuations* or *counting noise*.

We also wish to determine the composition of the target as a function of depth. Indeed the energy of detected recoil nuclei depends on the *energy loss* experienced by impinging ions (also called projectiles) before collision events and by recoiling nuclei afterward. This energy loss results from interactions of moving ions with electrons and nuclei in the target. It is a complex stochastic process composed of many individual interactions. Thus the mean value of the energy loss can be described by the simple notion of a *stopping power* of the target for projectiles and recoils, but what is really observed is a statistical distribution of energy loss around the mean value defined by the stopping power. The energy spread due to energy loss fluctuations is called *energy straggling*, and it also requires a thorough examination.

Some other phenomena are still involved. As a consequence of the large number of interactions with electrons and nuclei in the target, ions are deflected from ideally straight trajectories before and after the scattering event that is the object of the ERDA experiment. The result of many small deflections experienced by recoils or scattered projectiles is globally described as *multiple scattering*. It consists in both *lateral* and *angular* particles *spreading*. This dispersion deserves much attention in what follows.

The detection system itself introduces additional straggling and multiple scattering in the eventual absorber foil. The detector response to the energy of detected particles also presents some dispersion, known as *detector resolution*. The detector response may detect undesirable pulses of physical and electronic origins, referred to as *background* or *background noise*.

Before beginning the detailed study of elastic recoil spectrometry, we provide an overview of this book and the history and present status of ERDA in ion beam analysis (IBA) techniques.

1.2.   OBJECTIVES

When L'Ecuyer and coworkers wrote their short paper entitled "An accurate and sensitive method for the determination of the depth distribution of light elements in heavy materials" in 1976,[1] they could not guess that 20 years later ERDA would become popular in the ion beam analysis field, as shown by the recent Twelfth Ion Beam Analysis conference in Tempe (22–26 May 1995).[2]

Our main ambition in writing this book is to give any ion beam user in the world a practical guide to ERDA as it exists for Rutherford backscattering spectrometry (RBS), proton induced X-ray emission (PIXE), or nuclear reaction analysis (NRA). Our dream is to have this book become as helpful to the whole ion beam community as previously published books, such as:

Backscattering Spectrometry[3]

Introduction to Radioanalytical Physics[4]

PIXE: A Novel Technique for Elemental Analysis[5]

Ion Beam Handbook for Material Analysis[6]

Ion Beams for Materials Analysis[7]

Fundamentals of Surface and Thin Film Analysis[8]

Handbook of Modern Ion Beam Materials Analysis[9]

In contrast to RBS, an ERDA measurement is relatively difficult to carry on, essentially because of geometrical constraints and sample surface requirements (minimum roughness, lateral homogeneity, and low contamination conditions). On the other hand, when an ERDA spectrum is recorded, the information being sought (hydrogen depth distribution, for example) is not too difficult to extract, as in the case of a backscattering spectrum from a multielemental target. However, simplifications currently used in RBS for the cross section near 180° or the validity of the Rutherford assumption are generally not applicable for ERDA. Therefore, we sincerely think that the challenge of writing a book on the multiple aspects of elastic recoil spectrometry is worth the effort.

1.3. TOPICS

This book is divided into 15 chapters, including the introduction. Appendixes dealing with some key parameters for ERDA investigation, such as elastic cross-section data or solid-angle calculation formula follow. Chapter 2 discusses the basic principles of elastic recoil spectrometry: description of the two-body collision and development of fundamental equations. Chapter 3 is divided in two parts. The first part deals with elastic recoil scattering cross-section formalism: The second part is devoted to multiple scattering. We provide a detailed review of these phenomena and we discuss the basic equations.

Analytical characteristics of elastic recoil spectrometry are described in the first part of Chapter 4, for example the notions of recoil yield, selectivity, mass resolution, and the energy–depth relationship. The second part of Chapter 4 discusses practical requirements for investigating real targets with elastic recoil spectrometry. Various types of analytical approaches to ERDA are introduced in Chapters 5–9. Conventional elastic recoil spectrometry in reflexion or transmission geometry is described in Chapter 5. Time-of-flight recoil spectrometry is presented Chapter 6.[*] The use of an *ExB* filter is developed in Chapter 7.[*] The telescope technique is discussed in Chapter 8. Elastic recoil coincidence spectroscopy is presented in Chapter 9.[*]

We dedicate Chapter 10 to the description of various experimental devices required for elastic recoil investigation. Chapter 11 presents mathematical procedures available for performing quantitative interpretation of recoil spectra. The applications of elastic recoil spectrometry in both fundamental and applied research are given and abundantly illustrated in Chapter 12. Chapter 13 gives applications of ERDA for determining light-element distributions in solids (hydrogen isotopes excepted). The use of heavy ions in ERDA measurements is also discussed in Chapter 13.

Chapter 14 is devoted to a detailed analysis of secondary effects induced by ion irradiation in the target specimen: thermal effects, electronic effects and bond breaking, ballistic damage, and charge accumulation effects. Some particularly important points for elastic recoil spectrometry (for example temperature increase and elemental losses) are studied in detail, as well as the choice, preparation, and stability of standard samples. In Chapter 15 we discuss resonant nuclear reactions and elastic recoil spectrometry for hydrogen depth profiling in the near-surface region of solids.

For each chapter a detailed bibliography is given, and additional references are sometimes listed. A general conclusion presents the real position of elastic recoil spectrometry among the most common IBA methods, such as RBS, NRA, PIXE, and PIGE. Further technical developments and new application fields are then briefly introduced.

---

[*]These chapters contain important contributions due to the kind cooperation of Nick Dytlewski for TOF-ERDA, Guy Ross for ERDA *ExB*, and Hans Hofsäss for transmission and coincidence ERDA.

1.4. HISTORICAL BACKGROUND

In 1976, L'Ecuyer and coworkers described for the first time an analytical method based on detecting forward angles of light elements recoiling from a thin target after elastic collisions with an incident ion beam (25–40 MeV ^{35}Cl).[1] They investigated lithium-containing thin layers (LiF or LiOH) separated by a copper layer (thickness of 30–150 nm) deposited on a carbon or copper backing (Figures 1.3 and 1.4). As shown in these figures, contributions of different layers to the spectra are clearly separated, and contaminants are unambiguously identified. The achieved depth resolution is approximately 30 nm, and the estimated ultimate sensitivity is as small as 10^{14} at. cm^{-2}. Other relevant papers from the same group were published at the end of the seventies.[10,11]

Doyle and Peercy first proposed using a Van de Graaff with a low terminal voltage, such as 2.5 MV, to carry out ERDA experiments in 1979.[12] Using 2.4-MeV ^{4}He$^+$ with

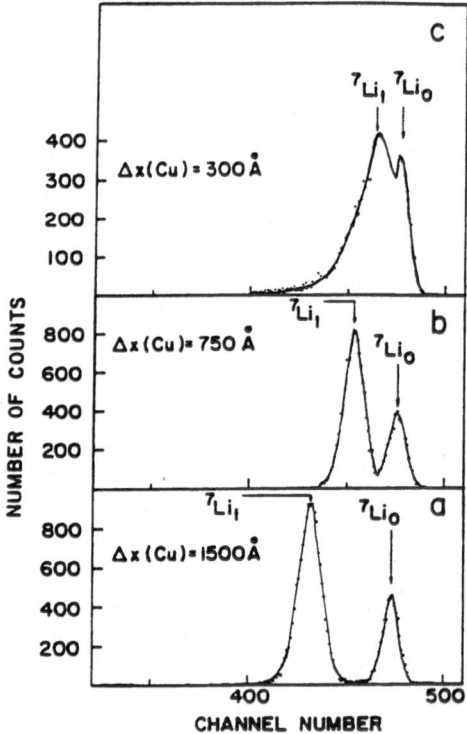

Figure 1.3. Energy spectra of ^{7}Li recoils observed when targets containing ^{7}LiF are bombarded by a 35-MeV ^{35}Cl beam. Each target was made of two thin layers of ^{7}LiF separated by a copper layer of thickness Δx; ^{7}Li$_0$ corresponds to the first layer and ^{7}Li$_1$ to the second layer. In (a) and (b) the target surface was perpendicular to the beam direction; in (c) it was tilted at 30° with the beam direction. (Data from Ref. 1.)

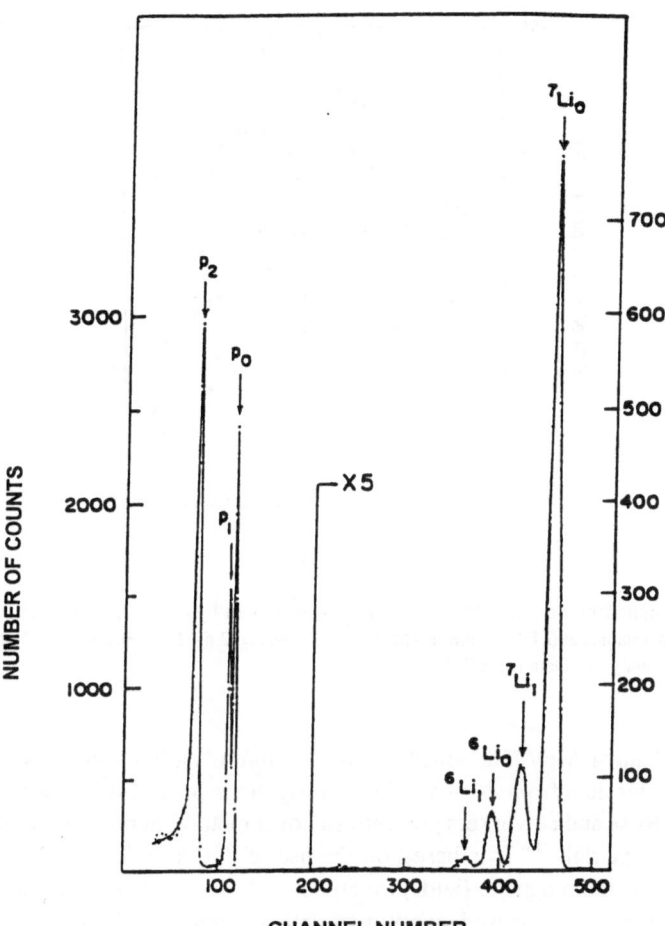

Figure 1.4. Spectra of particles emitted from a target containing two layers of LiOH separated by 1000 Å of copper. Particles coming from the first and second layer are labeled 0 and 1, respectively. Peak p_2 comes from moisture accumulated on the back surface of the backing. (Data from Ref. 1.)

an incidence angle of 75°, they measured the hydrogen distribution in Si_3N_4 layers, then compared these results with both 12-MeV ^{12}C elastic recoil spectrometry and 6.4-MeV ^{19}F resonant nuclear reaction profiling (Figure 1.5). They achieved a depth resolution of 70 nm and a sensitivity around 0.1 at. %.

The TOF principle was first applied to elastic recoil detection in 1983 by Groleau and coworkers[13] and by Thomas and coworkers.[14] This system, composed of a thin carbon foil and a microchannel plate detector, allows us to distinguish the contribution of different recoil species by measuring their velocity. In a recent paper, Gujrathi and Bultena reported results on hydrogen depth profiling within polymer films using both

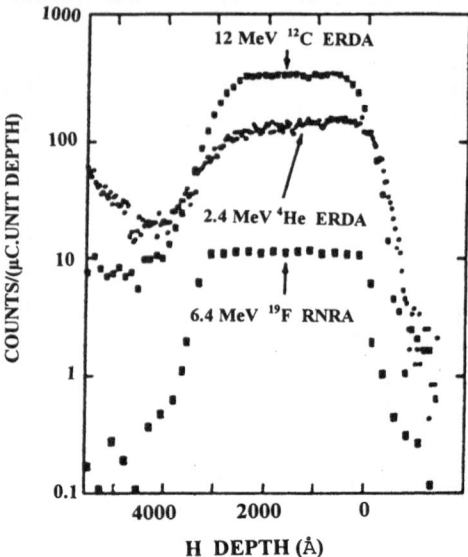

Figure 1.5. Comparison between ^{1}H profiles in Si$_3$N$_4$ measured by nuclear reaction analysis using the ^{1}H(^{19}F,$\alpha\gamma$) ^{16}O resonance at 6.4-MeV and 12-MeV ^{12}C ERDA and 2.4-MeV ^{4}He ERDA. RNRA = resonant nuclear reaction analysis. (Data from Ref. 12.)

30-MeV ^{35}Cl$^+$ and 4-MeV ^{4}He$^+$ ions[15] with a 10-µg/cm^2 carbon foil as a start detector. Figure 1.6 shows data from 4-MeV ^{4}He$^+$ investigations on a Mylar target.

In 1984 Ross and coworkers presented a novel method of profiling hydrogen and deuterium in materials.[16] It is based on the use of 350-keV ^{4}He$^+$ and an *ExB* filter (crossed electric and magnetic fields) to separate ^{1}H, ^{2}H, and ^{4}He contributions. Since 1984 many efforts have focused on optimizing their device.[17-19] Figure 1.7 illustrates the performances of *ExB* filtering for hydrogen and helium isotopes profiling in Be.

The analytical capabilities of ERDA in transmission geometry using 1.8–3 MeV ^{4}He$^+$ ions were first described in 1990 by Tirira and coworkers.[20] The typical analyzed depth is about 6 µm, and the depth resolution at the target surface is around 35 nm. These authors have applied this particular experimental configuration to irradiated polymer films and volcanic glasses.[21-24] Figures 1.8 and 1.9 illustrate performances of transmission ERDA for hydrogen profiling within krypton-irradiated 12.5-µm polyimide films.[22]

Hofsäss and coworkers introduced in 1990 the use of coincidence spectroscopy applied to ERDA.[25] Their approach is based on the pioneering works of Cohen,[26] Smidt and Pieper,[27] and Moore[28] in which coincidence counting was used to reject unwanted scattering events from other components of the target material. Hofsäss and coworkers proposed to relate each pair of energies of scattered and recoiled atoms to the depth where collision occurred and to the mass of the recoiled nucleus. Thus thin

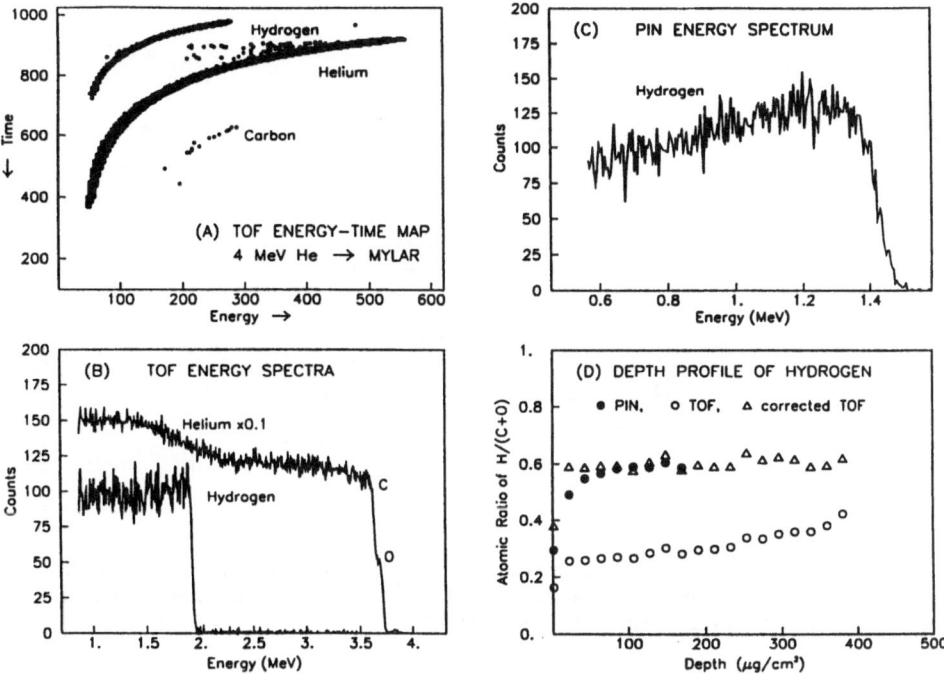

Figure 1.6. Results of 4-MeV ^{4}He on a Mylar target: (a) energy–time map, (b) mass-gated silicon surface barrier detector (SSBD) spectra (TOF) of He and H , (c) photodiode detector (PIN) hydrogen spectrum, and (d) depth profiles from PIN (closed circles), TOF (opened circles), and efficiency corrected TOF (triangles). (Data from Ref. 15.)

polycarbonate films (< 2 μm) have been investigated with 2-MeV ^{4}He$^+$ ions (Figure 1.10), and carbon and oxygen have been profiled over a depth range of 1 μm.

Further application examples have been published by the same group.[29] All modifications made in classical ERDA to use coincidence spectroscopy lead to the apparition of two groups of coincidence techniques. The first, denoted as coincident elastic recoil detection analysis (CERDA), is based on the adjustment of both scattering and recoil angles.[30,31] The second, denoted as scattering recoil coincidence spectroscopy (SRCS), or elastic recoil coincidence spectroscopy (ERCS), is based on the simultaneous measurement of scattered and recoiled energies.[32]

Very recently the use of ΔE-E solid-state telescopes was developed, allowing the discrimination of H, D, and T isotopes.[33,34] This consists of measuring, along with total particle energy, its velocity through its energy loss ΔE in a thin silicon surface barrier detector (10–20 μm).

The whole evolution of ERDA since 1976 clearly shows that it has become a well-recognized IBA technique.

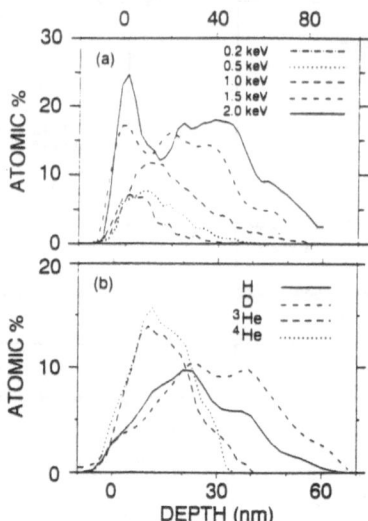

Figure 1.7. (a) Depth profiles of H implanted at different low energies in Si as measured by the ERDA *ExB* method; (b) depth profiles of H, D, ³He, and ⁴He implanted at 1 keV in Be (at different concentrations), as measured by the ERDA *ExB* method. (Data from Ref. 18.)

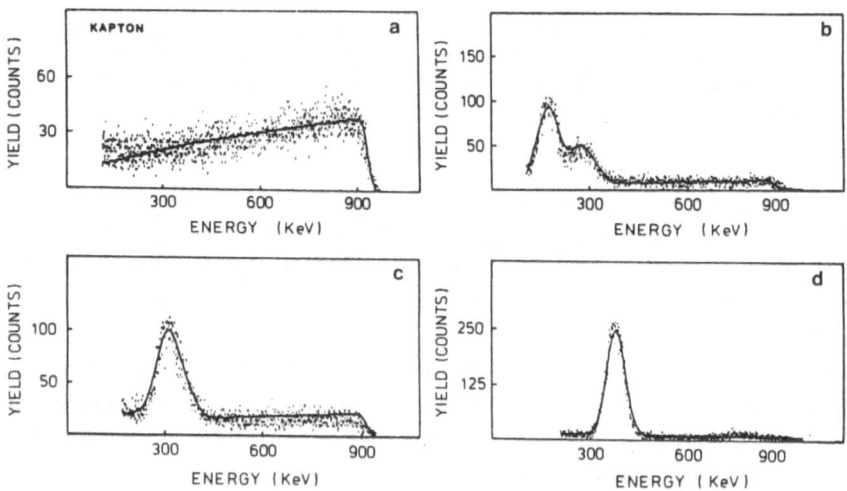

Figure 1.8. Experimental recoil spectra for 50-MeV ^{32}S irradiated polyimide film (thickness = 12.5 μm, ^{4}He energy = 2.05 MeV, time = 500 s, beam current = 1 nA, beam spot area = 100 μm^2). (a) Unirradiated sample, (b) fluence = 2.8 10^{13} S/cm^2, (c) fluence = 1.4 10^{14} S/cm^2, (d) fluence = 1.3 10^{15} S/cm^2. The dotted line corresponds to experimental data, and the solid line corresponds to a simulated fit. (Data from Ref. 22.)

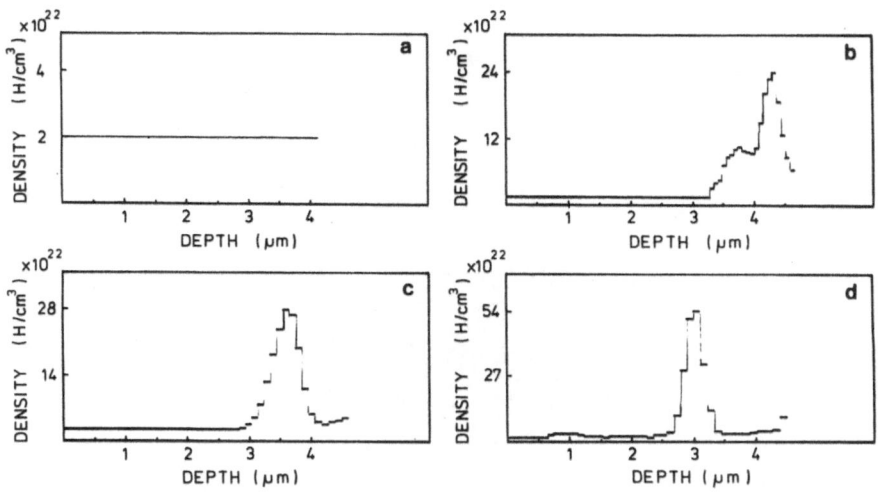

Figure 1.9. Hydrogen atomic density profiles derived from spectra in Figure 1.8 and calculated using the computer code GABY. (Data from Ref. 22.)

1.5. EXTENSION OF THE ERDA METHOD IN IBA LABORATORIES WORLDWIDE

The development of ERDA in IBA laboratories worldwide is continuously progressing since 1976, as seen from reading the proceedings of the Eleventh IBA conference held in Balatonfüred, Hungary.[35] More than 50 laboratories in the world currently use ERDA. In this section we give an overview of the groups involved in ERDA investigations. In the following section, we distinguish three types of ERDA works: MeV ^{4}He$^+$ induced recoil spectrometry, high-energy ^{4}He-ion-induced recoil spectrometry, and heavy-ion-induced recoil spectrometry.

In Europe two countries can be considered leaders in performing ERD: the Netherlands and Germany. In both countries several types of accelerators are used for this purpose: single-ended Van de Graaff, tandem accelerators and cyclotrons. The three types of ERDA measurements given above are thus applied.

In the Netherlands the most famous groups are located in Eindhoven and Utrecht. Klein, Rijken, Mutsaers, and coworkers perform high-energy ^{4}He ion induced ERDA and heavy-ion-induced ERDA measurements using the cyclotron from Eindhoven University of Technology.[30,31] Habraken, Arnoldbik, and coworkers carry on HI-ERDA investigations using the 6.5-MV tandem from Utrecht University.[33,36] A group at Philips Research Laboratories in Eindhoven is also involved in elastic recoil measurements on thin films.[37]

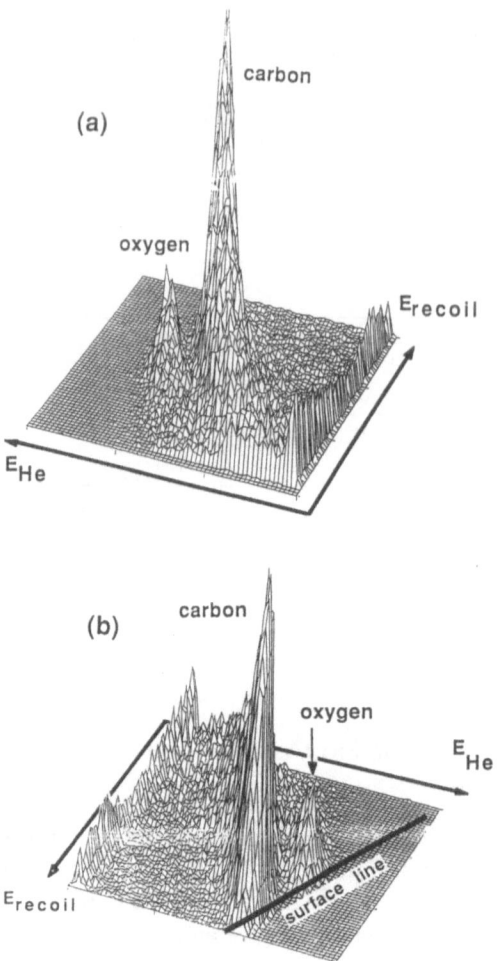

Figure 1.10. Examples of ERDA analysis of a thin polycarbonate film using the coincidence method. (Data from Ref. 25.)

In Germany a large number of laboratories are involved in recoil spectrometry. The HI-ERDA is currently performed at Gesellschaft für Schwer Ionenforschung Darmstadt by Fischer and coworkers.[38] The Hahn Meitner Institut in Berlin, with Goppelt, Gebauer, and coworkers, specializes in ultrahigh energy recoil spectrometry by using the high energy heavy ion accelerator.[39-41] In Garching specific detection devices are used in HI-ERDA experiments, such as ionization chambers by Assmann and coworkers[42] or magnetic spectrometer by Dollinger.[43] Neelmeijer and his group in Dresden are involved in thin-film recoil characterization using 30-MeV $^{35}Cl^{6+}$ ion beams supplied by the Rossendorf 5-MV tandem.[44] Hofsäss and coworkers at the

University of Konstanz are the leaders in using coincidence spectroscopy[25,29] through a collaboration with the group at Chapel Hill. Other laboratories are also involved, for example the European Institute for Transuranium Elements in Karlsruhe with Matzke,[45] Institut für Kernphysik der Universität Frankfurt,[46] Friedrich-Schiller-Universität Jena,[47] Leipzig Universität,[48] the Institute of Physics of the University of Erlangen,[49] Max-Planck-Institut für Festkörperforschung Stuttgart,[50] Freiburg Universität,[51] and Forschungszentrum Jülich.[52]

In France only four IBA laboratories are currently involved in ERDA experiments. Thomas, Chevarier, and coworkers from Institut de Physique Nucléaire de Lyon first developed heavy-ion-induced recoil spectrometry[14] and also MeV ^{4}He$^+$ recoil spectrometry using ExB filter.[53] The Centre de Spectrométrie Nucléaire et de Spectrométrie de Masse in Orsay,[54] the Laboratoire Pierre Süe in Saclay,[23,24] and the Groupe de Physique des Solides de l'Université Paris VII[55] regularly perform ERDA measurements with 2–4 MeV ^{4}He$^+$.

Some other countries in Europe host IBA laboratories undertaking recoil spectrometry experiments with MeV ^{4}He$^+$: in Hungary the Central Research Institute for Physics in Budapest,[56] in Belgium the Laboratoire d'Analyse par Réactions Nucléaires in Namur,[57] and in Italy the University of Padova, the University of Trento,[58] and the University of Catania.[59] The HI-ERDA is also performed with tandem accelerators at the Lund Institute of Technology (Sweden) by Whitlow and coworkers[60] and at the University of Helsinki by Räisänen, Rauhala, and coworkers.[61] Interesting papers have also been published by the Institute of Physics and Astronomy of Aarhus (Denmark),[62] by the Technical Research Center in Espoo (Finland),[63] and the Joint Institute for Nuclear Research in Dubna (Russia).[64]

In Canada ERD investigations are usually performed with either low-energy ^{4}He$^+$ ions or high-energy heavy ions. Ross and coworkers from Université du Québec in Varennes are the leading group in the world for using ExB devices.[16–19] L'Ecuyer and his group from the Laboratoire de Physique Nucléaire de l'Université de Montréal were the first team to propose using HI-ERDA in transmission geometry for light-element profiling.[1] In Ontario the Accelerator Laboratory of McMaster University in Hamilton with Siegele and Davies and the Chalk River Laboratories with Forster recently published papers on HI-ERDA.[65] The Department of Physics of the Queen's University in Kingston is also involved in recoil spectrometry.[66]

In the United States at least 15 IBA laboratories involved in ERDA measurements can be mentioned. At the Sandia National Laboratories in Albuquerque, Doyle and coworkers are currently involved in MeV recoil spectrometry[12] as well as HI-ERDA.[67] The Department of Physics and Astronomy at the University of North Carolina in Chapel Hill has published many important papers on the subject.[32] Wielunski, Benenson, Lanford, and coworkers from the State University of New York at Albany first demonstrated the use of 3.5–5 MeV ^{3}He or ^{4}He ions to determine hydrogen distribution through thin metal foils in transmission geometry.[68] At the Department of Physics at Idaho State University in Pocatello, Knox and his group developed a fruitful collaboration with the Finnish team from Helsinki.[61] Many other laboratories must be mentioned: Alabama A&M University[69]; Vanderbilt University in Nashville, where Arps and coworkers have presented recoil data using 300-keV ^{4}He$^+$ ions[70]; U.S. Naval Research Laboratory in Washington[71]; Cornell University in Ithaca[72]; IBM Almaden Research Center in San Jose[73]; IBM Thomas Watson Research Center in Yorktown Heights[74]; Southwest Missouri State University in Springfield[75]; Harvard University in Cambridge[76]; and Universal Energy Systems, Inc. in Dayton, Ohio.[77] Chu and coworkers from the Texas Center for Superconductivity at the University of Houston have been involved for a long time in ERDA investigations of thin films.[78] Some interesting papers have also been published by the University of Pennsylvania in Philadelphia,[79] Brookhaven National Laboratory in Upton,[80] and the University of New Mexico in Albuquerque.[81]

In South America two groups are involved in ERDA research: the Departamento de Fisica (Pontifica Universidade Catolica do Rio de Janeiro), equipped with a 4-MV Van de Graaff,[82] and the Departamiento de Fisica del CNEA in Buenos Aires, which develops HI-ERDA with 50-MeV sulfur ion beam.[83]

In China several IBA laboratories include physics groups in which ERDA measurements are carried out. Recoil spectrometry induced by MeV ^{4}He$^+$ is currently used by groups from Fudan University with a 3-MV Pelletron tandem from NEC[84] and by groups from East China Normal University in Shanghai,[85] the University of Lanzhou,[86] and Sichuan University in Chengdu.[87] The Department of Technical Physics at the University of Beijing[88] and the Institute of Physics in Beijing,[89] respectively, equipped with a tandem and a 1.7-MV Pelletron tandem are involved in ERDA measurements induced by medium mass ion beams (from ^{12}C to ^{35}Cl) at intermediate energies (< 1 MeV/amu).

Many laboratories currently perform ERDA measurements in Japan: Osaka University works with heavy-ion beams[90]; Nagoya University combines recoil spectrometry with low-energy ^{4}He$^+$ and HI-ERDA.[91] High-energy ^{35}Cl^{7+} and medium-energy ^{19}F^{3+} ion beams are used at the Tokyo Institute of Technology to characterize the composition of thin films.[92] Two groups from Ritsumeikan University in Kyoto[93] and Tohoku University in Sendai[94] perform ERDA measurements with 1–3 MeV ^{4}He$^+$. The Institute of Chemical and Physical Research in Wakoh is involved in HI-ERDA measurements with 50-MeV Ar ion beams.[95] A last group must be mentioned in Asia: the Nuclear Science and Technology Development Center at the National Tsing Hua University in Taiwan uses 3-MeV ^{4}He$^+$ for ERDA investigations on thin films.[96]

In Australia several groups have published data on ERDA studies. The ANSTO Lucas Heights[97] and the Department of Electronics and Engineering of the National University in Canberra[98] currently carry out HI-ERDA investigations on semiconductor multilayers. The CSIRO in North Ryde uses MeV ^{4}He$^+$ induced recoil spectrometry to characterize inclusions trapped in minerals and glasses.[99]

In South Africa a group from the National Accelerator Centre in Faure has recently started ERDA measurements using a 6-MV single-ended accelerator. This group described the use of a ΔE-E telescope with a 4-MeV ^{4}He$^+$ ion beam.[34]

We have selected only one representative paper from each group listed here to illustrate its contributions to ERDA investigations. Some of these papers are discussed in more details in Chapters 6–9, 12 and 13, devoted to examples of ERDA applications. A much larger number of papers are referenced at the end of each chapter.

1.6. CONCLUSION

The ERDA is a young but mature technique. It has seen much development in the last 20 years, and this development is still in progress. The ERDA appears to be particularly well-suited for profiling light elements in solids. Its many variants allow various applications to all sorts of materials. The time has now come to give an extended review of its physical principles and its various capabilities.

The ERDA is often combined with RBS to perform simultaneous analysis on major elements with hydrogen and other light elements. Experiments are generally conducted with ^{4}He millibeams (a few millimeters in diameter) on single-ended machines or cyclotrons or with heavy-ion millibeams (^{12}C, ^{28}Si, ^{35}Cl, ^{127}I, ^{197}Au) on tandems.

At the same time nuclear microprobe applications of ERDA have also been developed since the pioneering work of Tirira and coworkers[20–24] (see also Refs. 38, 67), and Sie[99] presented some interesting results at the Fourth International Conference on Nuclear Microprobe Technology and Applications in Shanghai.

REFERENCES

1. L'Ecuyer, J., Brassard, C., Cardinal, C., Chabbal, J., Deschênes, L., Labrie, J. P., Terreault, B., Martel, J. G., and Saint Jacques, R., An accurate and sensitive method for the determination of the depth distribution of light elements in heavy materials, *J. Appl. Phys.* **47**, 381 (1976).
2. Proc. of the Twelfth International Conference on Ion Beam Analysis (Tempe, Arizona, 22–26 May 1995), to be published in a special issue of *Nucl. Instrum. Methods Phys. Res. Sect. B.*
3. Chu, W. K., Mayer, J. W., and Nicolet, M. A., *Backscattering Spectrometry* (Academic, New York, 1978).
4. Deconninck, G., *Introduction to Radioanalytical Physics* (Elsevier Scientific Publishing, Amsterdam, 1978).
5. Johansson, S. A. E., and Campbell, J. L., *PIXE: A Novel Technique for Elemental Analysis* (Wiley, Chichester, UK, 1988).
6. Mayer, J. W., and Rimini, E., *Ion Beam Handbook for Material Analysis* (Academic, New York, 1977).
7. Bird, J. R., and Williams, J. S., *Ion Beams for Materials Analysis* (Academic, Sydney, Australia, 1989).
8. Feldman, L. C., and Mayer, J. W., *Fundamentals of Surface and Thin-Film Analysis* (North Holland, New York, 1986).
9. Tesmer, J. R., and Nastasi, M., eds., *Handbook of Modern Ion Beam Materials Analysis* (Materials Research Society, Pittsburgh, 1995).
10. Terreault, B., Martel, J. G., Saint Jacques, R. G., and L'Ecuyer, J., Depth profiling of light elements in materials with high-energy ion beams, *J. Vac. Sci. Technol.* **14**, 492 (1977).
11. L'Ecuyer, J., Brassard, C., Cardinal, C., and Terreault, B., The use of ^{6}Li and ^{35}Cl ion beams in surface analysis, *Nucl. Instrum. Methods* **149**, 271 (1978).

12. Doyle, B. L., and Peercy, P. S., Technique for profiling ^{1}H with 2.5-MeV Van de Graaff accelerators, *Appl. Phys. Lett.* **34**, 811 (1979).
13. Groleau, R., Gujrathi, S. C., and Martin, J. P., Time-of-flight system for profiling recoiled light elements, *Nucl. Instrum. Methods* **218**, 11 (1983).
14. Thomas, J. P., Fallavier, M., Ramdane, D., Chevarier, N., and Chevarier, A., High-resolution depth profiling of light elements in high atomic mass materials, *Nucl. Instrum. Methods* **218**, 125 (1983).
15. Gujrathi, S. C., and Bultena, S., Depth profiling of hydrogen using the high-efficiency ERD-TOF technique, *Nucl. Instrum. Methods Phys. Res. Sect. B* **64**, 789 (1992).
16. Ross, G. G., Terreault, B., Gobeil, G., Abel, G., Boucher, C., and Veilleux, G., Inexpensive, quantitative hydrogen depth profiling for surface probes, *J. Nucl. Mater.* **128/129**, 730 (1984).
17. Ross, G. G., and Terreault, B., ERD measurement of the mean ranges and variances of 0.75–2.0 keV deuterium ions in Be, C and Si, *Nucl. Instrum. Methods Phys. Res. Sect. B* **45**, 190 (1990).
18. Ross, G. G., Terreault, B., Pageau, J. F., and Gollier, P. A., Nuclear microanalysis by means of 350-keV Van de Graaff accelerator, *Nucl. Instrum. Methods Phys. Res. Sect. B* **66**, 17 (1992).
19. Ross, G. G., and Leblanc, L., Depth profiling of hydrogen and helium isotopes by means of the ERD ExB technique, *Nucl. Instrum. Methods Phys. Res. Sect. B* **62**, 484 (1992).
20. Tirira, J., Trocellier, P., and Frontier, J. P., Analytical capabilities of ERDA in transmission geometry, *Nucl. Instrum. Methods Phys. Res. Sect. B* **45**, 147 (1990).
21. Tirira, J., Trocellier, P., Frontier, J. P., Massiot, P., Costantini, J. M., and Mori, V., 3D hydrogen profiling by elastic recoil detection analysis in transmission geometry, *Nucl. Instrum. Methods Phys. Res. Sect. B* **50**, 135 (1990)
22. Trocellier, P., Tirira, J., Massiot, P., Gosset, J., and Costantini, J. M., Nuclear microprobe study of the composition degradation induced in polyimides by irradiation with high-energy heavy ions, *Nucl. Instrum. Methods Phys. Res. Sect. B* **54**, 118 (1991).
23. Mosbah, M., Clocchiatti, R., Tirira, J., Gosset, J., Massiot, P., and Trocellier, P., Study of hydrogen in melt inclusions trapped in quartz with a nuclear microprobe, *Nucl. Instrum. Methods Phys. Res. Sect. B* **54**, 298 (1991).
24. Tirira, J., Trocellier, P., Mosbah, M., and Metrich, N., Study of hydrogen content in solids by ERDA and radiation induced damage, *Nucl. Instrum. Methods Phys. Res. Sect. B* **56/57**, 839 (1991).
25. Hofsäss, H. C., Parikh, N. R., Swanson, M. L., and Chu, W. K., Depth profiling of light elements using elastic recoil coincidence spectroscopy (ERCS), *Nucl. Instrum. Methods Phys. Res. Sect. B* **45**, 151 (1990).
26. Cohen, B. L., Fink, C. L., and Degnan, J. H., Nondestructive analysis for trace amounts of hydrogen, *J. Appl. Phys.* **43**, 19 (1972).
27. Smidt, F.A., Jr., and Pieper, A. G., Studies of the mobility of helium in vanadium, *J. Nucl. Mater.* **51**, 361 (1974).
28. Moore, J. A., Mitchell, I. V., Hollis, M. J., Davies, J. A., and Howe, L. M., Detection of low-mass impurities in thin films using MeV heavy-ion elastic scattering and coincidence detection techniques, *J. Appl. Phys.* **46**, 52 (1975).
29. Hofsäss, H. C., Parikh, N. R., Swanson, M. L., and W. K. Chu, W. K., Elastic recoil coincidence spectroscopy (ERCS), *Nucl. Instrum. Methods Phys. Res. Sect. B* **58**, 49 (1991).
30. Klein, S. S., Separate determination of concentration profiles for atoms with different masses by simultaneous measurement of scattered projectile and recoil energies, *Nucl. Instrum. Methods Phys. Res. Sect. B* **15**, 464 (1986).
31. Rijken, H. A., Klein, S. S., and de Voigt, M. J. A., Improved depth resolution in CERDA by recoil time of flight measurement, *Nucl. Instrum. Methods Phys. Res. Sect. B* **64**, 395 (1992).
32. Chu, W. K., and Wu, D. T., Scattering recoil coincidence spectrometry, *Nucl. Instrum. Methods Phys. Res. Sect. B* **35**, 518 (1988).
33. Arnoldbik, W. M., de Laat, C. T. A. M., and Habraken, F. H. P. M., On the use of a ΔE-E telescope in elastic recoil detection, *Nucl. Instrum. Methods Phys. Res. Sect. B* **64**, 832 (1992).

34. Prozesky, V. M., Churms, C. L., Pilcher, J. V., Springhorn, K. A., and Behrisch, R., ERDA measurement of hydrogen isotopes with a ΔE-E telescope, *Nucl. Instrum. Methods Phys. Res. Sect. B* **84**, 373 (1994).

35. Gyulai, J., Pászti, F., Lohner, T., and Battistig, G., Proc. of the Eleventh International Conference on Ion Beam Analysis, *Nucl. Instrum. Methods Phys. Res. Sect. B* **85** (1994).

36. Habraken, F. H. P. M., Light-element depth profiling using elastic recoil detection, *Nucl. Instrum. Methods Phys. Res. Sect. B* **68**, 181 (1992).

37. Oostra, D. J., RBS and ERD analysis in materials research of thin films, *Philips J. Res.* **47**, 315 (1993).

38. Klein, S. S., Mutsaers, P. H. A., and Fischer, B. E., Mass selection and depth profiling by coincident recoil detection for nuclei in the middle mass region, *Nucl. Instrum. Methods Phys. Res. Sect. B* **50**, 150 (1990).

39. Goppelt, P., Gebauer, B., Fink, D., Wilpert, M., Wilpert, Th., and Bohne, W., High-energy ERDA with very heavy ions using mass- and energy-dispersive spectrometry, *Nucl. Instrum. Methods Phys. Res. Sect. B* **68**, 235 (1992).

40. Gebauer, B., Fink, D., Goppelt, P., Wilpert, M., and Wilpert, Th., Multidimensional ERDA measurements and depth profiling of medium-heavy elements, *Nucl. Instrum. Methods Phys. Res. Sect. B* **50**, 159 (1990).

41. Goppelt, P., Biersack, J. P., Gebauer, B., Fink, D., Bohne, W., Wilpert, M., and Wilpert, Th., Investigation of thin films by high-energy ERDA, *Nucl. Instrum. Methods Phys. Res. Sect. B* **80/81**, 142 (1993).

42. Assmann, W., Hartung, P., Huber, H., Staat, P., Steffens, H., and Steinhausen, Ch., Setup for materials analysis with heavy ion beams at the Munich MP tandem, *Nucl. Instrum. Methods Phys. Res. Sect. B* **85**, 726 (1994).

43. Dollinger, G., Faestermann, T., and Maier-Komor, P., High-resolution depth profiling of light elements, *Nucl. Instrum. Methods Phys. Res. Sect. B* **64**, 422 (1992).

44. Neelmeijer, Ch., Grötzschel, R., Klabes, R., Kreissig, U., and Sobe, G., Study of carbon and oxygen incorporation in reactively sputtered Cr-Si-Al films using ERDA, *Nucl. Instrum. Methods Phys. Res. Sect. B* **64**, 461 (1992).

45. Matzke, Hj., Della Mea, G., Dran, J. C., Rigato, V., and Bevilacqua, A., Chemical and physical modifications in waste glasses ion implanted at different temperatures. in *Modifications Induced by Irradiation in Glasses* (P. Mazzoldi, ed.). (North Holland, Amsterdam, 1992), pp. 25–31.

46. Wagner, W., Rauch, F., Ottermann, C., and Bange, K., Analysis of tungsten oxide films using MeV ion beams, *Nucl. Instrum. Methods Phys. Res. Sect. B* **68**, 262 (1992).

47. Rottschalk, M., Bachmann, T., and Witzmann, A., Investigation of proton exchanged optical waveguides in $LiNbO_3$ using elastic recoil detection, *Nucl. Instrum. Methods Phys. Res. Sect. B* **61**, 91 (1991).

48. Ascheron, C., Lehmann, D., Neelmeijer, C., Schindler, A., and Bigl, F., Hydrogen depth profiling in belleved proton-implanted semiconductors by ^4He ERDA and ^{19}F NRA, *Nucl. Instrum. Methods Phys. Res. Sect. B* **63**, 412 (1992).

49. Nölscher, C., Brenner, K., Knauf, R., and Schmidt, W., Elastic recoil detection analysis of light particles (^1H - ^{16}O) using 30 MeV sulphur ions, *Nucl. Instrum. Methods Phys. Res.* **218**, 116 (1983).

50. Kruse, O., and Carstanjen, H. D., High-depth resolution ERDA of H and D by means of an electrostatic spectrometer, *Nucl. Instrum. Methods Phys. Res. Sect. B* **89**, 191 (1994).

51. Bruder, F., and Brenn, R., Measuring the binodal by interdiffusion in blends of deuterated polystyrene and poly(styrene-co-4-bromostyrene), *Macromolecules* **24**, 5552 (1991).

52. Wang, M., Schmidt, K., Reichelt, K., Jiang, X., Hübsch, H., and Dimigen, H., The properties of W-C : H films deposited by reactive rf sputtering, *J. Mater. Res.* **7**, 1465 (1992).

53. Roux, B., Chevarier, A., Chevarier, N., Wybourn, B., Antoine, C., Bonin, B., Bosland, P., and Cantacuzene, S., High-resolution hydrogen profiling in superconducting materials by ion beam analysis (ERD-E x B), Institut de Physique Nucléaire de Lyon Report LYCEN 9417 (1994).

54. Pivin, J. C., Stehle, J. L., Piel, J. P., and Allouard, M., Correlation between gradients of composition and dielectric properties in oxinitride or diamond like films on Si by means of spectroscopic ellipsometry and ion beam analysis, *Philos. Mag. B* **64**, 1 (1991).

55. Quillet, V., Abel, F., and Schott, M., Absolute cross-section measurements for H and D elastic recoil using 1–2.5 MeV ^{4}He ions, and for the ^{12}C(d, p)^{13}C and ^{16}O(d,p$_1$)^{17}O nuclear reactions, *Nucl. Instrum. Methods Phys. Res. Sect. B* **83**, 47 (1993).

56. Pászti, F., Kótai, E., Mezey, G., Manuaba, A., Pocs, L., Hildebrandt, D., and Strusny, H., Hydrogen and deuterium measurements by elastic recoil detection using alpha particles, *Nucl. Instrum. Methods Phys. Res. Sect. B* **15**, 486 (1986).

57. Tirira, J., Bodart, F., Serruys, Y., and Morciaux, Y., Optimization algorithm for elastic recoil spectra simulation, *Nucl. Instrum. Methods Phys. Res. Sect. B* **79**, 527 (1993).

58. Antoni, V., Bagatin, M., Buffa, A., Della Mea, G., Freyre F., Jr., Mazzoldi, P., and Romanato, F., Investigation of the ETA-BETA II plasma edge by surface analysis of collector probes, *Nuovo Cimento Soc. Ital. Fis.* **13**, 435 (1991).

59. Compagnini, G., Calcagno, L., and Foti, G., Properties of fully implanted amorphous Si$_x$C$_{1-x}$: H alloys, *Nucl. Instrum. Methods Phys. Res. Sect. B* **80/81**, 978 (1993).

60. Whitlow, H. J., Possnert, G., and Petersson, C. S., Quantitative mass and energy dispersive elastic recoil spectrometry resolution and efficiency considerations, *Nucl. Instrum. Methods Phys. Res. Sect. B* **27**, 448 (1987).

61. Räisänen, J., Rauhala, E., Knox, J. M., and Harmon, J. F., Non-Rutherford cross sections in heavy ion elastic recoil spectrometry : 40–70 MeV ^{32}S ions on carbon, nitrogen and oxygen, *J. Appl. Phys.* **75**, 3273 (1994).

62. Stensgaard, I., Surface studies with high-energy ion beams, *Rep. Prog. Phys.* **55**, 989 (1992).

63. Ronkainen, H., Likonen, J., and Koskinen, J., Tribological properties of hard carbon films produced by the pulsed vacuum arc discharge method, *Surf. Coat. Technol.* **54/55**, 570 (1992).

64. Hrubcin, L., Huran, J., Sandrik, R., Kobzev, A. P., and Shirokov, D. M., Application of the ERD method for hydrogen determination in silicon (oxy)nitride thin films prepared by ECR plasma deposition, *Nucl. Instrum. Methods Phys. Res. Sect. B* **85**, 60 (1994).

65. Siegele, R., Davies, J. A., Forster, J. S., and Andrews, H. R., Forward elastic recoil measurements using heavy ions, *Nucl. Instrum. Methods Phys. Res. Sect. B* **90**, 606 (1994).

66. Forster, J. S., Leslie, J. R., and Laursen, T., Depth profiling of hydrogen isotopes in thin, low-Z films by scattering recoil coincidence spectrometry, *Nucl. Instrum. Methods Phys. Res. Sect. B* **66**, 215 (1992).

67. Doyle, B. L., and Wing, N. D., The Sandia nuclear microprobe, Sandia Report SAND82-2393 (1982).

68. Wielunski, L. S., Benenson, R. E., and Lanford, W. A., Helium-induced hydrogen recoil analysis for metallurgical applications, *Nucl. Instrum. Methods Phys. Res.* **218**, 120 (1983); see also Berning, P. R., and Benenson, R. E., An ERD based study of the electromigration of hydrogen in V, Ta, Nb and Ti, *Nucl. Instrum. Methods Phys. Res. Sect. B* **94**, 130 (1994).

69. Ila, D., Jenkins, G. M., Holland, L. R., Thompson, J., Evelyn, L., Hodges, A., Zimmerman, R. L., and Dalins, I., Measurement of accumulated contaminants in glassy carbon by RBS, ERD and NRA, *Nucl. Instrum. Methods Phys. Res. Sect. Sect. B* **64**, 439 (1992).

70. Arps, J. H., and Weller, R. A., Medium energy elastic recoil analysis of surface hydrogen, *Nucl. Instrum. Methods Phys. Res. Sect. B* **79**, 539 (1993).

71. Gossett, C. R., Use of a magnetic spectrometer to profile light elements by elastic detection, *Nucl. Instrum. Methods Phys. Res. Sect. B* **15**, 481 (1986).

72. Shull, K. R., Dai, K. H., Kramer, E. J., Fetters, L. J., Antonietti, M., and Sillescu, H., Diffusion by constraint release in branched macromolecular matrices, *Macromolecules* **24**, 505 (1991).

73. Russell, T. P., The characterization of polymer interfaces, *Ann. Rev. Mater. Sci.* **21**, 249 (1991).

74. Grill, A., Patel, V., and Meyerson, B. S., Temperature and bias effects on the physical and tribological properties of diamondlike carbon, *J. Electrochem. Soc.* **138**, 2362 (1991).

75. Wang, Y., Mohite, S. S., Bridwell, L. B., Giedd, R. E., and Sofield, C. J., Modification of high temperature and high performance polymers by ion implantation, *J. Mater. Res.* **8**, 388 (1993).
76. Gordon, R. G., Hoffman, D. M., and Riaz, U., Low-temperature atmospheric pressure chemical vapor deposition of polycrystalline tin nitride thin films, *Chem. Mater.* **4**, 68 (1992).
77. Bhattacharya, R. S., Rai, A. K., and McCormick, A. W., Ion-beam-assisted deposition of Al_2O_3 thin films, *Surf. Coat. Anal.* **46**, 155 (1991).
78. Liu, J. R., Zheng, Z. S., Zhang, Z. H., and Chu, W. K., RBS and ERD analysis using lithium ions, *Nucl. Instrum. Methods Phys. Res. Sect. B* **85**, 51 (1994).
79. Wallace, W. E., Zhong, Q., Genzer, J., Composto, R. J., and Bonnell, D. A., On the use of ion scattering to examine the role of hydrogen in the reduction of TiO_2, *J. Mater. Res.* **8**, 1629 (1993).
80. Hu, B., Cholewa, M., Rivers, M. L., Smith, J. V., and Jones, K. W., Determination of hydrogen content of phlogopite in garnet peridotites using elastic recoil detection, *Nucl. Instrum. Meth. Phys. Res. Sect. B* **47**, 97 (1990).
81. Arnold, G. W., Westrich, H. R., and Casey, W. H., Application of ion beam analysis (RBS and ERD) to the surface chemistry of leached minerals, *Nucl. Instrum. Methods Phys. Res. Sect. B* **64**, 542 (1992).
82. Freire, F. L., Jr., Achete, C. A., Graeff, C. F. O., Chambouleyron, I., and Mariotto, G., Study of annealed amorphous hydrogenated films by elastic recoil detection analysis, *Nucl. Instrum. Methods Phys. Res. Sect. B* **85**, 55 (1994).
83. Alurralde, M., Garcia, E., Abriola, D., Filevich, A., Garcia-Bermudez, G., Aucouturier, M., and Siejka, J., Identification and depth profiling of light elements elastically recoiled by heavy ion beams. in *Proc. of High-Energy and Heavy-Ion Beams in Materials Analysis Workshop* (J. R. Tesmer, C. J. Maggiore, M. Nastasi, J. C. Barbour, and J. W. Mayer, eds.) (Materials Research Society, Albuquerque, New Mexico, 14–16, June 1989), pp. 119–127.
84. Wang, Y., Chen, J., and Huang, F., The calculation of the differential cross sections for recoil protons in 4He-p scattering, *Nucl. Instrum. Methods Phys. Res. Sect. B* **17**, 11 (1986).
85. Wang, Y., Huang, B., Cao, D., Cao, J., Zhu, D., and Shen, K., Analysis of hydrogen in oxygen-doped polysilicon by $^4He^+$-H elastic recoil detection, *Nucl. Instrum. Methods Phys. Res. Sect. B* **84**, 111 (1994).
86. Wang, Y., Liao, C., Yang, S., and Zheng, Z., A convolution analysis method for hydrogen concentration profiles by elastic recoil detection, *Nucl. Instrum. Methods Phys. Res. Sect. B* **47**, 427 (1990).
87. Long, X., Peng, X., He, F., Liu, M., and Lin, X., The hydrogen concentration of diamondlike carbon films by elastic recoil detection analysis, *Nucl. Instrum. Methods Phys. Res. Sect. B* **68**, 266 (1992).
88. Xiting, L., Zonghuang, X., Kungang, Z., Changwen, J., Xihong, Y., Hongtao, L., Dongxing, J., and Yanlin, Y., A new method for measuring stopping powers by ERD, *Nucl. Instrum. Methods Phys. Res. Sect. B* **58**, 280 (1991).
89. Liu, J., Peiran, Z., Feng, Y., Juinshi, Z., Houxian, Z., and Zhang Q., An overview of the ion beam laboratory at Beijing Institute of Physics, *Nucl. Instrum. Methods Phys. Res. Sect. B* **56/57**, 1005 (1991).
90. Oura, K., Naitoh, M., Morioka, H., Watamori, M., and Shoji, F., Elastic recoil detection analysis of coadsorption of hydrogen and deuterium on clean Si surfaces, *Nucl. Instrum. Methods Phys. Res. Sect. B* **85**, 344 (1994).
91. Matsunami, N., A study of carbon thin films by ion beams, *Nucl. Instrum. Methods Phys. Res. Sect. B* **64**, 800 (1992).
92. Arai, E., Zounek, A., Sekino, M., Takemoto, K., and Nittono, O., Depth profiling of porous silicon surface by means of heavy-ion TOF ERDA, *Nucl. Instrum. Methods Phys. Res. Sect. B* **85**, 226 (1994).
93. Kido, Y., Miyauchi, S., Takeda, O., Nakayama, Y., Sato, M., and Kusao, K., Precise determination of H recoil cross sections for 1.5-3.0 MeV He ions, *Nucl. Instrum. Methods Phys. Res. Sect. B* **82**, 474 (1993).
94. Yamada, Y., Kasukabe, Y., Nagata, S., and Yamaguchi, S., Spontaneous hydrogenation of Ti films evaporated on NaCl substrates I, *Jpn. J. Appl. Phys.* **29**, L1888 (1990).

95. Nagai, H., Hayashi, S., Aratani, M., Nozaki, T., Yanokura, M., and Kohno, I., Reliability, detection limit, and depth resolution of the elastic recoil measurement of hydrogen, *Nucl. Instrum. Methods Phys. Res. Sect. B* **28**, 59 (1987).

96. Niu, H., Wu, S., Huang, S., Lin, J., and Deng, R., Hydrogen depth profiling of SiN_x films by the detection of recoiled protons, *Nucl. Instrum. Methods Phys. Res. Sect. B* **79**, 536 (1993).

97. Martin, J. W., Cohen, D. D., Dytlewski, N., Garton, D. B., Whitlow, H. J., and Russell, G. J., Materials characterisation using heavy ion elastic recoil time of flight spectrometry, *Nucl. Instrum. Methods Phys. Res. Sect. B* **94**, 277 (1994).

98. Williams, J. S., Proc. of the Conference on Ion Beam Modification of Materials (Canberra, Australia, 7–11, Feb. 1995), to be published in a special issue of *Nucl. Instrum. Methods Phys. Res. Sect. B*.

99. Sie, S., Suter, G. F., Chekhmir, A., and Green, T. H., Microbeam recoil detection for the study of hydration of minerals, *Nucl. Instrum. Methods Phys. Res. Sect. B* **104**, 261 (1995).

Basic Physical Processes of Elastic Spectrometry

2.1. INTRODUCTION

Forward recoil scattering can be completely described using the very well-known basic principles of charged particle interaction with matter, but charged particles moving through matter interact with both electron shells of atoms and nuclei. Thus, beside the scattering of ions by target nuclei, we have to consider energy loss resulting from ionizations and excitations of a dense medium.

If the total kinetic energy is conserved in the scattering event itself, and the internal energy of particles plays no role, this scattering process is called *elastic collision*. In the opposite case, both kinetic and internal energy are involved in the scattering, and atomic or nuclear excitations occur. In such interactions, called *inelastic collisions*, new particles are eventually formed during the interaction process.

Physical concepts of two-body elastic scattering are the basis of several nuclear methods for elemental material characterization. The kinetics of the collision and its probability of occurrence, which is quantified by means of the scattering cross section, have thus been the object of detailed studies. In particular RBS and ERDA are exploited using a simple projectile–target elastic collision. Rutherford backscattering basic theory, mathematical models, as well as results and applications have been extensively treated for 17 years (e.g., Chu et. al.,[1] Bird and Williams[2]). Although elastic recoil spectrometry[3–6] is quite similar to RBS, important and subtle differences exist; hence it is worthwhile to provide a distinct and detailed analysis.

Chapter 2 first deals with kinematics and geometrical parameters. Key expressions and a discussion of relationships are given in Sections 2.2 and 2.3. Fundamentals of the energy loss process along the track of particles are briefly considered in Section 2.4; indeed details are sufficiently discussed in a large number of reviews and reports (especially for protons and helium ions in solids). The important role of this energy loss process in the forward recoil yield is presented in Chapter 4. Statistical fluctuations of the energy loss (energy straggling) is discussed with particular attention in Section 2.5.

2.2. KINEMATICS OF ELASTIC COLLISION

2.2.1. Mechanics of the Collision

The elastic collision between two particles is physically simple, particularly when they are considered as two hard spheres. Figure 2.1 shows a schematic representation of an elastic collision. The kinematics of elastic collision between two particles can be completely described by applying the principles of conservation of energy and momentum (e.g., Landau and Lifshitz[7]). In the case of elastic scattering, the internal energy of particles is not taken into account as long as kinematics alone are considered. Indeed elastic nuclear reactions and resonances can be included in this treatment even though the internal state of particles is ignored.

Let M_1 be the mass of the incident particle and E_0 and $\vec{V}_0$ its energy and velocity in the laboratory system (Figure 2.1) while the target mass M_2 is at rest before the collision. After collision energy is transferred from the projectile to the stationary mass. Then E_1, $\vec{V}_1$ and E_2, $\vec{V}_2$ are the energies and velocities of projectile and target atom, respectively. To establish the relation between these kinetic parameters, we first consider the center-of-mass system C, which allows a very simple description of the collision because total momentum is zero. Since we are interested only in the laboratory system L, this description must be transformed, which is easy enough.

In the C system, velocities of particles before the collision are related to the velocity $\vec{V}_0$ in the laboratory system by:

$$\vec{V}_{10} = \frac{M_2 \vec{V}_0}{M_1 + M_2} \qquad \vec{V}_{20} = \frac{M_1 \vec{V}_0}{M_1 + M_2} \tag{2.1}$$

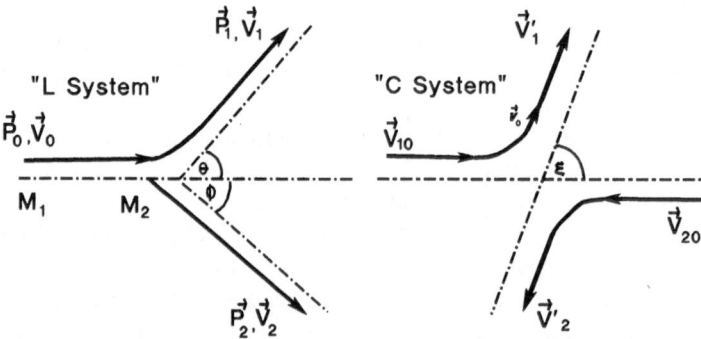

Figure 2.1.   Elastic collision between two particles (a) in the laboratory system, where the scattering angle θ and the recoil angle ϕ are defined as positive between the directions of motion after the collision and the incident axis (beam direction), (b) in the center-of-mass-system, where ξ is the scattering angle in this system.

In the C system the collision simply rotates the velocities; in fact they remain opposite in direction and unchanged in magnitude. Then velocities $\vec{V}_1$ and $\vec{V}_2$ of the two particles after collision in the C system are given by:

$$\vec{V}_1 = \frac{M_2\vec{V}_0}{M_1 + M_2} \qquad \vec{V}_2 = \frac{-M_1\vec{V}_0'}{M_1 + M_2} \qquad (2.2)$$

with $\vec{V}_0 = V_0\,\vec{v}_0$, $|\vec{V}_0| = |\vec{V}_0'|$, and $\vec{v}_0$ is the unit vector in the direction of the velocity of particle M_1 in the C system after collision.

Since $M_1\,\vec{V}_0/(M_1 + M_2)$ is the velocity of the center of mass in the laboratory system, we can write the velocities of particles in the laboratory system after the collision as:

$$\vec{V}_1 = \vec{V}_1 + \frac{M_1\vec{V}_0}{M_1 + M_2} \qquad (2.3a)$$

$$\vec{V}_2 = \vec{V}_2 + \frac{M_1\vec{V}_0}{M_1 + M_2} \qquad (2.3b)$$

Multiplying these equations by M_1 and M_2, respectively, we have the momenta after collision, $\vec{P}_1$ and $\vec{P}_2$. Adding these momenta (see Eq. 2.4), we obtain $\vec{P}_1$ and $\vec{P}_2$ as a function of the reduced mass $\mu = M_1M_2/(M_1 + M_2)$ and the momentum $\vec{P}_0$ of particle M_1 before the collision.

$$\vec{P}_1 + \vec{P}_2 = \mu\vec{V}_0 + \frac{M_1\vec{P}_0}{M_1 + M_2} \qquad (2.4)$$

We use this representation to obtain a simple geometrical interpretation of the collision (see Ref. 7). In fact if we draw a circle of radius $\mu\,\vec{V}_0$, we have a vectorial construction for momenta $\vec{P}_1$ and $\vec{P}_2$, where the vector **AB** is equal to the momentum $\vec{P}_0$ before collision. Figure 2.2 shows this collision diagram for both $M_1 < M_2$ and $M_1 > M_2$. In this representation angles θ and ϕ are defined as positive between the directions of motion after the collision and the incidence axis (beam direction). In particular we call ϕ the *recoil angle* (or forward angle) and θ the *scattering angle*. Furthermore, angles ϕ and θ can be expressed in the center-of-mass angle system as a function of angle ξ (see Figure 2.1) by

$$2\phi = \pi - \xi \qquad \tan\theta = \frac{M_2\sin\xi}{M_1 + M_2\cos\xi} \qquad (2.5)$$

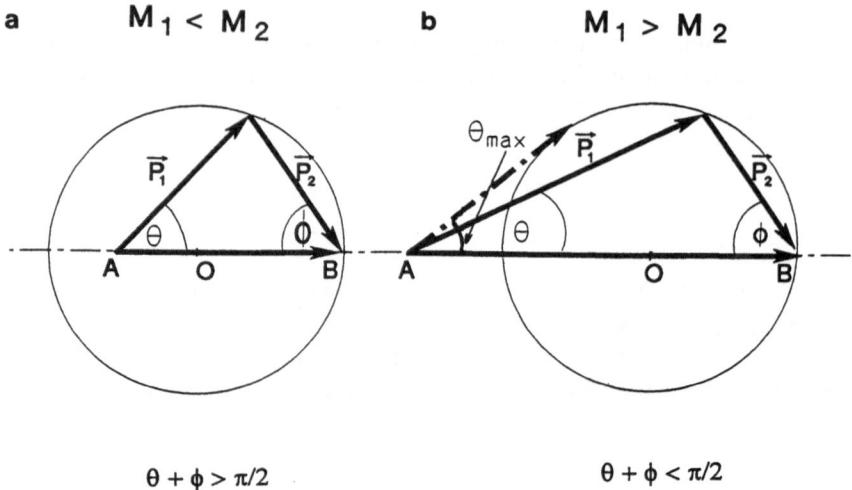

$$\theta + \phi > \pi/2 \qquad\qquad \theta + \phi < \pi/2$$

Figure 2.2. Interpretation diagram of the elastic collision, where the vectorial construction shows $\overline{OB} = \mu \, \vec{V}_0$ and $\overline{AB} = \vec{P}_1 + \vec{P}_2, = \vec{P}_0$, (a) when $M_1 < M_2$; (b) when $M_1 > M_2$.

From Eq. 2.5 the recoil angle can be expressed as a function of the scattering angle by the following expression:

$$\cos^2 \phi = \frac{(M_1 + M_2)^2 - \left(M_1 \cos\theta \pm \sqrt{M_2^2 - M_1^2 \sin^2 \theta}\,\right)^2}{4M_1M_2} \qquad (2.6a)$$

Equation 2.6a is an adequate representation for $\cos \phi$, which gives an explicit form (both signs in Eq. 2.6a) of both possible relations between angles ϕ and θ. The variation of scattering and recoil angles from Eq. 2.6a is shown in Figure 2.3 as a function of the mass ratio M_2/M_1. Another typical representation between scattering angle θ and recoil angle ϕ is given by Eq. 2.6b, which can be used to represent in simplified form analytical recoil equations:

$$\tan \theta = \frac{\sin(2\phi)}{(M_1 / M_2) - \cos(2\phi)} \qquad (2.6b)$$

However this representation does not show in explicit form the double-valued relation between these angles. An example of applications of these relations is given in Chapter 9. For practical coincidence recoil problems, different possible experimental configurations can be quickly inferred from Eq. 2.6.

If $M_1 < M_2$, the velocity of M_1 after collision may have any direction from $0-\pi$. Then only the plus sign holds in Eq. 2.6. In this case it is evident that $\theta + \phi > \pi/2$ (Figure 2.2a).

If $M_1 > M_2$ (as for ^1H target and ^4He beam), the M_1 particle cannot be deflected from the incident beam direction through an angle exceeding θ_{max}. This maximum value of θ corresponds to the tangent to the circle (Figure 2.2b). Then

$$\sin(\theta_{max}) = \frac{M_2}{M_1} \tag{2.7}$$

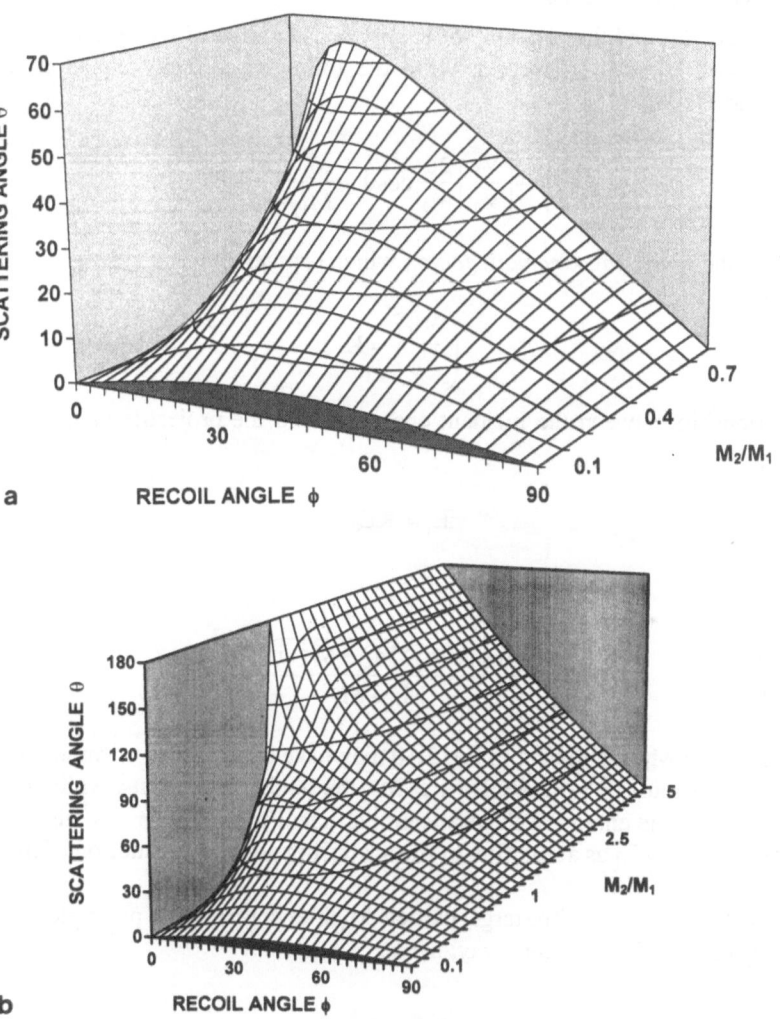

Figure 2.3. Relation between scattering and recoil angles as a function of mass ratio (a) for $M_2/M_1 < 0.8$, with an accent on the double-valued behavior of ϕ, and (b) for any M_2/M_1.

For example in scattering ^{4}He from ^{1}H, ^{2}H, and ^{3}H, the maximum scattering angles θ_{max} are 14.47°, 30°, and 48.59°, respectively. Furthermore $\theta + \phi < \pi/2$ ($\theta + \phi = \pi/2$ for $M_1 = M_2$). Note that for each scattering angle $\theta < \theta_{max}$, two different recoil angles ϕ exist as a consequence of the double sign in Eq. 2.6 (see Figures 2.2b and 2.3b). For example if the scattering angle θ tends to 0, the recoil angles tend to 0 or $\pi/2$.

2.2.2. Kinematic Factors

From Eq. 2.4 and the principle of energy conservation of the system expressed by $M_1 V_0^2 = M_1 V_1^2 + M_2 V_2^2$ we can write the following velocity equation for the incident particle M_1:

$$\left(\frac{V_1}{V_0}\right)^2 - \left(\frac{V_1}{V_0}\right)\frac{2\mu}{M_2}\cos\theta - \frac{M_1 - M_2}{M_1 + M_2} = 0 \qquad (2.8)$$

K is called the *scattering kinematic factor,* which is given by:

$$\left(\frac{V_1}{V_0}\right)^2 = K \qquad (2.9a)$$

The relationship between the incident energy E_0 and the projectile energy E_1 after scattering is

$$E_1 = KE_0 \qquad (2.9b)$$

with

$$K = \left(\frac{M_1 \cos\theta \pm \sqrt{M_2^2 - M_1^2 \sin^2\theta}}{M_1 + M_2}\right)^2 \qquad (2.10)$$

When $M_1 > M_2$, the radical in Eq. 2.10 may have either a plus or minus sign, but for $M_1 < M_2$, it must be positive. If $M_1 > M_2$, the double sign exists, which means a double value for the energy of the scattered particle. Figure 2.4 shows the variation of scattered energy E_1 as a function of the scattering angle as a function of the mass ratio.

To obtain the energy of the target particle M_2 after collision, the principle of energy conservation is used in a similar way; hence:

$$K' \equiv 1 - K \qquad (2.11a)$$

The energy of the recoil particle as a function of the incident energy is obviously:

$$E_2 = K'E_0 \qquad (2.11b)$$

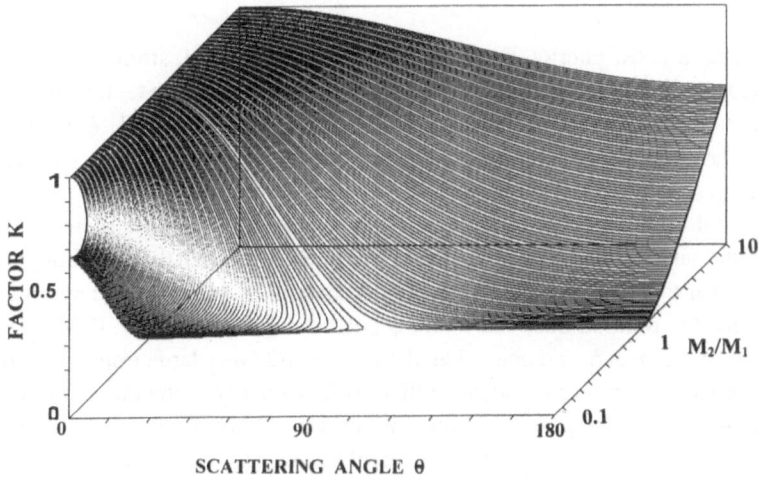

FACTOR K

0

90   180

SCATTERING ANGLE θ

Figure 2.4. Scattered energy E_1 is represented as a function of angle and mass ratio M_2/M_1. Note that for $M_2/M_1 < 1$, the energy of the scattered particle has a double value.

Using Eq. 2.11a, the *recoil kinematic factor* K' is related to the recoil angle ϕ in the following terms:

$$K' = \frac{4M_1M_2 \cos^2 \phi}{(M_1 + M_2)^2} \tag{2.12}$$

Figure 2.5 shows the dependence of recoil energy according to Eq. 2.12 as a function of recoil angle and mass ratio M_1/M_2. The energy E_1 of the projectile after

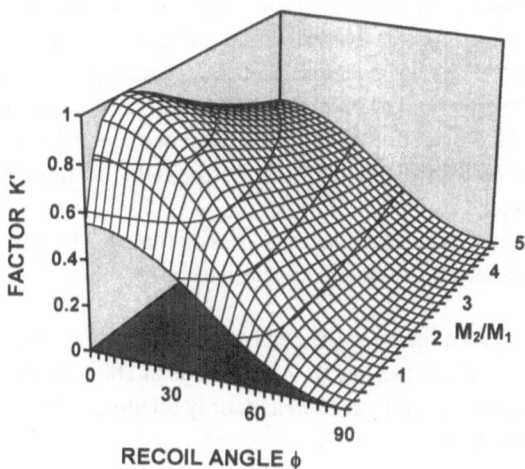

RECOIL ANGLE ϕ

Figure 2.5. Variation of recoil energy as a function of recoil angle and mass ratio. Note that the recoil energy has a single value.

collision and kinetic energy E_2 transferred to the recoiled atom are completely determined by kinematic factors K and K', respectively. Both these factors depend only on the ratio of masses and on the scattering angle θ and recoil angle ϕ, respectively. Moreover these factors contain essential information about how elastic spectrometry enables us to separate different masses.

A typical application of these relationships appears when the incident particle follows in the same direction (scattering angle θ goes to zero); in this case both possible expressions for the angle ϕ must be taken into account. If $M_1 < M_2$, the recoil angle ϕ goes to zero (the relation between θ and ϕ is a monotone function). If $M_1 > M_2$, we have both recoil angles ϕ, and one value of ϕ goes to $\pi/2$ (very large impact parameter), while the other goes to zero (head-on collision). The energy of the target particle after the collision then has two possible values. When $\phi = \pi/2$, the recoil energy is zero, and when $\phi = 0$, the energy transferred to the recoil has its greatest value E_{2max}, which can be obtained in a collision if the target particle is initially at rest:

$$E_{2max} = \frac{4M_1M_2E_0}{(M_1 + M_2)^2} \tag{2.13}$$

Note that no further information about the collision can be obtained from the laws of conservation of momentum and energy. The kinematic results previously discussed can be interpreted geometrically, and they are sufficient to carry out significant experiments in elastic spectrometry.

2.3. GEOMETRIC CONSIDERATIONS

The kinetic energy E_2 transferred to the target at rest and the energy E_1 of the projectile after scattering contain the essential information for carrying out both backscattering and recoil spectrometry. In fact the preceding concepts of kinematics are the basis for the development of several recent techniques, i.e., conventional elastic recoil detection[3,6] (ERDA), scattering recoil coincidence spectroscopy (SRCS)[4,5] and CERDA, elastic recoil by TOF-ERDA,[8-10] ExB cross field,[11,12] and elastic recoil with a magnetic spectrometer.[13]

Taking into account only the concepts of kinematics, the practical geometric configuration can be inferred for elastic spectrometry.

Backscattering spectrometry[1] ($M_1 < M_2$): Angles near 180° are of special interest. In fact K always has its lowest value at 180°. As the scattering angle θ moves away from 180°, K increases only quadratically with the difference $\pi - \theta$, and we have

$$K \approx 1 - 4\frac{M_1}{M_2} + (\pi - \theta)^2\left(\frac{M_1}{M_2}\right)^2 \tag{2.14}$$

Consequently backscattering projectile energy varies slowly for $\theta \cong \pi$. Furthermore when a target contains two types of atoms that differ in mass by a small amount ΔM_2, the difference of energy ΔE_1 after collision is largest when $\theta = \pi$ (see Section 4.1.5 on mass resolution). Therefore the vicinity of this scattering angle is the preferred location for a detector.

Forward spectrometry: Both the scattered ion and the recoiled particle can be detected in the forward direction independently of the mass ratio. Evidently when $M_1 > M_2$, the scattering angle is at most θ_{max}. According to Eq. 2.9 and Figure 2.5, the difference ΔE_1 after collision decreases with the decreasing scattering angle θ. The compromise of geometrical configuration is chosen according to beam-target parameters (see Chapters 5 and 9).

Recoil spectrometry: The energy of recoil particles is maximum for $\phi = 0°$, and a small variation of ϕ around $0°$ has only a slight influence on recoil energy E_2, since:

$$K' \cong \frac{4M_1M_2(1 - \phi^2)}{(M_1 + M_2)^2} \tag{2.15}$$

Furthermore the largest separation ΔE_2 between two signals arising from neighboring atoms in the target is desirable (see Section 4.2.5). The largest separation is reached for $\phi = 0°$ (Figure 2.5). Therefore the recoil angle $\phi = 0°$ is a favorable location for the detector.

Note that a special compromise must be found between different target–beam parameters to carry out an elastic spectrometry experiment with success. Indeed in recoil spectrometry using transmission geometry (ϕ around $0°$), special attention must be paid to target thickness (see Section 4.3.2). In coincidence recoil spectrometry, scattering and recoil angles are in general chosen large enough to avoid flooding the detector (see Chapter 9).

2.4. ENERGY LOSS

As particles penetrate matter, they slow down, and their kinetic energy decreases. The energy loss depends basically on projectile velocity, projectile–target masses, state of electronic charge, and target composition. The processes of *energy loss of a projectile* in a dense medium is also called *stopping power of a dense medium* for a particular projectile. Since the early days of nuclear physics, experimental determinations of energy loss have been performed for all projectile atoms over a wide range of energies.

Four principal methods were used extensively to measure stopping powers in materials:

Transmission: Energy loss is measured directly through self-supporting foils.

Backscattering: The difference between ion energy before and after collision is determined.

Gamma resonance shift: Resonance energy shift is measured as a function of depth.

Recoiled ions: Glancing geometry is used to determine energy loss directly through the target.

Figure 2.6 shows the typical dependence of the stopping power on projectile energy.

In a similar way several theories were developed in the past. The theory of energy loss in dense media began with the work of Bohr,[14,15] and it was continued by Bethe,[16–18] Bloch,[19] and Lindhard *et al.*[20–23] These theoretical treatments, reviewed by Bohr,[24] Firsov,[25] Fano,[26] Jackson,[27] Sigmund,[28] Andersen *et al.*,[29] Ziegler *et al.*,[30,31] Littmark *et al.*,[32] Montenegro *et al.*[33] will continue to be a source of research.

It is important to establish some concepts used to describe how a particle penetrates matter. Let dE/dx be the energy deposited through interaction processes with respect to the distance x traveled by an ion along its nominal actual path, where the energy ΔE lost per traversed Δx is given by:

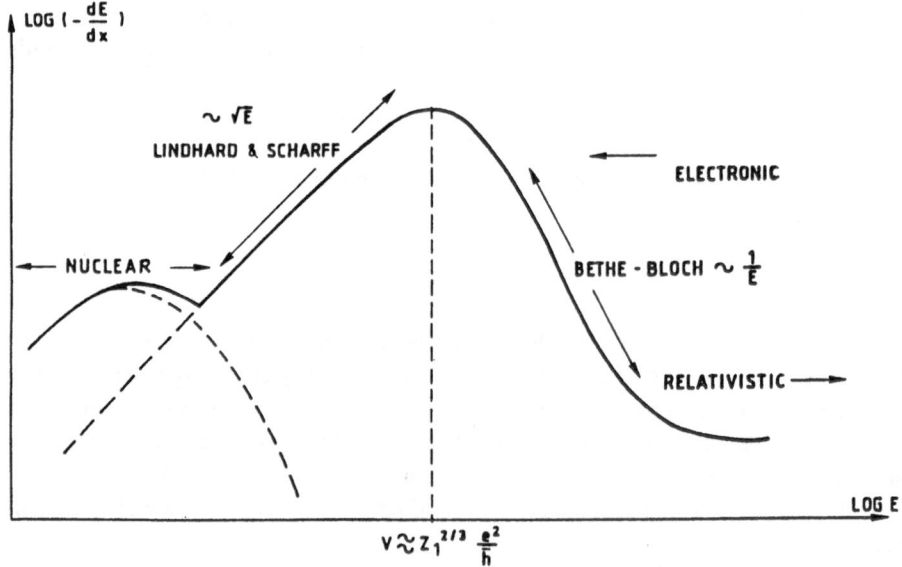

Figure 2.6. Typical variation of stopping power as a function of projectile energy.

$$\Delta E = \int_0^{\Delta x} \frac{dE}{dx}\, dx \qquad (2.16)$$

The energy loss per unit path is usually defined as *stopping power*, and it is represented by $S \equiv dE/dx$, where the conventional unit for S is $eV \cdot cm^2/g$, or alternatively the *stopping cross section* ε, for which the unit is $eV \cdot cm^2/atom$. The specific stopping power S(E) is generally known as a function of the energy E of an ion, and not of the depth x. Thus to find x(E), we must integrate over $(dE/dx)^{-1}$, and this procedure is frequently carried out by iterative approximation. Equation 2.16 is the starting point for energy loss calculations.

Following Bohr's work, several theoretical advances were made in understanding stopping powers, and complex formulations are used to describe precisely energy loss of particles in matter. Theoretical analysis considers the two processes of energy loss as independent interactions: the electronic interaction of moving ions with free electrons, called *electronic stopping power*, and the nuclear interaction of incidents ions with nuclei of target atom, called *nuclear stopping power*. Theoretical calculations assume that particles loose energy continuously from electronic interactions and by discrete amounts in nuclear scattering (elastic energy transfer), which is particularly important for low energies. Stopping powers and ion range distributions are active fields of investigation. Although around one hundred articles have been published, this subject is still a source of analysis. However essentially the same *basic energy loss theory* is commonly applied.

For completeness we present the well-known *Bethe–Bloch formula*[16-19] for the electronic stopping power S_{el}. This formula is largely applied to calculate energy loss in problems related to IBA. For high nonrelativistic energies, or more precisely when the velocity of the projectile is $V_0 \gg v_0 Z_1^{2/3}$, where $v_0 = 2.2 \times 10^8\ cm \cdot s^{-1}$ is the Bohr velocity and $V_{TF} \equiv v_0 Z_1^{2/3}$ is called *Thomas–Fermi* velocity, the electronic stopping power S_{el} is given by:

$$S_{el} = \frac{4\pi e^2 N Z_2 Z_1^2}{m_e V_0^2} L \qquad (2.17)$$

where Z_1 and Z_2 are the incident and target atomic numbers, respectively; N is the atomic density in the target; m_e is the electronic mass; and $L = \ln(2m_e V_0^2/\langle I \rangle)$ is called the *stopping number*, which varies slowly with ion energy; $\langle I \rangle$ is the average straggling ionization potential,[26] defined by the weighted average of the energy transfer and ionization energy I_i for electrons of the ith suborbital for which $2m_e V_0^2 > I_i$. In practice using the virial theorem for Coulomb potentials, the mean velocity of target electrons $\langle V_e \rangle$ is related to the mean ionization potential by $m_e \langle V_e \rangle^2 = 2\langle I_0 \rangle$.

Figure 2.6 shows the Bethe–Bloch region; the maximum of this stopping power curve lies approximately in the vicinity of Thomas–Fermi velocity. At low energies $(V_0 < v_o Z_1^{2/3})$, the Bethe–Bloch equation breaks down, but the stopping power can be approximated by a simple velocity–proportional dependence, described by using the LSS Lindhard–Scharff–Schiott[23] theory. At very low energies $V_0 \ll v_0 Z_1^{2/3}$, nuclear energy loss is the dominating process. For instance the electronic stopping power for He^+ in silicon is 239 keV/μm at 1 MeV and 42 keV/μm at 10 keV, while the nuclear stopping power is 0.37 keV/μm at 1 MeV and 8.2 keV/μm at 10 keV.

To solve the statistical problem of ion range distribution in matter, several numerical solutions of the transport equation have been found that calculate the moments of the energy distribution. Hence this procedure is a direct computer step from the universal stopping power's concept to final moments (mean range, straggling, etc.) of the distribution. This analysis has been attempted by many authors in two ways:

> *Integral equations method* for energy deposited in atomic processes that seeks analytical solutions to obtain an entire distribution profile of deposited energy. This type of solution is presented for example by Brice et al.[34] and Gibbons et al.[35] who tabulate theoretical predictions of the projected range and range statistics based on the moments method. Both use the measured electronic stopping power of Eisen et al.[36] to correct the Lindhard et al.[22] stopping power formula.

> *Monte Carlo method*, currently adopted in numerical calculations using different models for energy loss; this method is used by Biersack et al.[37] and Ziegler et al. [31]. In particular the electronic stopping power calculation for low energy is based on the Lindhard et al.[22, 23] treatment and for high energies, on the Bethe–Bloch formulation. For evaluating at very low energies, these researchers use the Molière[38] interatomic potential.

Indeed when the ion energy E is relatively high, so that the stopping power is assumed to be described only by electronic interactions, energy ΔE lost per Δx traversed can be estimated in a simple way if $\Delta E \ll E$. In fact the Bethe–Bloch formula for stopping power is valid for this energy range, so the approximation $S \propto 1/E$ can be used (Eq. 2.17). Then energy loss ΔE (Eq. 2.16) is easily evaluated. We add that this last approach is frequently applied (see Chapter 9) when we are seeking a rapid estimation of total energy loss.

Note that we need elaborate tables of ion stopping powers and ranges for elastic spectrometry. The first extensive tables were compiled by Northcliffe and Shilling.[39] Ziegler and Andersen[29,30] compiled data on proton and helions for all energies and almost all elemental targets ($\approx$ 13400 experimental data). They found an approximate analytic equation with the shape of the curves for elements for which many measurements have been made. This equation has limited theoretical basis; however it represents the best fit of stopping power values. It can also be seen from data compiled by Ziegler and Andersen[29,30] that in the vicinity of maximum stopping power (the 30–600 keV energy range) discrepancies exist, so new studies must be carried out to attempt to match theory with experimental data (e.g., Bauer,[40] Mertens[41]).

To calculate stopping powers in materials several programs, for example PRAL of TRIM[31,37] or PYROLE,[42] use the Ziegler–Andersen compilation.*

A great number of papers on stopping powers have been written during the last decade. We present only a brief review of this important problem and refer the reader to the original papers for more details, especially when a precise evaluation is required.

2.5. STRAGGLING

Energy loss of a particle in a dense medium is *statistical in nature* due to the large number of individual collisions between the particle and matter. Thus the evolution of an initially monoenergetic and monodirectional beam leads to dispersion of energy and direction. The resulting statistical distribution of energy is called *energy straggling*. In other words the initially narrow energy distribution changes with additional penetration into the target due to energy loss fluctuations. Figure 2.7 shows the propagation of the energy-straggling distribution through an absorber foil, where Tschalär and Maccabee[43] energy-straggling data are plotted as a function of depth in the material. Note that at high energies the stopping power increases with energy decrease, so that we expect the distribution to be broadened with the propagation. But at energies below the stopping power maximum, stopping power decreases again; hence the stopping power tends to reduce straggling, and at low enough energies, distribution finally shrinks. Furthermore since the straggling effect is due to differences in energy loss rate, it *depends* on other processes, for example angular and lateral spreading by multiple scattering (see Chapter 3). In Chapter 2 we are interested only in the energy-straggling distribution due to energy loss. However the combination of straggling with other processes causing energy spreading requires an adequate procedure; this is discussed in Sections 3.3.4 and 4.2.4.1.

2.5.1. Theories of Energy Straggling

It is convenient to divide the discussion of energy-straggling distribution into three domains, depending on the ratio $\overline{\Delta E}/\overline{E}$, where $\overline{\Delta E}$ is the mean energy loss and $\overline{E}$ is the average energy of the particle along the trajectory.

2.5.1.1. Very Low Fractional Energy Loss

For very thin films (small path lengths), where $\overline{\Delta E}/\overline{E} \leq 0.01$, Landau[44] and Vavilov[45] derived an energy loss distribution that includes contributions from infrequent single collisions with large energy transfers. The tabulation of this distribution,

---

*Chapter 4 discusses the role of energy loss on the forward recoil yield, and Chapter 11 shows the application of stopping powers in interpreting experimental spectra. For completeness additional information on the latest experimental energy loss measurements is provided in Appendix A, "Data Reference," in particular for protons and helions in light elements.

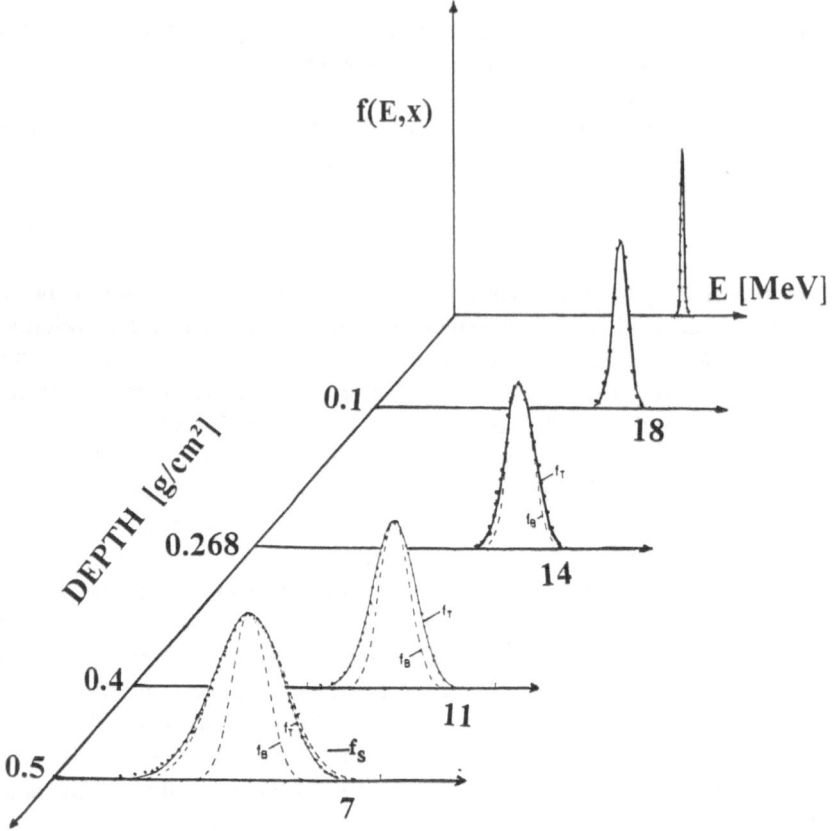

Figure 2.7. Propagation of the energy-straggling distribution through matter. The measured data distribution points from Ref. 43 represent the energy-straggling distribution in an Al foil for protons of 19.6 MeV. Models of the different distribution functions (f_B : Bohr, f_S : Symon, f_T : Tschalär) are explained in the text.

which gives a good account of experimental values in the first region, is presented in Ref. 46.

2.5.1.2. Medium Fractional Energy Loss

For regions where $0.01 < \overline{\Delta E} / \overline{E} \leq 0.2$, Bohr's model can be applied with success. In fact to calculate energy straggling, in 1915 Bohr[15] used a classical model based on electronic interactions. This basic model calculates the amount of energy straggling in terms of the areal density of electrons traversed by the beam. Bohr's model predicts a Gaussian energy distribution, which assumes that the number of collisions is large and follows a Poisson distribution. The standard deviation Ω_B^2 of the energy distribution is given by:

$$\Omega_B^2 = 4\pi(Z_1 e^2)^2 N Z_2 \Delta x \tag{2.18}$$

where $Z_2 N \Delta x$ is the number of electrons per unit area over the path length increment Δx.

In Bohr's model energy straggling is not a function of particle energy. This is a consequence of the assumption that individual energy transfers take place between free stationary electrons (ion velocity is much greater than the outer orbital electron velocity of target atoms) and the fully ionized projectile of charge $Z_1 e$. Furthermore, these individual energy transfers are small compared to total energy loss over the ion path length.

Several authors have proposed corrections to Bohr's model. Particularly useful is the modification proposed by Lindhard and Scharff[20] for low- and medium-particle energies. They divided the atomic electron cloud into two independent components, outer electrons with Thomas–Fermi velocity $V_{TF} \le V_0$ and inner electrons with $V_{TF} \ge V_0$. Using asymptotic expressions for both $V_0 >> V_{TF}$ and $V_0 << V_{TF}$, these researchers arrived at a remarkably simple expression (Eq. 2.19) related to Ω_B^2, which is a function of an ion velocity term $X = V_0^2/(v_0^2 Z_2)$, where V_0 is the initial ion velocity and v_0 is the Bohr velocity.

$$\Omega_{LS}^2 = \Omega_B^2 L(X)/2 \qquad X \le 3$$

$$\Omega_{LS}^2 = \Omega_B^2 \qquad X > 3 \tag{2.19}$$

where $L(X) = 1.36 X^{1/2} - 0.016 X^{3/2}$ is the stopping number of the target material; when the energy is given in MeV, M_1 in amu, and we can write $X \approx 40 E_0/(Z_2 M_1)$.

Improvements have been made by Bonderup and Hvelplund[47] and Chu[48] using more realistic Hartree–Fock–Slater atomic wave functions, but comparison of these theoretical results with experimental values is not conclusive; in fact energy-straggling measurements of C, N, and He ions in solid targets by Tukahashi et al.[49] and Belery et al.[50] show straggling values several times the Bohr values. However, all these theories produce estimates less than or of the order of magnitude of Bohr's energy straggling even for fractional energy $\overline{\Delta E}/\overline{E} < 0.15$. Results of these calculations differ only slightly from each other; furthermore straggling data determined by Harris and Nicolet[51] and Kido[52] for light ions (H, He) in Al, Ni targets support Lindhard and Scharff,[20] Bonderup, Hvelplund,[47] and Chu[48] theories.

Another interesting approach is the Bethe–Livingston approximation,[53] where Bohr's model is modified using quantum mechanical considerations according to:

$$\Omega_{B-L}^2 = \Omega_B^2 \left(\frac{Z'}{Z_2}\right)\left[1 + \frac{2\langle V_e \rangle^2}{3V_0^2} \ln\left(\frac{2mV_0^2}{\langle I_1 \rangle}\right)\right] \tag{2.20}$$

where $\langle V_e \rangle$ is the mean velocity of the target electrons and Z' the number of target electrons that participate in the straggling process. Furthermore like Bohr's model, this approximation considers a fully stripped ion.

Measurements of energy straggling of ^{4}He, ^{12}C, and ^{16}O in Al foils by Sofield et al.[54] and of ^{15}N in targets with $6 \leq Z_1 \leq 79$ by Briere and Biersack[55] show that the sum of the Bethe–Livingston approximation and the straggling effect of fluctuations in the charge state as ions traverse targets reconciles this theory with experimental data. To evaluate the straggling charge effect, Vollmer et al.[56] proposed a function that considers the capture cross section and the stopping power for the respective ionization states of the projectile. Besenbacher et al.,[57] who analyzed straggling for some gases at particle energies below 1 MeV/amu, assume a contribution from atomic correlation and the charge state, which is expressed as a function of the stopping power and an effective atomic section. However similar deviations are also found for energy straggling in solids (e.g., Friedland et al.[58,59]), where charge fluctuations are thought to be negligible.

2.5.1.3. Large Fractional Energy Loss

For fractional energy loss in the region of $0.2 < \overline{\Delta E} / \overline{E} \leq 0.8$, the energy dependence of stopping powers causes the energy loss distribution to differ from Bohr's straggling function. The Bohr theory must then break down for thick targets with large energy loss. Various theoretical advances were made in understanding energy straggling in this case.[43,60–64] Symon[60] proposed an expression for straggling in the region of $0.2 < \overline{\Delta E} / \overline{E} \leq 0.5$ that is a function of the moments $\mathbf{M}_i$ (i.e., $\mathbf{M}_1$ represents stopping power, $\mathbf{M}_2$ describes the variation in straggling with depth). Tschalär et al.[43,61] derived a straggling function solving the differential equation $d\Omega^2/dx = S(E) \, d\Omega^2/dE$. Figure 2.7 shows different models of the energy-straggling distribution of 19-MeV protons, where curves f_B are theoretical predictions of Bohr, f_S of Symon, and f_T of Tschalär. The Tschalär expression generally agrees well for protons in solid materials; nevertheless great discrepancies are seen for heavy incident ions (e.g., Antolak et al.[65]). L'Hoir and Schmaus[62] derived a similar theoretical expression with the condition that the energy spectrum is nearly Gaussian, and they check the validity of their assumptions using the straggling data of polyester. The Tschalär's expression,[61] which is valid for nearly symmetrical energy loss spectra, is

$$\Omega_T^2 = S^2[\overline{E(x)}] \int_{\overline{E(x)}}^{E_0} \frac{\sigma^2(E)dE}{S^3(E)} \tag{2.21}$$

where $\sigma^2(E)$ represents energy straggling per unit length, or the variance of energy loss distribution per unit length for particles of energy E and E(x) is the mean energy at depth x. The variation of $\sigma^2(E)$ with E for light ions in several matrices is very slow for energies around a few MeV/amu. In particular we can take $\sigma^2(E) \approx \Omega_B^2/\rho x$ (Bohr's formula, Eq. 2.18) and write[65]

$$\frac{\Omega_T^2}{\Omega_B^2} = \frac{S^2(\overline{E(x)})}{\rho x} \int_{\overline{E(x)}}^{E_0} \frac{dE}{S^3(E)} \qquad (2.22)$$

For example Cohen and Rose[66] analytically tested Eq. 2.22 using only straggling measurements in Kapton. They show that straggling in a Kapton foil can differ by a factor of 5 or more from Bohr's prediction for a relative energy loss above 60%. Equation 2.22 is a satisfactory expression of straggling for a large energy loss fraction.

2.5.2. Practical Straggling Evaluation

Attempts were made to find a general analytical expression for energy straggling, but nothing could be done for a complete energy range. Unfortunately few reliable energy-straggling data are available, and large discrepancies are often observed in verifying different theories. Discrepancies could not be resolved in terms of target nonuniformities (e.g., Sofield et al.,[54] Andersen et al.[67]). Agreement of a certain number of experimental data (e.g., Briere et al.[55]) with the Bethe–Livingston approximation for a fully ionized projectile suggests that nearly the whole projectile nuclear charge contributes to the energy-straggling process. Furthermore agreement between the measured straggling data and Bohr's model in the velocity regime (low energies), where electrons must be considered as bound to atoms, and not free and stationary as assumed in Bohr's theory, must be taken as coincidental. In general the theories do not take into account the effect of charge state in broadening energy straggling.

For practical elastic spectrometry, the various results described suggest that for small $\overline{\Delta E}/\overline{E} \le 0.01$, the variance of straggling can be calculated by the Landau–Vavilov method. For medium fractional energies, $0.01 < \overline{\Delta E}/\overline{E} \le 0.2$, the Gaussian energy distribution describes straggling satisfactorily. In fact energy straggling can be calculated using improvements to Bohr's model, as previously discussed, or the Bethe–Livingston model (Ω_{B-L}^2) for fully stripped ions. For large energy loss, $0.2 < \overline{\Delta E}/\overline{E} \le 0.8$, the energy distribution is still nearly Gaussian, but the total energy straggling may be several times the Bohr values. In this case straggling expressions, as surveyed by Cohen and Rose[66] (from Tschälär et al.,[61] L'Hoir et al.[62] and Schwab et al.[63]) can be used; nevertheless supplementary experiments are useful for a more complete test of these theories. Furthermore in particular cases, the charge state effect can be introduced. This effect is significant at the stopping power maximum, where charge exchange is highly likely.

2.6. CONCLUSION

Kinematics of elastic spectrometry are fully described by applying the principles of energy and momentum conservation. The energy of scattered particles and recoiled atoms is expressed by a simple two-body collision. Geometrical considerations often used in ERDA are deduced from Eq. 2.6. Note that the mass ratio is a determining factor; indeed if $M_2 < M_1$, the energy of scattered particles has a double value, which requires a choice for recoil and scattered angles in a particular experimental situation. Section 2.3 gives the key expressions often used to carry out experiments in elastic recoil spectrometry.

The process of energy loss of a particle in a dense medium is briefly treated in Section 2.4, since a great number of reviews exist where details of these processes are thoroughly discussed.

The statistical process of energy loss is discussed in Section 2.5. Analysis depends on the ratio $\overline{\Delta E}/\overline{E}$ (where $\overline{\Delta E}$ is the mean energy loss and $\overline{E}$ is the average energy of the particle along the trajectory). Thus we divide the discussion into three areas. For thin films, $\overline{\Delta E}/\overline{E} \leq 0.01$, the Landau and Vavilov calculation can be applied. For medium path length, $0.01 < \overline{\Delta E}/\overline{E} \leq 0.02$, Bohr's model can be applied (Eq. 2.18). For thick targets, $0.2 < \overline{\Delta E}/\overline{E} \leq 0.8$, the full integral (Eq. 2.22) must be used. Despite straggling measurements should still be performed for certain types of target–particle systems, nevertheless the various theoretical approaches discussed in Chapter 2 are quite reasonable.

REFERENCES

1. Chu, W. K., Mayer, J. W., and Nicolet, M. A., *Backscattering Spectrometry* (Academic, New York, 1978).
2. Bird, J. R., and Williams, J. S., *Ion Beams for Materials Analysis* (Academic, Sydney, 1989).
3. Doyle, B., and Peercy, P., Technique for profiling ^{1}H with 2.5–MeV Van de Graaff accelerators, *App. Phys. Lett.* **34**, 811 (1979).
4. Chu, W. K., and Wu, D. T., Scattering recoil coincidence spectrometry, *Nucl. Instrum. Methods Phys. Res. Sect. B* **35**, 518 (1988).
5. Hofsäss, H. C., Parikh, N. R., Swanson, M. L., and Chu, W. K., Elastic recoil coincidence spectroscopy (ERCS), *Nucl. Instrum. Methods Phys. Res. Sect. B* **58**, 49 (1991).
6. Tirira, J., Trocellier, P., Frontier, J. P., and Trouslard, Ph., Theoretical and experimental study of low-energy ^{4}He-induced ^{1}H elastic recoil with applications to hydrogen behaviour in solids, *Nucl. Instrum. Methods Phys. Res. Sect. B* **45**, 203 (1990).
7. Landau, L. D., and Lifshitz, E. M., *Mechanics* (Pergamon, New York, 1960).
8. Thomas, J. P., Fallavier, M., Ramdane, D., Chevarier, N., and Chevarier, A., High-resolution depth profiling of light elements in high atomic mass materials, *Nucl. Instrum. Methods Phys. Res.* **218**, 125 (1983).
9. Houdayer, A., Hinrichsen, P. F., Gujrathi, S., Martin, J., Monaro, S., Lessard, L., Oxorn, K., Janicki, C., Brebner, J., Belhadfa, A., and Yelon, A. N., Trace element and surface analysis at the University of Montreal, *Nucl. Instrum. Methods Phys. Res. Sect. B* **24/25**, 643 (1987).
10. Groleau, R. Gujrathi, S. C., and Martin, J. P., Time-of-flight system for profiling recoiled light elements, *Nucl. Instrum. Methods Phys. Res.* **218**, 11 (1983).
11. Ross, G., Terreault, B., Gobeil, G., Abel, G., Boucher, C., and Veilleux, G., Inexpensive, quantitative hydrgen depth profiling for surface probes, *J. Nucl. Mat.* **128/129**, 730 (1984).
12. Serruys, Y., and Tirira, J., ERDA analysis from hydrogen to oxygen using a modified ExB filter, submitted to *Nucl. Instrum. Methods Phys. Res. Sect. B*.
13. Gosset, C. R., Use of a magnetic spectrometer to profile light elements by elastic recoil detection, *Nucl. Instrum. Methods Phys. Res. Sect. B* **15**, 481 (1986).
14. Bohr, N., On the theory of the decrease of velocity of moving electrified particles on passing through matter, *Philos. Mag.* **25**, 10 (1913).

15. Bohr, N., On the decrease of velocity of swiftly moving electrified particles in passing through matter, *Philos. Mag.* **30**, 581 (1915).
16. Bethe, H. A., Zur theorie des Durchgangs schneller Korpuskularstrahlen durch Materia, *Ann. Phys.* **5**, 325 (1930).
17. Bethe, H. A., Bremsformel für elektronen relativistischer Geschwindigkeit, *Z. Phys.* **76**, 293 (1932).
18. Bethe, H. A., and Heitler, W., On the stopping of fast particles and on the creation of positive electrons, *Proc. R. Soc. London* A **146**, 83 (1934).
19. Bloch, F., Zur Bremsung rasch bewegter Teilchen beim Durchgang durch Materia, *Ann. Phys.* **16**, 285 (1933); Bremsvermögen von Atomen mit mehreren Elektronen. *Z. Phys.* **81**, 363 (1933).
20. Lindhard, J., and Scharff, M., Energy loss in matter by fast particles of low charge, *Mat. Fys. Medd. Dan. Vid. Selsk.* **27**(18), 1 (1953).
21. Lindhard, J., On the properties of a gas of charged particles, *Mat. Fys. Medd. Dan. Vid. Selsk.* **28**(8), 1 (1954).
22. Lindhard, J., and Scharff, M., Energy dissipation by ions in the keV region, *Phys. Rev.* **124**, 128 (1961).
23. Lindhard, J., Scharff, M., and Shiott, H. E., Range concepts and heavy ion ranges. (Notes on atomic collisions, II), *Mat. Fys. Medd. Dan. Vid. Selsk.* **33**(14), 1 (1963).
24. Bohr, N., The penetration of atomic particles through matter, *Mat. Fys. Medd. Dan. Vid. Selsk.* **18**(8), (1948).
25. Firsov, O. B., A qualitative interpretation of the mean electron excitation energy in atomic collisions, *Sov. Phys. JETP.* **36**, 1076 (1959).
26. Fano, U., Penetration of protons, alpha particles and mesons, *Ann. Rev. Nucl. Sci.* **13**, 1 (1963).
27. Jackson, J. D. C., *Classic Electrodynamics* (Wiley, New York, 1975), chap. 13.
28. Sigmund, P., *Radiation Damage Process in Materials* (Noordhoff, Leyden, 1975), chap. 1.
29. Andersen, H. H., and Ziegler, J. F., *Hydrogen Stopping Powers and Ranges in All Elements*, vol. 3 (Pergamon, New York, 1977).
30. Ziegler, J. F., and Andersen, H. H., *Helium Stopping Powers and Ranges in All Elements*, vol. 4 (Pergamon, New York, 1977).
31. Ziegler, J. F., Biersack, J. P., and Littmark, U., *Stopping Powers and Range of Ions in Solids* (Pergamon, New York, 1985).
32. Littmark, U., and Ziegler, J. F., *Handbook of Range Distribution for Energetic Ions in All Elements* (Pergamon, New York, 1980).
33. Montenegro, E. C., Cruz, S. A., and Vargas–Aburto, C., A universal equation for the electronic stopping of ions in solids, *Phys. Lett.* A **92**, 195 (1982).
34. Brice, D. K., *Ion Implantation Range and Energy Deposition Distribution* (Plenum Press, New York, 1975).
35. Gibbons, J. F., Johnson, W. S., and Mylroie, S. W., *Projected Range Statistics* (Dowden, Hutchinson, and Ross, Stroundsburg, PA 1975).
36. Eisen, F. H., Welch, B., Westmoreland, J., and Mayer, J. W., in *Atomic Collision Processes in Solids* (Palmer, D. W., Thompson, M. W., and Townsend, P. D., eds.) (Pergamon, London, 1970).
37. Biersack, J. P., and Haggmark, L. G., A Monte Carlo computer program for the transport of energetic ions in amorphous targets, *Nucl. Instrum. Methods Phys. Res.* **174**, 257 (1980).
38. Molière, G., Theorie der Streuung schneller geladener Teilchen I, *Z. Naturforsch.* A **2**, 133 (1947).
39. Northcliffe, L. C., and Schilling, R. F., Range and stopping-power tables for heavy ions, *Nucl. Data Tables* A **7**, 233 (1970).
40. Bauer, P., How to measure absolute stopping cross section by backscattering and transmission methods. Part I. Backscattering, *Nucl. Instrum. Methods Phys. Res. Sect.* B **27**, 301 (1987).
41. Mertnes, P., How to measure absolute stopping cross section by backscattering and transmission methods. Part II. Transmission, *Nucl. Instrum. Methods Phys. Res. Sect.* B **27**, 315 (1987).
42. Trouslard, P., Pyrole : un logard an service des analyses par faisuan d'ions, Rapport CEA-R-5703 (1995).

43. Tschalär, C., and Maccabee, H. D., Energy straggling measurements of heavy charged particles in thick absorbers, *Phys. Rev.* B **1**, 2863 (1970).
44. Landau, L., On the energy loss of fast particles by ionization, *J. Phys. JETP* **8**, 201 (1944).
45. Vavilov, P. V., Ionization losses of high-energy heavy particles, *S. Phys. JETP.* **5**, 749 (1957).
46. Deconninck, G., and Fouille, Y., Energy spreading calculations and consequences. in *Ion Beam Surface Layer Analysis* (Meyer, O., Linker, G., and Käppeler, F., eds.) (Plenum, New York, 1976), pp. 87–97.
47. Bonderup, E., and Hvelplund, P., Stopping power and energy straggling for swift protons, *Phys. Rev.* A **4**, 562 (1971).
48. Chu, W. K., Calculation of energy straggling for protons and helium ions, *Phys. Rev.* A **13**, 2057 (1976).
49. Tukahashi, T., Awaya, Y., Tonuma, T., Kumagai, H., Izumo, K., Hashizume, A., Uchiyama, S., and Hitachi, A., Energy Straggling of C and He ions in metals foils, *Nucl. Instrum. Methods Phys. Res.* **166**, 587 (1979).
50. Belery, P., Delbar, T., and Gregoire, G., Multiple scattering and energy straggling of heavy ions in solid targets, *Nucl. Instrum. Methods Phys. Res.* **179**, 1 (1981).
51. Harris, J. M., Chu, W. K., and Nicolet, M.-A., Energy straggling of ^{4}He below 2 MeV in Pt, *Thin Solid Films*. **19**, 259 (1973).
52. Kido, Y., Energy and Z_2 dependencies of energy straggling for fast proton beams passing through solids, *Phys. Rev.* B **34**, 73 (1986).
53. Livingston, M. S., and Bethe, H. A., Nuclear dynamics, Experimental, *Rev. Mod. Phys.* **9**(3), 245 (1937).
54. Sofield, C. J., Cowern, N. E. B., and Freeman, J. M., The role of charge-exchange in energy-loss straggling, *Nucl. Instrum. Methods Phys. Res.* **191**, 462 (1981).
55. Briere, M. A., and Biersack, J. P., Energy loss straggling of MeV ions in thin solid films, *Nucl. Instrum. Methods Phys. Res. Sect. B* **64**, 693 (1992).
56. Vollmer, O., Der Einfluss von Ladungsfluktuationen auf die Energieverlustverteilung geladener Teilchen, *Nucl. Instrum. Methods Phys. Res.* **121**, 373 (1974).
57. Besenbacher, E., Andersen, J. U., and Bonderup, E., Straggling in energy loss of energetic hydrogen and helium ions, *Nucl. Instrum. Methods Phys. Res.* **168**, 1 (1980).
58. Friedland, E., and Kotze, C. P., Energy loss straggling of protons, deuterons and α-particles in copper, *Nucl. Instrum. Methods Phys. Res. Sect. B* **191**, 490 (1981).
59. Friedland, E., and Lombaard, J. M., Energy loss straggling of alpha particles in Al, Ni, and Au, *Nucl. Instrum. Methods Phys. Res.* **168**, 25 (1981).
60. Symon, K. R., Ph.D. diss. Harvard University (1948); see Symon's theory discussion in Payne, M. G., Energy straggling of heavy particles in thick absorbers, *Phys. Rev* **185**, 611 (1969).
61. Tschalär, C., Straggling distribution of large energy losses, *Nucl. Instrum. Methods Phys. Res. Sect. B* **61**, 141 (1968).
62. L'Hoir, A. L., and Schmaus, D., Stopping power and energy straggling for small and large energy losses of MeV protons transmitted through polyester films, *Nucl. Instrum. Methods Phys. Res. Sect. B* **4**, 1 (1984).
63. Schwab, Th., Geissel, H., Armbruster, P., Gillibert, A., Mitting, W., Olson, R. E., Witerborn, K. B., Wollnik, H., and Münzenberg, G., Energy and angular distribution for Ar ions penetrating solids, *Nucl. Instrum. Methods Phys. Res. Sect. B* **48**, 69 (1990).
64. Wilken, B., and Fritz, T. A., Energy distribution functions of low energy ions in silicon absorbers measured for large relative energy losses, *Nucl. Instrum. Meth. Phys. Res.* **138**, 331 (1976).
65. Antolak, A. J., Handy, B. N., Morse, D. H., and Pontau, A. E., Energy loss and Straggling measurements of ions in solid absorbers, *Nucl. Instrum. Methods Phys. Res. Sect. B* **59/60**, 13 (1991).
66. Cohen, D. D., and Rose, E. K., Large energy loss straggling of protons and He ions in Mylar foils, *Nucl. Instrum. Methods Phys. Res. Sect. B* **64**, 672 (1992).
67. Andersen, H. H., Besenbacher, F., and Goddiksen, P., Stopping power and straggling of 80–500 keV Lithium ions in C, Al, Ni, Cu, Se, Ag, and Te, *Nucl. Instrum. Methods Phys. Res.* **168**, 75 (1980).

Elastic Scattering
Cross-Section and Multiple Scattering

3.1. INTRODUCTION*

The advent of ERDA as an analytical technique has produced renewed investigations of elastic collisions. The principles of elastic spectrometry are generally well-known but not completely enough to describe entirely the elastic yield from light atoms. To determine thoroughly the scattering yield, it is necessary to examine scattering cross sections. For usual Rutherford backscattering spectrometry, the effective cross section is straightforward. However, for ERDA, when light particles (e.g., ^1H, ^2H, ^3H) recoil after scattering heavier impinging ions (e.g., ^4He), the collision process cannot be considered according to the simple Rutherford theory.[1,2] Hence the cross section remains an open question.

The elastic scattering problem can be developed according to the following structure:

Elastic atomic collision: Interaction involves the screened Coulomb potential for a large distance of closest approach R. This means that $R \gg a_1 + a_2$ (sum of screening radii). Typically this is important at very low energies, for example 10 keV for Kr ions.

Elastic Coulomb nuclear collision: Unscreened Coulomb potential describes the interaction process (Rutherford scattering). This is relevant if the distance of closest approach is of the order of the screened radii of the ion and atom nuclei. In this case the incident energy of projectiles is below the Coulomb barrier E_{co} (i.e., the energy of an ion for the distance of closest approach to equal the limit of incipient nuclear interaction).

Elastic nuclear collision: Projectiles have enough energy to approach the target nucleus and the nuclear force field of targets becomes important as well as

*The authors wish to acknowledge Peter Sigmund and Georges Amsel for their critical and stimulating reviewing of the multiple scattering section.

the unscreened potential deviates from the Coulomb potential. This is the case for energies larger than the Coulomb barrier energy. In practice, we must often define a threshold energy less than E_{co} at which deviations from the Coulomb collision begin to appear.

Interactions involving nuclear forces can include excitation of the nucleus and transfer of one or more nucleons. However in the present development we consider only collisions where total kinetic energy is conserved and the energy release Q equals zero. In the following we are mainly interested in the initial and final states of a binary collision, so intermediate excited states are not treated herein. Depending on the kind of elastic collision, we can treat the problem using a Coulomb potential,[3] a weak deviation from a Coulomb potential,[4] and special potentials, which can be used to describe the interaction process. Indeed significant deviations from Rutherford theory are quite evident for elastic nuclear reactions induced by ^{4}He ions on hydrogen isotopes,[1,2,5] since such nuclear particles effectively penetrate the repulsing Coulomb barrier.

To complete our review of elastic scattering, we present an overview of multiple scattering (MS) which occurs whenever particles passing through matter undergo successive scattering events; MS is thus a stochastic process. Angular and lateral spreads due to MS are very important in ERDA. One reason is that both MS of scattered projectiles and recoiled light atoms are responsible for a decreased counting rate at low energies (see Section 9.6), corresponding to scattering at large depth. Multiple scattering also leads to energy spread that causes a deterioration of the depth resolution (see Chapter 4).

In Chapter 3 we first discuss the elastic scattering cross section in Section 3.2, and the MS distribution is analyzed in Section 3.3. As we are particularly concerned with hydrogen isotopes, the following discussion of cross section goes as far as available theoretical models and experimental data permit. Both new measurements of cross-section values and some theoretical reviews are presented.[5–8] We then consider the problem of cross section in the energy range (E_{He}) 1–4 MeV, which is the most common domain for recoil spectrometry. To include recoil spectrometry using other ion beams (i.e., coincidence recoil spectrometry, light elements profiling using heavy ions) for which collisions can be interpreted as usual Coulomb interactions, we must first introduce the physical concepts of *classical Rutherford scattering* and next the threshold energy for a limit of Rutherford behavior. Chapter 3 continues with a complete discussion of multiple scattering.

3.2. ELASTIC CROSS SECTION

It is usually necessary to measure cross sections at specific energy values and recoil angles useful for IBA. For the purposes of forward recoil spectrometry, compilations of experimental scattering cross-section data are difficult to handle. Not only for this reason, but also for the purpose of improving theoretical concepts, attempts were made to match theoretical expressions or semiempirical fits with experimental data. This necessarily involves some preliminary selection among available data. These analytical expressions may be much more suitable for practical calculations in spectra analysis than a compilation of individual values. Differential cross-section functions have been the concern of several studies in the past, as discussed in the following sections. In particular special attention is paid to understanding collisions of helium with hydrogen isotopes, because helium beams are largely used to analyze these elements.

The absolute cross section for hydrogen isotopes from ^{4}He ions has been extensively studied in the past.[1,2,6] These investigations were carried out to study nuclear levels of light nuclei (e.g., ^{5}Li, ^{6}Li, ^{7}Li). Although there are four angular observables, only the differential cross section and polarization have been determined in the medium energy range $E_{He} > 3$ MeV. Energy-dependent phase shifts analyses already exist in this energy range. However for recoil spectrometry, it is very important to know these scattering parameters for energies generally below and up to $E_{He} = 3$ MeV. In particular an accurate knowledge of the absolute cross sections is often important for correctly interpreting spectra in such studies as profiling hydrogen isotopes in metals and minerals.

3.2.1. Rutherford Scattering

In Chapter 2 individual two-body collisions are treated from the kinematic point of view. In practice we examine the scattering of an ion beam by one target atom. Although this scattering is merely composed of two-body collisions with well-determined kinematics, the question of the number of scattering events arises. Thus the concept of a scattering cross-section must be introduced. This parameter is defined as the ratio of the number of particles scattered through angles between θ and $\theta + d\theta$ per time unit and the total number of particles traversing a unit of area per time unit. In other words the scattering cross-section is the effective area presented to incoming ions by each interaction center; this parameter is expressed in units of cm^2 or barn ($=10^{-24}$ cm^2). We add that the cross section is completely determined by the type of interaction potential V(R) associated with the collision, and the cross section represents the most important parameter of a collision process. This discussion is based on the classical description for the simplified case of a central potential.[3] Although the scattering cross-section problem can be resolved in the laboratory system, we begin the discussion using the center-of-mass coordinates C, then transform this solution into the laboratory system.

A beam of identical particles impinging with uniform velocity V_0 on the scattering center is scattered through different angles ξ in the center-of-mass system (Figure 2.1). Since the scattering angle ξ is a monotonically decreasing function of the impact parameter, only particles with impact parameters between $\rho(\xi)$ and $\rho(\xi) + d\rho(\xi)$ are scattered at angles between ξ and $\xi + d\xi$. The number of particles scattered per unit time through angles between ξ and $\xi + d\xi$ is equal to $2\pi\rho(\xi)d\rho(\xi)N_0$, where N_0 is the number of particles passing per unit time through an unit area of the beam cross section. The effective scattering cross-section in the C system is therefore:

$$d\dot{\sigma}_c = \frac{\rho(\xi)}{\sin \xi} \frac{d\rho(\xi)}{d\xi} d\Omega \tag{3.1}$$

where $d\Omega$ is the solid angle between cones with angles ξ and $\xi + d\xi$.

The scattering of charged particles in a Coulomb potential is the most classical case of elastic collision. For a Coulomb potential $V(R) = \alpha/R$, the impact parameter as a function of angle ξ is given by:

$$\rho^2(\xi) = \frac{\alpha^2 \cot^2(\xi/2)}{\mu^2 V_0^4} \tag{3.2}$$

where μ is the reduced mass (see Eq. 2.4).

Using Eqs. (3.1) and (3.2), the differential scattering cross-section $d\sigma_c$ can be written in the following form:

$$d\sigma_c = \left(\frac{\alpha}{2\mu V_0^2 \sin^2(\xi/2)} \right)^2 d\Omega \tag{3.3}$$

This is the classical Rutherford formula in the center-of-mass system. This famous expression is the basis of several physical applications involving projectile collisions. For a Coulomb repulsion potential $\alpha = Z_1 Z_2 e^2$, Eq. (3.3) can be written as:

$$\left(\frac{d\sigma}{d\Omega} \right)_C = \left(\frac{Z_1 Z_2 e^2}{4E_c \sin^2(\xi/2)} \right)^2 \tag{3.4}$$

here Z_1 and Z_2 are the atomic numbers of the projectile and target atom, respectively, and E_c is the energy before scattering of the projectile with reduced mass μ in the C system. Note that for Coulomb scattering, both classical and quantum mechanical calculations give this same well-known result.

In practice using a general experimental set-up for elastic spectrometry, we often detect particles within a finite solid angle $\Delta\Omega$ (a few msr). Thus we measure an average cross section σ_T rather than a differential cross section:

$$\sigma_T \equiv \frac{1}{\Omega} \int \frac{d\sigma}{d\Omega} d\Omega \tag{3.5}$$

The generic appelation scattering cross-section is currently used for values obtained from Eq. (3.5).

3.2.2. Cross Section in the Laboratory System

Equation 3.4 gives the differential scattering cross-section $(d\sigma/d\Omega)_C$ in the frame of reference in which the center-of-mass of colliding particles is at rest. To obtain the cross-section formula in the laboratory frame $(d\sigma/d\Omega)_L$ for incidents projectiles or particles initially at rest (called recoils), we must consider the following transformation relationships:

- The total number of scattered particles in the solid angle $(d\Omega)_C$ is equivalent to the number measured in the solid angle $(d\Omega)_L$. Hence the conservation relation for recoils:

$$\left(\frac{d\sigma}{d\Omega}\right)_C \sin \xi \, d\xi = \left(\frac{d\sigma}{d\Omega}\right)_L \sin \phi \, d\phi \qquad (3.6)$$

The same equation holds for projectiles with θ instead of angle ϕ.

- Using the transformation between recoil and center-of-mass angles $(2\phi = \pi - \xi)$, we obtain the following expression for the *differential recoil cross-section in the laboratory system* $(d\sigma/d\Omega)_\phi$:

$$\left(\frac{d\sigma}{d\Omega}\right)_\phi = \left(\frac{d\sigma}{d\Omega}\right)_C 4\cos \phi \qquad (3.7)$$

3.2.3. Rutherford Recoil Cross Section

Using Eq. 3.7 and the adequate transformation of the energy between C and L systems, we can write the differential recoil cross-section in the laboratory system given by:

$$\left(\frac{d\sigma}{d\Omega}\right)_\phi = \left(\frac{Z_1 Z_2 e^2}{2E_0}\right)^2 \left(\frac{M_1}{M_2} + 1\right)^2 \frac{1}{\cos^3 \phi} \qquad (3.8)$$

where E_0 is given in MeV, $e^2 \approx 1.4398 \ 10^{-13}$ MeV.cm, and $(d\sigma/d\Omega)_\phi$ is given in cm^2.

Figure 3.1 shows variations of $(d\sigma/d\Omega)_\phi$ as a function of recoil angle and mass ratio M_2/M_1. The functional dependence of classic recoil scattering (Eq. 3.8) can be summarized as follows:

- $(d\sigma/d\Omega)_\phi$ exhibits a minimum value for $\phi = 0°$ ($\propto \cos^{-3}\phi$).

- $(d\sigma/d\Omega)_\phi$ tends to infinity for $\phi \to \pi/2$, which corresponds to large impact parameters. At these distances the interaction between the projectile and the target nuclei is not completely Rutherford. In fact we must include electron screening in the scattering field (e.g., see Ref. 9).

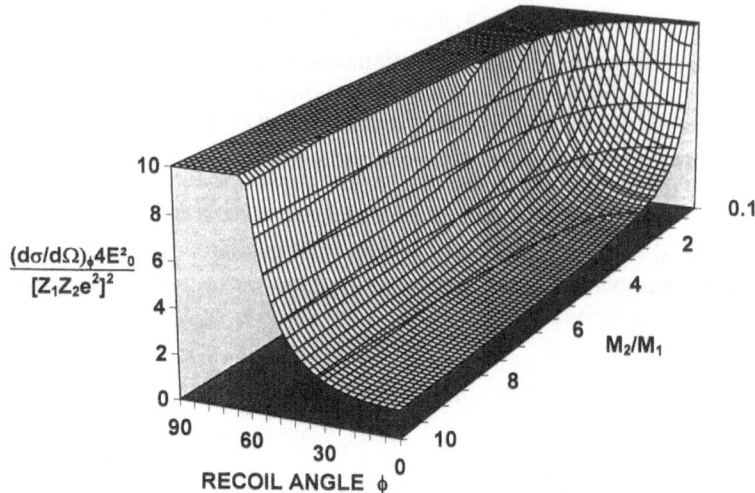

Figure 3.1. Rutherford recoil cross section given by Eq. (3.8) as a function of recoil angle ϕ and mass ratio M_2/M_1.

- The yield of recoil particles rises rapidly with decreasing incident energy ($\propto E^{-2}$).

- When $M_2 \gg M_1$, $(d\sigma/d\Omega)_\phi$ is approximately independent of the mass ratio and almost the same cross-section values can be assigned to different light atoms colliding with a heavy projectile.

In ERDA, a Rutherford behavior is observed predominantly in the case of heavy ions colliding with light atoms (some examples are shown in Chapters 9 and 13), and for collisions below 1 MeV, where the Coulomb field alone may be considered.

3.2.4. Rutherford Projectile Cross Section

The study of recoil spectroscopy sometimes needs the analysis of the cross section for scattered incident particles, (for example in the case of coincidence spectrometry). Thus, starting from Eq. 3.4 and using the adequate transformations from the center-of-mass to the laboratory system (equation 3.6) the differential cross-section for projectiles $(d\sigma/d\Omega)_\phi$ in a Rutherford collision is given by:

$$\left(\frac{d\sigma}{d\Omega}\right)_\theta = \left(\frac{Z_1 Z_2 e^2}{2E_0}\right)^2 \frac{(\{1 - [(M_1/M_2) \sin \theta]^2\}^{1/2} + \cos \theta)^2}{\{1 - [(M_1/M_2) \sin \theta]^2\}^{1/2}} \frac{1}{\sin^4 \theta} \qquad (3.9)$$

Figure 3.2 shows the variation of $(d\sigma/d\Omega)_\theta$ as a function of angle θ and mass ratio M_2/M_1. Furthermore we can summarize the main characteristics of this differential cross-section (Eq. 3.9) as follows (see Ref. 10 for more details):

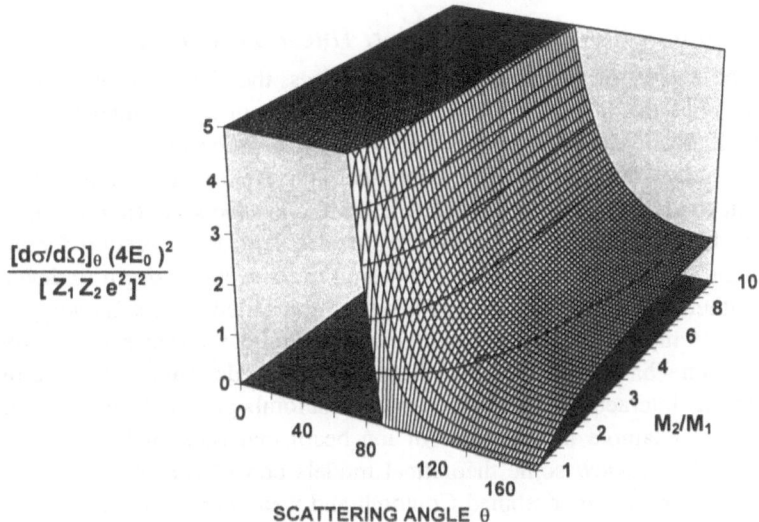

$$\frac{[d\sigma/d\Omega]_\theta \, (4E_0)^2}{[\,Z_1 Z_2 e^2\,]^2}$$

SCATTERING ANGLE θ

Figure 3.2. Rutherford scattering cross-section for incident particles given by Eq. (3.9) as a function of angle θ and mass ratio M_2/M_1.

- $d\sigma/d\Omega_\theta$ is inversely proportional to the square of the projectile energy. Hence the scattered yield increases rapidly with decreasing energy.

- $d\sigma/d\Omega_\theta$ is approximately proportional to $\sin^{-4}(\theta/2)$ for $M_1 \ll M_2$. This implies that the backscattering yield increases rapidly as the scattering angle is diminished.

- Deviations from Rutherford expression do exist for $\theta \to 0$ which corresponds to the recoil angle $\phi \to \pi/2$ (see preceding discussion). In such cases scattering cross-section must be derived, including electron screening in the potential (see Everhard *et al.*[9]).

- $d\sigma/d\Omega_\theta$ has its lowest value at 180°. Furthermore near this angle, the differential cross section varies slowly with the scattering angle.

Depending on the projectile–target nucleus combination and projectile energy, the scattering process may still be elastic. Indeed heavy incident beams with energies higher than 200 MeV have been used for elastic recoil spectrometry of several atoms.[11] This type of experiment is discussed in Chapter 13. In the medium energy range (1–4 MeV for ⁴He colliding on light elements), short-range nuclear forces influence the scattering process, so the differential cross-section deviates from Rutherford.[4,6] However to know the limit of validity of the Rutherford theory, we must examine scattering cross-section as a function of incident energy to define an incident energy limit for which elastic scattering begins to deviate from a purely Coulomb interaction.

3.2.5. Limit of Rutherford Behavior: Threshold Energy

As the energy of incident particles increases, the simple Rutherford collision theory can no longer be used to describe a classical elastic scattering using a Coulomb potential. Incident particles have enough energy to probe the nuclear force of the interacting target nucleus. Several theoretical and experimental studies[4,12–15] were carried out to understand how nuclear forces affect elastic scattering for high-energy ion beams. In other words these studies attempted to develop a procedure for predicting the incident energy at which elastic scattering cross sections begin to deviate from classic Rutherford theory. This energy limit is often called *threshold energy* E_{th}.

As an attempt to match theory with experimental data, some specific projectile–nucleus systems have been studied. Values of E_{th} were determined by assuming that the scattering interaction is described by a Coulomb potential and a perturbating potential. For example in the frame of ion beam materials analysis, Bozoian and co-workers[4,13,14] review some theoretical models considering that scattering takes place in the presence of combined Coulomb and weak-perturbing nuclear potential, for example a Yukawa-like potential. Measurements of energy E_{th} were made for the elastic scattering cross-section of some projectile–target couples[12,15] that deviate by about 4% from Rutherford values in the angular range used in backscattering spectrometry ($\theta > 150°$). Recently, Räisänen and Rauhala[15,16] extended the E_{th} study; they propose a polynomial fit to calculate this energy E_{th}. From another point of view, Goppelt and co-workers[11] calculate energy E_{th} on the basis of the optical potential model, using the condition that the elastic cross-section deviates from Rutherford scattering only by 1%, and compare E_{th} energies with Coulomb barrier energies. In fact it is a currently used convention to define the *experimental threshold energy* as the energy at which the cross section deviates from the pure Rutherford formula by a few percent. A survey of literature shows that a theoretical closed solution based on the interaction potential is complex; it is not discussed here. However an approach is given for some projectile– nucleus systems, and a simple expression is proposed for practical problems. To estimate the maximum incident energy below the Coulomb barrier E_{co} causing only slight deviations from the pure elastic collision, we can use the classic model of Coulomb scattering. This non-Rutherford threshold energy E_{th} can be considered a first approximation of the practical high-energy limits applicable to ERDA. When the distance of closest approach R_0 is used as a free parameter, incident energy E_{th} in the laboratory system can be written in the following form[4,14]:

$$E_{th} = \frac{Z_1 Z_2 e^2}{2R_0} \left(1 + \frac{M_1}{M_2} \right) \left(1 + \frac{1}{\cos\phi} \right) \tag{3.10}$$

where R_0 represents the closest approach corresponding to the non-Rutherford threshold energy, it is expressed as a function of the mass numbers A_1 and A_2 of projectile and target, respectively. Indeed based on extensive literature survey of over 100 experimental data points Räisänen et al.[15] propose the following linear fit for the

Table 3.1. Typical E_{th} Values (MeV) for Some Usual Elastic Collisions[a]

Target→	^{12}C	^{14}N	^{16}O
ION ↓			
^{35}Cl	57.37	59.24	61.12
^{81}Br	215.51	218.20	221.03
^{127}I	463.84	466.56	469.60
^{197}Au	984.96	986.37	988.38

[a]From Eq. 3.10.

parameter R_0 as a function of $A_1^{1/3} + A_2^{1/3}$ (which is the typical approximation of nuclear radii):

$$R_0 = C_1 + C_2 \left(A_1^{1/3} + A_2^{1/3} \right) \tag{3.11}$$

with $C_1 = 6.003$ fm, $C_2 = 0.864$ fm, and $5.2 < A_1^{1/3} + A_2^{1/3} < 8.6$.

Note that the aim of some authors[13] is to obtain an expression as simple as possible to calculate this threshold energy, which can be used to evaluate an experimental limit of incident energy for cross sections to be certainly Rutherford. However it appears necessary to check some of the expressions in the literature in light of experimental data.

For completeness Table 3.1 gives some E_{th} values calculated from Eq. 3.10 for typical projectile–target pairs currently used in recoil spectrometry and for a recoil angle $\phi = 30°$. Figure 3.3 shows typical incident energy E_{th} values for representative incident ion/target nucleus couples and two different recoil angles.[11] We see that in the case of ^{127}I, it is possible to profile elements between ^1H and ^{28}Si with an incident energy below 470 MeV (under the condition that the deviation is $\approx 1\%$ and for $\phi = 50°$), so that the Coulomb barrier is greater than 500 MeV.[*] Calculations for two-body collisions, other than those represented in Figure 3.3 require taking into account relevant nuclear shell effects.

3.2.6. Non-Rutherford Scattering

The Rutherford formula cannot be applied for elastic scattering when the interaction potential deviates from a simple Coulomb potential (see Eq. 3.3, which is based on the assumption that incident particles are scattered in the presence of a Coulomb potential). Moreover some elastic scattering of light particles[1,2,5] (i.e., ^1H, ^2H, ^3H, ^4He . . .) by alpha particles over different energy ranges [e.g., greater than 1 MeV for

---

[*]Evidently this is only a theoretical result, which must be correlated with experiments.

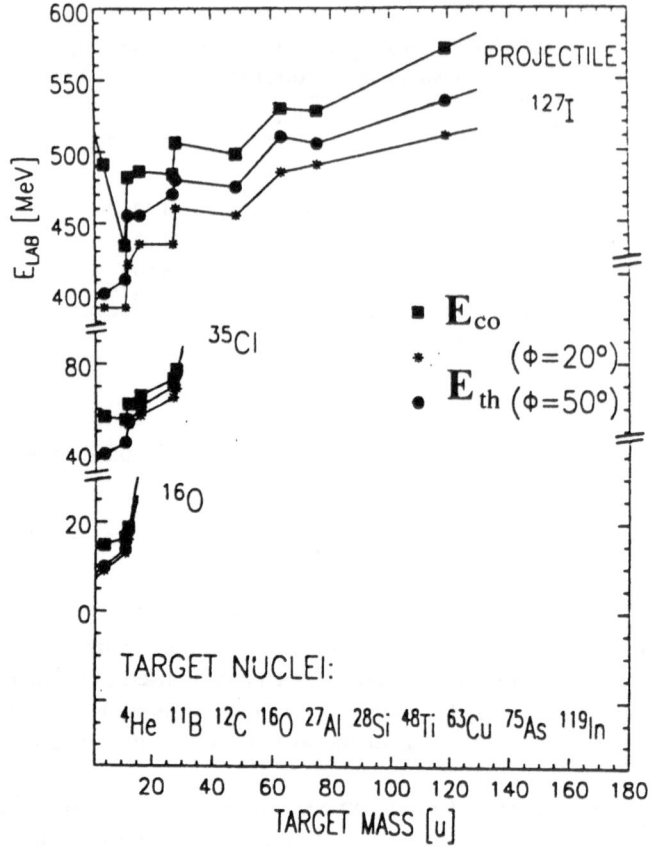

Figure 3.3. Comparison between Coulomb barrier energy (E_{co}) and incident energy (E_{th}) as determined from the optical model in the case of elastic collision for ^{16}O, ^{35}Cl, and ^{127}I at recoil angles of 20 and 50°. All parameters are expressed in the laboratory frame (data from Ref. 11). Connections between data points are a graphical artefact.

alpha particles in an ^{1}H(^{4}He,^{4}He)^{1}H reaction] are distinctly non-Rutherford, and the collision in such elastic scattering involves a non-Coulomb potential. Experiments for studying this type of elastic cross-section have been performed for different reactions,[1,2,17,18] and data have been analyzed in terms of nuclear phase shifts. The following sections present the differential recoil cross-section in hydrogen isotopes, which have a priority in our development. Following the usual convention a collision is called *direct collision* when the projectile is lighter than the target atom and *inverse collision* when the projectile is heavier than the target atom.

3.2.7. Alpha-Proton Collisions up to 4 MeV

The scattering of protons by ^{4}He target particles (direct collision) is extensively investigated both theoretically and experimentally in early research works (e.g., Freier *et al.*,[1] Miller *et al.*[2]). Resonances in the energy range of $E_p = 1-20$ MeV, which were of primary interest in studying $P_{1/2}$ and $P_{3/2}$ levels in the ^{5}Li nucleus, have been extensively studied.[19] Energy-dependent analyses were also performed at medium energies up to 16 MeV for protons.[4,18,20] These early published phase shifts analyses recreate the physical observables as the scattering cross-section and polarization for medium energies. However as observed in recent years, measurements of scattering cross-section using the inverse collision (^{4}He-proton scattering) in the energy range $E_{He} = 1-4$ MeV, are not consistent with early theoretical analyses. A great number of reviews and reports have been recently written on these scattering cross sections at low energy. In this section we first review theoretical calculations, then a data selection. The reader who is not interested in the details of cross-section theory will find polynomial fits to evaluate differential cross-sections in Section 3.2.7.4.

3.2.7.1. Differential Cross-Section Calculation

The analysis of scattering cross-section follows the usual line: considering the incident beam as an infinite plane wave in the center-of-mass coordinates (C) and expanding in eigenfunctions of orbital angular momentum.[2,21] In the presence of spin orbit effects, waves are scattered according to the relative orientation of spin and orbital momentum. Note that the transformation of scattering cross-section values between center-of-mass and laboratory systems is very easy for recoil atoms (see Eq. 3.7).

The differential scattering cross section $(d\sigma^H/d\Omega)_C$ for the helium-4 proton collision in the center-of-mass system is the average of the absolute squares of the scattered amplitudes summed over spin components. In particular for energies up to $E_{He} = 4$ MeV (for protons $E_p \approx 1$ MeV in the laboratory system), we use only terms involving $l = 0$ and $l = 1$; when S and P partial waves are used in the derivation, the following expression for $(d\sigma^H/d\Omega)_C$ can be written[2]

$$\left[\frac{d\sigma^H(\xi, E_p)}{d\Omega}\right]_c = \frac{1}{K^2} \left| \frac{-\eta}{2} \csc^2(\xi/2) \exp\{i \ln[\csc^2(\xi/2)]\} + \exp(i\delta_0) \sin \delta_0 + \right.$$

$$\left. + \cos\xi[2 \exp(i\delta_1^+) \sin \delta_1^+ + 2 \exp(i\delta_1^-) \sin \delta_1^-]\exp(i\alpha_1) \right|^2$$

$$+ \sin^2 \xi + \sin^2(\delta_1^+ - \delta_1^-) \tag{3.12}$$

with

$$K = \frac{\mu}{\hbar}\left(\frac{2E_p}{M_p}\right)^{1/2} = 0.1754 \, E_p^{1/2} \qquad \eta = \frac{2\mu}{\hbar^2 K^2} = 0.3161 \, E_p^{1/2}$$

where

E_P is the energy of proton in the laboratory system

$\alpha_1 = 2 \arctan(\eta)$

δ_0 is the phase shift for $S_{1/2}$ *partial waves*

δ_1^+, δ_1^- are phase shifts for P partial waves ($1 \pm 1/2$)

Note that Eq. 3.12 gives the effective cross section when the center-of-mass of colliding particles is at rest, and it uses E_p as an explicit variable. The transformation into the laboratory system for particles initially at rest (proton or alpha particles) can be made using adequate transformations for energies and angles.[21] To evaluate Eq. 3.12, we define the following quantities R_i and ζ:

$$R_1 = \eta \csc^2(\xi/2) \cos\{\ln[\csc^2(\xi/2)]\}$$

$$R_2 = 1 - \eta \csc^2(\xi/2) \cos\{\ln[\csc^2(\xi/2)]\}$$

$$R^2 = R_1^2 + R_2^2$$

$$\frac{R_1}{R_2} = \operatorname{tg}(\zeta + \alpha_1)$$

We also define ρ, β, γ:

$$\rho \cos\beta = 2\cos(2\delta_1^+) + \cos(2\delta_1^-) - 3$$

$$\rho \sin\beta = 2\sin(2\delta_1^+) + \sin(2\delta_1^-)$$

and

$$\delta_0 = \frac{(\alpha_1 + \gamma)}{2}$$

Using the previously defined variables, the cross-section equation (Eq. 3.12) can be written as:

$$\left[\frac{d\sigma(\xi, E_p)}{d\Omega}\right]_c = \frac{1}{4K^2}\{R^2 - 2R[\cos(\gamma - \zeta) + \cos\xi \cos(\beta - \zeta)] + 2\rho \cos\xi \cos(\gamma - \beta)$$

$$+ 0.5\rho^2(3\cos^2\xi - 1) - 3\rho \cos\beta \sin^2\xi + 1\} \tag{3.13}$$

The starting point in cross-section calculations (Eq. 3.13) is evaluating nuclear phase shifts. In fact phase shifts can be obtained by data reduction using different models: analysis based on the optical potential, the R-matrix formulation for elastic scattering, the effective range expansion of phase shifts.

An optical model for scattering can be interpreted as a convenient parameterization of data in which energy-dependent phase shifts are characterized by a relatively small number of parameters. These phase shifts can then be used to interpolate between energies of available measurements. However introducing a potential model implies that a physical interpretation can be given to the potential as for example optical model potentials for nucleon scattering (^{4}He - H), such as the Woods–Saxon[19] form with a spin orbit coupling term used by Satchler et al.[19]

The R-matrix theory was extensively reviewed by several authors; for elastic scattering the theory simplifies so that the R-matrix reduces to an R-function. This function describes scattering phase shifts in terms of model states of the nucleus. For example the R-matrix formulation for proton-alpha collision for energies up to 20 MeV was studied by Stammbach et al.[18,22]

Another method that can be applied for energy-dependent phase shifts analysis is the effective range theory,[6,20] in which the phenomenological repulsion potential is taken into account by introducing either a very strong but finite repulsion or a hard core at a certain radius. It is a common feature of this type of analysis that parameters in the potential take different values for each angular momentum state.

A universal method that agrees with experiments for any energy range does not exist. Various analyses may represent accurately data over a small energy interval within the large energy range. The test carried out by Brandan et al.[23] on difference sets of phase shifts between E_p = 2 and 9 MeV shows that the Arndt et al.[6,20] analyses using effective range theory are the best ones for their measurements. Phase shifts tables established by Dodder et al.[22] (from R-matrix) give values similar to Arndt's analyses[6,20]; however neither set of phase shifts agrees with recent cross-section values (Appendix A). In the light of new experimental data, Tirira et al.[7,24] extend previous phase shifts studies using only measurements up to 4 MeV. The parameterization employed in their analysis uses the effective range expansion of phases. In particular they retained the scattering length and effective range in each phase shift expansion for $S_{1/2}$, $P_{1/2}$, and $P_{3/2}$ partial waves. This energy dependent analysis permits to calculate the scattering cross section as a function of energy up to E_{He} = 4 MeV, using as input their optimized solutions for the set of phase shifts parameters.

3.2.7.2. Phase Shift Analyses

Effective range parameterization, used for nuclear phase shifts in the alpha-proton analyses can be written in the following form when the Coulomb correction has been made[6,20]

$$C_l^2(\eta)\left[\text{ctg}\delta_l + \frac{2\eta H(\eta)}{C_0^2(\eta)}\right] = K^{-2l-1}(-A_l^{-1} + 0.5\,\Gamma_l K^2 + 0.25\,P_l K^4) \qquad (3.14)$$

with

$$H(\eta) = \eta^2 \sum \frac{1}{S(S^2 + \eta^2)} - \ln(\eta) - \gamma$$

$$\gamma = 0.577216 \text{ (Euler's constant)}$$

$$C_0^2 = \frac{2\pi\eta}{\exp(2\pi\eta) - 1}$$

$$C_l^2 = C_{l-1}^2 \left(1 + \frac{\eta^2}{l^2}\right)$$

where l is the orbital angular momentum; A_l, Γ_l, and P_l represent the scattering length (fm), effective range (fm), and shape parameter (fm^3), respectively.

Using this effective range expansion, it is possible to define an acceptable set of phase shifts (δ_0, δ_1^+, δ_1^-) for one set of data at a given energy (single energy analyses) as one for which the minimum of a χ^2 function is reached. Note that in spin 0–spin 1/2 scatterings there are four angular observables (cross section, polarization, spin rotation parameters) at each energy. However only scattering cross-section has been measured for the alpha-proton collision at energies up to E_{He} = 4 MeV (with the exception of polarization reported by Ad'jasevich et al.[25]). Hence the quality of data fitting is indicated by:

$$\chi^2 = \sum \left[\frac{\sigma_{exp}(\phi_i) - \sigma_{th}(\phi_i)}{\Delta\sigma_{exp}(\phi_i)}\right]^2 \tag{3.15}$$

where $\sigma_{exp}(\phi_i)$ and $\sigma_{th}(\phi_i)$ are the measured and predicted cross sections, respectively, at angle ϕ_i in the laboratory system. Equation (3.13) is transformed into $\sigma_{th}(\phi_i)$ in the laboratory system by means of Eq. (3.7); $\Delta\sigma_{exp}(\phi_i)$ is the error associated with experimental data. The minimization was made separately for different energies and then combined for all energies; the least-squares-method was used to fit the ensemble of phase shifts with just three variables A_l, Γ_l, and P_l for each phase.

3.2.7.3. Data Selection

Experimental studies of elastic scattering of alpha particles by protons have now been performed over a wide range of energies. In particular the elastic recoil cross section has recently been measured up to E_{He} = 4 MeV in the laboratory system and for recoil angles up to $\phi = 40°$. In a collection of experimental cross-section measurements, there are always inconsistencies. Inconsistent data are most safely identified by simultaneously considering all data over the selected energy range, then eliminating experiments that are obviously inconsistent with the entire body of data. A careful

Table 3.2. Parameters Obtained from Ref. 7 for A_l, Γ_l
in Proton–Alpha Collision up to Ep = 1 MeV

	A (fm)	Γ (fm)
$S_{1/2}$	3.35 ± 0.1	6.45 ± 0.15
$P_{1/2}$	−17.8 ± 0.8	−0.43 ±0.05
$P_{3/2}$	−37.6 ±1.5	−0.62 ±0.03

reading of experimental cross-section papers is essential. A complete list of references is given in Appendix A. Available data in the energy range of E_{He} = 1–4 MeV are compiled over the last 8 years (measurements are rarely performed at energies E_{He} smaller than 1.5 MeV).

In an attempt to match theoretical predictions with experimental data, several authors compared their calculations to their data set with more or less success. We see that calculated values of $d\sigma^H/d\Omega$ agree well with each other, because they all use similar phase shifts values, for example the tables established by Dodder et al.[22]; however experimental data vary widely (i.e., up to 35% from one author to another for the same conditions). Tirira et al.[7] surveyed the literature and selected data (A_1, A_4, A_6–A_9 in Appendix A) consistent with each other within broad data uncertainties. They compare their calculations to experimental data and conclude that the formulation of phase shifts as shown herein (Eq. 3.14) gives a good account of the observed cross sections. Moreover cross-section evaluations appear in Ref. 7 to converge with Arndt's values at energies higher than Ep = 1 MeV. Their phase shifts studies are given in Table 3.2 for A_l, Γ_l, and P_l parameters, where P_l (for $P_{3/2}$) is neglected, since P_l values consistent with experiment are erratic, and P_l contributions to p-phases are small. Figure 3.4 compares measured data and calculated cross-section values, as discussed in this section.

For ERDA interest in cross-section values centers predominantly on $E_{He} \leq 3$ MeV and $\phi \leq 30°$ because these are the most frequently used conditions for the analyzing beam. In Figure 3.5 scattering cross-section behavior is shown as a function of the energy of ^{4}He projectiles and the recoil angle. We can then summarize the functional dependence of non-Rutherford cross section $d\sigma^H/d\Omega$ in the laboratory system in the following terms:

- $d\sigma^H/d\Omega$ exhibits a minimum value in the energy range $2.4 < E_{min}(\phi) < 3$ MeV for recoil angles from 0°–30°. As the angle increases from 0°, the minimum moves toward larger energies.

- $d\sigma^H/d\Omega$ always has its lowest value at $\phi = 0°$ for $E_{He} < E_{min}$ (E_{min} at 0° is approximately equal to 2.41 MeV). However this cross section also has its highest value at $\phi = 0°$ for $E_{He} > E_{min}$.

- $d\sigma^H/d\Omega$ appears to reach the same constant value around $E_{He} = 3$ MeV for 0° $< \phi < 40°$.

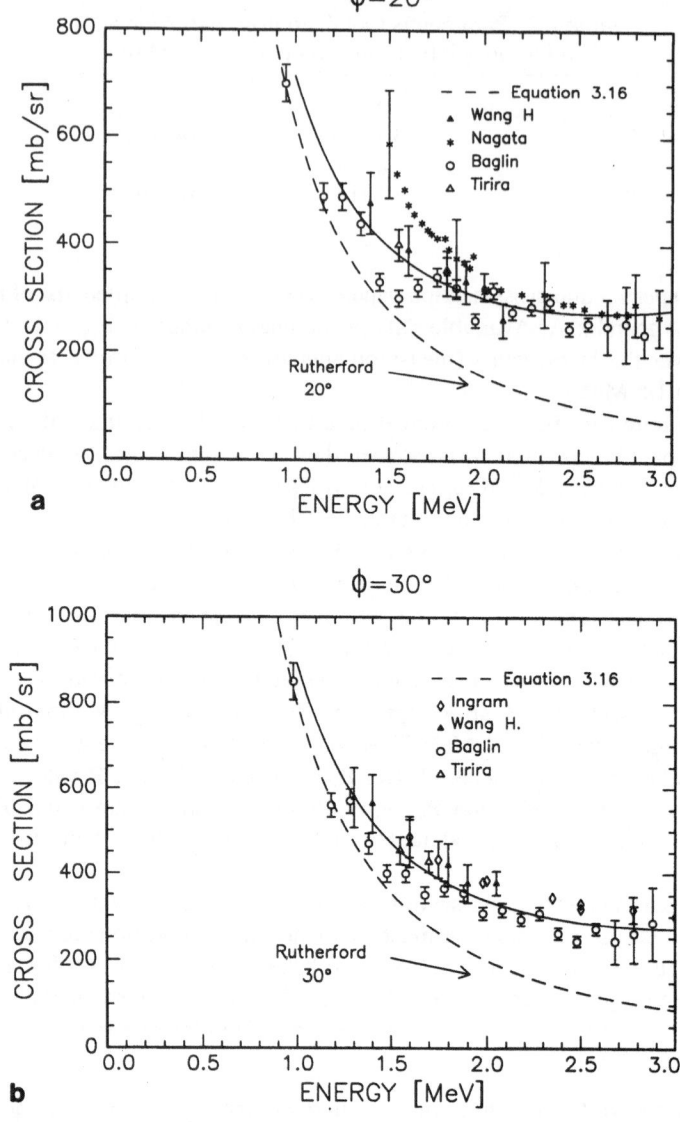

Figure 3.4. Comparison between selected data and calculation as discussed herein. The solid line represents Eq. (3.16). (a) $\phi = 20°$; (b) $\phi = 30°$.

- $d\sigma^H/d\Omega$ near 0° varies very slightly with respect to the ion beam energy E_{He}, (i.e., for $\phi = 0° \pm 3°$, $d\sigma^H/d\Omega$ varies less than 2% from 1.5–2.5 MeV).

- $d\sigma^H/d\Omega$ varies slightly from 2–3 MeV in the $\phi = 0°$ to 30° range. This dependence gives quasi-constant recoil yields as the recoil angle varies.

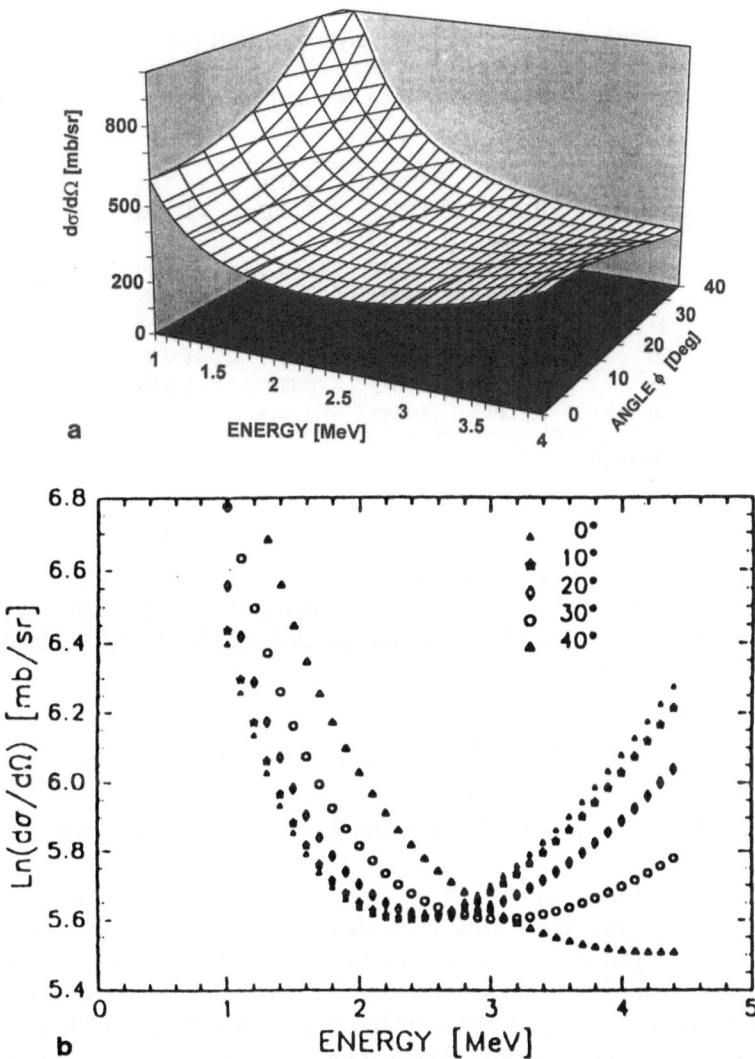

Figure 3.5. Differential cross-section $d\sigma^H/d\Omega$ (lab.) for helium proton collision is shown as a function of (a) both the incident energy of ^{4}He ions and the angle ϕ; (b) the incident energy at various selected angles.

- $d\sigma^H/d\Omega$ is quasi-Rutherford for $E_{He} < 1$ MeV. Hence until more experimental data are available, the Rutherford recoil cross section (Eq. 3.8) is the most appropriate expression for $E_{He} < 1$.

3.2.7.4. Semiempirical Evaluation

To obtain a simple evaluation of $d\sigma^H/d\Omega$ in the laboratory system, some authors proposed semiempirical models that appear to give a good account of observed values.

Table 3.3. Coefficients of Polynomial Fit (Eq. 3.16) for $d\sigma^H/d\Omega$ in the Laboratory System (mb/sr) in the Energy Range from E_{He} = 1 and 4 MeV[a]

ϕ (deg)	A_1	A_2	A_3	A_4
0	0.7651	1.7201	5.6116	−1.7011
5	0.7581	1.7321	5.6302	−1.7148
10	0.7366	1.7716	5.6797	−1.7527
15	0.6994	1.8492	5.7417	−1.8049
20	0.6449	1.9807	5.7890	−1.8568
25	0.5732	2.1840	2.7888	−1.8906
30	0.4779	2.4758	5.7117	−1.8897
35	0.3651	2.8682	5.5349	−1.8408
40	0.2349	3.3687	5.2445	−1.7350

[a]From Ref. 7.

In particular Tirira *et al.*[7,24] proposed a polynomial fit obtained from their theoretical phase shifts analyses and experimental compilation. In fact this function is the least-squares interpolation that fits the theoretical $d\sigma^H/d\Omega$ (Eq. 3.13 in the laboratory system) using the following form:

$$\ln\left[\frac{d\sigma^H(E_{He},\phi)}{d\Omega}\right] = A_1 E_{He} + A_2 + A_3 E_{He}^{-1} + A_4 E_{He}^{-2} \qquad (3.16)$$

where the angle ϕ is an implicit variable; coefficients of this function are given in Table 3.3 for each recoil angle.*

Using Eq. 3.16 some authors[8,26] produce different sets of parameters A_i. Separately these Refs. show an excellent interpolation, with each measurement using a different set of polynomial coefficients. One can remark that coefficients presented herein are the result of analysis using theoretical and selected experimental compilation data. Obviously this type of interpolation models is very useful in a quick evaluation context. However, a complete solution is found using Eq. 3.13 for a differential cross-section and Eq. 3.14 for phase shifts with A_l, Γ_l parameters.

3.2.8. Alpha-Deuteron Collision up to 2.5 MeV

Deuteron scattering by ^{4}He target atoms (direct collision) has been studied in the past. For obtaining information on ^{6}Li nucleus levels, differential cross-sections of

---

*The subroutine for recoil spectrometry of the well-known RUMP software (see Chapter 11) uses this expression for evaluating recoil scattering cross section.

^{2}H(^{4}He, ^{4}He)^{2}H elastic collisions were measured in different energy ranges. In particular a narrow resonance of width $\approx$70 keV exists near 2.13 MeV for ^{4}He ions.[27-29] Data on deuteron scattering by ^{4}He particles seems quite incomplete in comparison to proton-^{4}He collision. Using ERDA as an analytical technique has generated renewed interest in experimental measurements of this cross section $d\sigma^D/d\Omega$. Energy-dependent phase shifts analyses were made in the past, particularly near the resonance energies[27-29]. No renewed analyses have been carried out to match theory with experimental data in light of new experimental values; nevertheless some reviews have been written on scattering cross-section in the helium energy range up to 2.5 MeV. The discussion of $d\sigma^D/d\Omega$ is conducted herein as far as available measurement data allow, using only semiempirical models. A complete list of references is given in Appendix A.

3.2.8.1. Data Selection

Scattering cross-section measurements of the ^{2}H-^{4}He collision (direct collision) have been carried out using either gaseous ^{4}He targets[27-29] or a solid hydrogenated target (inverse reaction) by elastic recoil spectrometry (Appendix A). Almost all available experimental data show primary interest in elastic resonance occurring near deuteron energy $E_D = 1.065$ MeV (which corresponds to $E_{He} \approx 2.128$ MeV in the laboratory system). In fact few scattering cross-section measurements extend to energies well below this deuteron-^{4}He resonance.

To compare ^{2}H-^{4}He data, the direct scattering studied must be converted into more recently studied inverse-scattering data (as already noted, direct or inverse collisions are equivalent if the incident energy of the projectile is adequately transformed). Moreover absolute values of $d\sigma^D/d\Omega$ must be deduced from relative differential cross-section measurements as reported by Besenbacher et al.[30] Selected data are consistent with each other within broad uncertainties (a graphical display is shown in Figure 3.6). However some differences exist at the top of the resonance. Maximum cross-section values measured by Besenbacher et al.[30] are systematically larger than those of Galonsky et al.[28,29] and Quillet et al.[8] An obvious explanation for this discrepancy is not possible, since such cross-section measurements often depend strongly on the energy stability of ^{4}He ion beams, recoil angle, and target composition stability. Furthermore it is difficult to compare cross-section values absolutely with each other. Thus until more experimental data are available, the semiempirical analysis that follows may be used as a good approach to evaluating these cross sections.

Data analysis reveals characteristic dependence of $d\sigma^D/d\Omega$ in the laboratory system for energies in the 1–2.5 MeV range (Figure 3.6):

- $d\sigma^D/d\Omega$ exhibits a fairly narrow (70 keV wide) elastic resonance near 2.128 MeV.

- $d\sigma^D/d\Omega$ always has its highest value at $\phi = 0°$ for an energy resonance peak with a value of $\approx 8.4 \pm 10\%$ b/sr. The resonance peak decreases with an increasing recoil angle; for $\phi = 35°$ the peak essentially disappears.

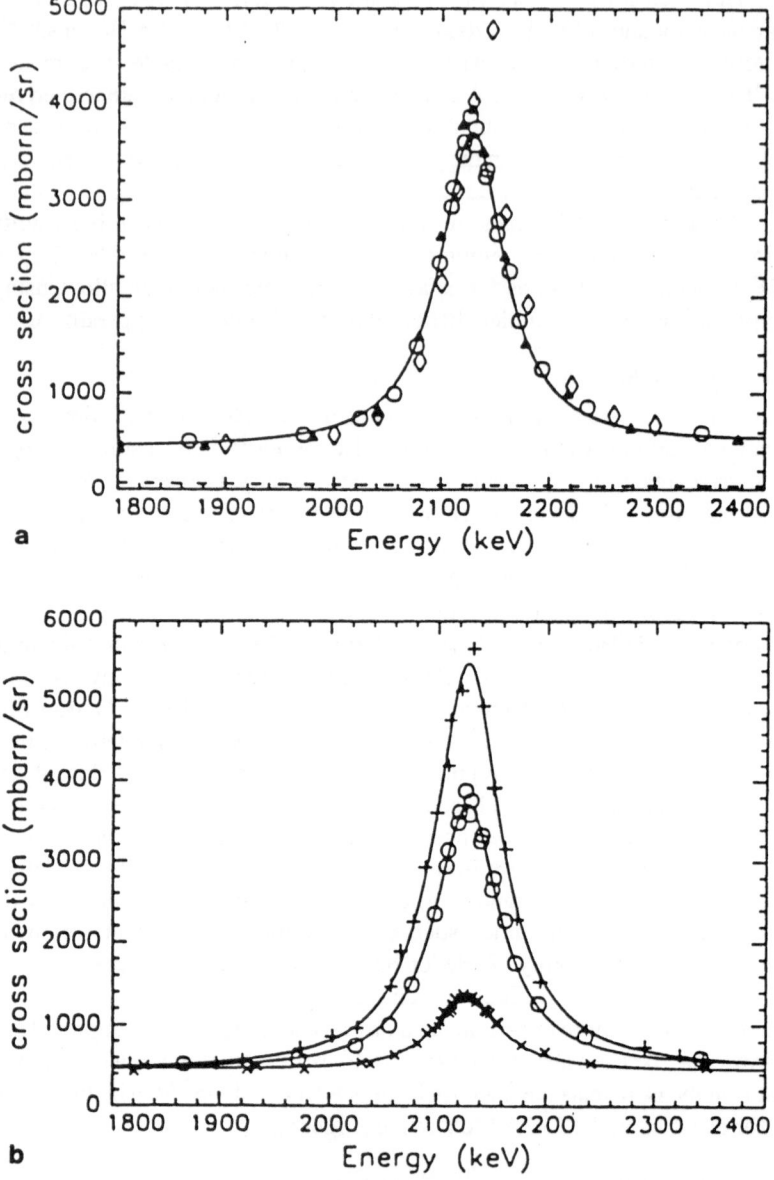

Figure 3.6. (a) Graphical display of selected differential cross-section values $d\sigma^D/d\Omega$ for $\phi = 20°$ where (O) = Ref. 8, (Δ) = Ref. 29 and 30. (b) Comparison between experimental values for $\phi = 10°$ (+), $\phi = 20°$ (o), $\phi = 30°$ (x), and semi empirical calculation according to Eq. (3.17) (solid line, from Ref. 8). (c) $d\sigma^D/d\Omega$ as a function of both incident energy of ^{4}He ions and angle ϕ.

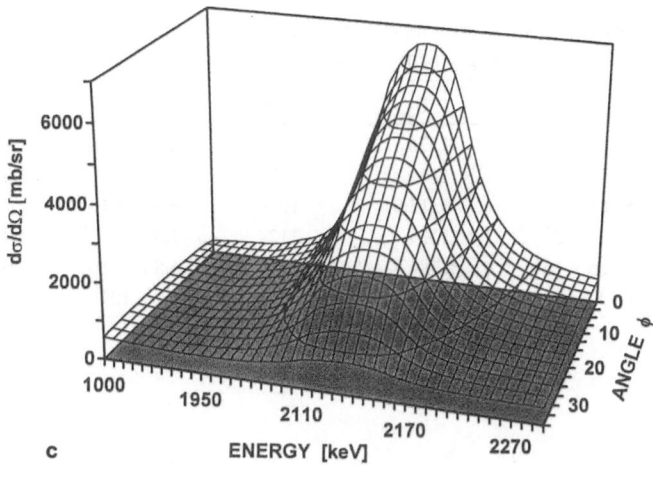

Figure 3.6. Continued.

- $d\sigma^D/d\Omega$ always has both its lowest value at $\phi = 0°$ for $E_{He} < 1.5$ MeV and its highest value for $1.8 < E_{He} < 2.5$ MeV.

- $d\sigma^D/d\Omega$ varies slightly from $1-1.7$ MeV for recoil angle $\phi = 0°$. This dependence gives quasi-constant recoil yields as the incident energy varies.

- $d\sigma^D/d\Omega$ appears to reach the same constant value (≈ 480 mb/sr) around $E_{He} = 1.75$ MeV for $0° < \phi < 30°$.

- $d\sigma^D/d\Omega$ is not yet Rutherford for $0.6 < E_{He} < 1$ MeV. Until more experimental data are available, Eq. (3.8) can be used even though its accuracy is not granted for estimating this parameter when $E_{He} < 0.6$.

3.2.8.2. Semiempirical Evaluation

Existing energy-dependent phase shifts analyses are not complete for the deuteron-helium-4 collision in the energy range up to $E_{He} = 3$ MeV; neither are the earliest phase analyses for the highest energies, which cannot be directly applied to energies generally used in recoil spectrometry. Different approaches exist in the literature; these consist of interpolating experimental data with an analytical function. Indeed a particular function is chosen to reflect as well as possible the dependence of the differential cross-section as a function of energy and recoil angle. The least-squares procedure is an attempt to match data with a semiempirical function. For an ^{4}He-^{2}H collision, when the energy of ^{4}He ions is taken as a free parameter, the differential cross-section can be written as the summation of a second-degree polynomial and a Lorentzian function[8]:

Table 3.4. Interpolation Parameters of $d\sigma^D/d\Omega$ (mb/sr) Using Eq. 3.17 in the Energy Range from E_{He} = 1 and 2.5 MeV[a]

ϕ (deg)	B_1	B_2	B_3	B_4	B_5
5	7.845×10^6	1376.5	421.25	2.125×10^{-2}	-1.13×10^{-5}
10	6.77×10^6	1340	785	-3.86×10^{-1}	1.03×10^{-4}
15	5.565×10^6	1312.5	1019.5	-6.387×10^1	1.72×10^{-4}
20	4.23×10^6	1294	1124	-7.37×10^{-1}	1.96×10^{-4}
25	2.765×10^6	1284.5	1099.25	-6.807×10^{-1}	1.08×10^{-4}
30	1.17×10^6	1284	945	-4.7×10^{-1}	1.08×10^{-4}

[a]From Ref. 8.

$$\frac{d\sigma^D(E_{He}, \phi)}{d\Omega} = \frac{B_1}{(E_{He} - E_\Gamma)^2 \, 10^6 + B_2} + B_3 + B_4 \, E_{He} 10^3 + B_5 \, E_{He}^2 \, 10^6 \qquad (3.17)$$

where the recoil angle ϕ is used in implicit form, E_{He} is in MeV, E_Γ is the energy resonance equal to 2.128 MeV, B_i are angular coefficients given in Table 3.4, and $d\sigma^D/d\Omega$ is given in mb/sr. These coefficients represent only a fit that gives a good account of experimental data measured by Quillet *et al.*[8] Since data are not abundant on the whole, it is not a simple task to eliminate data that are inconsistent with each other. Hence these coefficients can be optimized in the light of experimental values.

3.2.9. Alpha-Triton Collision up to 3 MeV

The ^{3}H-^{4}He collision was studied in the past using the elastic reaction ^{3}H(^{4}He,^{4}He)^{3}H; however there is little data for this scattering cross-section. Energy levels of ^{7}Li were studied[31,32] by scattering tritium ions from helium gaseous targets (direct collision) for E_T > 1.7 MeV ($E_{He} \approx 2.3$ MeV). Moreover using tritium gas (inverse reaction), cross sections were measured for E_{He} above 3 MeV.[17,33] Owing to the high energy loss of ^{4}He particles in absorber windows in this method, no attempt was made to measure differentials cross-section $d\sigma^T/d\Omega$ at energies lower than 3 MeV.*

3.2.9.1. Data Selection

Energy-dependent phase shifts analyses were carried out in early works[14,34] for E_{He} energy higher than 3 MeV. A survey of the literature shows that the structure of ^{7}Li can be described using only S and P waves in phase shifts analyses.† For ERDA interest in $d\sigma^T/d\Omega$ values centers predominantly on energies lower than E_{He} = 3 MeV,

---

*Several elastic resonances exist for this reaction out of the energy range of interest for recoil spectrometry.
†No data comparison has been published for this cross section in the energy range up to E_{He} = 3 MeV.

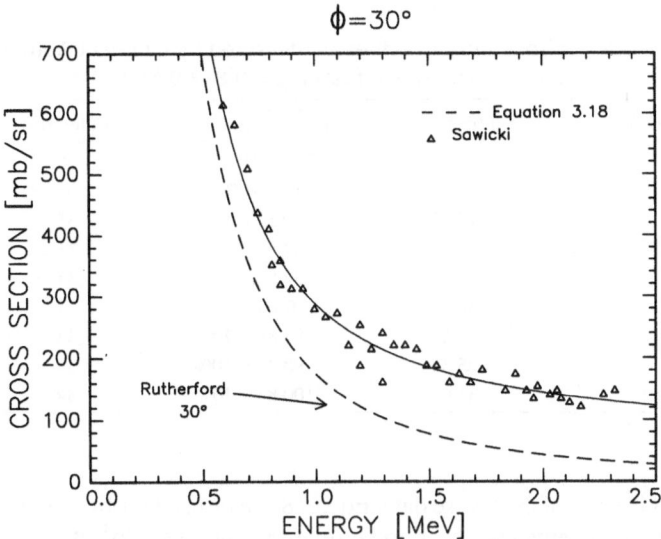

$\phi = 30°$

Figure 3.7. Selected cross-section values of $d\sigma^T/d\Omega$ for $\phi = 30°$ and a comparison with a semiempirical calculation.

the most frequently used in elemental tritium characterization. Furthermore in this energy region, nuclear reactions of other nuclei with ^{4}He are generally absent. Among all these data (Appendix A) only measurements by Sawicki[35] in the range of $E_{He} = 0.5$ to 2.5 MeV were carried out using recoil scattering. We thus select these $d\sigma^T/d\Omega$ values (Appendix A) and available experimental data below 3 MeV from earlier literature (Appendix A) that are consistent with each other. Figure 3.7 shows selected data for ϕ = 30° as a function of E_{He} energy.

3.2.9.2. Semiempirical Evaluation

We do not ignore the theoretical treatment of ^{3}H-^{4}He collisions in previous works for $E_{He} > 3$ MeV; we simply cannot extend theory to low energies while new data are not available. For ERDA we use only data as selected before and carry out a simple semiempirical analysis. Thus we propose a polynomial fit that reflects as well as possible the analytical form of experimental data. Available data seem to show a simple deviation from the Rutherford cross section (Eq. 3.8). Hence it turns out that a parameterization function in the laboratory system, according to Eq. (3.18) gives a good account of data from Sawicki[35] for a recoil angle of 30°.

$$\left[\frac{d\sigma^T(E_{He}, 30°)}{d\Omega}\right] = \left[\frac{d\sigma^T(E_{He}, 30°)}{d\Omega}\right]_{Ruth} + 124.59 - 11.6E_{He} \qquad (3.18)$$

where the energy E_{He} is in MeV and $d\sigma^T/d\Omega$ is given in mb/sr.

Table 3.5. $d\sigma^T/d\Omega$ (mb/sr) Values as Measured by Different Authors in the Energy Range from $E_{He} = 0.8–2.9$ MeV

E (MeV)	θ (deg)	$d\sigma^T/d\Omega$ (mb)	Reference
0.8	15	330 ± 10%	35
1.4	15	110 ± 10%	35
2.0	15	90 ± 10%	35
2.28	20	152.4	31
2.28	30	234.5	31
2.86	30	107.04	31
2.84	21.5	97.88 ± 7%	34
2.84	25.35	98.32 ± 10%	34
2.84	31.6	100.84 ± 10%	34

Note that in spite of the sparse data, Eq. 3.18 can be used for first approximations; however new coefficients may have to be revised using new $d\sigma^T/d\Omega$ measurements. For completeness some values of cross-section-scattering differentials are listed in Table 3.5. The reader is referred to the original works for details.

The functional dependence of $d\sigma^T/d\Omega$ in the laboratory system cannot be completely specified; however the following comments can be added for energies from 0.5–3 MeV (Figure 3.7):

- $d\sigma^T/d\Omega$ does not exhibit resonances in this energy range.
- $d\sigma^T/d\Omega$ always has its lowest value at $\phi = 0°$ for $E_{He} < 3$ MeV.
- $d\sigma^T/d\Omega$ varies very slightly with ion beam energy (E_{He}) in the 1.8–2.5 MeV range (i.e., for $\phi = 30°$ $d\sigma^T/d\Omega$ varies less than 4%).
- $d\sigma^T/d\Omega$ presents a quasi-Rutherford behavior in the 0.5–2.5 MeV range for $\phi = 30°$, as shown in Figure 3.7.

3.2.10. ^{3}He-Proton Collision up to 3 MeV*

In contrast to the extensive research on ^{4}He-hydrogen isotopes collisions, only a few studies involving ^{3}He have been done to profiling hydrogen. The ^{4}He ions are preferred as projectiles because below 3 MeV ^{4}He ions induce elastic reactions with the target atoms almost exclusively. From another point of view, proton scattering from ^{3}He has been studied to investigate levels of ^{4}Li in the energy range above 1 MeV for protons.[36,37] Indeed measurements of differential scattering cross-section of ^{3}He(^{1}H,^{1}H)^{3}He elastic reaction have been made and subsequent phase shifts analyses carried out to complete these studies.[37] However in this section we are principally

---

*The authors wish to thank Guy Terwagne from LARN, Namur, Belgium, for having communicated his measurements of helium-3 - proton scattering cross sections and authorized their presentation in this section.

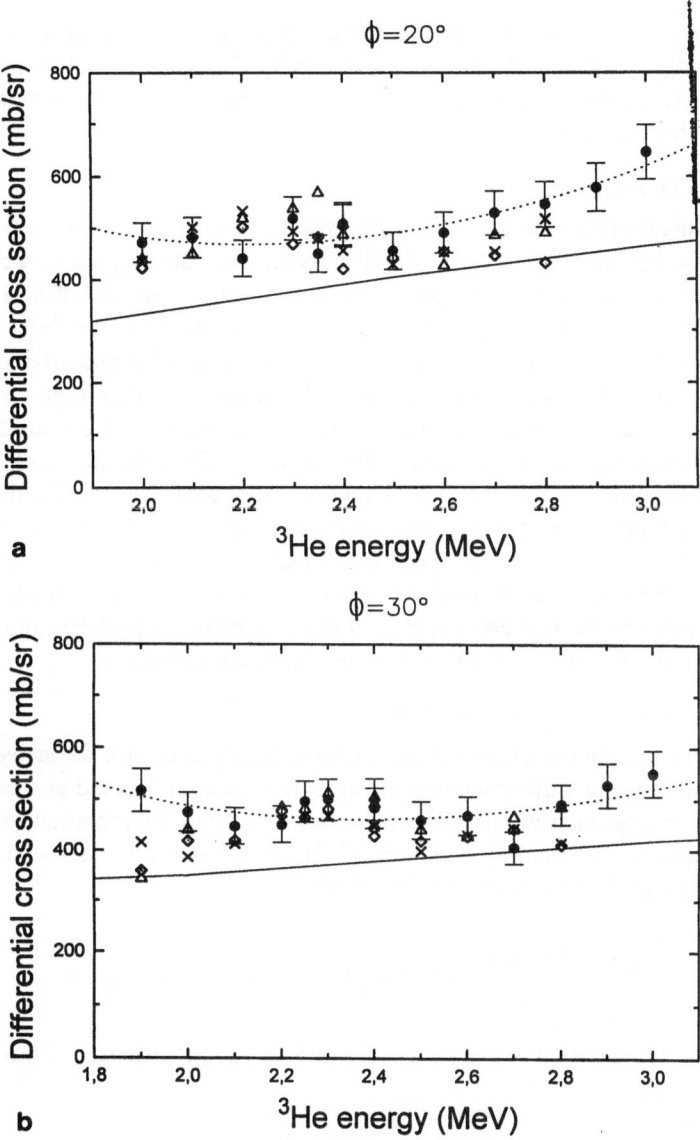

Figure 3.8. Differential cross-section for the ^{1}H(^{3}He,^{3}He)^{1}H elastic reaction measured by Terwagne from the LARN-Namur Laboratory (*private communication*). These cross-section values are performed relative to the gold and carbon content in the multilayer target sample. The full line represents theoretical data evaluated by Benenson *et al.*,[5] and the dotted line is the polynomial fit using Eq. 3.19. Symbols used represent data measured relative to (•) RBS gold signal, (Δ)^{12}C(^{3}He,p$_0$)^{14}N nuclear reaction, (♦) ^{12}C(^{3}He,p$_1$)^{14}N nuclear reaction, and (+) ^{12}C(^{3}He,p$_2$)^{14}N nuclear reaction; (a) for recoil angle φ = 20° and (b) for recoil angle φ = 30°

interested in elastic recoil from ^{3}He-proton collisions used to profile hydrogen, for which the energy range usually chosen is less than 3 MeV for ^{3}He ions. For this interval of energy, data are rare and theoretical analysis has been done with extrapoled data from earlier high-energy studies.[5]

3.2.10.1. Data Selection

Surveying the literature we cannot find cross-section data at energies up to 3 MeV for ^{3}He ions. For completeness in our discussion on scattering cross-section, we introduce data measured by Terwagne[38] that have not yet been published. He measured the recoil cross-section of ^{1}H(^{3}He,^{3}He)^{1}H elastic reaction between $E_{^3He}$ = 1.9 and 3 MeV at two recoil angles: $\phi = 20°$ and 30°. The target is a C:H thin layer deposited on a silicon wafer. A thin gold layer (2.86×10^{15} at/cm^2) was deposited on the C:H layer. The nominal hydrogen content was 18.7×10^6 H/cm^2. The stability of the hydrogen content was controlled during bombardment. Cross-section data were obtained by measuring the recoil hydrogen counts normalized using the gold RBS signal. To check a second normalization was carried out using the nuclear reaction induced by ^{3}He on the C content. An adequate cross section for ^{3}He-gold and ^{3}He-carbon collisions were used, respectively. Figure 3.8 shows differential cross-section values, where full circles with errors bars ($\approx 10\%$) represent normalized gold data and triangles and crosses (with incertitude > 15%) show normalized carbon data.

3.2.10.2. Semiempirical Evaluation

Since cross-section data are scarce, theoretical analysis cannot be carried out. An extrapolation of phase shifts data as made by Benenson et al.[5] could give hazardous values. In first approach we propose to use a polynomial interpolation, which fits experimental data measured by Terwagne.[38] These cross-section data seem to show the following simple form (similar to Eq. 3.16):

$$\ln\left[\frac{d\sigma^H(E_{^3He},\phi)}{d\Omega}\right] = D_1 E_{^3He} + D_2 + D_3 E_{^3He}^{-1} + D_4 E_{^3He}^{-2} \qquad (3.19)$$

where the recoil angle ϕ takes two values, 20° and 30°, and coefficients of this function are given in Table 3.6.

Table 3.6. Coefficients of Polynomial Fit (Eq. 3.19) for $d\sigma^H$ ($E_{^3He},\phi$)/$d\Omega$ in the Laboratory System (mb/sr) in the Energy Range from $E_{^3He}$ = 1.9–3 MeV[a]

ϕ (deg)	D_1	D_2	D_3	D_4
20	1.3657	0.0269	7.0730	−0.4577
30	1.0705	0.5529	8.4432	2.89949

[a]From Ref. 38.

The functional dependence of the differential cross-section for an ^{3}He-H collision cannot be completely given; however some comments can be added for energies from 1.9–3 MeV (Figure 3.8):

- $d\sigma^H(^3\text{He})/d\Omega$ does not exhibit resonances in this energy range.[38]
- $d\sigma^H(^3\text{He})/d\Omega$ has its lowest value at recoil angles close to 0° for $E_{He} < E_{min}$ [5] (E_{min} is a function of a recoil angle; for $\phi = 0°$ $E_{min} \approx 2$ MeV). However this cross section also has its highest values at $\phi = 0°$ for $E_{^3He} > E_{min}$.[5]
- $d\sigma^H(^3\text{He})/d\Omega$ varies very slightly for $\phi = 30°$ with respect to the ion beam energy in the range from 1.9–3 MeV (i.e., $d\sigma^H(^3\text{He})/d\Omega$ varies less than 4%). However for $\phi = 20°$, the cross section varies around 18% in the same interval.[38]

Due to the fact that measured data of this cross section are very scarce, behavior as quoted herein may have to be revised in light of new scattering cross-section measurements.

3.2.11. Cross-Section Summary

The evaluation of scattering cross-sections is a difficult and old problem. For each elastic collision, we must consider the energy range to which a particular solution can be applied. Different analyses may accurately represent data over a small part of the large energy range for which data exist. Our preceding discussion progressed as far as available experimental data. In general there are more theoretical treatments at high energies than measured values for energies regarded in ERDA. Although we could give a complete analysis for a large energy interval, we presented some key expressions to be used for practical IBA problems. Note that since semiempirical expressions are obtained from experimental and semitheoretical analysis, they cannot be extrapolated outside the energy region in which they were deduced. Furthermore for some collisions, we gave expressions for scattering cross section for particular recoil angles. Their applicability is thus limited, and new measurements must be carried out for other conditions.

3.3. MULTIPLE SCATTERING

When an ion beam penetrates matter, particles undergo successive scattering events and deviate from the original direction. Initially well-collimated ions do not have the same direction after passing through a thickness Δx of a random medium. As a result both angular and lateral deviations from the initial direction occur. These increase with path length, and they are subject to fluctuations. Figure 3.9 gives a schematic representation of the geometry of the beam over its path in matter. This process is called MS, and it is statistical in nature due to the large number of collisions. Figure 3.10 shows the propagation of angular spread distribution through a dense medium where we have superimposed an exact angular distribution (representing a

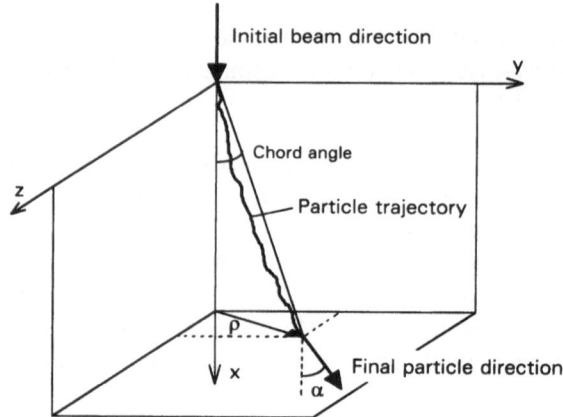

Figure 3.9. Multiple-scattering scheme where the ion beam is directed in the x direction. Lateral displacement perpendicular to the beam direction is $\rho(y,z)$, and α is the total angular deviation after a penetrated depth x.

typical distribution as it is experimentally obtained) and a Gaussian approximation. We stress that the MS distribution is *never* Gaussian, even when the traversed thickness is large. The MS distribution of ions traversing matter is one of the basic processes of interest in experiments involving energetic projectiles. As a matter of fact, MS adds an important contribution to energy dispersion of detected particles, and this contribution is difficult to account for exactly.

The angular distribution of a beam is often the investigated quantity in the study of MS. *The lateral distribution is closely related to the angular one but secondary to it, since lateral displacement is a consequence of angular divergence.*[39–41] Lateral distribution represents the beam profile in the matter. Note that when the combined lateral–angular MS is considered, the relation between both lateral and angular MS distributions must be adequately treated, since both processes are *dependent on* each other.

Note that angular MS of charged particles emerging from a target is still a current problem although an old one. Surveying the current literature we observe that the improvement of the theoretical treatment is still underway and available experimental data are much less abundant than for stopping powers or energy straggling.

The analysis of MS was started by Bothe[42] and Wentzel[43] in the twenties using the well-known small-angle approximation. The physics of energy straggling and MS was developed quite far by Williams[44–46] from 1929 to 1945. In particular Williams devised a theory based on a now currently used method. It consists of fitting the MS distribution as a *Gaussian-like portion* due to small scattering angles and the *single collision tail* due to the large angles (Figure 3.10). Goudsmit and Saunderson[47] give

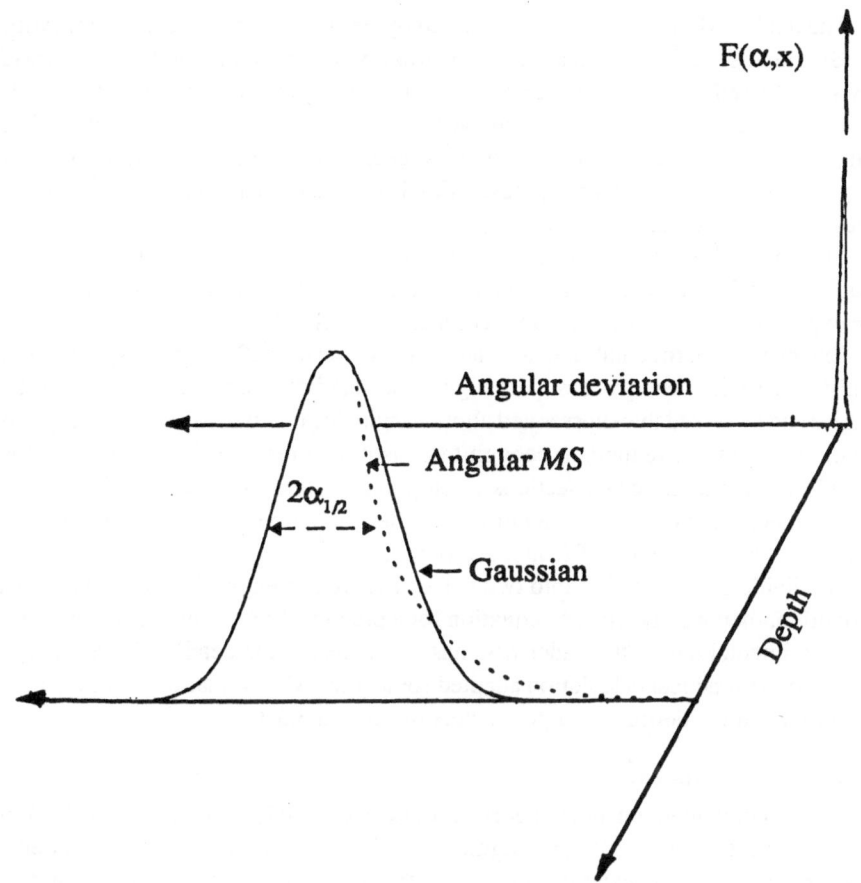

$F(\alpha,x)$

Angular deviation

Angular *MS*

$2\alpha_{1/2}$

Gaussian

Depth

Figure 3.10. Propagation of the MS angular distribution through matter. The half-width of the angular MS distribution is called $\alpha_{1/2}$, and it generally characterizes the width to the distribution.

a more complete treatment of MS, including large angles. The angular distribution from Coulomb scattering has been studied in detail by Molière[48] and further continued by Marion and coworkers.[49] Scott[39] presents a complete review of basic theory, mathematical methods, as well as results and applications. A comparative development of MS at small angles is presented by Meyer,[50] based on a classical calculation of single scattering cross section. Sigmund and Winterbon[40,51–54] reevaluate Meyer's calculation to extend it to a more general case. Marwick and Sigmund[41] carried out development on lateral spreading by MS, which resulted in a simple scaling relation with the angular distribution.

The double MS profile was first outlined by Fermi[55] in the diffusion approximation. Sigmund et al.[51] show that the correlation between angular and lateral spreads allows us to reduce their combination to an expression containing only angular distributions; in recent years some new works were presented on this subject.[52,56] For example, the combined MS distribution was recently discussed in detail by Amsel et al.[57,58] The effect of MS on energy resolution is treated by Szilágyi et al.,[59] who also discuss the projected distribution.

The energy loss effect in MS—the so-called Bothe–Landau formula—was first introduced by Sigmund and Winterbon,[40,52] then discussed in more detail in relation with experimental measurements by Schmaus and L'Hoir.[60]

Extensive experimental data are necessary to verify MS theory. Available data mainly concern low-energy ions. Although these data[51,61–72] cover a broad interval of energies and traversed thicknesses and allow convincing comparisons with theory, they are not abundant nor redundant enough to constitute a real data base, such as those available for scattering cross-sections or stopping powers. In particular comparison between theory and experiment require more data about compounds; moreover data for large traversed amounts of matter are very rare.

We divide our discussion into two parts: first we present basic equations for MS distribution, then we give the key equation for a practical MS evaluation and compare it to experimental data. The reader who has no interest in the details of MS theory is referred to the practical calculation adapted for a quick MS evaluation (Section 3.3.3) after looking at the definition of parameters in Section 3.3.1.1.

3.3.1. Basic Equations

The general mathematical theory is quite well-established by Bothe,[42] Williams,[44–46] Scott,[39] Meyer,[50] and Sigmund.[40,51] Several approaches exist to attempt to understand MS, for example the transport equation, Monte Carlo method, or collision summation. Therefore discussion of general equations presented herein is brief. In this section we begin with the theory of Meyer–Sigmund and Winterbon which is largely used in MS studies and progress to comments on more recent works. The reader is invited to see original papers for details.[39–41,50,51]

To begin we define the following variables:

$\alpha =$ total deflection spatial angle
$\alpha_{1/2} =$ half-width at half-maximum of the angular distribution (cf. Figure 3.10)
$\rho =$ total lateral displacement
$\rho_{1/2} =$ half-width at half-maximum of the lateral distribution
$\theta =$ single-collision laboratory angle
$\tilde{\theta} =$ reduced laboratory scattering angle
$\tau =$ reduced thickness
$\varepsilon =$ reduced energy
$\tilde{\alpha} =$ reduced deflection angle

$\tilde{\alpha}_{1/2}$ = reduced half-width of the distribution defined as the angle where intensity has dropped to half its value at $\tilde{\alpha} = 0$

$\tilde{\rho}$ = reduced lateral displacement

$\tilde{\rho}_{1/2}$ = reduced half-width of the distribution defined as the lateral displacement where intensity has dropped to half its value at $\tilde{\rho} = 0$

$d\Omega$ = solid angle element around α

$d\sigma(\theta)$ = differential scattering cross-section

$J_0(x)$ = zero-order Bessel function

3.3.1.1. Neglected Energy Loss Effects

In theoretical models basic MS equations rely on the following assumptions:

1. Scattering centers in the medium are distributed at random i.e., collisions can be considered as independent; thus channeled ions are excluded.
2. Energy loss of ions over their path is very small (less than 10% of E_0).
3. Only scattering cross-sections with cylindrical symmetry are considered.
4. Only small scattering angles are taken into account ($\sin \theta \sim \theta \Rightarrow \theta < 12°$), assuming that large scattering angles make a small contribution to the total deflection. Furthermore, traveled path length and penetrated depth are indistinguishable.
5. Spatial distributions are considered to be cylindrically symmetric around the initial beam direction (this follows directly from Assumptions 1 and 3).

Taking into account these assumptions and the classical scattering theory based on the impact parameter concept, we can write the cylindrically symmetric *spatial distribution* F(x,α) of particles having traversed a distance x; this function is often called the Bothe formula.

$$F(x,\alpha)d\Omega = \frac{d\Omega}{2\pi} \int_0^\infty dk k J_0(k\alpha) \exp\left\{-Nx \int_0^\infty d\sigma(\theta)\left[1 - J_0(k\theta)\right]\right\} \qquad (3.20)$$

where the dummy variable k results from the two-dimensional Fourier tranform.

Note that Assumption 4 becomes central in the solution: it requires the cross-section integral at small angles to make a strong enough contribution, so that large scattering angles make an insignificant contribution. Thus the integration interval can be extended to infinity.

In the small-angle approximation the differential cross-section obtained for a screened Coulomb interaction can be used; it can be written as (e.g., Ref. 73):

$$d\sigma(\theta) = \frac{\pi a^2}{\varepsilon^2} \frac{g(\varepsilon\theta)}{\theta^2} d\theta \qquad (3.21)$$

This expression has been widely used; it assumes collisions to be two-body interactions, which is a particular case compatible with Assumption 3. A more realistic scattering cross-section describing eventually three- or many-body interactions could in principle be substituted in what follows, but it may be mathematically untractable. In Eq. 3.21 reduced energy ε is given by:

$$\varepsilon = \frac{Ea}{2Z_1Z_2e^2} \tag{3.22a}$$

where $\mathbf{a} = 0.8853a_0/(Z_1^{2/3} + Z_2^{2/3})^{1/2}$ is the screening radius, a_0 the Bohr radius, and $g(\varepsilon\theta)$ the scattering function.

Using the following reduced variables:

the reduced laboratory scattering angle

$$\tilde{\theta} = \varepsilon\theta \tag{3.22b}$$

the reduced thickness

$$\tau = \pi\mathbf{a}^2Nx \tag{3.22c}$$

the reduced angle

$$\tilde{\alpha} = \varepsilon\alpha \tag{3.22d}$$

We obtain the following simple form of Eq. 3.3.20:

$$F(x,\alpha)d\Omega = \tilde{\alpha}\,d\tilde{\alpha}\,f_1(\tau, \tilde{\alpha}) \tag{3.23}$$

The function $f_1(\alpha,\tau)$ defined by Eq. (3.23) is the *universal MS* distribution first introduced by Meyer.[50] This function can be written in the following form using the new variable $y = k/\varepsilon$:

$$f_1(\tau,\tilde{\alpha}) = \int_0^\infty dy\, y\, J_0(\tilde{\alpha}y)\exp\left\{-\tau \int_0^\infty d\tilde{\theta}\, \frac{g(\tilde{\theta})}{\tilde{\theta}^2}\left[1 - J_0(y\tilde{\theta})\right]\right\} \tag{3.24}$$

Using Eq. 3.21 the universal MS has been evaluated, so tabulated values exist. Particularly calculations by Sigmund and coworkers[40] are the most largely used as reference data. Note that Eq. 3.24 represents a cylindrically symmetric *spatial distribution* that describes the absolute angular deviation from the original beam direction. In other words this distribution is measured for example when the scattered beam is detected at angles close to 0°. However, when the beam is impinging onto the target surface with a grazing angle and scattered particles are also detected at a grazing angle, we must consider the spatial distribution projected onto the collision plane. Hence a general transformation of the spatial distribution must be applied to obtain the projected distribution. An adequate transformation has been discussed in detail by Scott et al.[39] and more recently by Amsel and coworkers.[57,58,74]

In particular for a very thin target foil (zero-limit thickness), we obtain from Eqs. (3.20) or (3.24):

$$\lim_{x \to 0} F(x, \alpha) \, d\Omega = Nx \, d\sigma(\alpha)$$

which corresponds to a single scattering with $\alpha \neq 0$.

In the formalism developed here above, one of the most important scaling parameters is the reduced energy ε, which allows a simple formulation of the scaled angular distribution. This formulation is largely used to calculate MS distributions; indeed it is often applied to a particular case of a power potential $V(R) \propto R^{-1/m}$ between collision partners, where the cross section also takes the power form for small angles, so that for $g(\tilde{\theta}) \propto \tilde{\theta}^{1-2m}$ for $0 \leq m \leq 1$. Thus Sigmund and Winterbon[40,53] show that when the cross section takes the power form, the reduced half-width $\tilde{\alpha}_{1/2}$ of the MS distribution can be written as a function of thickness:

$$\tilde{\alpha}_{1/2} = \kappa \tau^p \tag{3.25}$$

where $p = 1/2m$ as originally defined by Sigmund.

This expression has been checked for different intervals of reduced thickness values. Indeed for thin targets, Andersen et al.[62,64] find good agreement between their measurements and Sigmund's theory for some target projectile couples. For $\tau < 20$ Spahn and coworkers[67] and independently Hooton and coworkers[75] find overall agreement with the Meyer–Sigmund theory. Anne and coworkers[72] have experimentally determined half-width angles, and they confirm the consistency of Eq. 3.25 with their data for $40 \leq \tau \leq 10^6$ (Figure 3.11).

Note that the most important result is the simple correlation between τ and $\tilde{\alpha}_{1/2}$ that appears as a universal relation, with a change of slope within different regions of reduced thickness.

Marwick and Sigmund[41] show similar scaling properties for the lateral distribution $G(x,\rho)$ as the angular distribution $F(x,\alpha)$ when the power cross-section is used. Moreover Winterbon[53] extended this analysis to show that power cross-sections are a good approximation of the MS distribution calculated with more exact cross-section expressions. In particular the two distributions exhibit nearly identical shape using the following scale factors[41]:

$$\alpha = \frac{\Gamma_m}{x} \rho \tag{3.26}$$

Furthermore the reduced half-widths of both distributions are *related* by:

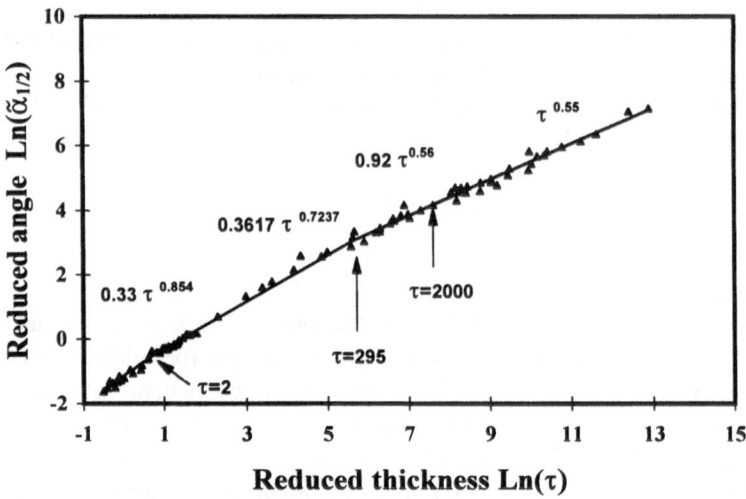

Figure 3.11. Measured reduced half-width of MS is plotted as a function of the reduced thickness. Data for $30 < \tau < 10^6$ from Ref. 72 represent $\alpha_{1/2}$ measured in the energy range from 20–90MeV/uma for ^{16,17}O, ^{36}K, ^{40}Ar, ^{86}Kr, and ^{100}Mo heavy ions. Data for $\tau < 20$ from Ref. 67 represent $\alpha_{1/2}$ measured in the energy range from 1–11 MeV for H, He, N, and Ne ions. Data for $\tau < 15$ from Refs. 62 and 64 represent measurements with low energy ions (200–1000 keV). The curve represents the fit of data according to the subdivision into four domains of reduced thickness, as discussed in the text.

$$\tilde{\alpha}_{1/2} = \frac{\Gamma_m}{x} \tilde{\rho}_{1/2} \qquad (3.27)$$

with

$$\Gamma_m = (1 + 2m)^{1/2m} \qquad (3.28a)$$

The parameter Γ_m can also be evaluated using Eq. 3.28b, which was derived by Marwick and Sigmund[41] as a function of reduced thickness and simplified as proposed by Bird and Williams[76]:

$$\Gamma = 1.77 + 0.172\tau^{-0.335} \qquad (3.28b)$$

Marwick shows that these scaling relations relating angular and lateral MS distributions are accurate to a few percent. Particular deviations from simple scaling are expected to occur in the tails of the distributions. The combination of both MS distributions to build the total energy dispersion by MS requires special attention, since both distributions have a non-Gaussian form.

3.3.1.2. Energy Loss Effects

In the basic MS equations just discussed, initial particle energy was considered quasi-constant during the path in the matter; hence the energy loss process of particles was negligible. However in many cases this last condition may not be satisfied, so that different parameters for the differential cross-section and obviously the angular and lateral distributions must be calculated more exactly. Several authors[52,56,60] have developed models to describe this process using different approaches but with the common assumption of a screened Coulomb interaction.

As before considering each scattering event to be independent of previous ones, an MS distribution can be calculated by including the energy variation $E(x)$ as a function of penetration depth x directly into the scattering cross-section function. Thus we can rewrite Eq. 3.20 as[40]:

$$Nx\sigma(k) \rightarrow N \int_0^x dx' \, \sigma[k, E(x')] \tag{3.29}$$

To evaluate the MS distribution as a function of energy, we choose an adequate model to describe the stopping power $S[E(x)]$. For example Valdés and coworkers[56] calculated the MS distribution by assuming a simple power velocity dependence of stopping power. This supposition has been tested for a reduced thickness τ lower than 1, and it shows an effective value $\hat{E}$ of ion energy can be used for practical problems (see Section 3.3.3 for more details).

For more general cases Schmaus and coworkers[60] developed a procedure to evaluate Eqs. 3.20 and 3.29. For example in the case of power potentials, they introduce a new scale factor $\Lambda(x)$ for large energy losses, so that we can rewrite Eq. 3.24[60] as:

$$N \int_0^x dx' \{\sigma[k, E(x')]\} = Nx\sigma(k, E_0) [\Lambda(x)]^{2m} \tag{3.30}$$

where

$$\Lambda(x) = \left\{ \frac{1}{x} \int_0^x dx' \left[\frac{E(x')}{E_0} \right]^{-2m} \right\}^{1/2m} \equiv \frac{E_0}{\hat{E}} \tag{3.31}$$

The reduced scattering angle is actually given by:

$$\alpha(\hat{E}) = \frac{E_0}{\hat{E}} \alpha(E_0) \tag{3.32}$$

where the ratio $E_0/\hat{E}$ is obviously larger than 1. This ratio is easily calculated by making suitable assumptions. In this particular case when the power potential is taken as $V(R) \propto R^{1/m}$ for $0 \le m \le 1$, if the stopping power takes the form $S(E) \propto E^{-n}$, Schmaus and L'Hoir[60] show that the energy loss scale factor can be written as:

$$\Lambda(x) = \left[\frac{1 - r^{n+1-2m}}{(1 - r^{n+1})(1 - 2m/(n+1))} \right]^{1/2m} \quad \text{for } \frac{2m}{n+1} \neq 1 \qquad (3.33)$$

and

$$\Lambda(x) = \left[\frac{\ln(r^{-n-1})}{(1 - r^{n+1})} \right]^{1/2m} \quad \text{for } \frac{2m}{n+1} = 1 \qquad (3.34)$$

where $r = E(x)/E_0$ and $E(x)$ is the energy at depth x. For example when $S(E) \propto E^{-1}$ and $V(R) \propto R^{-2}$ (m = 0.5), Eq. 3.33 gives the effective energy equal to the *arithmetical mean value* $\{\hat{E} = [E_0 + E(x)]/2\}$, which is often used to take into account the effect of energy loss as proposed by Sigmund.[52] In most practical cases, this arithmetical average agrees well with experimental results.[64,65]

Note that Eqs. 3.33 and 3.34 constitute a useful procedure for evaluating MS parameters as a function of mean energy. Nevertheless suitable expressions for the power potential and the stopping power must be chosen, so that both parameters m and n be well-defined for a practical problem.

3.3.2. Multielemental Targets

The theories on MS just outlined apply to a monoatomic target. However it is possible to introduce the case of polyatomic targets by using a convenient transformation. For example Schmaus and L'Hoir[60] extended the Bothe formulation (Eq. 3.20), writing:

$$F(x, \alpha)d\Omega = \tilde{\alpha}\, d\tilde{\alpha} \int_0^\infty dy [y\, J_0(\tilde{\alpha}y)] \exp\left\{ -\sum_i \tau_i \int_0^\infty d\tilde{\theta}\, \frac{g(\tilde{\theta})}{\tilde{\theta}^2} \left[1 - J_0\!\left(\frac{y}{\varepsilon_i} \tilde{\theta} \right) \right] \right\} \qquad (3.35)$$

where each type of scattering center is labeled with the index i and indexed quantities are defined for each atomic kind.

Note that this last formulation has been checked[60] against some experimental data, in particular for a polyester target using a proton beam of various energies from 1.2–2 MeV. The average discrepancy between theory and these experimental data is reported to be about 6%.[60]

Meanwhile a simplified treatment for polyatomic targets has been proposed, which consists of taking a mean atomic number $\hat{Z}$ so that angular distribution can be evaluated by assuming a monoatomic target. In this case reduced variables are calculated directly by using $\hat{Z}$ in Eqs. 3.22a–c. In particular for a power potential interaction, Schmaus and L'Hoir[69,71] show that the mean $\hat{z}$ of a polyatomic target can be written as:

$$\hat{Z} = \frac{1}{\hat{a}^{(1/m)-1}} \left(\sum_i \frac{N_i}{N_T} Z_i^{2m}\, a_i^{2-2m} \right)^{1/2m} \qquad (3.36)$$

where $\mathbf{a}_i$ is the screening radius for species i, $\hat{\mathbf{a}}$ is the average screening radius, N_T and N_i the total number of scattering centers and the number of scattering centers of atomic species i, respectively. Equation 3.36 has been tested on a few materials,[69,71] but its general applicability to compounds is not demonstrated. It is most certainly justified for small differences in Z, but there are few reasons to assume that the parameter m can be taken as a constant for larger differences.

3.3.3. Practical Evaluation

Different methods can be used to evaluate MS. On the one hand a complex exact evaluation can be calculated using the equations just outlined. On the other hand we can apply scaling properties of the MS distribution, then use a simple polynomial fit to evaluate some reduced scale parameters.

The following key equations are used for a practical evaluation: the Reduced Energy equation Eq. 3.22a:

$$\varepsilon = \frac{1.625 \times 10^4 E}{Z_1 Z_2 (Z_1^{2/3} + Z_2^{2/3})^{1/2}} \tag{3.37}$$

with E in MeV; and the Reduced Thickness equation 3.22c:

$$\tau = \frac{41.5 \times 10^3 t}{M_2 (Z_1^{2/3} + Z_2^{2/3})} \tag{3.38}$$

where t is given in mg·cm^{-2}.

Using Eqs. 3.22d and 3.37, we write the half-angle as:

$$\alpha_{1/2} = 6.153 \times 10^{-5} \frac{\tilde{\alpha}_{1/2}}{E} Z_1 Z_2 \left(Z_1^{2/3} + Z_2^{2/3} \right)^{1/2} \tag{3.39}$$

with E in MeV.

To evaluate $\tilde{\alpha}_{1/2}$ we take as a guideline experimental data for finding parameters κ and p in Eq. 3.25. Sigmund et al.[51] determined the dependence between angular and lateral MS distributions for very low reduced thicknesses in gaseous targets ($\tau = 0.036–0.082$). Andersen et al.[62,64,65] measured the MS of low-energy ions (200–1000 keV) in targets with small reduced thicknesses ($\tau = 0.5–15$). Below 100 keV and for $\tau = 0.5–20$, Högberg et al.[61] and Meyer and Krygel[63] determined angular distributions in different target materials and similarly Hooton et al.[66] and Spahn and Groeneveld[67] at energies about 10 MeV for $\tau = 1–15$. L'Hoir et al. measured MS distributions for oxide targets,[68,69,71] and they used MS to study thickness heterogeneities in self-supporting thin films.[70] Anne and coworkers[72] measured MS distributions for very high-energy ions (20–90 MeV/amu oxygen, argon, krypton, and molybdenum ions) from $\tau = 40$ to $\tau = 10^6$.

For example we graphically obtain parameters κ and p from Eq. 3.25 from Figure 3.11, which shows a compilation of selected data from references.[62,64,67,72] The data have the universal form given by Eq. 3.25. Analyzing the data as represented in Figure 3.11, we see that continuity exists between low and high τ, with a change of slope as a function of reduced thickness. In particular we can obtain expressions representing a fit from the original data for different intervals of reduced thickness τ.[*]

For $2 \times 10^3 < \tau < 10^6$, $\tilde{\alpha}_{1/2} = 1.0\tau^{0.55}$, as shown by Anne and coworkers,[72] which is not far from the asymptotic form proposed by Sigmund et al.[40]

For $1000 < \tau$, Sigmund theory gives values 25% less than Anne et al.[72] measurement. We propose to use the following function as fitted from Anne's original data; thus: for $295 < \tau \le 2.10^3$, $\tilde{\alpha}_{1/2} = 0.92\tau^{0.56}$.

Similarly, using data from both Anne et al. and Spahn et al.,[67] we obtain the following relationship, which deviates from Sigmund's theoretical values about 10%: for $2 < \tau \le 295$, $\tilde{\alpha}_{1/2} = 0.3617\tau^{0.7237}$; for $0.5 < \tau \le 2$, $\tilde{\alpha}_{1/2} = 0.33\tau^{0.854}$, according to Andersen et al.[62,64]

Some discrepancies between theoretical and experimental data are not completely explained. Some may be caused by correlated scattering in crystalline targets.

The preceding four relations constitute a continuous fit over the whole range within a few percent. Note that a similar analysis was carried out by Bird and Williams.[76] However in their analysis, there is some overlap between intervals, and parameters κ and p are not continuous between one interval and the next. (Note that some typewriting errors exist in the MS equations used by Bird and Williams). Szilágyi et al.[59] proposed a polynomial fit to evaluate these parameters when projected MS distributions are considered. This procedure represents an interesting method of calculating projected half-width angles; however several experimental tests should be performed to check the validity of this method. We can add the recent evaluation of projected MS distribution carried out by Amsel et al.,[57,77] which represents a new approach to evaluating MS parameters accurately.

To evaluate the energy that must be used in Eq. 3.38, we assume that if energy loss is neglected, the initial energy before traveling in the matter can be used, for example incident energy E_0 for projectiles or recoil energy E_2 just after the collision. When the energy lost by the particles (ΔE) is relatively large (for example $\Delta E/E_0 > 5\%$), we can use the median energy, then apply the following criteria.

For low energies when particle velocities are $v_1 < v_0 Z_1^{2/3}$ (where $v_0 = 2.2 \times 10^8$ cm.s^{-1} is the Bohr velocity), stopping powers can be approximated by a simple velocity proportional dependence. In this case Valdés and coworkers[56] show that an account of the energy loss effect on angular distribution is given by inserting a mean energy in

---

[*]Thomas–Fermi potential is applicable for thick targets while Lenz–Jensen potential is more adapted for these targets but these two approaches converge.

the Meyer-Sigmund[40,50] theory. In fact they check various models for this mean ion energy to show that the model in good agreement with earlier theories has the following form:

$$\bar{\varepsilon} = \frac{\varepsilon_0}{4}\left[1 + \left(\frac{\varepsilon_1}{\varepsilon_0}\right)^{1/2}\right]^2 \tag{3.40}$$

where ε_0 and ε_1 are the initial and final reduced energies, respectively Eq. 3.22.

For the region where the stopping power cannot be taken as proportional to the velocity, another mean energy is used, for example Eqs. 3.33 and 3.34, which approximate this energy. To calculate the mean energy using these equations, we must define values for parameters n (from the stopping powers) and m (from the power potential). In particular for energies at which $v_1 \gg v_0 Z_1^{2/3}$, the electronic stopping power S_{el} (Bethe–Bloch Eq. 2.17) can be used. Then in this energy interval, $S(E) \propto E^{-1}$, so that $n = 1$, which can be used in Eqs. 3.33 or 3.34. However it is also well-known that the Bethe–Bloch equation cannot be applied to any energy interval; thus we must use the adequate stopping power law for each ion-beam specific problem.

There are different approaches for determining the parameter m. As noted before Winterbon shows that power cross-sections are a good approximation to MS distributions calculated with more exact cross sections. However the power must be chosen according to the target thickness. For example Sigmund shows that m = 0.191 gives a good agreement with the Lenz–Jensen potential for $\tau < 0.2$; m = 0.31 for $0.2 < \tau < 1$ and m = 0.5 for $1 < \tau < 5$. Furthermore Warwick and Sigmund[41] determined m as a function of reduced thickness for both Thomas–Fermi and Lenz–Jensen potentials. One can add that a theoretical calculation was developed by Schmaus et al.[70] to determine the parameter m as a function of the reduced thickness τ. Furthermore it is clear that as the MS distribution becomes approximately a Gaussian function, m is close to 1. The reader is referred to Refs. 41 and 70 for details.

One can also obtain m values in another way: in Eq. 3.25 parameter p is equal to 1/2m; one can directly deduce m values from $\tilde{\alpha}_{1/2}$ fitted functions or other measured values. For example when $2.10^3 < \tau < 10^6$, $m \approx 0.909$.

In the treatment of multielemental targets, we can use for example the $\hat{Z}$ approximation, then apply a procedure equivalent to that for a monoatomic target. For a polyester target, we obtain from Eq. 3.36 $\hat{Z} = 5.13$. This procedure is described in detail in Refs. 68 and 71, and it has been tested using both experimental data and the more exact solution from Eq. 3.35. These authors claim that both methods are equivalent (within less than 3%,) but as previously mentioned validity of the $\hat{Z}$ method is not demonstrated for any material.

Example. For a 12.6-MeV oxygen ion beam impinging on a 45-µg/cm^2 Ni foil, we obtain from Hooton et al.[75] the measured value of $\alpha_{1/2} = 2.8 \ 10^{-3}$ rad. This result is a guideline for estimating $\alpha_{1/2}$ in the following terms.

From Eq. 3.38 the reduced thickness is $\tau = 2.41$, and for this value we have $\tilde{\alpha}_{1/2}$ $= 0.3617\tau^{0.7237}$; then $\tilde{\alpha}_{1/2} = 0.6836$, and $m = 0.69$. Using Eq. 3.39, there are two cases: (1) Using the neglected energy loss approach; $\alpha_{1/2} = 2.7 \ 10^{-3}$ rad; (2) considering energy loss, we can use Eq. 3.33 to obtain $E_0/\hat{E} = 1.014$ and $\alpha_{1/2} = 2.74 \ 10^{-3}$ rad. Thus in this particular case, the deviation is about 1.5%; however for higher values of the relative energy loss, the deviation may become significant. Thus a comparison between the values as obtained from the preceding procedure and those measured by Hooton et al.[75] shows good agreement within error bars.

3.3.4. Combined Lateral–Angular Multiple Scattering

A Gaussian approximation is often used when evaluating the angular and lateral MS distributions to calculate each MS width. However MS distributions cannot be approximated with a Gaussian shape in every case, and great deviations may appear, as discussed by many authors. Nevertheless when reduced thickness is larger than a few thousands, *in first approximation a Gaussian function can be used although an* MS *distribution is never a Gaussian function.*

Moreover it is well-known that lateral spread is a consequence of angular divergence. Thus the angular and lateral MS cannot be considered independently of each other. As a consequence it is not justifiable to apply a summation in quadrature of both distribution widths, even in the Gaussian approximation.

Amsel and coworkers[57,77] carried out a systematic study of MS contributions and proposed an accurate evaluation of the resulting MS. In particular they evaluated the projected angular distribution from the related *spatial distribution* given by Eq. 3.24, and they also determined the projected half-width angle. Szilágyi et al.[59] recently showed that angular MS can be well-fitted by means of a Pearson VII type distribution. They also showed that the convolution of two different MS distributions described by the Pearson VII type distribution is not Pearson VII type, and they propose a summation method to compose two MS distributions.

Note that during the past years, a simple Gaussian convolution composition has often been used to calculate the final width, a method that can introduce errors in MS evaluations. To determining the total energy resolution (defined and discussed in Chapter 4), the resulting MS width must be evaluated correctly for example according to the procedure developed by Szilágyi in Ref. 59.

3.3.5. Multiple Scattering Summary

Multiple scattering is still a current problem although it has been studied for a long time. Several theoretical approaches exist, particularly when the total energy loss of particles over their path is very small. Obviously such approaches may be used for only very thin foils. When energy loss cannot be neglected, theoretical treatments are few. Although Landau[78] in 1944 gave a theoretical formulation, no universal procedure can be applied to evaluate MS; furthermore surveying the literature we observe that there are many theoretical studies rather than abundant experimental data. Hence it is difficult to check the validity of models that can be applied to a practical problem.

We presented an overview of the MS problem that must be considered a brief and simple outline. For a more accurate evaluation, the reader is referred to original papers. In particular when combined lateral–angular MS is considered, both lateral and angular MS distributions must be adequately summed, for example using the procedure of Amsel *et al.*[57,58,74]

Note that a Gaussian function is often used to represent an MS distribution, which is a very rough approximation, particularly for thin targets. In fact MS *distributions are never Gaussian functions* even when the thickness is very large. However *in first approximation* a Gaussian function can be used when the reduced thickness is larger than a few thousands.

When grazing geometry is used, we much use the projected distribution instead of the spatial distribution as given in the preceding discussion. A procedure for calculating projected distributions is described in detail in Refs. 57, 59, and 77.

It is well-known that theoretical advances were made in understanding MS; however the various theories are complex, so it is not easy to carry out a practical evaluation in a simple way. Nevertheless several values were tabulated, and some approaches were presented. These procedures may be sufficient for calculating MS in problems related to IBA.

REFERENCES

1. Freier, G., Lampi, E., Sleator, W., and Williams, J. H., Angular distribution of 1 to 3.5 MeV protons scattered by ^{4}He, *Phys. Rev.* **75**, 1345 (1949).
2. Miller, G. R., and Phillips, G. C., Scattering of protons from helium and level parameters in Li^5, *Phys. Rev.* **112**, 2043 (1958).
3. Landau, L., and Lifshitz, E., *Course of Theoretical Physics, Vol. 1: Mechanics* (Pergamon, Oxford, 1960).
4. Bozoian, M., Hubbard, K. M., and Nastasi, M., Deviations from Rutherford scattering cross section, *Nucl. Instrum. Methods Phys. Res. Sect. B* **51**, 311 (1990).
5. Benenson, R. E., Wielunski, L. S., and Lanford W. A., Computer simulation of helium induced forward recoil proton spectra for hydrogen concentration determinations, *Nucl. Instrum. Methods Phys. Res. Sect. B* **15**, 453 (1986).
6. Arndt, R. A., Roper, L. D., and Shotwell, R. L., Analyses of elastic proton-alpha scattering, *Phys. Rev.* **3**, 2100 (1971).
7. Tirira, J., and Bodart, F., Alpha-proton elastic scattering analyses up to 4 MeV, *Nucl. Instrum. Methods Phys. Res. Sect. B* **74**, 496 (1993).
8. Quillet, V., Abel, F., and Schott, M., Absolute cross-section measurements for H and D elastic recoil using 1–2.5 MeV ^{4}He ions, and for the ^{12}C(d,p)^{13}C and ^{16}O(d,p$_1$)^{17}O nuclear reaction, *Nucl. Instrum. Methods Phys. Res. Sect. B* **83**, 47 (1993).
9. Everhard, E., Stone, G., and Carbone, R. J., Classical calculation of differential cross section for scattering from a Coulomb potential with exponential screening, *Phys. Rev.* **99**, 1287 (1955).
10. Chu, W. K., Mayer, J. W., and Nicolet, M. A., *Backscattering Spectrometry* (Academic, New York, 1978).

11. Goppelt, P., Gebauer, B., Fink, D., Wilpert, M., Wilpert, Th., and Bohne, W., High energy ERDA with very heavy ions using mass and energy dispersive spectrometry, *Nucl. Instrum. Methods Phys. Res. Sect. B* **68**, 235 (1992).

12. Martin, J. A., Nastasi, M., Tesmer, J. R., and Maggiore, C. J., High energy elastic backscattering of helium ions for compositional analysis of high temperature superconductor thin films, *Appl. Phys. Lett.* **52**, 2177 (1988).

13. Bozoian, M., A useful formula for departures from Rutherford backscattering, *Nucl. Instrum. Methods Phys. Res. Sect. B* **82**, 602 (1993).

14. Bozoian, M., Threshold of non-Rutherford nuclear cross section for ion beam analysis, *Nucl. Instrum. Methods Phys. Res. Sect. B* **56/57**, 740 (1991).

15. Räisänen, J., Rauhala, E., Hnox, J., and Harmon, J. F., Non-Rutherford cross section in heavy ion elastic recoil spectrometry: 40–70 MeV 32S ions on carbon, nitrogen, and oxygen. *J. Appl. Phys.* **75**, 3273 (1994).

16. Räisänen, J., and Rauhala, E. J., Angular distribution of ^{12}C, ^{14}N, and ^{16}O ion elastic scattering by sulfur near the Coulomb barrier and the high-energy limits of heavy-ion Rutherford scattering, *J. Appl. Phys.* **77**, 1761 (1995).

17. Spiger, R. J., and Tombrello, T. A., Scattering of ^{4}He by ^{3}He and of ^{3}He by tritium, *Phys. Rev.* **163**, 964 (1967).

18. Stammbach, Th., and Walter, R., R-matrix formulation and phase shifts for n-^{4}He scattering for energies up to 20 MeV, *Nucl. Phys. A* **180**, 225 (1972).

19. Satchler, G. R., Owen, W., Elwyn, A. J., Morgan, G. L., and Walte, R., An optical model for the scattering of nucleons from ^{4}He at energies below 20 MeV, *Nucl. Phys. A* **112**, 1 (1968).

20. Arndt, R. A., Long, D. D., and Roper, L. D., Nucleon-alpha elastic-scattering analyses, *Nucl. Phys. A* **209**, 429 (1973).

21. Critchfield, C. L., and Dodder, D. C., Phase shifts in proton-alpha scattering, *Phys. Rev.* **76**, 602 (1949).

22. Dodder, D. C., Halle, G. M., Jarmie, N., and Witte, K., Tables of Phase Shifts and Experimental Observables for 4He Elastic Scattering, Report LA-6389-MS (Los Alamos, NM, 1976), pp. 1–58.

23. Brandan, M. E, Plattner, G. R., and Haeberli, W., A test of p-^{4}He phase shifts between 2 and 9 MeV, *Nucl. Phys. A* **263**, 189 (1976).

24. Tirira, J., Trocellier, P., Frontier, J. P., and Trouslard, P., Theoretical and experimental study of low-energy ^{4}He-induced ^{1}H elastic recoil with application to hydrogen behavior in solids, *Nucl. Instrum. Methods Phys. Res. Sect. B* **45**, 203 (1990).

25. Ad'jasevich, B. P., Antonenko, V. G., Polunin, Y. P., and Fomenko, D. E., Elastic scattering of polarized protons by helium at low energies, *Sov. J. Nucl. Phys.* **5**, 665 (1967).

26. Baglin, J. E., Kellock, A. J., Crockett, N. A., and Shih, A. H., Absolute cross section for hydrogen forward scattering, *Nucl. Instrum. Methods Phys. Res. Sect. B* **64**, 469 (1992).

27. Lauritsen, T., Huus, T., and Nilson, S. G., Scattering and deuterons in helium, *Phys. Rev.* **92**, 1501 (1953).

28. Galonsky, A., and McEllistrem, M. T., Energy levels of Li^6 from the deuteron-helium differential cross section, *Phys. Rev.* **98**, 590 (1955).

29. Galonsky, A., Douglas, R. A., Haeberli, W., McEllistrem, M. T., and Richards, H. T., Deuteron-helium differential scattering cross section, *Phys. Rev.* **98**, 586 (1955).

30. Besenbacher, F., Stensgaard, I., and Vase, P., Absolute cross section for recoil detection of deuterium, *Nucl. Instrum. Methods Phys. Res. Sect. B* **15**, 459 (1986).

31. Hermmendinger, A., Cross section for T-^{4}He scattering, *Bull. Am. Phys. Soc. Ser. II* **1**, 96 (1956).

32. Allen, R., and Jarmie, N., Triton reaction cross section, *Phys. Rev.* **111**, 1129 (1958).

33. Ivanovitch, M., Young, P. G., and Ohlsen, G. G., Elastic scattering of several hydrogen and helium isotopes from tritium, *Nucl. Phys. A* **110**, 441 (1968).

34. Chuang, L. S. The elastic scattering reaction ^{4}He(t,t)^{4}He and ^{4}He(τ,τ)^{4}He near 2 MeV, *Nucl. Phys. A* **174**, 399 (1971).

35. Sawicki, J. A., Measurements of the differential cross section for recoil tritons in ^{4}He-^{3}T scattering at energies between 0.5 and 2.5 MeV, *Nucl. Instrum. Methods Phys. Res. B* **30**, 123 (1988).

36. Tombrello, T. A., Miller-Jones, C., Phillips, C. G., and Weil, J. L., The scattering of protons from ^{3}He, *Nucl. Phys.* **39**, 541 (1962).

37. Meyerhof, W. E., and Tombrello, T. A., Energy levels of light nuclei A = 4, *Nucl. Phys. A* **109**, 1 (1968).

38. Terwagne, G., Cross section measurements of the ^{1}H(^{3}He,^{1}H)^{3}He elastic reaction between 1.9 and 3 MeV, *Nucl. Instrum. Methods Phys. Res.*, to be published.

39. Scott, W. T., The theory of small angle multiple scattering of fast charged particles, *Reports Mod. Phys.* **35**, 231 (1963).

40. Sigmund, P., and Winterbon, K. B., Small-angle multiple scattering of ions in the screened Coulomb region. I, *Nucl. Instrum. Methods Phys. Res.* **119**, 541 (1974).

41. Marwick, A. D., and Sigmund, P., Small-angle multiple scattering of ions in the screened Coulomb region. II, *Nucl. Instrum. Methods Phys. Res.* **126**, 317 (1975).

42. Bothe, W., Das allgemeine Fehlergesetz, die Schwankungen der Feldestärke in einem Dielektrikum und die Zertreuung der α-Strahlen, *Z. Phys.* **5**, 63 (1921).

43. Wentzel, G., Zur Theorie der Streuung von β-Strahlen, *Ann. Phys.* (Leipzig) **69**, 335 (1922).

44. Williams, E. J., The straggling of β particles, *Proc. Roy. Soc. A* **125**, 420 (1929).

45. Williams, E. J., Multiple scattering of fast electrons and alpha particles, and curvature of cloud tracks due to scattering, *Rev. Mod Phys.* **17**, 292 (1940).

46. Williams, E. J., Application of ordinary space-time concepts in collision problems and relation of classical theory to Born's approximation, *Phys. Rev.* **58**, 217 (1945).

47. Goudsmit, S., and Saunderson, J. L., Multiple scattering of electrons, *Phys. Rev.* **57**, 24 (1940).

48. Molière, G., Theorie der Streuung schneller geladener Teilchen, *Z. Naturforsch.* **2a**, 133 (1947); *Z. Naturforsch.* **3a**, 78 (1948).

49. Marion, J. B., and Young, F. C., *Nuclear Reaction Analysis* (North-Holland, Amsterdam, 1968)

50. Meyer, L., Plural and multiple scattering of low-energy heavy particles in solids, *Phys. Status Solidi B* **44**, 253 (1971).

51. Sigmund, P., Heinemeier, Besenbacher F., Hvelplund, P., and Knudsen, H., Small-angle multiple scattering of ions in the screened Coulomb region. III. Combined angular and lateral spread, *Nucl. Instrum. Methods Phys. Res.* **150**, 221 (1978).

52. Sigmund, P., in *Interaction of Charged Particles with Solids and Surfaces* (Gras–Marti, A., Urbassek, H., and Arista, N., eds.) (Plenum, New York, 1991), pp. 73–144.

53. Winterbon, K. B., Finite-angle multiple scattering, *Nucl. Instrum. Methods Phys. Res. Sect. B* **21**, 1 (1987).

54. Winterbon, K. B., Finite-angle multiple scattering: Revised, *Nucl. Instrum. Methods Phys. Res. Sect. B* **43**, 146 (1989).

55. Fermi, E., referred to by Scott, W. T., *Phys. Rev.* **97**, 12 (1949).

56. Valdés, J. E., and Arista, N. R., Energy loss effects in multiple scattering angular distribution of ions in matter, *Phys. Rev.* **49**, 2690 (1994).

57. Amsel, G., Battistig, G., Pászti, F., and Szilágyi E., Projected small-angle multiple scattering and lateral spread distribution and their combination, Part 2: Analytical approximation, *Nucl. Instrum. Methods Phys. Res.*, to be published.

58. Battistig, G., Amsel, G., d'Artemare, E., and L'Hoir A., Multiple scattering induced resolution limits in grazing incidence resonance depth profiling, *Nucl. Instrum. Methods Phys. Res. Sect. B* **85**, 572 (1994).

59. Szilágyi, E., Pászti, F., and Amsel, G., Theoretical approximation for depth resolution calculation in IBA methods, *Nucl. Instrum. Methods Phys. Res. Sect. B* **100**, 103 (1995).

60. Schmaus, D., and L'Hoir, A., Lateral and angular spread up to large energy losses for MeV protons transmitted through polyester, *Nucl. Instrum. Methods Phys. Res. Sect. B* **4**, 317 (1984).

61. Högberg, G., Nordén, H., and Berry, H. G., Angular distribution of ions scattered in thin carbon foils, *Nucl. Instrum. Methods Phys. Res.* **90**, 283 (1970).

62. Andersen, H. H., and Bøttiger, J., Multiple scattering of heavy ions of keV energies transmitted through thin carbon films, *Phys. Rev. B: Condens. Matter* **4**, 2105 (1971).

63. Meyer, L., and Krygel, P., The determination of screening parameters from measurements of multiple scattering of low energy heavy particles, *Nucl. Instrum. Methods Phys. Res.* **98**, 381 (1972).

64. Andersen, H. H., Bøttiger, J., Knudsen, H., and Møller-Petersen, P., Multiple scattering of heavy ions of keV energies transmitted through thin films, *Phys. Rev. A: Gen. Phys.* **10**, 1568 (1974).

65. Knudsen, H., and Andersen, H. H., Multiple scattering of MeV gold and carbon ions in carbon and gold targets, *Nucl. Instrum. Methods Phys. Res.* **136**, 199 (1976).

66. Hooton, B. W., Freeman, J. M., and Kane, P. P., Small angle multiple scattering of 12-40 MeV heavy ions from thin foils, *Nucl. Instrum. Methods Phys. Res.* **124**, 29 (1975).

67. Spahn, G., and Groeneveld, K. O., Angular straggling of heavy and light ions in thin solid foils, *Nucl. Instrum. Methods Phys. Res.* **123**, 425 (1975).

68. Schmaus, D., Abel, F., Bruneaux, M., Cohen, C., L'Hoir, A., Della Mea, G., Drigo, A. V., Lo Russo, S., and Bentini, G. G., Multiple scattering of MeV light ions through thin amorphous anodic SiO_2 layers formed on silicon single crystals, *Phys. Rev. B: Condens. Matter* **19**, 558 (1979).

69. Schmaus, D., and L'Hoir, A., Multiple scattering of MeV light ions transmitted through thin Al_2O_3 films: detailed analysis of angular distributions, *Nucl. Instrum. Methods Phys. Res.* **194**, 75 (1982).

70. Schmaus, D., L'Hoir, A., and Cohen, C., Ion multiple scattering: a tool for studying the thickness topography of self supporting targets, *Nucl. Instrum. Methods Phys. Res.* **194**, 81 (1982).

71. Schmaus, D., and L'Hoir, A., Multiple scattering angular distribution of MeV ^4He ions transmitted through Ta_2O_5 targets, *Nucl. Instrum. Methods Phys. Res. Sect. B* **2**, 187 (1984).

72. Anne, R., Herault, J., Bimbot, R., Gauvin, H., Bastin, G., and Hubert, F., Multiple angular scattering of heavy ions ($^{16,17}O$, ^{36}Kr, ^{40}Ar, ^{86}Kr and ^{100}Mo) at intermediate energies (20–90 MeV/u), *Nucl. Instrum. Methods Phys. Res. Sect. B* **34**, 295 (1988).

73. Lindhard, J., Nielsen, V., and Scharff, M., Approximation method in classical scattering by screened Coulomb fields, *Mat. Fys. Medd.* **36**(10), 1 (1968).

74. Dieumegard, D., Dubreuil, D., and Amsel, G., Analysis and depth profiling of deuterium with the D(3He,p)4He reaction by detecting the protons at backward angles, *Nucl. Instrum. Methods Phys. Res.* **166**, 431 (1979).

75. Hooton, B. W., Freeman, J. M., and Kane, P. P., Small-angle multiple scattering of 12–40 MeV heavy ions from thin solids, *Nucl. Instrum. Methods Phys. Res.* **124**, 29 (1975).

76. Bird, J. R., and Williams, J. S., *Ion Beams for Materials Analysis* (Academic, Sydney, 1989), pp. 620–623.

77. Amsel, G., Battistig, G., and L'Hoir, A., Projected small-angle multiple scattering and lateral spread distribution and their combination, Part 1: Basic formulae and numerical results, *Nucl. Instrum. Methods Phys. Res.*, to be published.

78. Landau, L., On the energy loss of fast particles by ionization, *J. Phys. JETP* **8**, 201 (1944).

4

Elastic Spectrometry
Fundamental and Practical Aspects

4.1. INTRODUCTION

The scattering of energetic ions by elemental target atoms has proved to be a useful tool in studying materials (e.g., thin films, surfaces, bulk solids). The parameters accessible to measurement by an elastic interaction can be divided into two groups: physicals parameters attached to the ion trajectory and the collision process, for example, cross sections, stopping powers, ranges and path lengths, resonances, etc.; the inventory of the target sample, for instance, target thickness, roughness of target surface, elemental constituents, depth profiles, elemental impurities, etc.

Elastic recoil detection is an old technique introduced and described by Marsden[1] in 1914. However this technique received considerable development in recent years using as a guideline works on backscattering spectrometry. In particular spectrum shape and height were investigated for backscattering spectrometry (RBS) in both theoretical and experimental studies in the past. Indeed theoretical analysis of spectra was given by Wenzel et al.[2] as early as the fifties, and several different theoretical versions have been proposed to give an adequate explanation of the backscattering yield.[3-8] These different models have a similar conceptual development, and they are the subject of continuous progress (e.g., Williams,[9] Chu et al.,[10] Lewis[11]), with the introduction of more precise functions for such secondary effects as straggling, multiple scattering, etc. One can note that the complete review presented in Chu et al.[10] has a special place in the world of backscattering spectrometry. This review shows in a simple way the physical concepts and the mathematical formalism of RBS. In another sense, the study of elastic recoil spectrometry is a relatively recent problem, and several new approaches are proposed to interpret recoil yields. Thus renewed studies have been carried out to obtain formulae for recoil spectrometry based on early backscattering works (e.g., Doyle et al.,[12] Benenson et al.,[13] Hofsäss et al.,[14] Tirira et al.,[15] Pászti et al.,[16] Szilágyi et al.[17]). In fact different versions of theoretical treatment of the elastic yield have been presented. Nevertheless almost all approaches are based on the same conceptual development.

In Chapter 4 we discuss the fundamentals of recoil spectrometry and summarize the practical aspect of the experimental problem. In Section 4.2 we present some fundamental theoretical expressions that are useful for ERDA spectrometry, implicitly assuming both an *ideal* experimental set-up and an *ideal* target sample. In Section 4.3 we discuss in detail problems arising from the usual imperfections in *real* set-ups and targets. In many cases of practical interest, one may wish to analyze data from elastic scattering of both the incident particle and the recoil. Thus looking for a general development of analytical expressions, one has to analyze basic relations that apply to general elastic spectrometry (projectile and recoil scattering).

4.2.　FUNDAMENTALS OF RECOIL SPECTROMETRY

As a starting point, we assume that dispersion phenomena—straggling, MS, and detector resolution—are absent, because if they can be ignored, elastic spectrometry can be described as a perfectly deterministic process. A sequence of simple arguments then leads from scattering collision to the yields and shapes of spectra generated by projectiles elastically scattered from atoms. It is thus a traditional approach to describe in a first step the principles for ideal straight trajectories and to assume that energy loss is completely defined by particle energy and target composition. This *ideal* behavior represents a *mean* behavior in both lateral and angular spreads and energy straggling. Then we introduce spatial and energetic deviations from this *ideal* behavior that can be superimposed on this simple description in a second step. We proceed in this way, considering only *mean* trajectories through Section 4.2.3.3. First we must describe trajectories of both incoming and outgoing particles, then derive the depth scale, applying energy conservation.

Note that the shape of the detected recoil spectrum is related to the type of detection system used in the experiment. In Chapters 5–9 we discuss different types of observed recoil spectra. However the transition from conventional set-up (an absorber foil in front of a classic silicon detector) to other detection types (time-of-flight, telescope, etc.) essentially affects the shape of the output signal, but not physical processes inside the target. Moreover the output signal is finally converted into one or several yields versus energy spectra, and thus the following discussion of energy–depth relations applies to all experimental set-ups. Thus in Chapter 4 the discussion of recoil spectrometry is based on the classic recoil arrangement; the adaptation of this analysis of conventional ERDA to other arrangements is given in Chapters 6–9.

To evaluate energy loss of particles in target materials, we first assume that the target sample is laterally uniform and constituted by a monoisotopic element. This assumption allows a simple description of energy–depth scales and elastic scattering yield. Afterward we extend the treatment to a multielemental target by introducing the combination of stopping powers, which are generally assumed to obey a linear additivity rule (*Bragg's rule*).

This section first describes the relationship between incoming and exiting paths and the depth at which the collision occurs, using the geometric factor of the collision. Next we describe how to relate the energy of the detected particle (scattered incident ion or recoil) to the depth at which the collision occurs.

4.2.1. Depth Scale

Figure 4.1 shows the inward and outward paths for mean trajectories. A more realistic aspect of trajectories appears in Figure 4.7.

To calculate easily incoming and outgoing paths as a function of the collision depth, it is convenient to consider first the case shown in Figure 4.1 that corresponds to the simplest arrangement for forward recoil scattering: In this case both trajectories are in a plane perpendicular to the target surface, and incoming and outgoing paths are the shortest possible ones for a given collision depth and given scattering and recoil angles.

Impinging ions reach the surface, making an angle θ_1 with the inward-pointing normal to the surface; some of these are scattered by atoms at rest. After collision their velocity makes an angle θ_2 with the outward surface normal; the atom initially at rest recoils, making an angle θ_3 with this normal. Detection is possible if one of these angles is such that the particle crosses the target surface.

Paths of the particles are related to the collision depth x, measured along a normal to the surface. For the impinging ion the incoming path is

$$L_1 = \frac{x}{\cos \theta_1} \qquad (4.1)$$

The outgoing path of the scattered projectile is

$$L_2 = \frac{x}{\cos \theta_2} \qquad (4.2)$$

and the outgoing path of the recoil

$$L_3 = \frac{x}{\cos \theta_3} \qquad (4.3)$$

In this simple case of a collision plane perpendicular to the target surface, the scattering angle of the impinging ion is $\theta = \pi - \theta_1 - \theta_2$, and the recoil angle is equal to $\phi = \pi - \theta_1 - \theta_3$.

The general case can easily be described if we keep the plane containing the incoming and outgoing paths (collision plane) as reference and tilt the sample around

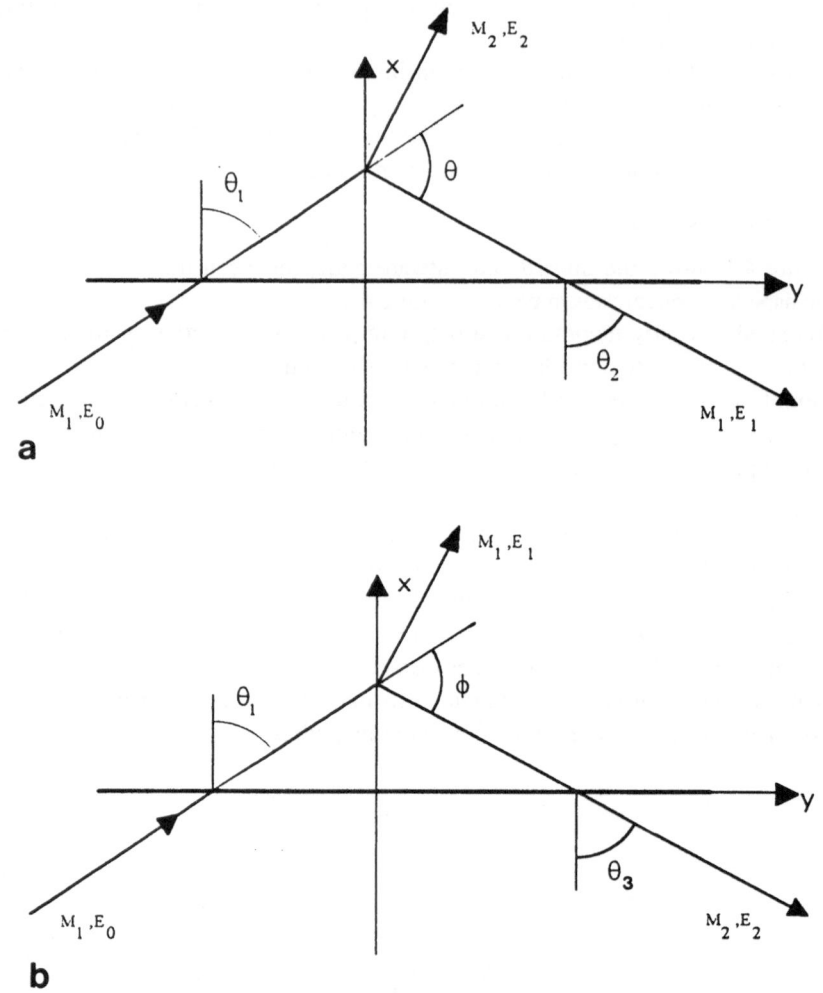

Figure 4.1. Planar representation of collisions in the target when both incoming and outgoing paths are in a plane perpendicular to the target surface. Depending on the impact parameter of the collision, the outgoing particle may be (a) the scattered projectile or (b) the recoil.

the y-axis of Figure 4.1. The target normal now makes an angle α (cf. Figure 4.2) with the collision plane, and each path is augmented by a factor $1/\cos\alpha$.

   For the purpose of converting the energy of outgoing particles into collision depth or vice-versa, it is useful to define *geometrical factors of the collision* as follows. For the recoil $R(\phi,\alpha)$ is defined by:

$$\sin L_3 = R(\phi,\alpha)\, L_1 \qquad R(\phi,\alpha) = \frac{\cos\theta_1 \cos\alpha}{\sin\phi\,\sqrt{\cos^2\alpha - \cos^2\theta_1} - \cos\theta_1 \cos\phi} \tag{4.4}$$

For forward scattering of the projectile $R(\theta,\alpha)$ by:

$$L_2 = R(\theta,\alpha)\, L_1 \qquad R(\theta,\alpha) = \frac{\cos\theta_1 \cos\alpha}{\sin\theta\,\sqrt{\cos^2\alpha - \cos^2\theta_1} - \cos\theta_1 \cos\theta} \tag{4.5}$$

Now there is somewhat of a conflict between choosing the simplest reference system and building a convenient target holder. Frequently we prefer to have a reference system bound to the target holder rotation axes rather than to the collision plane. In this case the plane defined by the impinging beam and the normal to the target surface is taken as a reference instead of the collision plane, so the outgoing path can be described using angle $\psi^{(18)}$ or angle $\beta^{(15)}$ of Figure 4.2. Expressions of the geometrical factor for the recoil are then:

$$R(\phi,\beta) = \frac{\sqrt{1 - \sin^2\theta_1 \cos^2\beta}}{\cos\beta\,\sin\theta_1\,\sin\phi - \cos\phi\sqrt{1 - \sin^2\theta_1 \cos^2\beta}} \tag{4.6}$$

$$R(\phi,\psi) = \frac{\cos\theta_1\,(\cos^2\theta_1 + \sin^2\theta_1 \cos^2\psi)}{\cos\psi\,\sin\theta_1\,\sqrt{\cos^2\theta_1 \sin^2\psi + \cos^2\psi \sin^2\phi} - \cos\phi\,\cos\theta_1} \tag{4.7}$$

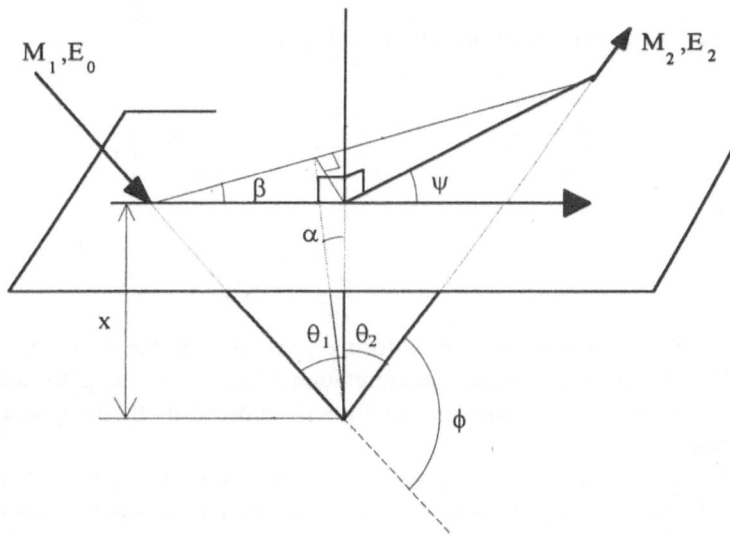

Figure 4.2. Typical geometrical configuration of recoil spectrometry. Paths of scattered particles are L_1 for incident particles, L_2 for scattered incident particles, and L_3 for recoiled atoms.

and for the scattered projectile $R(\theta,\beta)$ and $R(\theta,\psi)$ by merely changing ϕ to θ.

In following sections, we refer to $R(\theta,\alpha)$ and $R(\phi,\alpha)$; it is sometimes preferable to substitute expressions as functions of β or ψ, depending on the experimental arrangement under consideration.

4.2.2. Energy-Depth Relationship

To relate the energy $E_0(x)$ of the incident particle before the collision to its initial energy E_0 and the depth x at which the scattering event occurs, we use the following expression:

$$E_0(x) = E_0 - \int_0^{x/\cos\theta_1} S(E)\, dL_1 \tag{4.8}$$

Similarly we can write the energy $E_1(x)$ for emergent scattered incident particles and $E_2(x)$ for recoiled atoms as:

$$E_1(x) = KE_0(x) - \int_0^{x/\cos\theta_2} S(E)\, dL_2 \tag{4.9}$$

$$E_2(x) = K'E_0(x) - \int_0^{x/\cos\theta_3} S_r(E)\, dL_3 \tag{4.10}$$

Using Eq. (2.16) and energy conservation (cf. Figure 4.3):

$$L_1 = \int_{E_0(x)}^{E_0} \frac{dE}{S(E)} \qquad L_2 = \int_{E_1(x)}^{E_{01}(x)} \frac{dE}{S(E)} \qquad L_3 = \int_{E_2(x)}^{E_{02}(x)} \frac{dE}{S_r(E)} \tag{4.11}$$

where

$$E_{01}(x) \equiv KE_0(x) \qquad E_{02}(x) \equiv K'E_0(x)$$

and $S(E)$ and $S_r(E)$ are stopping powers for the projectile and the recoil in the target material. Figure 4.3 gives a graphic interpretation of these path integrals and shows that the energy scale as a function of depth is determined by the reciprocal of the stopping power.

Because we are looking for a general expression for recoil particles, only the trajectory of the incoming particle and the outgoing recoil scattered nucleus are considered in this section. A conversion to scattered projectiles can always be made from the development that follows by substituting θ for ϕ, K for K', L_2 for L_3, E_1 for E_2, and corresponding to the outward path, S for S_r.

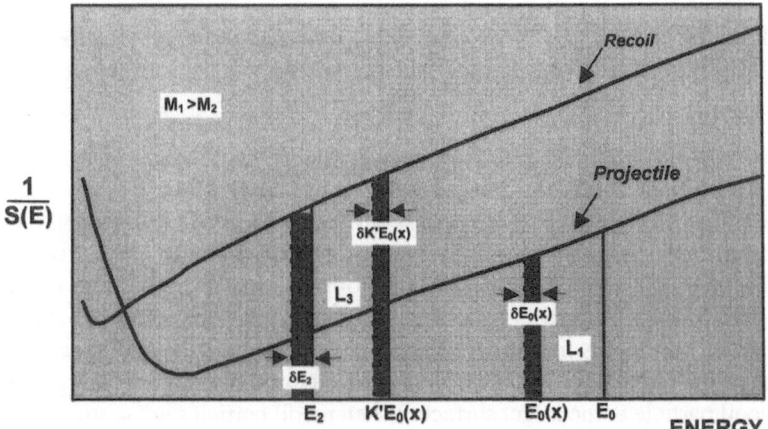

Figure 4.3. Schematic representation of reciprocal stopping powers as a function of energy. An interpretation of path integrals is shown for a small change δE_2 in energy E_2.

To obtain the energy–path scale, we must evaluate the variation of energy δE_2 around the outgoing energy E_2 at the target surface for an increment δx of the collision depth while E_0 remains fixed. Evidently this causes corresponding changes in the paths L_1 and L_3. Figure 4.3 shows a graphical display of path integrals variations (Eq. 4.11) for a small change δE_2 of emergent energy. In fact a variation of path around the collision point x can be related to the corresponding variation in energy before scattering by:

$$\delta L_1 = \frac{\delta E_0(x)}{S[E_0(x)]}$$

(4.12)

Immediately after scattering energy of the recoil is $K'E_0(x)$ [$\equiv E_{02}(x)$], and a variation of $E_0(x)$ results in a variation of recoil energy $E_2(x)$. Moreover particles with slightly different energies after the scattering at depth x undergo slightly different energy losses on their outgoing path. Then the change δL_3 of the path length L_3 can be written as:

$$\delta L_3 = - \frac{\delta[K'E_0(x)]}{S_r[K'E_0(x)]} + \frac{\delta(E_2)}{S_r(E_2)}$$

(4.13)

where the two terms represent the contribution to the variation of path due to the variation of energy just after the collision and the variation of energy loss along the outward path. Figure 4.3 shows an interpretation of paths variation, where the black areas under the reciprocal stopping power curves represent changes in paths (Eqs. 4.12 and 4.13).

Solving Eqs. 4.12 and 4.13 with $\delta x \to 0$ for the derivative dL_1/dE_2 with $L_3 = R(\phi,\alpha) L_1$, yields

$$\frac{dL_1}{dE_2} = \frac{1}{\{S_r(E_2)/S_r[K'E_0(x)]\}\{R(\phi,\alpha)S_r[K'E_0(x)] + K'S[E_0(x)]\}} \tag{4.14}$$

For a fixed geometry $[R(\phi,\alpha) = $ constant$]$, Eq. 4.14 shows that a small change in the recoil energy E_2 at the target surface is governed almost exclusively by the reciprocal of $S_r(E_2)$, since the contribution from the change in the ratio of stopping powers $S[E_0(x)]/S_r[K'E_0(x)]$ is generally less important. This effect is enhanced for very large outward paths for which $R(\phi,\alpha) \gg 1$. Equation 4.14 also shows that path length at which scattering occurs can be evaluated directly using the energy variation of the recoil particle at the target surface. When recoil particles are scattered from the top of a surface layer, the energy $E_0(x)$ goes to E_0 and Eq. 4.14 becomes

$$\frac{\delta L_1}{\delta E_2} \approx \frac{1}{K'S(E_0) + R(\phi,\alpha)S_r(K'E_0)} \tag{4.15}$$

where the ratio $S_r(E_2)/S_r(KE_0)$ is almost unity in this approximation and δE_2 can be interpreted as the energy difference $K'E_0 - E_2$. This approximation leads to a linear relation between the path length at which scattering occurs and the energy difference δE_2.

Equation 4.15 represents the *surface energy approximation* (superscript ^{0}) where stopping powers for incident and recoil particles are evaluated at energies E_0 and $K'E_0$, respectively. This means that when stopping powers are assumed constant along the way, the path–energy scale relationship may be considered linear. If α tends to zero, scattering particles and the surface normal are all contained in one plane, so we can rewrite Eq. 4.15 in the following simple form:

$$\frac{\delta L_1^0}{\delta E_2} \approx \frac{1}{[S^0] \cos \theta_1} \tag{4.16}$$

with

$$[S^0] = \frac{K'S(E_0)}{\cos \theta_1} + \frac{S_r(K'E_0)}{\cos \theta_3} \tag{4.17}$$

In elastic spectrometry, the term $[S]$ is usually called *energy loss factor*, which is given by:

$$[S] = \frac{K'S(E(x))}{\cos \theta_1} + \frac{S_r(K'E(x))}{\cos \theta_3} \tag{4.18}$$

Otherwise using the atomic density N of the target material and the *stopping cross section* defined by $\varepsilon(E) \equiv S(E)/N$, we can use the $[\varepsilon]$ factor, called *stopping cross-section factor*:

$$[\varepsilon] = \frac{K'\varepsilon(E(x))}{\cos \theta_1} + \frac{\varepsilon_r (K'E(x))}{\cos \theta_3} \tag{4.19}$$

The surface energy approximation is used in practical cases for a quick evaluation of energy loss and a rough estimate of the energy–depth scale.

As previously pointed out these expressions for the energy–depth scale were developed by taking the recoil particles as a reference. Therefore to introduce the scale for incident particle scattering (e.g., backscattering spectrometry), we need only replace in Eq. 4.14 the equivalent stopping powers of incident particles for the outward paths and the geometric factor $R(\theta,\alpha)$.

In previous discussions we derived formulae that allow energy and depth to be converted into each other. In practical problems of depth profiling, these simple expressions often prove useful for interpreting recoil energy spectra (Chapter 11).

4.2.3. Forward Recoil Yield

The scattering yield is the signal that must be analyzed to extract information about different parameters of the projectile–target system. Indeed stopping power values, scattering cross section, and elemental composition of the target can be extracted from the yield of interaction products.

4.2.3.1. Thin Target Sample

The number of recoil particles elastically scattered (number of counts) from depths from x to $x + \delta x$ through the recoil angle ϕ into the solid angle $\Delta\Omega$ is

$$\frac{dN(x)}{dx} \delta L_1 = N \, Q \, \delta\sigma[\phi, E_0 (x)]\Delta\Omega \, \delta L_1 \tag{4.20}$$

where N (atoms cm^{-3}) is the target atomic density, Q (particles cm^{-2}) is the total number of incident particles, $\delta\sigma[\phi, E_0(x)] \equiv d\sigma[\phi, E_0(x)]/d\Omega$ (barn atom^{-1} sr^{-1}) is the differential cross-section, which must be a specific function for each interaction type, and $\Delta\Omega$ (sr) is the solid angle subtended by the particle detector. If the yield $Y_r[E_0(x), E_2]\delta E_2$ is the number of recoil particles (Figure 4.4) observed by the detector having an efficiency ξ, with energy from E_2 to $E_2 + \delta E_2$, where δE_2 is the energy width of a channel in the multichannel analyzer, and the influence of energy broadening on the spectrum is neglected, we can write:

$$Y_r[E_0 (x), E_2] \, \delta E_2 = \frac{dN(x)}{dx} \, \xi \, \frac{dL_1}{dE_2} \, \delta E_2 \tag{4.21}$$

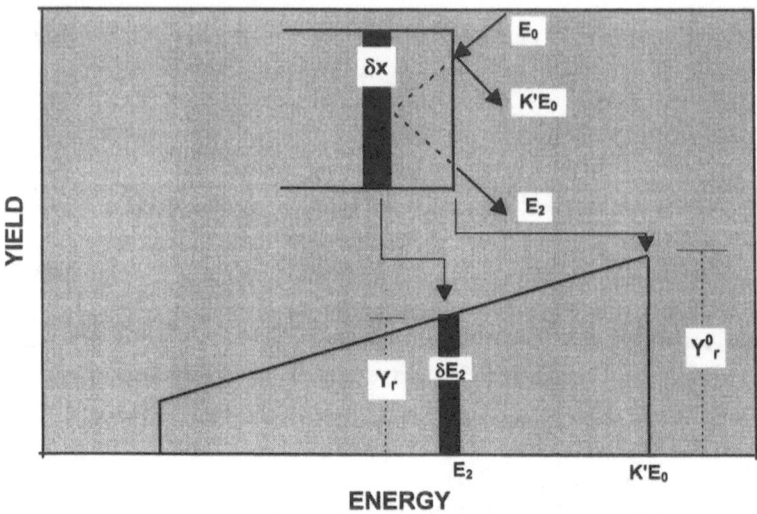

Figure 4.4. Elastic recoil spectrum as measured by a particle detector. Schematic showing the correspondence between the recoil process at depths between x and x + *dx* and the recoil yield observed with energies between E_2 and $E_2 + \delta E_2$.

Replacing the derivative dL_1/dE_2 from Eq. 4.14, the recoil yield (Eq. 4.21) is given by:

$$Y_r[E_0(x),E_2]\delta E_2 = \frac{NQ\xi \, \delta\sigma[\phi,E_0(x)]\Delta\Omega \, \delta E_2}{\{S_r(E_2)/S_r[K'E_0(x)]\} \, \{R(\phi,\alpha)S_r[K'E_0(x)] + K'S[E_0(x)]\}} \qquad (4.22)$$

Equation 4.22 represents the height of an energy recoil spectrum from a δx layer of an elemental target sample (Figure 4.4), when only two-body collisions are considered. The effects of MS and energy spreading are treated below. In practice for silicon detectors, ξ can generally be taken as unity. Equation 4.22 is a key expression in interpreting a recoil spectrum as a function of energy. In fact the variation of incident energy $E_0(x)$ of projectiles induces changes on both stopping powers and scattering cross section, so that the spectrum height may either decrease or increase following the ratio of their values. Indeed if the cross section varies slowly with energy (for example alpha-proton collision in the range of 2–3 MeV) and the ratio of stopping powers $S[E_0(x)]/S_r[K'E_0(x)]$ changes only slightly with the energy (for example a Si target, where hydrogen is considered an impurity), then the height of the hydrogen recoil yield varies almost exclusively as the reciprocal of stopping powers $S_r(E_2)$. This means that if the stopping power is low, the recoil yield is large, and vice versa. However in other cases the variation in scattering cross section and stopping powers at intermediate energies can play an important role, so that the form of the spectrum may be governed by variations in the ratio $\delta\sigma[\phi,E_0(x)]/S_r(E_2)$.

Note the dominant role played by the geometry of the experimental arrangement in determining relative strengths of stopping powers and scattering cross section. In particular the form of the spectrum is strongly influenced by variation in the geometrical factor R. Indeed for a fixed geometry (R = constant) and an homogeneous target, we showed a recoil spectrum that increases as energy increases (Figure 4.4). This effect is due to the fact that scattering cross section varies slowly with energy (for ^{4}He-hydrogen isotopes outside resonance regions), and the stopping power increases as energy decreases.

The recoil yield at the top of the surface layer when α tends to zero becomes

$$Y_r^0(E_0,E_2)\delta E_2 \approx \frac{N\ Q\ \xi\ \delta\sigma(\phi,E_0)\Delta\Omega\delta E_2}{[S^0]\cos\theta_1} \tag{4.23}$$

Equation 4.23 shows that the recoil yield at the surface is fundamentally proportional to the cross-section/stopping factor ratio evaluated at incident energy E_0. In this approximation the imaged thickness of the sample is given by a slab of width τ_0 ($\equiv \delta E_2/[S^0]$) from which all recoil particles are measured in the channel i.

In the case of a target containing more than one element, particles loose energy as a result of interacting with several constituents of the target. Therefore the principal problem becomes evaluating the stopping power in a compound sample. To evaluate at *first approximation* the recoil yield of a multielemental target, where F_j is the atomic fraction of the jth element, we can suppose that:

- Bragg's rule is a valid concept; thus we can apply the linear additivity of stopping cross-section. Using this approximation the compound stopping power is

$$S^c(E) \approx S^c_{Bragg}(E) = \sum_j F_j S_j(E) \tag{4.24}$$

  where S_j is the stopping power for an elemental target and the superscript c is related to the compound target. It is important to stress that this concept is not valid for certain compounds where significant deviations are quite evident (see Section 4.3.3).

- The height of the signal produced by each element (or each isotope) can be treated independently (linear additivity of signals).

Hence the individual recoil yield for one constituent (sometimes called a *contribution* of this constituent) becomes

$$Y_{jr}^c[E_0(x),E_2]\delta E_2 = \frac{NF_jQ\xi_j\delta\sigma_j[\phi,E_0(x)]\Delta\Omega\ \delta E_2}{\{S_r^c(E_2)/S_r^c[K_j'E_0(x)]\}\{R(\phi,\alpha)S_r^c[K_jE_0(x)] + K_jS^c[E_0(x)]\}} \tag{4.25}$$

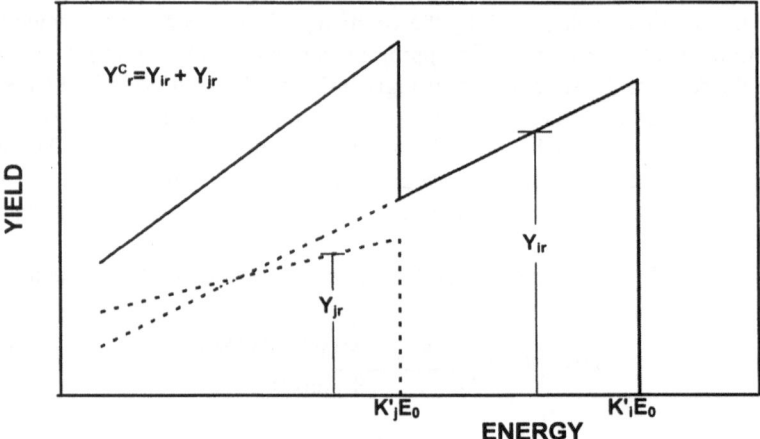

Figure 4.5. Recoil spectrum from a target sample composed of two elements, where E_0 is the energy of incident particles and $K_j E_0$ is the edge of element j in the spectrum.

where subscript j indicates that the parameters are related to the jth element in the compound target. The energy $K_j E_0(x)$ of the recoil particles immediately after collision is different for each one, since the kinematic factor is different for each element. Figure 4.5 shows the schematic representation of an energy recoil spectrum from a target composed of two elements.

To study the recoil yield of the jth element in more detail, it is useful in some cases to normalize this recoil yield. Thus we normalize Eq. 4.25 in terms of the ith yield arising from the compound target when α tends to zero, as:

$$\frac{Y_{jr}^c[E_0(x),E_2]}{Y_{ir}^c[E_0(x),E_2]} = \frac{F_j}{F_i} \frac{\xi_j}{\xi_i} \frac{\delta\sigma_j[\phi,E_0(x)]}{\delta\sigma_i[\phi,E_0(x)]} \frac{S_i^c[E_0(x)]}{S_j^c[E_0(x)]} \frac{S_r^c[K_j E_0(x)]}{S_r^c[K_i E_0(x)]} \qquad (4.26)$$

where the indices *i* and *j* in the stopping factor characterize both kinematic factors. The ratio of detection efficiency ξ_j/ξ_i is almost unity. Equation 4.26 shows that the simple ratio of experimental heights gives important information about the atom faction in the target sample if stopping powers and cross sections are well known. Strictly taken it is difficult to resolve the atomic fraction problem when elements constituting the target are not known. However using an iteration procedure, we can obtain a good approximation of the atomic fractions (cf. Section 11.4). Equation 4.26 can be used in a similar way when comparing results arising from a standard sample of known composition under identical beam conditions.

4.2.3.2. Thick Target Sample

In the preceding section, two effects are not mentioned: the intermediate energy influence on the stopping power and cross-section calculation and the effect of energy

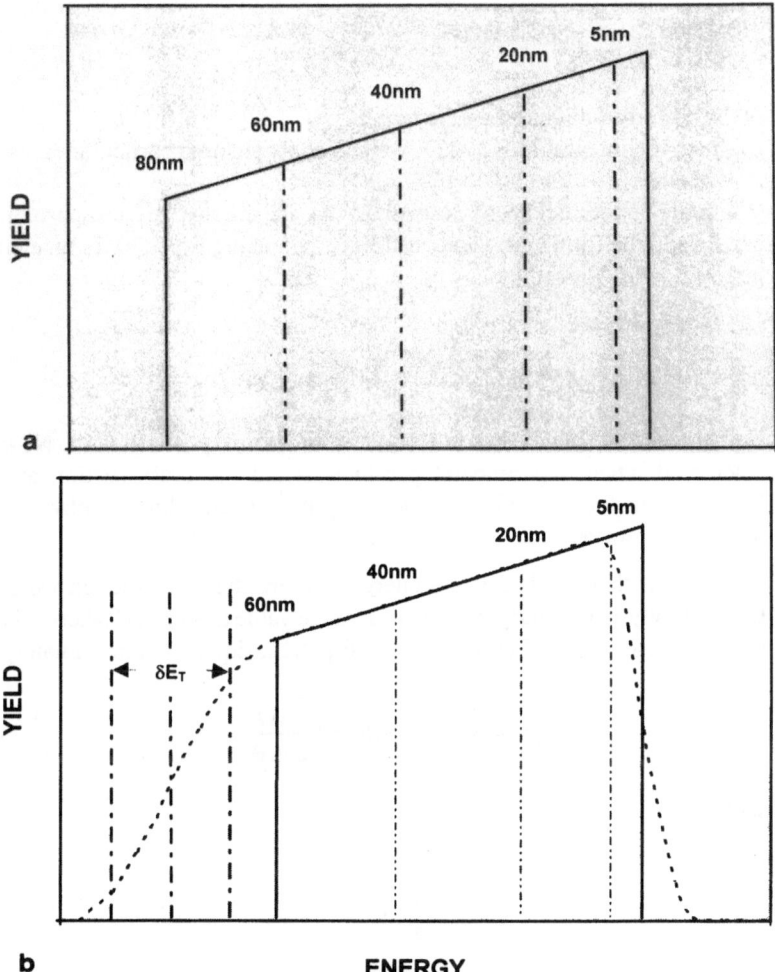

Figure 4.6. Total recoil yield from a monoelemental thick target (upper) without energy broadening and (lower) with energy broadening.

broadening on recoil spectra, treated in Section 4.2.3.3. These effects are particularly important for some targets: those thick enough (yet still quite a thin layer) so that energy of particles cannot be considered a constant and so-called thick samples, which means that the target thickness is larger than the range of the incident particles.

When energy broadening is not considered, the observed recoil yield (Figure 4.6) can be written as:

$$\mathscr{Y} = \int_0^t \frac{dN(x)}{dx} \frac{dx}{\cos\theta_1}$$ (4.27)

where t is the traversed thickness in the target.

To evaluate the integral in Eq. 4.27, we can analyze some particular cases.

- When the target thickness t is so small that the energy surface approximation can be applied (thin homogeneous film), then using Eq. 4.20 for an elemental target, Eq. 4.27 becomes

$$\mathscr{Y}^0 \approx N t Q \xi \, \delta\sigma(\phi,E_0) \frac{\Delta\Omega}{\cos\theta_1}$$ (4.28)

Equation 4.28 shows that the number of atoms per unit area Nt can be calculated when experimental parameters and the cross section are well-known, with the exception of the stopping power, which is obviously not needed in a surface energy approximation.

- For intermediate thickness and energies where the cross section varies relatively slowly, particularly taking the average value as $\sigma(\phi, \overline{E})$ [where $\overline{E}$ is equal to an average energy $E_0(x)$] and using Eq. 4.11, Eq. 4.27 approximately gives

$$\mathscr{Y} \approx N t Q \xi \, \delta\sigma(\phi,\overline{E}) \frac{\Delta\Omega}{\cos\theta_1}$$ (4.29)

with

$$t = \int_{E_t}^{E_0} \frac{dx}{dE} \, dE$$

If we have a thick target, t is the range of incident particles.

- For thick targets, it is difficult in general to resolve the recoil yields in a closed form. Nevertheless we can apply some method to approach a solution. Using the theorem of the average value, it is possible to find one analytical value Ξ of the energy in the interval (E_0, E_t) such that Eq. 4.27 for an elemental target gives

$$\mathscr{Y} = \frac{N Q \xi \, \delta\sigma(\phi,\Xi) \, \Delta\Omega}{S(\Xi)} (E_0 - E_t)$$ (4.30)

Then the essential problem in this solution is to find the analytical value Ξ, which cannot be determined experimentally.

Another way of calculating this type of yield is to carry out the integral process using analytical expressions for stopping power and cross section. Thus an analytical procedure can be applied in simulation methods (cf. Chapter 11). Note that if the atomic density has a concentration gradient $N(x)$, a numerical evaluation is almost the only solution.

4.2.3.3. Influence of Energy Broadening on a Recoil Spectrum

In the preceding development energy spreading is not taken into account. For brevity in the following, we call the yield resulting from the preceding expressions under the assumption of zero energy spreading *ideal yield* and the resulting spectrum *ideal spectrum*. Despite their abstract character, both these notions are quite useful because they present the advantage of a completely determined energy–depth relation. Nevertheless it is well-known that straggling, MS, and experimental apparatus may make a significant contribution to the shape of elastic recoil spectra. Figure 4.6b shows the effect of energy broadening on the shape of the spectrum. Energy straggling and MS are discussed in detail in Chapters 2 and 3, and these effects must be introduced into the recoil yield. Since the different effects of energy broadening are statistical in nature, we can treat these effects by introducing the probability distribution that implicitly takes into account both energy straggling and MS for a specific case. Thus let $P_1[E_0, E_0(x), L_1] dE_0(x)$ be the probability that the incident particle energy is found between $E_0(x)$ and $E_0(x) + \delta E_0(x)$ when the path length is L_1 ($\equiv x/\cos \theta_1$) if the energy at the surface was E_0. Similarly let $P_2[K'E_0(x), E_2(x), L_3] dE_2(x)$ be the probability of finding the energy of the recoil particle in the interval between $E_2(x)$ and $E_2(x) + \delta E_2(x)$ when the path is L_3 [$\equiv R(\phi, \alpha)x/\cos \theta_1$] if the energy just after the collision is $K'E_0(x)$. To calculate the influence of these expressions on the recoil yield, we choose a distribution that satisfies the following conditions[19]:

$$\int_0^{E_0} P_1 [E_0, E_0(x), L_1] \, dE_0(x) = 1 \qquad (4.31)$$

with

$$P_1 [E_0, E_0(x), 0] = \delta[E_0 - E_0(0)]$$

$$P_1 [E_0, E_0(x), L_1] = 0 \qquad \text{if} \qquad \begin{bmatrix} E_0 < E_0(x) \\ E_0(x) < 0 \end{bmatrix}$$

Hence the *real* recoil yield $Y_r[E_0(x), E_2]$ can be written as:

$$Y_r[E_0(x), E_2] = \int_{E_0}^{E_0(x)} \int_{E_2}^{K'E_0(x)} P_1(E_0, E_0', L_1) \frac{dN(x)}{dx} \delta L_1 P_2(K' E_0', E_2', L_2) dE_0' dE_2' \qquad (4.32)$$

If a delta function is substituted for energy distributions P, we obtain the yield $Y_r^0(E_0,E_2)\delta E_2$, which is equivalent to Eq. 4.23 in the energy surface approximation, as discussed before. When a thick target is considered, we must include one integral over the variable t that affects both paths. As noted before evaluating such integrals (Eq. 4.32) is not a simple task. However numerical techniques can be applied to solve this problem (see Chapter 11).

For practical purposes it is quite usual to assume a Gaussian form of energy broadening. In fact the Gaussian form is a good approximation for broadening due to beam energy fluctuations, finite detector acceptance angle, and detector resolution. As concerns the straggling distribution, it is Gaussian only when Bohr's theory is applicable. For MS it is never really Gaussian even if it is fairly approximated by a Gaussian for large enough depth, as discussed in Section 4.2.4.1. However the Gaussian model is generally preferred for convenience, because it allows a straight-forward convolution of yields obtained from Eq. 4.25, while exact calculations of Landau–Vavilov or Tschälär distributions for straggling and of the Bothe distribution for MS are tedious and time consuming.

The Gaussian approximation for straggling and MS at low depths produces a nonphysical yield above the high-energy edge of spectra, since a Gaussian is nonzero for any energy.[20] Indeed a Gaussian is not compatible with the third condition expressed in Eq. 4.31. This unphysical high-energy tail still exists, albeit to a lesser extent, if the more appropriate Landau–Vavilov distribution is taken instead of Bohr's distribution near the surface. In practice since the width of the straggling distribution, even in Bohr's approximation, tends to zero at the target surface, this tail represents more a theoretical anomaly than a practical problem.

4.2.3.4. Statistical Fluctuations of the Recoil Yield

Since scattering events occur independently of each other and with a constant probability determined by the cross-section scattering, the number of detected events is a statistical variable ruled by a binomial distribution. Cross sections are small enough for this distribution to be approximated by a Poisson distribution, for which the standard deviation is the square root of the expected mean number of events. Thus recoil yields exhibit fluctuations of the order of the square root of the number of counts predicted by Eqs. 4.25 and 4.32.[21] These fluctuations are sometimes referred to as *counting fluctuations* or *counting noise*.

Counting noise is liable to mask small details in the spectra. It is also the origin of strong perturbations in numerical methods of data processing. The possibility of reducing the noise/signal ratio by increasing the number of counts is limited by the risk of damaging the target and by the fact that the noise/signal ratio decrease is proportional only to the square root of the signal increase. The influence of counting noise on data processing and a possible way of reducing this influence by signal-processing methods are discussed in Section 11.4.4.

Counting noise is clearly distinct from *background noise*, which has a completely different physical origin. Counting noise is due to the random distribution ofimpact parameters. Background noise is composed of spontaneous impulsions from the detector (electronic background) in the low-energy range and of a generally negligible number of external particles (cosmic rays, gamma rays from concrete, etc.) reaching the detector.

4.2.4. Depth Resolution

4.2.4.1. Theoretical Expression

An important parameter that characterizes recoil spectrometry is the depth resolution. This parameter may be defined as the ability of an analytical technique to detect a variation in atomic distribution as a function of depth. This is the capability to separate in energy the recoil signal arising from a small depth interval. We can naturally define the *depth resolution* δR_x ($\equiv \delta L_1$) from Eq. 4.14 using the following expression:

$$\delta R_x = \frac{\delta E_T}{\{S_r(E_2)/S_r[K'E_0(x)]\}\{[R(\phi,\alpha)S_r[K'E_0(x)] + K'S[E_0(x)]\}} \qquad (4.33)$$

where δE_T is the total *energy resolution* of the system.

The energy resolution of a system results from combining fluctuations of physical and experimental origins. Different studies have been carried out to attempt to estimate δE_T. In particular a complete approach was recently given by Szilágyi et al.,[17] who developed a theoretical study on the ion beam depth resolution.

To determine energy resolution we must define the geometry used to detect recoil particles. The typical configuration of recoil spectrometry, which is shown in Figure 4.2, can be taken as a starting point of our discussion. We notice that in Figure 4.2 fluctuations of both collision angles and particle trajectories are not shown. Taking into account these fluctuations and projecting angles and path lengths onto the collision plane, we obtain Figure 4.7. We present herein a summary of models often used; the following discussion is based on the theoretical study in Ref. 17. In this description the collision plane is considered to be perpendicular to the target surface ($\alpha = 0$).

- *Detector energy resolution* δE_d: A function of electronic conditions and the detected energy E_2; δE_d varies very slowly with E_2.

- Energy spreading *by the fluctuations of projectile energy* $\delta E_p(x)$ can be calculated as:

$$\delta E_p(x) = \frac{\partial E_{02}}{\partial E_0} \delta \left\{ E_0 - \frac{x\,S[E_0(x)]}{\cos\theta_1} \right\}$$

$$+ \frac{\partial E_{02}}{\partial\phi} \delta(\pi - \theta_1 - \theta_3) - \delta \left\{ \frac{x\,S[E_2(x)]}{\cos\theta_3} \right\} \qquad (4.34)$$

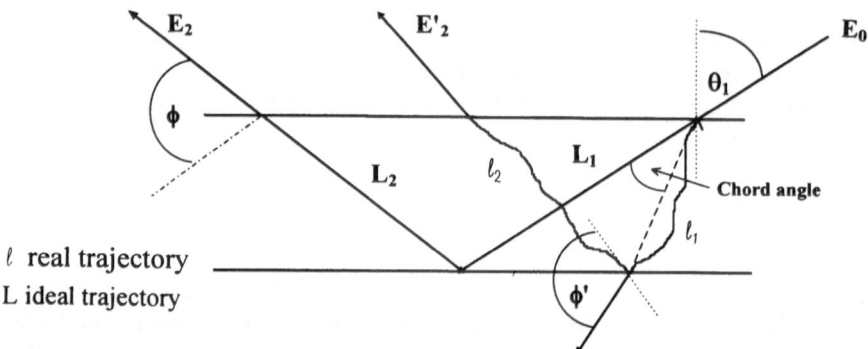

Figure 4.7. Schematic representation of incoming and outgoing paths for both ideal and real situations. Trajectories and angles are projected onto the collision plane. Paths exhibit a lateral displacement characterized by chord angles.

where E_{02} is the recoil energy just after the collision. This expression includes fluctuations in the beam energy and associated path-length fluctuations. Note that in thick targets, the energy loss fluctuation must be considered as induced by the path-length spread.

- *Effect of the beam angular spread* δE_{θ_1}: This term is a function of the angular width $\delta\theta_1$, which depends on accelerator optics. This term can be written as:

$$\delta E_{\theta_1} = \frac{dE_2(x)}{d\theta_1} \delta\theta_1 \qquad (4.35)$$

where

$$\frac{dE_2}{d\theta_1} \equiv \frac{\partial E_2}{\partial E_1}\frac{\partial E_1}{\partial \theta_1} + \frac{\partial E_2}{\partial \phi}\frac{\partial \phi}{\partial \theta_1}$$

Hence we can write

$$\delta E_{\theta_1} = [\tan\theta_1 K_{in} - 2E_0(x)\tan\phi]K'\frac{S_r(E_2)}{S_r(E_{02})}\delta\theta_1 \qquad (4.36)$$

where the factor K_{in}[17] is a function of the trajectory length and the stopping power of the projectile and $\delta\theta_1$ is obtained experimentally.

- *Geometrical energy spreading* δE_g arising from the particular detection type: This factor is a function of the finite beam size and the detector acceptance angle. Thus this term is given by:

$$\delta E_g(x) = \frac{dE_2(x)}{d\phi}\delta\phi_{det} \qquad (4.37)$$

where

$$\frac{dE_2}{d\phi} \equiv \frac{\partial E_2(x)}{\partial K'} \frac{\partial K'}{\partial \phi} + \frac{\partial E_2(x)}{\partial \theta_3} \frac{\partial \theta_3}{\partial \phi}$$

Then we can write

$$\delta E_g(x) = \left[\tan \theta_3 \, K_{out} - 2K' \, E_0(x) \tan \phi \, \frac{S_r(E_2)}{S_r(E_{02})} \right] \delta\phi_{det} \qquad (4.38)$$

where the factor K_{out}[17] is a function of the trajectory length and the stopping power of recoil atoms and $\delta\theta_{det}$ is the effective detector acceptance angle, which is approximated by:

$$\delta\theta_{det} \approx \delta\phi_{det} \approx \frac{1}{D} \sqrt{w^2 + \frac{d^2 \cos \theta_1}{\cos \theta_3}} \qquad (4.39)$$

where D is the detector target distance, w is the detector entrance width, and d is the incident beam width in the collision plane. Variation of the geometrical factor with x shows the dependence of this factor on the change of recoil energy between the scattering site and the target surface.

- Energy spreading by *energy straggling* δE_{st}: This factor takes into account the energy loss fluctuations δE_{st}^{in} and E_{st}^{out} that affect, respectively, the incoming and outgoing paths. These contributions can be calculated by:

$$\delta E_{st}^{in} = \frac{\partial E_2}{\partial E_{02}} \frac{\partial E_{02}}{\partial E_0} \Omega_{st}^{in} \qquad (4.40)$$

$$\delta E_{st}^{out} = \frac{\partial E_2}{\partial E_{02}} \Omega_{st}^{out} \qquad (4.41)$$

where we can take

$$\frac{\partial E_2}{\partial E_{02}(x)} \approx \frac{S(E_2)}{S[E_{02}(x)]} \quad \text{and} \quad \frac{\partial E_{02}}{\partial E_0} = K'$$

and Ω_{st}^{in} and Ω_{st}^{out} are adequate expressions for energy straggling. (See Section 2.5 in regard to the energy fraction $\overline{\Delta E}/\overline{E}$.)

- Energy spreading by *MS* δE_{mul}: This factor is used to evaluate the evolution from a monodirectional beam to one with an energy–direction distribution (see Section 3.3). Multiple scattering causes a strong deterioration of energy resolution, particularly when paths inside the target are long (especially for oblique angles). To calculate the MS contribution, it is useful to consider two contributions arising from *kinematic spreading*, where the full width half maximum (FWHM) of the angular distribution dispersion is $\delta\phi_{mul}$, which is twice the corresponding $\alpha_{1/2}$ of Section 3.3; and *lateral spreading*, where the path-length distribution is characterized by a width $\delta\rho$.

Discussion of these quantities and data for MS angles are given in Section 3.3 as a function of the reduced target thickness. Note that as stated in Section 3.3, lateral spread is closely related to angular divergence. Hence the corresponding contributions are not independent, and they must be combined in an adequate form to avoid large errors.

Energy fluctuations of incoming trajectories that affect the energy spread are given by:

$$\delta E_{mul}^{in,an} = \frac{\partial E_2}{\partial \phi} \, \delta\phi_{mul}^{in} \tag{4.42}$$

$$\delta E_{mul}^{in,la} = \frac{\partial E_2}{\partial L_1} \, \delta\rho_{mul}^{in} \tag{4.43}$$

where $\delta\phi_{mul}^{in}$ and $\delta\rho_{mul}^{in}$ are the FWHM for the angular (superscript ^{an}) and lateral (superscript^{la}) ingoing distributions.

Energy fluctuations of outgoing trajectories that affect the energy spread are given by:

$$\delta E_{mul}^{out,an} = \frac{\partial E_2}{\partial \phi} \, \delta\phi_{mul}^{out} \tag{4.44}$$

$$\delta E_{mul}^{out,la} = \frac{\partial E_2}{\partial L_2} \, \delta\rho_{mul}^{out} \tag{4.45}$$

where $\delta\phi_{mul}^{out}$ and $\delta\rho_{mul}^{out}$ are the FWHM for the angular and lateral outgoing distributions.

Using chain derivation Eq. 4.14 and the definition of MS as discussed in Section 3.3, we then have an analytical expression for MS broadening. The evaluation of these expressions is discussed in detail in Refs. 17 and 22.

• Spreading by *Doppler effect* δE_D: The motion of bound target atoms also affects their recoil energy measured by ERDA. If v_0 is the velocity of an atom bound in a solid, in first approximation, the Doppler energy shift can be given by:

$$\delta E_D \approx \frac{-2 \cos^2 \phi \, M_1 M_2 (M_1 - M_2) \vec{V}_0 \cdot \vec{v}_0}{(M_1 + M_2)^2} \tag{4.46}$$

For example for a bound proton ($v_0 \approx 1.1 \times 10^4$ m/s) recoiling after collision with an ^{4}He ion of 1.7 MeV ($V_0 \approx 9.1 \times 10^6$ m/s), the Doppler shift is at most 1 keV. So the Doppler broadening effect is small in comparison with other contributions. However the term must be taken into account for a complete evaluation of energy resolution. A more complete analysis of the Doppler effect is given in Refs. 23–24.

- Spreading by the *optional filter used in front of the detector* δE_{fil}: This parameter takes into account different systems for removing scattered ions or mass spectrometers, which can be used to separate recoil particles (e.g., absorber foil, *ExB* filter, TOF, etc.; see Chapters 5–8).

- Spreading due to the *roughness of the surface target* δE_{rou}: In some practical cases roughness parameters make a significant contribution to the recoil spectra. In fact roughness involves more than a depth resolution contribution, and it is preferable to dissociate its treatment from depth resolution, as discussed in Sections 4.3.2 and 11.4.5.1.

The total energy resolution of a system can be evaluated in the following way. Contributions considered to take a Gaussian character can be grouped together as an unique Gaussian distribution by quadratic summation of their standard deviations. In particular this is the case for fluctuations of experimental origin and for energy straggling (for large thicknesses); MS contributions cannot be summed in this way. Thus we ought to compute the convolution product between the Gaussian function and the total MS distribution. This task is not easy, and different procedures can be proposed; for example Szilágyi et al.[17] propose to add both distributions as a summation between two Pearson VII distributions, which is no obvious task. However although MS distributions cannot be represented by Gaussians in first approximation, when the target reduced thickness τ is larger than about 2000 we can assimilate the MS distribution to a Gaussian form. This last criterion has often been used in the IBA community to evaluate MS. Nevertheless this does not assume that the evaluation is completely wrong but simply that the MS contribution was under- or overevaluated. The reader is invited to see original papers for more details.

4.2.4.2. Practical Importance of Depth Resolution

The concept of depth resolution as previously presented represents the ability of recoil spectrometry to separate the energies of particles issued by scattering events that occurred at slightly different depths. Conversely this concept can be interpreted as a measure of scatter in energy of particles issued at the same depth.

For ERDA as well as other analytic techniques, δR_x is rather commonly interpreted as an absolute limit for determining concentration profiles. From this point of view features in concentration profiles separated by a depth interval of the order of magnitude of δR_x would be undistinguishable in the spectrum. Similarly it would be impossible with an accuracy better than δR_x to assign depth to a detail of the concentration profile producing a given feature in the spectrum. In fact the practical significance of depth resolution is more subtle, so that such an interpretation may be misleading. In particular the fact that signals corresponding to features of the concentration profile separated by less than δR_x strongly overlap in the spectrum does not necessarily mean that they are undistinguishable *in any case*, but only that attempting to distinguish them requires careful analysis and *may* finally appear as impossible. In the following discussion we examine the practical implications of depth resolution to give more insight into the limitations of depth profiling, taking into account the possibilities of data processing. This leads us to introduce the concept of *final* depth resolution, which describes the ability of the technique of resolving depth intervals after data processing. This discussion has application in Chapter 11, and it stresses the importance of thoroughly analyzing various physical processes contributing to depth resolution, as presented in the preceding section.

As a preliminary and somewhat elementary remark, one should not confound depth resolution with depth sensitivity. Very localized changes in target composition— thin layers, small concentration steps, etc.—are distinctly evidenced in a spectrum as soon as the corresponding yield is large enough.

The main ambiguity in interpreting depth resolution resides in the difference between identifying the depth at which *one* scattering event occurred, or the depth interval between *two* scattering events, and the resolution between *a number of* particles issued at two different depths. Before developing this distinction, two remarks illustrate this point of view:

- According to simple models of energy spread (e.g., Bohr's formalism for straggling), the energy distribution of particles issued at a given depth extends to infinity, which is obviously absurd, but indeed it extends over a very broad energy range. Thus it is nearly impossible to assign any depth to a detected scattering event. In other words considering a *single* event, any IBA technique is incapable of resolving depths.

- It should be kept in mind that there is a difference between *mixed* information and *lost* information. Due to energy broadening the spectrum is something like a cipher, but having analyzed in detail the various spreading effects enumerated in Section 4.2.4.1, we are in possession of a good part of the key to the cipher.

Let us consider energy distributions of recoils emerging from two slightly different depths. For the sake of mathematical simplicity, we can imagine them momentarily as Gaussian. Considering *a number of* particles, we first observe that their distribution is centered at a given energy corresponding to a well-defined depth. It is well-known that the sum of two Gaussian distributions with equal amplitude and equal standard deviation σ has two maxima when their barycenters are separated by more than 2σ and only one if this distance is less. But the sum looses its Gaussian character as soon as the barycenters do not coincide. In principle it is thus always possible to assign the resulting distribution to two distinct Gaussian distributions,[25] which can be separated analytically.[26] The position of both barycenters exactly determines the two different depths at which scattering has occurred. This discussion can be extended to any type of distribution, eventually to assymetric distributions, provided analytical tools for separating both distributions and finding their barycenters are at hand. If so the smallest detectable energy difference, namely, the channel width used to record the spectrum, determines the depth resolution; i.e., we could substitute the channel width into δE_T in Eq. 4.33 to obtain what we call the *final* depth resolution, or in other words, the depth resolution after data processing.

From another point of view, if the energy spread were *perfectly* described by some mathematical, energy-independent distribution, we could imagine obtaining a nearly unlimited resolution by performing a mathematical deconvolution, i.e., by dividing the Fourier transform of the experimental spectrum by the Fourier transform of the energy spread distribution, then taking the inverse Fourier transform of the result. [This result would be the ideal yield for which the energy-depth relation is unambiguous.] This is impossible in practice, because the energy spread distribution depends on energy and also because numerical deconvolution is extremely difficult to realize. But as shown in Section 11.4, there are other approaches to a similar, albeit necessarily approximate, result.

Schematically we can still give a more precise idea of the concept of *final* depth resolution by considering the difference between an experimental spectrum and a spectrum simulated by calculation. Assuming the energy spread distribution to be energy-independent, the experimental spectrum is the convolution product of an ideal yield $Y_{id}(E)$ by an energy-spreading distribution $P(E)$, as in Eq. 4.32. In fact this assumption is used for clarity, but any more appropriate operator can be substituted into the convolution product in what follows without changing the conclusions. Statistical fluctuations in the experimental yield (counting noise) can conveniently be represented by an additional term $b(E)$ without loss of generality. Starting with hypothetical concentration profiles and using equations available for elastic scattering and energy loss, we can calculate a simulated ideal yield $Y_{est}(E)$ that is an estimate of the ideal yield $Y_{id}(E)$ of a given target, and a simulated spectrum by convolution of $Y_{est}(E)$ by an estimated energy spread distribution $P_{est}(E)$ that represents knowledge available about energy dispersion. Then the difference between experimental and simulated spectra can be expressed as:

$$\Delta Y(E) = Y_{id}(E) \otimes P(E) + b(E) - Y_{est}(E) \otimes P_{est}(E) \qquad (4.47)$$

or equivalently by:

$$\Delta Y(E) = [Y_{id}(E) - Y_{est}(E)] \otimes P(E) + Y_{est}(E) \otimes [P(E) - P_{est}(E)] + b(E) \qquad (4.48)$$

Since ideal spectra are unambiguously related to depth profiles by a well-defined energy–depth relation (Eq. 4.14), finding the depth profile is equivalent to finding a simulated yield $Y_{est}(E)$ that exactly fits $Y_{id}(E)$. But practically we can only fit the experimental spectrum $Y_{id}(E) \otimes P(E) + b(E)$ by a simulated spectrum $Y_{est}(E) \otimes P_{est}(E)$, which consists in minimizing $\Delta Y(E)$. From Eq. 4.48 it appears that minimizing $\Delta Y(E)$ approximates a fit of the depth profile within an uncertainty determined by the second and third terms of the right side, i.e., by the difference between real and estimated energy spread distributions, $P(E) - P_{est}(E)$, and the counting noise $b(E)$.

The inaccuracy of energy spread models—not the energy spread itself—and counting noise thus appear as two major factors contributing to the *final* depth resolution. For instance a peak slightly broader than the estimated energy broadening should represent a layer of finite thickness. But it is impossible to attribute the excessive width to a finite layer thickness as long as excessive width is not definitely larger than the incertitude on the energy broadening. Similarly estimating the width of a noisy peak includes some incertitude that must be taken into account in interpreting peak width. Another term must also be considered as a consequence of these assumed simplifications. If it can be taken for granted that there is no approximation in kinematic formulae, incertitudes in stopping powers and cross sections necessarily cause some distortions in the depth scale of $Y_{est}(E)$ as compared to $Y_{id}(E)$ and thus influence the final depth resolution.

In conclusion it appears that despite the possibilities of using an appropriate analysis of data to surpass the apparent limits of depth resolution, there exists a finite *final* depth resolution resulting from both theoretical and experimental limitations. At the present state of knowledge, this *final* resolution is not amenable to a theoretical evaluation such as the classical depth resolution δR_x precisely because it results from three terms that escape theoretical estimation:

- Incertitude due to approximations of energy spread models

- Incertitude of data on stopping powers and cross sections

- Statistical fluctuations of recoil yield (counting noise)

The first term can be reduced by taking into account as exactly as possible the available knowledge of energy spreading or by improving theoretical models. A thorough evaluation of δR_x is thus a prerequisite to attaining the best possible *final* resolution. It is also clear that optimizing experimental conditions to minimize δR_x contributes to reducing the corresponding absolute incertitude and thus to improving the *final* resolution. Similar observations apply to the second term. Concerning counting noise, beside the limited possibilities of reducing it experimentally, discrimination by signal processing should be considered as a way of reducing its influence: This is the topic of Section 11.4.4.

4.2.5. Mass Resolution

In a similar way mass resolution is a parameter that characterizes the capability of recoil spectrometry to separate two signals arising from two neighboring elements in the target. The difference in the energy δE_2 of recoil atoms after collision when two types of atoms differ in their masses by a quantity δM_2 is

$$\frac{\delta E_2}{\delta M_2} = E_0 \frac{dK'}{dM_2} \qquad (4.49)$$

$$\frac{\delta E_2}{\delta M_2} = 4E_0 \frac{M_1(M_1 - M_2)\cos^2 \phi}{(M_1 + M_2)^2} \qquad (4.50)$$

where δE_2 is related to the energy resolution of the particular system under consideration. The left side of Eq. 4.50 is usually called mass resolution δM_R ($\equiv \delta E_2/\delta M_2$).

When ϕ tends to zero, Eq. 4.50 is proportional to $(1 - \phi^2)$. Thus the mass resolution is optimized with both the recoil angle going to zero and increasing incident energy. These considerations suggest that transmission geometry offers the optimal condition for mass resolution. Optimizing δM_R requires an optimal energy resolution. Therefore mass resolution is an intrinsic quantity that characterizes each detection system.

When mass resolution is determined from Eq. 4.49, we take into account only the kinematic process. However for detection systems where individual signals are measured for each mass, the mass resolution concept must be extended or even replaced by a concept of *mass selectivity*. For example in the case of TOF ERDA, an extension of the concept of mass resolution is necessary because the measured signal depends on energy, TOF, and flight path. If we assume incertitudes in energy and velocity to be uncorrelated, the mass resolution is given by:

$$\frac{\Delta M}{M} \approx \left[\left(\frac{\delta E}{E} \right)^2 + \left(\frac{2\delta t}{t} \right)^2 + \left(\frac{2\delta \ell}{\ell} \right)^2 \right]^{1/2} \qquad (4.51)$$

where δE and δt are the energy and time resolution, respectively, and $\delta \ell$ takes into account variations in the flight path. This expression represents the FWHM of the distribution of mass measurements around the true mass value.

For other systems such as the *ExB* filter or magnetic spectrometers for instance, particles with different masses are separated before energy measurement so that the energy difference of kinematic origin is no longer significant for mass separation. Then the concept of mass resolution as defined in Eq. 4.49 is no longer relevant, and it has to be replaced by the notion of mass selectivity, which is an intrinsic characteristic of the filter. More precisely mass selectivity is related to the ability of the detection system to separate masses independently of energy.

Note that in the case of TOF ERDA, concepts of kinematic (Eq. 4.49) and instrumental (Eq. 4.51) separation are complementary, so they must both be optimized. However in some mass spectrometers, kinematic separation as previously discussed may be of secondary importance or even contradict kinematic criteria for the choice of the projectile determined by other considerations. Discussions proper to each detection method are developed in Chapters 5–9.

4.2.6. Selectivity

To distinguish elemental atoms in a particular target, we must optimize the ion beam used to bombard the target. In others words we must know which ion beam to use to increase mass separation. In fact when only the kinematics in the target are considered, selectivity is determined by the difference in the energy $\delta E_2/\delta M_2$ (Eq. 4.50) of recoil atoms after collision related to the variation of mass δM_1 of incident particles, which is given by:

$$\frac{d(\delta M_R)}{dM_1} = \frac{4\,E_0\,\cos^2\,\phi(M_1^2 - 4\,M_1M_2 + M_2^2)}{(M_1 + M_2)^4} \tag{4.52}$$

Equation 4.52 shows that the maximum of the mass separation is obtained when $M_1 = 3.73\,M_2$. This means that when the projectile mass is around four times the mass of recoil atoms, the difference in energy of the recoil particles that differ in their masses by a small amount (e.g., isotopes) is maximum. This dependence gives a rapid diagnostic of the incident ion mass suitable for a particular target. For example an ^4He ion beam is suitable for analyzing hydrogen isotopes. When the detection system itself has the ability to separate masses (*ExB* filter, magnetic spectrometer), selectivity no longer depends essentially on this criterion.

4.3. PRACTICAL SPECTROMETRY OF REAL TARGETS

4.3.1. Ideal and Real Targets

In the preceding sections targets were implicitly considered to be ideal from different points of view.

- Their composition, and consequently also cross sections and stopping power, were assumed to be well-defined functions of the position along the paths of the impinging ion and recoil. Moreover the traveled path (in addition to any consideration of beam dispersion by MS) was assumed to depend only on the distance from the surface, which is itself considered well-defined (an even more idealized assumption). These assumptions depend on the invariance of target properties relative to a displacement parallel to a hypothetically plane target surface. We call this invariance lateral uniformity.

- It was also admitted for the simplicity of the presentation that the stopping power of compounds is given by Bragg's rule.

- Additionally any change in the target description during the experiment was disregarded.

- Finally targets were assumed to have no other interaction with ions and recoils than elastic scattering and energy loss, including straggling and MS.

Unfortunately such characteristics are only exceptionally encountered in real samples:

- Lateral nonuniformity is the most frequent and possibly the most disturbing deviation from the ideal situation. It may assume many forms (surface or interface roughness, porosity, texture, composition fluctuations, discontinuous coatings, etc.), and it has been widely discussed.[27,28]

- Bragg's rule is not perfectly satisfied in many materials, and it is even strongly violated in some.

- Target modification by the impinging beam may occur through different mechanisms.

- Electrostatic charge accumulation in insulating samples can reach high potentials and interact with the beam and recoils.

- Target crystallinity can modify the behavior of ions, and thus it can be regarded as another cause for unexpected beam–target interaction.

4.3.2. Lateral Nonuniformity

Lateral nonuniformity (also called lateral heterogeneity) is characterized by a variation in structure or composition in directions parallel to the target surface. Alternatively it may be defined as the impossibility of describing target properties by using the distance to the macroscopic surface as a unique parameter.

In such situations the behavior of ions and recoils depends on the impact point. If we imagine a beam composed of many very narrow beams, spectra generated by these partial beams differ from each other, and the spectrum obtained from the total beam is a superposition of these spectra.

Practically lateral nonuniformity must be taken into account in IBA when the following conditions are verified:

- The beam section is large enough for significant differences to occur between the paths of different ions.

- Variations in structure and composition are not so microscopic that no difference in the approximation of an homogeneous medium can be detected.

Nevertheless experimental conditions may exist that do not indicate lateral nonuniformity even when both of these conditions are satisfied (see, for example, the case of normal incidence in Figure 4.8).

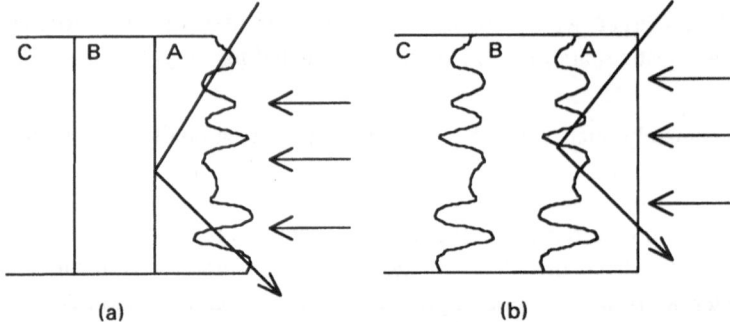

Figure 4.8. Equivalence of surface and interface roughness in the case of normal incidence. Situations (a) and (b) do not present any difference for normally impinging ions, but the ion at grazing incidence encounters several times the surface in (a); hence a broadening of the front edge of A appears, while it crosses several times the A–B interface in (b), which causes broadening of both A's back edge and B's front edge.

Roughness is a typical case of nonuniformity. It is frequently encountered because it is impossible to polish some materials or in most cases polishing would modify or destroy the region of interest. The path length from the real surface to a given depth relative to the apparent surface plane is variable. As a result, recoil energies from a given depth are dispersed. The importance of roughness becomes considerable at a grazing incidence because path differences are increased and also some ions may cross the target surface or interfaces more than once. Fluctuations in total target thickness play a role only in transmission arrangements. In principle they have the same effect as roughness, i.e., a dispersion in path lengths. Since small fluctuations in relative value can be large compared to usual roughness, they can be quite detrimental. A 1% variation in a 5-μm thick sample is equivalent to a 50-nm roughness, a large value; it causes path-length fluctuations that may be larger than the depth resolution. This statement also applies to transmission through the absorber foil. Multiple-scattering theory was applied by Valdés *et al.*[29] to correct energy loss in the case of rough foils.

Rough interfaces have similar effects: Figure 4.8 shows that ions at normal incidence and backscattered at 180° make no difference in surface and interface roughness. However this is not true for low-scattering angles and grazing incidence, because it is quite different for ions to cross the surface more than once or the interface several times. Porosity is also characterized by path-length variations for different ions. Except for large isolated voids, it is more easily treated as density fluctuations. Other typical lateral nonuniformities are discontinuous or fractured overlayers, interfacial voids and precipitates, columnar epitaxial layers. In the latter case stopping power and cross-section variations for different ions appear instead of path-length variations.

While lateral nonuniformity has been widely discussed in the frame of RBS analysis (Refs. 27, 28, 30, and 31), it is referred to only in general terms in ERDA literature, possibly because it presents more difficulties, mainly related to large

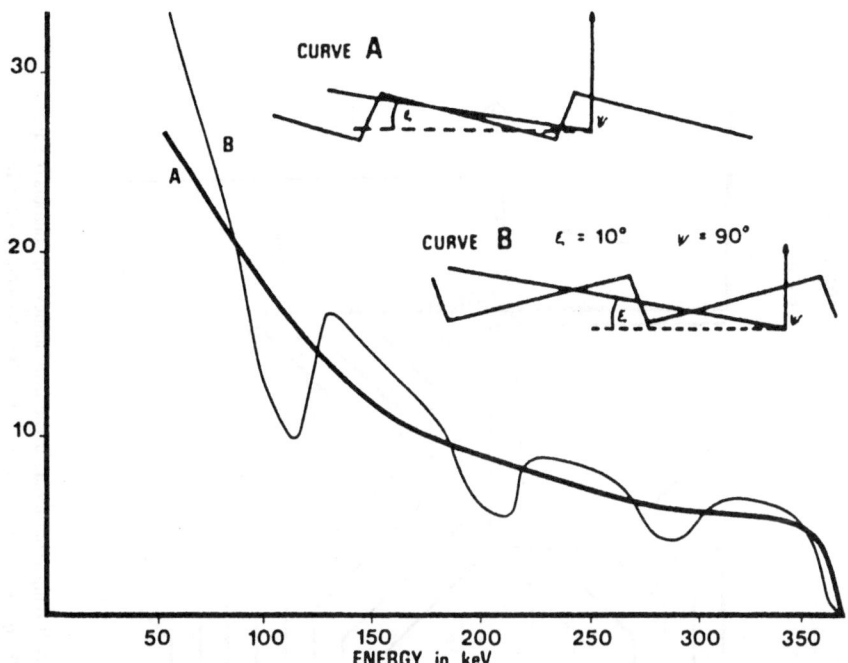

Figure 4.9. Backscattering from a surface with sawtooth topography. Two calculated spectra are compared for protons impinging in opposite directions with respect to the surface structure but with the same incidence angle relative to the average surface plane. (Reproduced from Ref. 32)

scattering angles and grazing incidence. It is possible to predict the effects of lateral nonuniformity in most cases and even to retrieve some of its characteristics from RBS spectra (cf. Section 11.4.5.1), but such treatments rely on an assumption of similitude of inward and outward paths (i.e., a complete determination of inward and outward path lengths once the impact point is given), which is justified only in the vicinity of a 180° scattering angle, hence not for ERDA analysis. Typically Edge and Bill[32] studied the case of periodic roughness at grazing incidence. Notwithstanding the academic character of the chosen example, they demonstrate in a spectacular way (Figure 4.9) how strongly incidence and exit angles influence the spectrum shape when impact and exit points are not equivalent. They also show the unexpected shape of contours of equal path length at grazing incidence on a rough surface (Figure 4.10). An exception is the treatment of thickness fluctuations of the absorber foil, which is traversed only once and at normal incidence.

As discussed in Section 11.4.5.1 a detailed interpretation of ERDA spectra of laterally nonuniform targets is quite difficult, so we are often bound to consider merely the lateral nonuniformity as a loss of resolution. The main concern then is to avoid misinterpretations resulting from an unjustified assumption of uniformity. Is it useful

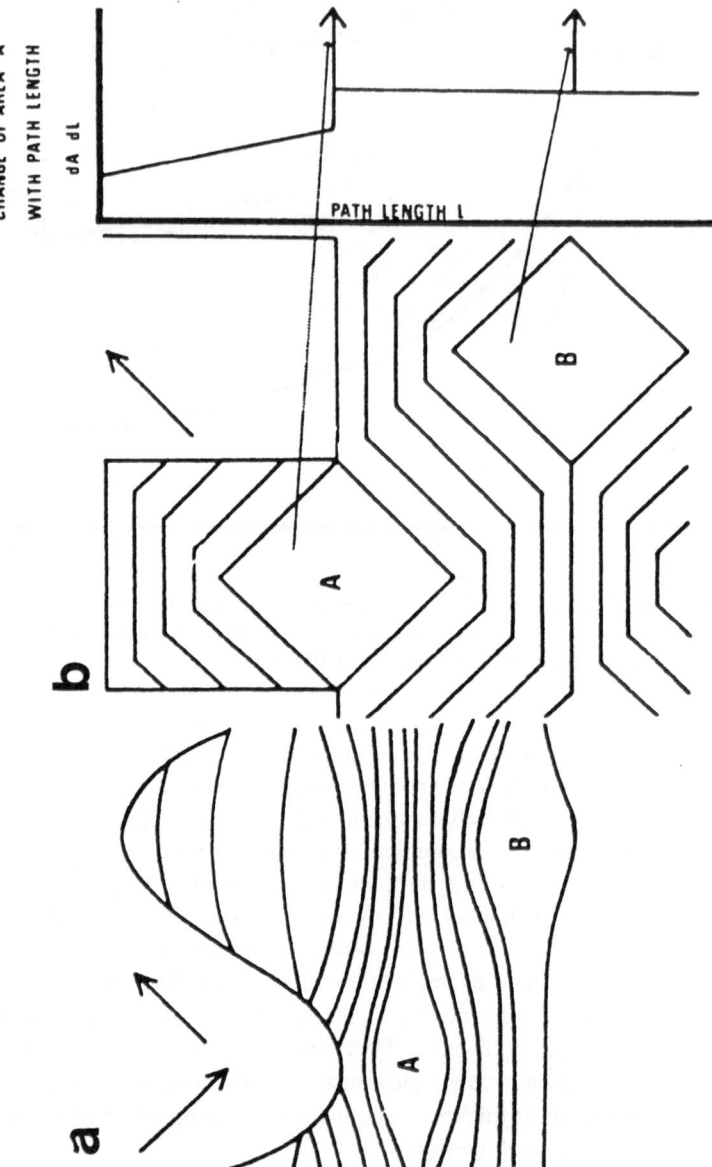

Figure 4.10. (a) Contours of equal path length for particles entering and leaving a sinusoidal surface at 45° angles. (b) Contours of equal path length for particles entering and leaving a square wave surface and (*right*) the rate of change of area between contours as a function of path length. (Reproduced from Ref. 32)

to say that a preliminary microscopic observation of samples is always a good precaution? In any case, hidden nonuniformities can exist (for example, interfacial precipitation). In such cases ERDA analysis sometimes enables us to detect nonuniformity through discrepancies that appear, assuming lateral uniformity, between spectra recorded with different tilt angles or beam energies.[31,33] This occurs because distortions caused by nonuniformity depend on stopping powers and of the geometrical factor.

4.3.3. Deviations from Bragg's Rule

Bragg's rule (Eq. 4.24) relies on the assumption that compounds can be treated as mixtures of pure elements; their stopping power is thus a sum of weighted atomic stopping cross-sections. This assumption has its origin in the idea that the energy loss is a random sequence of independent interactions with electrons and atomic cores without any influence from the surrounding atoms. Despite its attractive straightforward expression, this rule can be useful only if its accuracy is of the order of literature data accuracy for pure elements' stopping powers, i.e., a few percent. However abundant experimental literature on stopping power measurements departs from Bragg's rule for many compounds, which is not surprising, owing to the oversimplified character of this idea.

As can be expected good agreement is reported for metallic alloys and compounds in which free electron gas behavior justifies Bragg's assumption. Organic compounds, oxides, and more generally compounds with tightly bound electrons present noticeable deviations from Bragg's rule, especially at ion energies near and below the stopping power maximum. These deviations have been the topic of numerous publications in the last 20 years, which are extensively reviewed by Thwaites[34,35] and Powers.[36] For example the stopping power of solid oxygen for helium ions, derived from measurements in Al_2O_3 and SiO_2 assuming the validity of Bragg's rule, was reported to be 5–15% larger in the latter.[37] Similarly departures from additivity are evidenced in hydrogenated carbon films by discrepancies between estimations for hydrogen depletion from variations in peak integrals and widths.[38] Taking peak integrals of ERDA and proton-enhanced cross-section scattering as reference for the H:C ratio, experimental stopping powers of protons and helium according to peak widths is found to be about 30% larger than predicted by Bragg's rule. Very large departures have also been reported for AlB_2, TaB_2, and B_4Si.[39]

The considerable literature allows to judge for a large variety of compounds and mixtures whether it is reasonable to accept the approximation of Bragg's rule. Now where significant departures exist, it is generally not satisfactory to proceed to approximative corrections on the basis of these data, and it is preferable to find more suitable substitutes for Bragg's rule.

Stopping power data applicable to some compounds (e.g., those of Santry and Werner[37] for ^4He in Al_2O_3 and SiO_2) do exist, and these have been used with success by many authors. Similarly Reiter et al.[40] have derived a stopping cross-section for atomic nitrogen from measurements in O_2, N_2, NO, and N_2O. But to avoid a particular treatment for each compound, and to find a theoretical basis for the stopping power of compounds, attention was directed to less empirical approaches that consist of attempting to build an alternative rule using chemical bonds or molecular fragments as additive elements instead of constituents of the compound. This approach has its origin in the fact that valence electrons dominate energy loss at low energies; it has been developed by Sabin and Oddershede.[41] Their model is based on the kinetic theory of Sigmund,[42] which subdivides excitation energy into contributions from core and valence electrons accordingly to orbital weight factors (equivalent to occupation numbers) and valence velocity densities derived from experimental and theoretical Compton profiles. Bragg's rule is thus replaced by:

$$S_{compound}(v) = \overset{(bonds)}{\sum_{i}} S_i^{bond}(v) + \overset{(cores)}{\sum_{j}} S_j^{core}(v) \qquad (4.53)$$

In this model only the Bethe–Bloch domain is considered; Z_1^3 and Z_1^4 terms are neglected, being of opposite sign and significant only at low velocities. Cores are represented by their Hartree–Fock approximation and valence bonds by velocity densities from isotropic Compton profiles. Orbital weight factors and orbital values of mean excitation energies are taken from Refs. 43–45. The model predicts a 35% deviation below Bragg's rule at the stopping power peak for protons in methane, which is in excellent agreement with experiment (Figure 4.11).

Formerly an alternative approach consisting of adding a correction term to Bragg's rule for each bond was developed by the Köln group[39]; this approach has more recently adopted the functional group approach.[46,47] This latter was applied with success to hydrogenated carbon layers.[38] Bauer[48] found good agreement between experimental stopping power of protons in SiO_2 and calculations wherein SiO_2 is considered to be composed of two O^{4+} ions, one Si^{4+} and 12 valence electrons and the contribution from valence electrons is computed using a velocity distribution derived from an experimental Compton profile.

Following the cores-and-bonds approach and scaling heavy ion stopping powers to equivalent proton stopping powers, Ziegler and Manoyan[49] developed a calculation for stopping powers of protons and helium and lithium ions that fits experimental data for hydrocarbons within about 1%.

The cores-and-bonds theory is certainly the most promising approach presently available; nevertheless this method is based on an additivity rule similar to Bragg's rule. This means that when bond–bond interactions, ring strains, steric hindrance, and particularly strong polar or very electropositive or electronegative substituents are present, deviations may still occur; there is also some uncertainty about the effective charge of the impinging ion. Moreover isotropic Compton profiles are not always available for a valence bond. For all these reasons the cores-and-bonds model remains difficult to apply in general cases. Progress is still expected in the area of data necessary

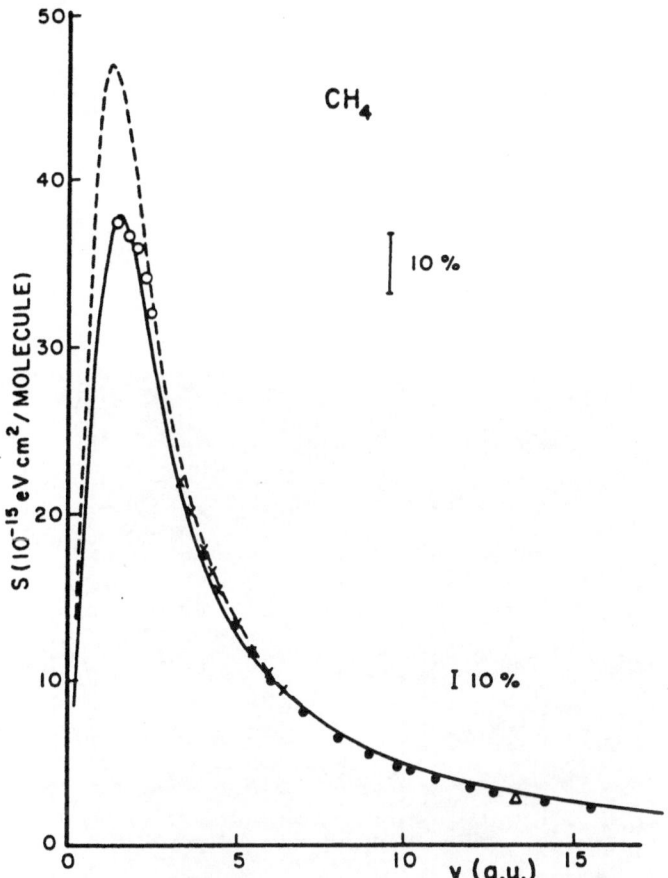

Figure 4.11. Stopping cross-section curve for protons in methane. *Dashed curve*: Bragg's rule; *solid curve*: calculation according to Ref. 41 using a theoretical velocity density. Experimental data are indicated by (Δ), (×) and (○). (Reproduced from Ref. 41)

for its application in cases for which it was developed and for extensions of the theory to more intricate cases.

4.3.4. Beam Damage

If the target is modified during analysis, whatever the physical nature of the beam-induced modification, may be a single spectrum no longer represents the material. Real targets are generally subject to beam damage—either through direct irradiation effects or processes induced by beam heating or electrostatic charge accumulation. Chapter 14 is dedicated to beam-induced damage and possible ways of reducing it.

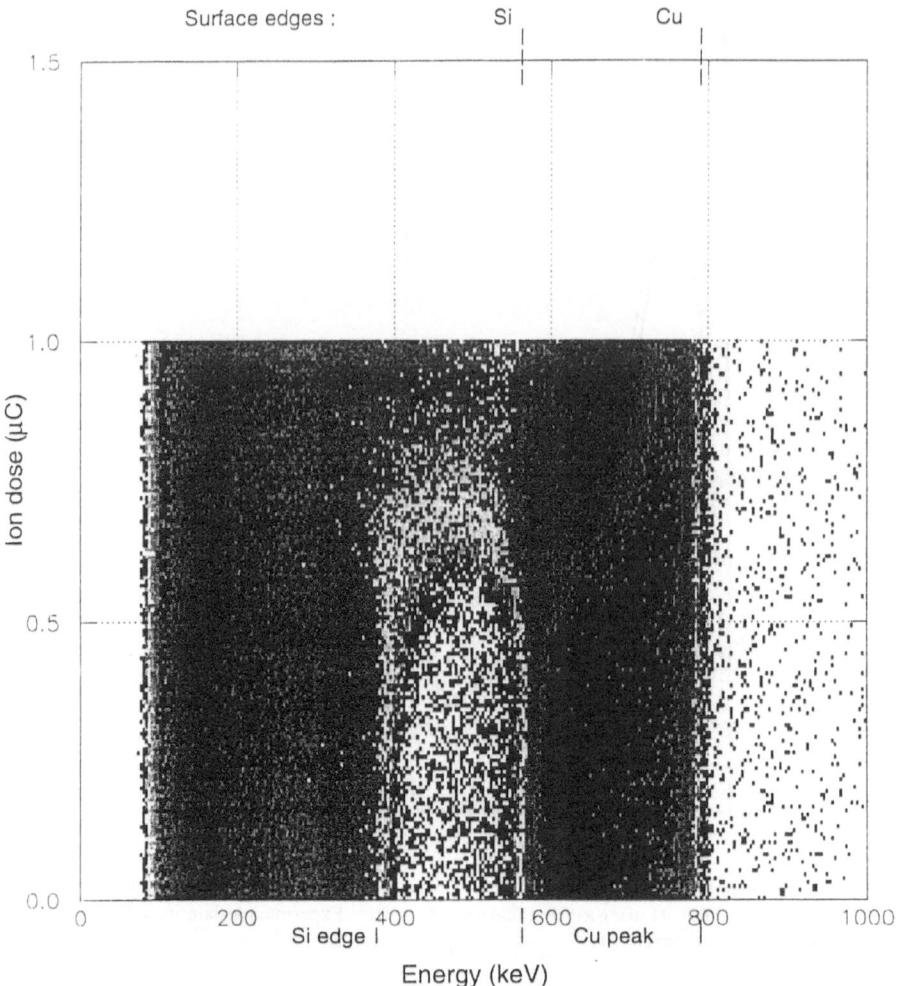

Surface edges :   Si   Cu

Figure 4.12. Event-by-event record of yield versus energy and ion flounce of a 200-nm Cu film on Si. Yields are represented by color levels as a function of energy and ion fluence. Initial copper region and silicon edge are labeled along the energy axis. (Reproduced from Ref. 50)

We mention here that changes in target composition may draw our attention for themselves and quote the interesting possibilities offered by event-by-event data acquisition,[50] a method of recording detection events not only as a function of energy (and eventually velocity) but also as a function of time, or more significantly of projectile fluence. Figure 4.12 shows this method for RBS analysis of a 200-nm Cu film deposited on Si using a ring detector with a 20-msr aperture. Evolution of the sample clearly appears as shifts in edge location and peak width variation on such a representation.[50] These shifts explain differences observed in global spectra recorded under the same ion fluence but with a very different beam focusing ($0.08-10 \, pA/\mu m^2$) and describe the dependence of the global amount of copper as a function of ion fluence. Wide solid angles are useful for recording a statistically significant number of events for small ion fluence increments. This is relatively easy for RBS using ring detectors but more difficult using ERDA, where wide solid angles result in resolution loss. Nevertheless large cross sections favor ERDA, and still unpublished hydrogen analyses have been performed with success.[51]

4.3.5. Insulators

Insulators are subject to charge accumulation in the near-surface region due to both ion implantation and secondary electron emission. This process can have several consequences—target damaging (dielectric breakdown) or modification (electromigration, changes in electrical and optical properties)—and since they are examined in Chapter 14, we overlook their consequences in the following discussion of analysis perturbations caused by charge accumulation.

Electric field buildup under irradiation is quite important and rapid. It is evidenced, for example by the repulsion of low-energy charged particles (electrostatic mirror effect).[52] Using the energy shift of alpha particles emitted by the proton–sodium resonant nuclear reaction, the potential rise of a glass sample submitted to a 600-keV proton beam was evaluated to about 10–15 kV.[53,54] This potential is reached after a few seconds despite the low beam current (a few nA/mm^2), and it requires in the given experimental geometry an electric field of at least 100 kV/cm in the near-surface region.

The field distribution inside and outside the target depends in a complex way on the balance between charge implantation, secondary emission, and conduction; it was discussed extensively by Cazaux.[55-58] Its quantitative prediction is difficult because the secondary electron yield is itself modified by the field buildup.

A first effect of charge accumulation is energy change in incident particles and recoils when traversing the field. Changes of several keV can be expected, and these are indeed observed,[30] but they are difficult to take into account exactly in the treatment of spectra (cf. Section 11.4.5.2).

Since particle trajectories are not perpendicular to equipotentials (except perhaps in transmission geometry), which are themselves not planes because of the finite beam width (see Figure 4.13), some angular deflection also occurs; it can scarcely be described precisely.

When the electric field exceeds the vacuum dielectric strength, superficial breakdowns can occur, these are accompanied by electron and photon emissions, which may reach the detector and add spurious counts to the spectrum. Moreover in such cases there is no permanent regime of charge accumulation, so energy shift and particle deflection become erratic.

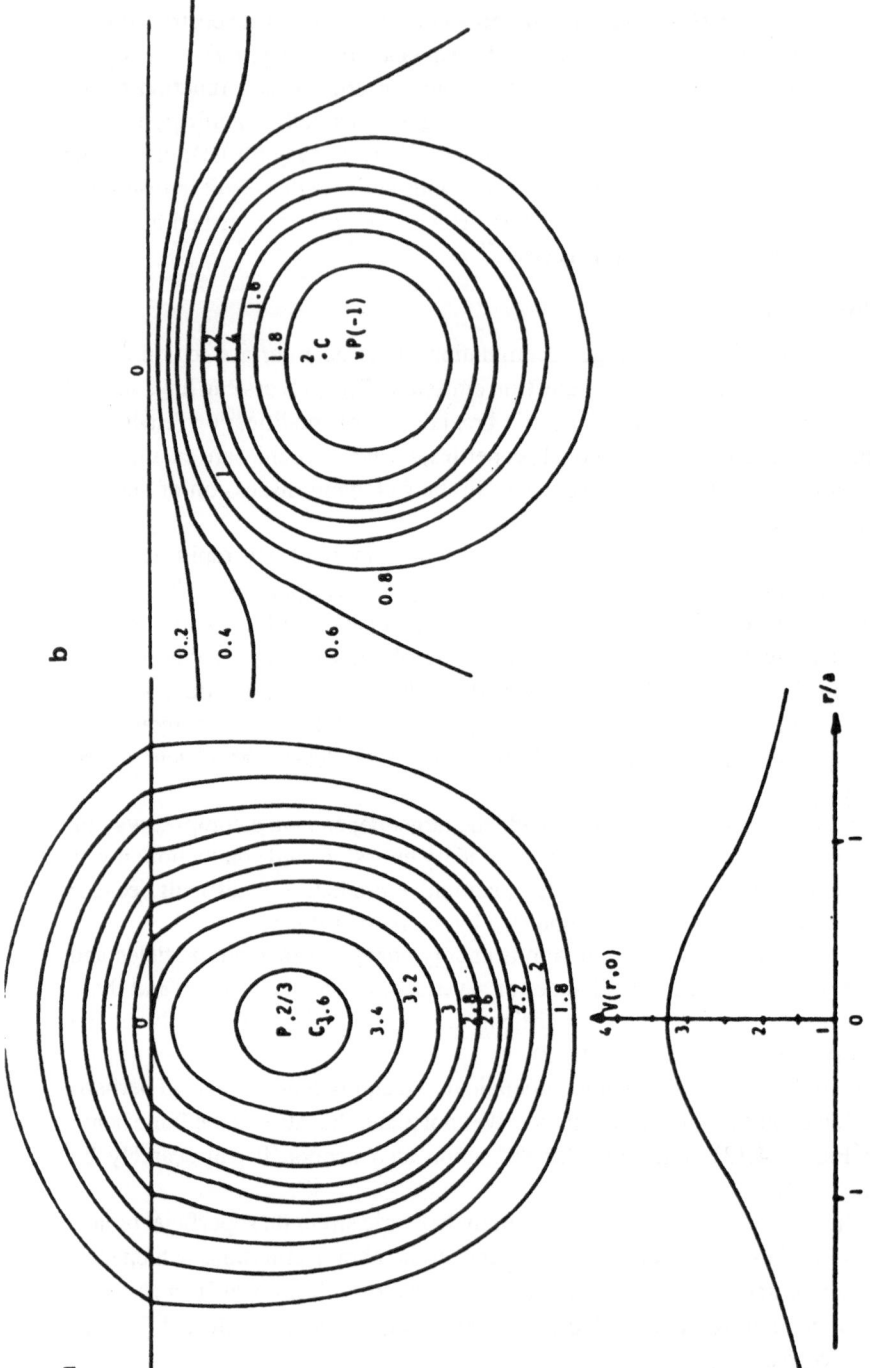

Figure 4.13. Equipotential lines around a uniformly charged sphere inside an insulator bounded by vacuum (a) or covered by a thin conducting layer (b). Curvature of the equipotentials build an electrostatic lens liable to deflect impinging ions. (From Ref. 56)

Another secondary effect of ion bombardment is the desorption of ionic or neutral species observed in halides and oxides. Quite recently Cazaux[59] discussed the mechanism of this effect under electron or x-ray irradiation; the principle should clearly apply to any ionizing beam. Desorption can be explained by the instability of ions in ionic or covalent crystals in the presence of electron vacancies left after electronic rearrangement following Auger processes. Halogen or oxygen loss in such materials can thus result in target modification during analysis; moreover the charge of ionic desorbed species is an additional term in the charge accumulation process.

Several solutions have been proposed for reducing or neutralizing charging effects, but these require some comment in light of what precedes. A grounded metallic coating fixes a null surface potential, but as seen in Figure 4.13, it repels the electric field toward the inside of the sample. The potential extremum is thus shifted away from the surface, which may reduce the deflection effect for shallow ion penetrations, but certainly does not warrant complete cancellation. Nevertheless the presence of such a charge carrier sink at the surface may be favorable, because it allows a more rapid elimination of charges by conduction and thus decreases the field for low-enough beam currents.[60] However, any conducting coating, even carbon, introduces an additional stopping power into the already complex data analysis. A grid applied on the surface is supposed to avoid this last inconvenience while providing a zero-potential plane as well, but it may cause too much shadowing at grazing incidences.

In principle the best way of neutralizing charging effects is to counterbalance the excess charge by electrons from a heated filament or a flood gun.[61] Note that the penetration depth must be quite close to the extent of the charge accumulation to perform an exact neutralization.[57]

An elegant solution consists in illuminating the sample with a broad ultraviolet beam.[58] If the energy of the photons is larger than the band gap of the insulator, electron pairs are formed, and these provide the necessary conductivity for driving charges toward a grounded layer surrounding the irradiated area. However such a solution is applicable only if the detection geometry prevents reflected photons from reaching the detector.

Whenever charge accumulation is unavoidable, we can possibly apply corrections in the data processing of spectra; this was attempted for RBS, and the possibility of adapting the retained solution is discussed in Section 11.4.5.2. Note that charge accumulation and particularly hot electron release in nonstationary regimes may be responsible for errors in charge integration, hence in the calibration of yields. Some of the means for reducing charging effects may also cause such errors or prohibit the use of target current as a measure for beam current.

4.3.6. Target Crystallinity

The enhanced penetration, or channeling effect, of ions into crystalline solids when the beam is almost parallel to atomic rows can cause considerable changes in the shape of IBA spectra. Although the ion–matter interaction in a crystal, and even a polycrystal, is never identical to what it would be in an amorphous system, whatever its crystal orientation, channeling is significant in only a very narrow angle interval around the main crystalline axes. It is thus rarely encountered unless the sample were deliberately oriented. However in transmission geometry, the growth mechanism of many superficial layers favors orientations capable of producing channeling at normal incidence. A tilt of 5–10° from a channeling orientation is generally enough to avoid channeling conditions. Otherwise a continuous azimuthal rotation at a constant and well-controlled tilt angle enables us to realize the condition of random orientation.[62] Since ERDA, unlike RBS, has been used only rarely for crystallographic studies taking advantage of this phenomenon,[63] it does not seem to require a more detailed discussion in this book.

4.3.7. Statistical Considerations

As a concluding remark on real targets, we point out that displacements in the target, beam rastering, or target azimuthal rotation are possible ways of minimizing the inconveniences of beam-induced phenomena—beam damage or charge accumulation—or avoiding channeling conditions, but these procedures are rather commonly suspected to arise difficulties caused by lateral nonuniformity. We stress that such a suspicion is not really relevant; indeed it is true that only a very small, immobile beam enables us to consider a possibly nonuniform sample as if it were uniform, but it also warrants a minimal statistical value of the recorded data. The broader the analyzed area, the more statistically representative the spectrum. Whether more than average parameters can be retrieved after accumulating data from nonequivalent target points is quite a different and certainly complex question into which some insight is given in Section 11.4.5.1.

REFERENCES

1. Marsden, E., The passage of particles through hydrogen, *Philos. Mag.* **27**, 824 (1914).
2. Wenzel, W. A., and Whaling, W., The stopping cross section of D_2O ice, *Phys. Rev.* **87**, 499 (1952).
3. Powers, D., and Whaling, W., Range of heavy ions in solids, *Phys. Rev.* **126**, 61 (1962).
4. Siritonin, E. I., Tulinov, A. F., Fiderkevich, A., and Shyskin, K. S., The determination of energy losses from the spectrum of particles scattered by a thick target, *Radiat. Effects* **15**, 149 (1972).
5. Tsurushima, T., and Tanoue, H. J., Spatial distribution of energy deposited by energetic heavy ions in semiconductors, *J. Phys. Soc. Jpn.* **31**, 1695 (1971).
6. Nicolet, M. A., Mayer, J. W., and Mitchell, I. V., Microanalysis of materials by backscattering spectrometry, *Science* **177**, 841 (1972).
7. Brice, D. K., Theoretical analysis of the energy spectra of backscattering ions, *Thin Solid Films* **19**, 121 (1973).
8. Jack, H. E., Jr., Some general features of random elastic scattering spectra, *Thin Solid Films* **19**, 267 (1973).
9. William, J. S., and Möller, W., On the determination of optimum depth resolution conditions for Rutherford backscattering analysis, *Nucl. Instrum. Methods Phys. Res.* **157**, 213 (1978).
10. Chu, W. K., Mayer, J. W., and Nicolet, M. A., *Backscattering Spectrometry* (Academic, New York, 1978).

11. Lewis, M. B., A deconvolution technique for depth profiling with nuclear microanalysis, *Nucl. Instrum. Methods Phys. Res.* **190**, 605 (1981).
12. Doyle, B. B., and Brice, D. K., The analysis of elastic recoil detection data, *Nucl. Instrum. Methods Phys. Res. Sect. B* **35**, 301 (1988).
13. Benenson, R. E., Wielunski, L. S., and Lanford, W. A., Computer simulation of helium-induced forward recoil proton spectra for hydrogen concentration determinations, *Nucl. Instrum. Methods Phys. Res. Sect. B* **15**, 453 (1986).
14. Hofsäss, H. C., Parikh, N. R., Swanson, M. L., and Chu, W. H., Elastic recoil coincidence spectroscopy (ERCS), *Nucl. Instrum. Methods Phys. Res. Sect. B* **58**, 49 (1991).
15. Tirira, J., Frontier, J. P., Trocellier, P., and Trouslard, Ph., Development of a simulation algorithm for energy spectra of elastic recoil spectrometry, *Nucl. Instrum. Methods Phys. Res. Sect. B* **54**, 328 (1991).
16. Pászti, F., Szilágyi, E., and Kótai, E., Optimization of the depth resolution in elastic recoil detection, *Nucl. Instrum. Methods Phys. Res. B* **54**, 507 (1991).
17. Szilágyi, E., Pászti, F., and Amsel, G., Theoretical approach of depth resolution in IBA geometry, *Nucl. Instrum. Methods Phys. Res. Sect. B* **100**, 103 (1995).
18. Chu, W. K., Mayer, J. W., and Nicolet, M. A., *Backscattering Spectrometry* (Academic, New York 1978), p. 203.
19. Chu, W. K., Mayer, J. W., and Nicolet, M. A., *Backscattering Spectrometry* (Academic, New York 1978), p. 323.
20. Chu, W. K., Mayer, J. W., and Nicolet, M. A., *Backscattering Spectrometry* (Academic, New York 1978), pp. 323–327.
21. Knoll, G. F., *Radiation Detection and Measurement* (Wiley, New York, 1979), pp. 111, 115, 127–30.
22. Amsel, G., L'Hoir, A., and Battistig, G., Projected small-angle multiple-scattering angular and lateral spread distribution and their combination, Part 1: Basic formulae and numerical results, submitted to *Nucl. Instrum. Methods Phys Res. Sect. B*.
23. Lanford, W. A., Analysis for hydrogen by nuclear reaction and energy recoil detection, *Nucl. Instrum. Methods Phys. Res. Sect. B* **66**, 65 (1992).
24. Zinke–Allmang, M., Kalbitzer, S., and Weiser, M., Nuclear reaction spectrometry of vibrational modes of solids, *Mat. Res. Soc. Symp. Proc.* **82**, 59 (1986).
25. Bachelard, G., *Essai sur la connaissance approchée* (Librairie Philosophique J. Vrin, Paris, 1968), p. 151.
26. Guye, C. E., *L'évolution physico-chimique* (Editions Chiron, Paris, 1922).
27. Campisano, S. U., Foti, G., Grasso, F., and Rimini, E., Determination of concentration profile in thin metallic films: applications and limitations of He^+ backscattering, *Thin Solid Films* **25**, 431 (1975).
28. Baglin, J. E. E., and Williams, J. S., in *Ion Beams for Materials Analysis* (J. R. Bird and J. S. Williams, eds.) (Academic, New York, 1989), pp. 132–41.
29. Valdés, J. E., Martínez-Tamayo, G., Arista, N. R., Lantschner, G. H., and Eckardt, J. C., The influence of foil roughness on low-energy stopping powers: calculations based on multiple-scattering theory, *J. Phys. Condens. Matter* **5**, A293 (1993).
30. Calmon, P., Contribution de l'analyse RBS à l'étude des effets d'irradiation sur la diffusion dans les verres d'oxydes, Rapport CEA-R-5560 (1991).
31. Marin, N., Serruys, Y., and Calmon, P., Extraction of lateral nonuniformity statistics from Rutherford backscattering spectra, *Nucl. Instrum. Methods Phys. Res. Sect. B* **108**, 179 (1996).
32. Edge, R. D., and Bill, U., Surface topology using Rutherford backscattering, *Nucl. Instrum. Methods Phys. Res.* **168**, 157 (1980).
33. Serruys, Y., and Bibić, N., unpublished work.
34. Thwaites, D. I., Review of stopping powers in organic materials, *Nucl. Instrum. Methods Phys. Res. Sect. B* **27**, 293 (1987).
35. Thwaites, D. I., Departures from Bragg's rule of stopping power additivity for ions in dosimetric and related materials, *Nucl. Instrum. Methods Phys. Res. Sect. B* **69**, 53 (1992).

36. Powers, D., An overview of current stopping power phenomena, measurements, and related topics, *Nucl. Instrum. Methods Phys. Res. Sect. B* **40/41**, 324 (1989).

37. Santry, D. C., and Werner, R. D., Energy loss of ^{4}He ions in Al_2O_3 and SiO_2, *Nucl. Instrum. Methods Phys. Res. Sect. B* **14**, 169 (1986).

38. Boutard, D., Möller, W., and Scherzer, B. M. U., Influence of H-C bonds on the stopping power of hard and soft carbonized layers, *Phys. Rev. B: Condens Matter* **38**, 2988 (1988).

39. Neuwirth, W., Pietsch, W., Richter, K., and Hauser, U., On the invalidity of Bragg's rule in stopping cross sections of molecules for swift Li ions, *Z. Physik A* **275**, 215 (1975).

40. Reiter, G., Baumgart, H., Kniest, N., Pfaff, E., and Clausnitzer, G., Proton and helium stopping cross sections in N_2, O_2, NO and N_2O, *Nucl. Instrum. Methods Phys. Res. Sect. B* **27**, 287 (1987).

41. Sabin, J. R., and Oddershede, J., Theoretical stopping cross sections of C-H, C-C and C=C bonds for swift ions, *Nucl. Instrum. Methods Phys. Res. Sect. B* **27**, 280 (1987).

42. Sigmund, P., Kinetic theory of particle stopping in a medium with internal motion, *Phys. Rev. A: Gen. Phys.* **26**, 2497 (1982).

43. Dehmer, J. L., Inokuti, M., and Saxon, R. P., Systematics of moments of dipole oscillator-strength distributions for atoms of the first and second row, *Phys. Rev A: Gen. Phys.* **12**, 102 (1975).

44. Inokuti, M., Baer, T., and Dehmer, J. L., Addendum: Systematics of moments of dipole oscillator-strength distributions for atoms of the first and second row, *Phys. Rev A: Gen. Phys.* **17**, 1229 (1978).

45. Inokuti, M., Dehmer, J. L., Baer, T., and Hanson, J. D., Oscillator-strength moments, stopping powers, and total inelastic-scattering cross sections for all atoms through strontium, *Phys. Rev A: Gen. Phys.* **23**, 95 (1981).

46. Kreutz, R., Neuwirth, W., and Pietsch, W., Electronic stopping cross sections of liquid organic compounds for 200-840-keV Li ions, *Phys. Rev. A: Gen. Phys.* **22**, 2598 (1980).

47. Kreutz, R., Neuwirth, W., and Pietsch, W., Analysis of electronic stopping cross sections of organic compounds, *Phys. Rev. A: Gen. Phys.* **22**, 2606 (1980).

48. Bauer, P., Stopping power of light ions near the maximum, *Nucl. Instrum. Methods Phys. Res. Sect. B* **45**, 673 (1990).

49. Ziegler, J. F., and Manoyan, J. M., The stopping of ions in compounds, *Nucl. Instrum. Methods Phys. Res. B* **35**, 215 (1988).

50. Boutard, D., and Berthier, B., Irradiation-Induced Modifications in Metal–Silicon Interfaces under MeV-focused helium beam (IBMM'95 Conference, Canberra, Australia, Feb. 1995).

51. Boutard, D., and Berthier, B., private communication.

52. Vigouroux, J. P., Duraud, J. P., Le Moël, A., Le Gressus, C., and Boiziau, C., Radiation induced charges in SiO_2, *Nucl. Instrum. Methods Phys. Res. Sect. B* **1**, 521 (1984).

53. Vigouroux, J. P., and Serruys, Y., Claquage, fractoemission, diffusion: rôle des défauts dans les diélectriques, *Vide, Couches Minces* **42**, 419 (1987).

54. Régnier, P., Serruys, Y., and Zemskoff, A., Electric field stimulated sodium depletion of glass and related penetration of environmental atomic species, *Phys. Chem. Glasses* **27**, 185 (1986).

55. Cazaux, J., Some considerations on the electric field induced in insulators by electron bombardment, *J. Appl. Phys.* **59**, 1418 (1986).

56. Cazaux, J., Electrostatics of insulators charged by incident electron beams, *J. Microsc. Spectrosc. Electron.* **11**, 293 (1986).

57. Cazaux, J., and Lehuede, P., Some physical descriptions of the charging effects of insulators under incident particle bombardment, *J. Electron Spectrosc. and Related Phenom.* **59**, 49 (1992).

58. Cazaux, J., in *Ionization of Solids by Heavy Particles*, vol. 306 (R. A. Baragiola, ed.) (NATO ASI Series B, 1993), *Physics*, (Plenum Press, New York), pp. 325–350.

59. Cazaux, J., The role of the Auger mechanism in the radiation damage of insulators, *Microsc. Microanal. Microstr*, in press.

60. Cazaux, J., Correlations between ionization radiation damage and charging effects in transmission electron microscopy, *Ultramicroscopy* **60**, 411 (1995).
61. Blanchard, B., Carriere, P., Hilleret, N., Marguerite, J. L., and Rocco, J. C., Utilisation des canons à ions pour l'étude des isolants à l'analyseur ionique, *Analusis* **4**, 180 (1976).
62. Williams, J. S., and Elliman, R. G., in *Ion Beams for Materials Analysis* (J. R. Bird and J. S. Williams, eds.) (Academic, New York, 1989), p. 303.
63. Janicki, C., Hinrichsen, P. F., Gujrathi, S. C., Brebner, J., and Martin, J.-P., An ERD / RBS / PIXE apparatus for surface analysis and channeling, *Nucl. Instrum. Methods Phys. Res. B* **34**, 483 (1988).

5

Conventional Recoil Spectrometry

With Hans Hofsäss

5.1. INTRODUCTION

Elastic recoil spectrometry ERDA has successfully been applied to depth-profiling light elements in a heavy matrix. As discussed in previous chapters, this analytical technique was first introduced in the seventies, using both transmission and reflection geometries.[1-4] These pioneering experiments were carried out on profiling hydrogen in thin targets[1] by means of transmission geometry. In addition concentration profiles of heavier elements like lithium, carbon, or oxygen were measured using appropriate projectiles like ^{16}O, ^{35}Cl, and ^{79}Br with energies in the range of 10–40 MeV.[2,3] Doyle and Peercy[4] introduced the widely used ERDA technique for hydrogen profiling with rather low energetic 2–3 MeV $^4He^+$ beams, which are readily available from many Van de Graaff accelerators commonly used for backscattering analysis.

Forward elastic-scattering spectrometry encounters two intrinsic difficulties: The recoil mass and the depth where a scattering event took place cannot be unmistakably determined (we describe this condition as *the mass–depth ambiguity*); in some cases there is a difficulty in distinguishing whether a recoiled target atom or a scattered projectile has been detected (we call this condition *recoil-scattered ion ambiguity* or *recoil–projectile ambiguity*).

The most common experimental set-up, which we call *conventional ERDA*, consists in detecting particles using a simple silicon barrier detector in reflection or transmission geometry. To resolve the recoil–projectile ambiguity in the case of conventional ERDA, an absorbing foil filters scattered projectiles in reflection geometry; it is optional in transmission. Other ERDA arrangements presented in Chapters 6–9 are essentially aimed at resolving the recoil–projectile ambiguity when it is impossible to use an absorbing foil for conventional ERDA ($M_1 < M_2$) and at resolving more efficiently the mass–depth ambiguity.

Conventional ERDA offers the possibility of multiple-element analysis, in particular for profiling different isotopes of hydrogen,[5-9] with a rather good depth resolution down to 10 nm.[10] However a reasonable depth resolution is obtained only for small-detector solid angles, which limits the sensitivity of the technique.

In Chapter 5 we first discuss the mass–depth and recoil–projectile ambiguities and solutions used to manage these effects (Section 5.2). In Sections 5.3 and 5.4 we treat ERDA capabilities in glancing and transmission geometries, respectively. In Section 5.5 we analyze mass resolution and detection limits in conventional ERDA. Applications are widely developed in Chapters 12 and 13.

5.2. MASS–DEPTH AND RECOIL-SCATTERED ION AMBIGUITIES

Depth profiling by conventional ERDA and also Rutherford backscattering spectroscopy are usually carried out by a rather simple scattering or recoil geometry. For projectiles of given mass M_1 and energy E_0 and a well-defined scattering angle θ or recoil angle ϕ, final energies of only one species, either scattered ions or recoiled target atoms E_1 or E_2, are measured. From data pairs (E_1, θ) or (E_2, ϕ) obtained from such an experiment, the mass of recoiled target atoms can be calculated from energy and momentum conservation, as described in Chapter 2. Depth information is obtained from energy loss in ions and recoil atoms in the target material. However conventional ERDA suffers from both mass–depth ambiguity and recoil–scattered ion ambiguities.

For given projectile mass and energy and a fixed recoil angle, the energy E_2 of recoiling atoms detected in an ERDA measurement is a function of the recoil mass M_2 and depth x where the scattering event has taken place. Therefore from the recoil energy E_2 alone, we cannot decide which fraction of the energy was transferred in the scattering process and how much energy was lost due to electronic or nuclear stopping in the material. Hence it is possible for recoils with different masses emerging from different depths to reach the detector with equal energies (see Figure 5.1). In other words we may not know which recoil mass was detected; we call this condition *mass–depth ambiguity*. In forward detection it is often impossible to determine from the detected energy alone whether a recoil target or a scattered ion was detected; we call this condition *recoil–scattered ion ambiguity*.

For $M_1 > M_2$ the most simple way of resolving the recoil–projectile ambiguity and to some extent the mass–depth ambiguity in conventional ERDA is to stop in an absorber foil all heavy particles that are moving in the direction of the detector; thus only the light elements of interest are transmitted. This is an absolute solution to mass–depth ambiguity only if the absorber stops all particles except the lightest isotope present in the target. This does not mean that the ambiguity cannot be resolved when several species, e.g., various hydrogen isotopes, are transmitted, but only that the interpretation of spectra will include the problem of separating the signal from different species, which may be quite intricate (cf. Chapter 11).

For $M_1 < M_2$ absorber foils are no longer a solution to either ambiguity. Moreover we cannot rely on interpretation methods to solve the recoil–projectile ambiguity because an abundance of scattered ions can overflood the detector and even damage it. In this case one possible approach is to discriminate scattered light projectiles, like high-energy alpha particles, from heavier recoils, like C, N, or O, by the pulse shape

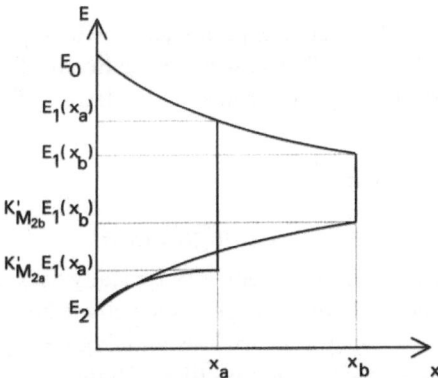

Figure 5.1. Energy versus depth shows the mass–depth ambiguity in reflexion geometry for two scattering centers of respective mass M_{2a} and M_{2b} located at different depths x_a and x_b. The slope of energy loss curves is schematic, depending on particle energy and target composition.

generated from particles penetrating the depletion layer of a semiconductor detector.[11,12]

In addition to conventional ERDA there are several ways of solving both difficulties. The general principle is to measure the momentum $P_2 = |\vec{P}_2| = M_2 |\vec{V}_2|$ of the recoiled atoms or the Z-dependent energy loss $\Delta E(Z)$ in a thin detector in addition to the data pair (E_2, ϕ). If we take such a measurement, we end up with data triplets (E_2, ϕ, P_2) or $(E_2, \phi, \Delta E)$ from which the mass M_2 of the recoiled atom as well as the depth x at which the scattering event took place can be derived.

Commonly used methods of measuring momentum include TOF spectroscopy,[13–21] discussed in Chapter 6, electric or magnetic spectrometers,[22–24] described in Section 10.5.1.4, or the *ExB* filter introduced by Ross *et al.*[25] and recently optimized by Serruys and Tirira.[26] This technique is discussed in Chapter 7.

The Z-dependent energy loss ΔE of high-energy heavy ions is usually measured with ΔE–E detector telescopes, consisting of a gas ionization chamber in combination with a silicon surface barrier detector.[27–29] Recently ΔE-E solid state telescopes using thin ($\approx$ 10-μm) transmission silicon surface barrier detectors were also developed, which are able to discriminate H, D and T isotopes recoiled by rather low energetic 4-MeV ^{4}He projectiles.[30–32] This method is discussed in Chapter 8.

Coincidence techniques (CERDA, SCRS) can also be used to solve simultaneously the depth–mass and recoil-scattered ion ambiguities.[34–36] Detecting the coincidence of both the recoiled target atom and the scattered projectile allows us to deduce the full information about a scattering event, i.e., E_1, E_2, θ, and ϕ, so that the recoil mass M_2 and the depth x can easily be determined. To assign a recoiled target atom and a scattered ion to the same scattering event, it is necessary to detect both particles in time coincidence. The CERDA and SCRS techniques are widely discussed in Chapter 9, including the use of position-sensitive detectors (PSD). However all these solutions involve more sophisticated experimental devices and for some of them, special interpretation techniques. This is why conventional ERDA remains a widely used arrangement for profiling light elements, particularly hydrogen isotopes.

5.3. GLANCING GEOMETRY

5.3.1. Experimental Reflection Configuration

Conventional ERDA in reflection geometry typically uses ^{4}He beams as primary ions with energies in the 1–3 MeV range, especially when measuring hydrogen isotopes.[4,7–10] Nevertheless heavy primary ions with high energy from 5 to 50 MeV have been used to measure concentration profiles of light elements H to N.[3,37,38]

A classic experimental configuration for ERDA in reflection geometry is shown in Figure 5.2. In this case the beam is directed onto the target surface at a glancing angle, so that recoiled atoms are detected from the same surface also at a glancing angle. Using this simple scattering geometry for projectiles of given mass M_1 and incident energy E_0 and for a well-defined scattering angle θ, the final energies E_2 of recoiled atoms are measured for a recoil angle ϕ. For a given projectile mass and energy and for a fixed recoil angle ϕ, the energy E_2 of detected recoils is a function of two parameters (see Chapter 2): the ratio of masses M_1/M_2 and the energy before the collision $E_0(x)$ that corresponds to the depth x at which scattering took place. As discussed before to solve the problem of recoil-scattered ion ambiguity in conventional ERDA using glancing geometry for $M_1 > M_2$, an absorber foil is placed in front of the detector. The absorber foil can be chosen with an appropriate thickness, so that only light elements are transmitted and scattered ions are stopped (Figure 5.3), which completely eliminates the recoil-scattered ion ambiguity. Furthermore the mass–depth ambiguity can be partially solved for an energy interval (see Figure 5.4) when individual signals can be separated. However a mathematical treatment of the recoil spectrum (cf. Chapter 11) may be needed to separate the signals completely from other recoil masses as a function of depth in the target.

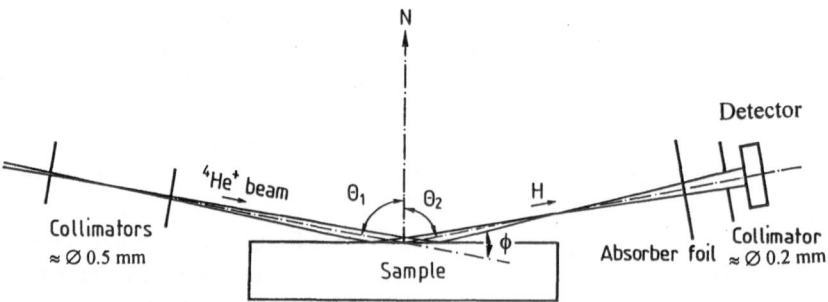

Figure 5.2. Experimental set-up for elastic recoil detection using reflection geometry. A second detector (*not shown on this figure*) can be used to detect simultaneously backscattered ions.

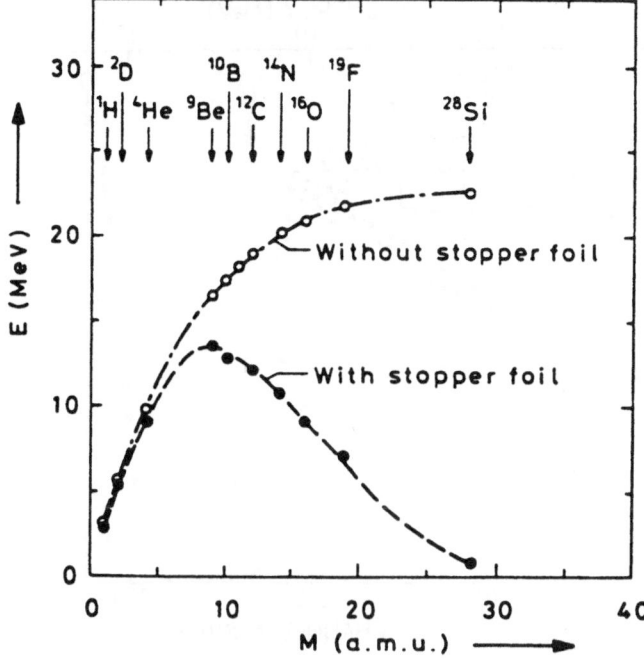

Figure 5.3. Typical variations of the surface recoil energy with and without an absorber foil in front of the detector. Calculation was made for $\phi = 30°$ using a 30-MeV Si ion beam and a 8.9-μm Mylar absorber foil. (Data from Ref. 38)

5.3.2. Analytical Capabilities of ERDA in Reflection

5.3.2.1. Multielemental Analysis

It is possible to carry out multiple-element analysis, in particular for profiling the different hydrogen isotopes.[7,9] In fact using an appropriate absorber thickness, for example an 11-μm-thick aluminium foil for a 2.5-MeV ^{4}He ion beam, hydrogen isotopes can be easily detected. Otherwise using heavy primary ions with high energy, we can analyze several light recoil elements.[37,38]

5.3.2.2. Depth Resolution

As discussed in Section 4.2.4.1 the depth resolution δR_x is one of the most important parameters characterizing an analytic technique. Although this parameter is very important, we cannot give a complete and general discussion for all combinations of incident ion beam, target sample, and experimental arrangement. In fact since depth resolution (Eq. 4.33) depends on many physical parameters—stopping power, straggling, MS, and experimental set-up—each analytic problem (ion, target, set-up) must

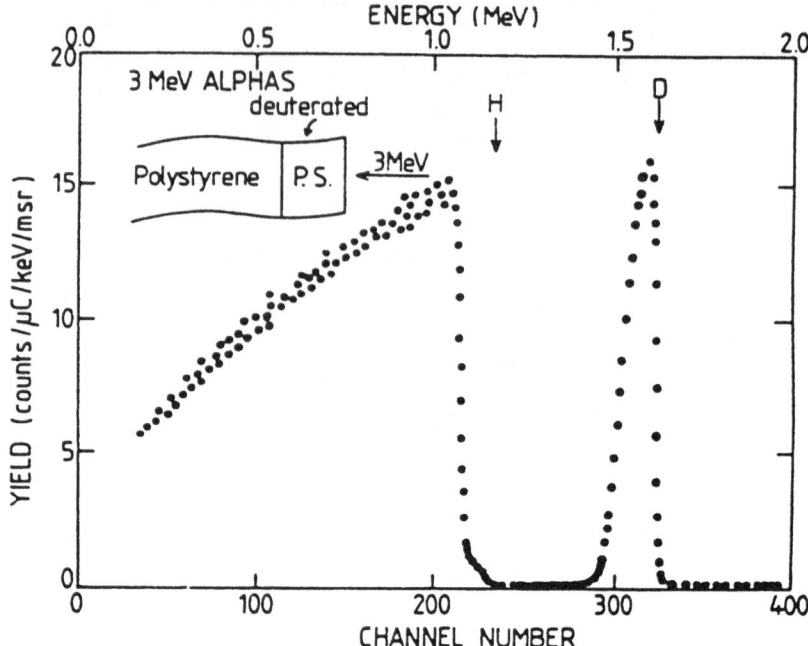

Figure 5.4. Elastic recoil spectrum from hydrogen recoil isotopes. (Data from Ref. 7)

be studied in detail. Hence the discussion herein offers only a partial view of this difficult problem. We emphasize that this treatment takes its inspiration primarily from earlier analyses of depth resolution for other IBA techniques—nuclear reactions,[39] Rutherford backscattering.[40] For conventional elastic recoil spectrometry, several theoretical approaches have been used and several experimental measurements have been carried out.[10,41–43] Surveying the literature we find that in hydrogen depth profiling, there are different opinions about achievable depth resolution. For example several authors (Nagata *et al.*,[42] Pászti *et al.*,[10] Tirira *et al.*[41]) analyzed this problem in detail and claimed values for δR_x from 10 nm at the surface to 40 nm in depth. This discrepancy probably results from differences in experimental configurations.

More generally the behavior of δR_x is described by Eq. 4.33, so we can optimize incidence and recoil angles for the best possible depth resolution. For example Pászti *et al.*[9,10] studied depth resolution systematically at different depths and energies for a pure silicon target. Figure 5.5 shows the variation of hydrogen depth resolution as a function of incident and recoil angles for a silicon matrix using a ^{4}He beam. In principle by simultaneously decreasing the recoil angle and taking an incidence angle larger than 80°, depth resolution can be optimized relatively; the only limitation is taking measurements. Thus in practice recoil angles with values around 20° and incidence

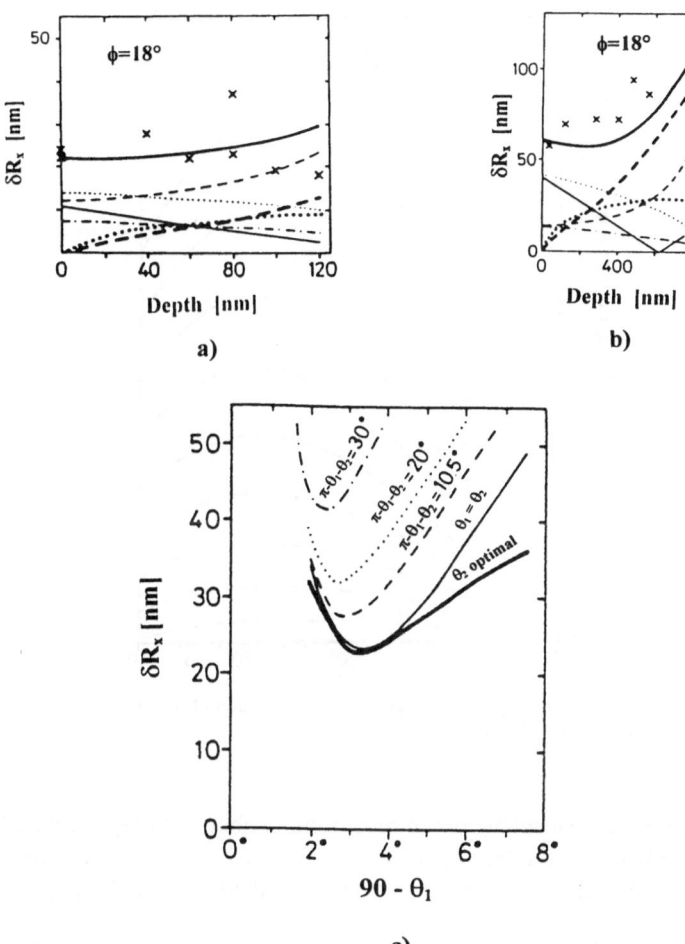

Figure 5.5. Depth resolution as a function of depth for a pure silicon target in conventional glancing geometry for recoil angle $\phi = 18°$ and incidence angle kept constant at 79.5°. (a) $E_0 = 1.2$ MeV and (b) $E_0 = 3$ MeV. (c) Depth resolution as a function of the incidence angle for $E_0 = 3$ MeV and x = 100 nm. (Data from Ref. 10)

angles with values around 80° offer a good compromise for an acceptable depth resolution (see Figure 5.5). Note that for an extreme case with $\theta_1 \approx 87°$ and $\phi \approx 12°$ in a hydrogenated silicon target the *theoretical* depth resolution as calculated by Pászti *et al.*[(10)] is better than 10 nm; however such a value is only an example of an ideal case, since such angles are acceptable only due to the exceptional smoothness of silicon surfaces. Depth resolution is strongly influenced by surface roughness (see Section 4.3.2), so a slight error in the recoil angle can lead to a relatively large error in

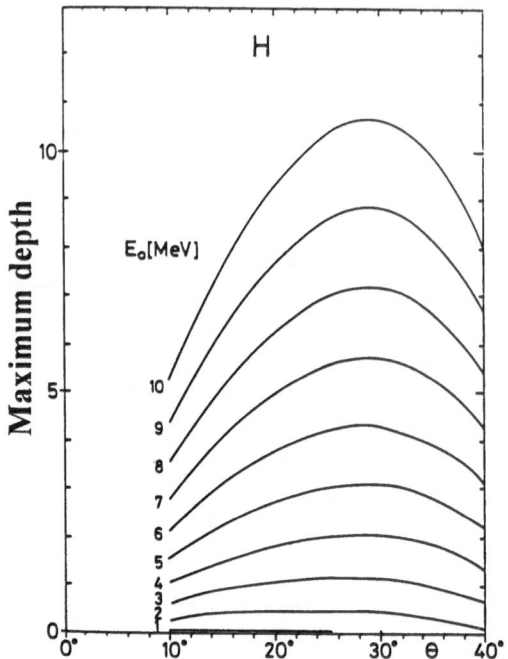

Figure 5.6. Maximum probing depth for a Si:H target as a function of recoil angle at different incident energies of ^{4}He ions. (Data from Ref. 9)

identifying depth. A reasonable depth resolution is obtained only for small detector solid angles ($\approx 10^{-4}$ sr), and this factor limits the sensitivity of the technique. These intrinsic limitations can be improved using other detection methods (see Chapters 6–9).

In calculating depth resolution (Eq. 4.33) as previously discussed, almost all authors summed in quadrature *all contributions* to total energy resolution (detector, geometry, straggling, MS, etc.); however as stressed in Section 3.3.4, this procedure must be considered a first approximation. A more appropriate treatment, particularly for MS, is given recently by Szilágyi *et al.*[43] Note that the theoretical evaluation of depth resolution represents only one aspect of the problem of depth resolution, since it does not define practical limits that can reach a suitable analysis of spectra, as explained in Section 4.2.4.2.

5.3.2.3. Maximum Analyzable Depth

The maximum probing depth in conventional reflection geometry does not generally exceed 1 μm, due to low glancing angles of incidence and detection. For example when profiling hydrogen using an incident 3-MeV ^{4}He beam, this maximum depth is generally not larger than 0.5 μm. To achieve the maximum probing depth, we must find the suitable incidence and recoil angles at a given energy. We use Eqs. 4.9 and 4.10 to calculate the variation of energy loss for incident ^{4}He ions and recoil atoms and obtain the optimum angles. Figure 5.6 shows the maximum probing depth for a Si:H target as a function of incident energy E_0 and recoil angle ϕ. However there is no general solution, because the maximum probing depth is a function of stopping powers, which differ for each target composition.

5.4. TRANSMISSION GEOMETRY

5.4.1. Experimental Transmission Configuration

An experimental configuration for ERDA in transmission geometry is shown in Figure 5.7. In this case the beam is incident onto the target surface under an angle $\theta_1 \approx 0$. This means that the beam is generally normal to the surface, and atoms recoiling from the back surface are detected from the other surface in a forward direction, often at angles close to 0°. Note that films must be uniformly thick, at least over an area given by the beam spot size (in fact larger than the spot size, because of MS). In general an absorber foil in front of the detector is not used if target thickness is large enough to stop scattered ions moving in direction of the detector.

5.4.2. Analytic Capabilities of ERDA in Transmission

Compared to the reflection geometry commonly used for ERDA, the transmission geometry offers several advantages. The main advantage is of course that coincident detection of both scattered and recoiled particles becomes possible (e.g., Hofsäss et al.[34,35]). A complete discussion of this method is developed in Chapter 9. The analytic capabilities of ERDA in transmission geometry are discussed by Tirira et al.[44] for the case of hydrogen profiling using an incident ^{4}He beam. Considering these works but not restricting ourselves to the case of ^{4}He-H elastic collision, we summarize specific features of ERDA in transmission geometry in the sections that follow.

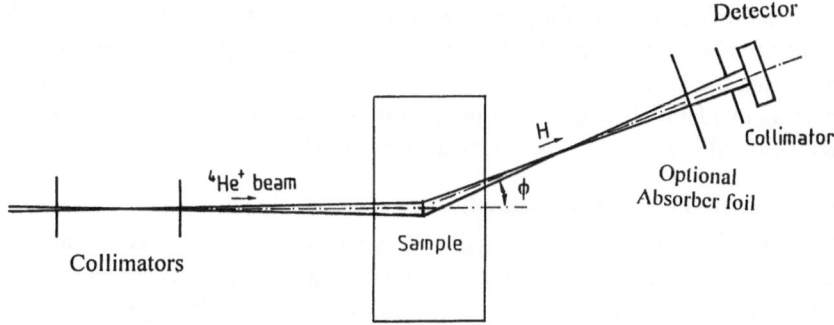

Figure 5.7. Experimental set-up for elastic recoil detection in transmission geometry.

5.4.2.1. Surface Roughness and Thickness Heterogeneity

Surface roughness is not a critical parameter, and less caution is needed for the sample surface alignment, since in transmission geometry the beam is incident perpendicularly to the sample front surface[2,44–46] or at large angles of incidence,[47] and recoils are detected preferably in a forward direction, often at 0°. On the other hand a *critical parameter* in transmission is the uniformity of thickness (cf. Section 4.3.2), at least over an area given by the beam spot size.

5.4.2.2. Detection near Zero Degree

In transmission geometry where recoils are detected downstream around 0°, large solid-angle detectors can be used to maximize the counting rate without affecting depth resolution. In fact for small recoil angles ϕ, the kinematic factor K' is almost independent of ϕ (see Eq. 2.14). Deviation from $K'(0°)$ is less than 1% for angles up to 5° and increases to about 5% for angles up to 12°. For example L'Ecuyer *et al.*[2] studied Li concentration profiles in thin films by transmission ERDA, where Li recoils from a 35-MeV ^{35}Cl incident ion beam were detected under 0° to the beam direction by a large solid-angle detector. In these experiments depth resolution was 30 nm, and the detection limit was estimated to be 10^{14} Li/cm^2. Transmission ERDA using large detectors at 0° was also applied to measure hydrogen concentration profiles in free-standing thin films (e.g., Wielunski *et al.*,[45] Tirira *et al.*[44]). With detector solid angles up to 0.05 sr, a detection limit of 10 ppm was reached (e.g., Wielunsky *et al.*[46]). For recoil angles close to 0°, the energy $E_2 = K'(0°)E_0$ transferred to the recoil atoms is largest, which is one reason for the increased probing depth in transmission ERDA measurements.

5.4.2.3. Maximum Analyzable Depth

In transmission geometry the maximum probing depth can be more than five times larger than in reflection geometry. For normal incidence and a 0° recoil angle, an analyzed depth up to 6 μm was achieved for an incident 3-MeV ^4He beam (e.g., Tirira *et al.*[44,48]). For a similar experimental set-up Wielunsky *et al.*[45] reached an even larger probing depth of up to 8 μm for profiling hydrogen by 3–5 MeV ^3He ions. The ^3He ions have an advantage over ^4He because of a larger energy transfer in the recoil process. From the kinematic factor K' given by Eq. 2.12, it follows that 64% of the energy of an incident ^4He and 75% of the energy of an incident ^3He are transferred to a proton recoiled in forward direction at an angle of 0°.

5.4.2.4. Recoil–Projectile Ambiguity

Since $\theta_{max} = \sin^{-1}(M_2/M_1)$ according to a suitable choice of M_1 ($> M_2$) and detection angle θ, signals arising from scattered projectiles can be completely suppressed in a transmission ERDA experiment. For example analyzing a polymer sample containing ^{16}O as the heaviest element with a ^{35}Cl ion beam, the detection recoil angle ϕ should be larger than $\theta_{max} = 27.2°$. Similarly in TOF-ERDA experiments by Gujrathi *et al.*,[16] a detection angle of 30° was chosen to study Al-coated polymer films. Except for ^{35}Cl scattered from ^{27}Al, only the light recoiled target atoms are seen in the mass spectrum. In coincidence ERDA experiments performed by Forster *et al.*[49,50] to profile ^{1}H and ^{2}H by an incident ^{4}He beam, one of the two detectors subtended an angular range ϕ from 33° to 69°, so that only recoiled ^{1}H and ^{2}H were recorded with this detector (see Section 9.1).

5.4.2.5. Depth Resolution

In transmission geometry the beam is nearly perpendicular to the sample surface, and recoil atoms are detected preferably in forward direction, often close to 0°. In this case the depth resolution δR_x is almost independent of the optimal incident and recoil angles. In fact we determine depth resolution keeping other experimental conditions similar to those in glancing geometry. Furthermore total energy resolution δE_T varies slowly as the projectile loses energy, and the stopping power increases as the beam penetrates the sample. Thus depth resolution improves with depth. For example in the case of a 25-μm thickness Kapton layer, δR_x varies from 38 nm at the surface to 29 nm at a 6-μm depth, using 3-MeV ^{4}He ions.

5.4.2.6. Microbeam Capabilities

It is possible to analyze an area as small as the incident beam spot size, which can be a few μm^2 in the case of a microbeam. Tirira *et al.*[48] demonstrated the three-dimensional mapping of hydrogen concentration in thin films (a possibility already pointed out in the early work of Cohen *et al.*)[1] by using a 3-MeV ^{4}He microbeam with a beam spot size of 100 μm^2, corresponding to a lateral resolution of about 10 μm. Along the beam direction, a depth range up to 6 μm can be analyzed with a resolution of 30–40 nm.

5.5. SENSITIVITY

To evaluate the sensitivity of an elastic recoil experiment, we must analyze the differential scattering cross section which is largely discussed in Section 3.2. In the case of Rutherford scattering, for small recoil angles ϕ the recoil cross section (Eq. 3.8) is smallest but almost independent of ϕ. This fact allows to use a larger acceptance angle in the detector and still obtain accurate values for the cross section close to 0°; this simplifies the quantitative analysis.[44,46] A high sensitivity is achieved because the smaller recoil cross sections are overcompensated by large detector acceptance angles. On the other hand for non-Rutherford elastic scattering, we can analyze two typical cases: ^{4}He-^{1}H and ^{4}He-^{2}H.

In the case of ^{4}He-^{1}H scattering (see Section 3.2.7.3), note that near 0° the cross section varies slowly as a function of the recoil angle, so we have two energy regions with different behaviors. On the one hand if the incident energy of ^{4}He ions is lower than approximately 2.4 MeV, the cross section is smallest close to 0°. Thus a larger acceptance angle can be used in a similar way. On the other hand this scattering cross section has its highest value at $\phi = 0°$ for incident energy higher than 2.4 MeV. Hence the highest sensitivity can be reached in this energy range.

In the case of ^{4}He-^{2}H scattering (see Section 3.2.8.1), the cross section exhibits a fairly narrow resonance near 2.128 MeV, and $d\sigma^D/d\Omega$ is highest at $\phi = 0°$. This fact can be used to improve the sensitivity of deuterium analysis by ERDA in transmission (besides $d\sigma^D/d\Omega$ varies slowly around 0° and always has its lowest value at $\phi = 0°$ for $E_{He} < 1.5$ MeV and its highest value for $1.8 < E_{He} < 2.5$ MeV). In the latter case sensitivity can be optimized by using large acceptance angles. The reader is referred to Section 3.2.8, for a detailed discussion of the scattering cross section for Rutherford and non-Rutherford collisions.

5.6. MASS RESOLUTION

The difference in energy δE_2 of recoil atoms after the collision when two types of atoms differ in mass by a quantity δM_2 characterizes the capability of recoil spectrometry to separate two signals arising from two neighboring elements in the target. This quantity is usually called mass resolution δM_R $(\equiv \delta E_2/\delta M_2)$ (see Section 4.2.5). Equation 4.50 shows that for a conventional ERDA measurement, δM_R is largest for recoil angles close to 0° and decreases with increasing recoil angles. The mass resolution of ERDA in transmission geometry is therefore optimized for recoil angles close to 0°. It also follows from Eq. 4.50 that for target masses M_2 comparable to

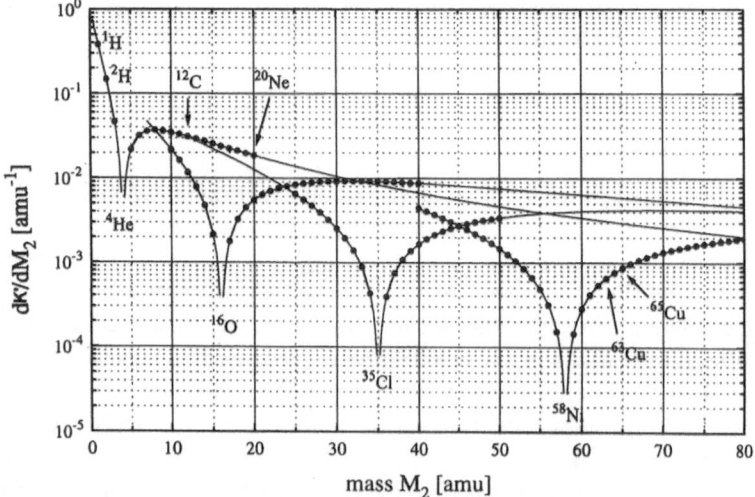

Figure 5.8. Variation of δM_R from Eq. 4.50 as a function of the mass M_2 of recoiled target atoms, calculated for four different projectile masses M_1 (^{4}He, ^{16}O, ^{35}Cl, and ^{58}Ni). Note that for target masses M_2 comparable to incident projectile mass M_1, mass resolution is poor and tends to 0.

incident projectiles mass M_1, the mass resolution is poor. To achieve a reliable mass resolution, a compromise must be found between both masses. Figure 5.8 shows δM_R for four projectiles as a function of the recoil mass; for example the best mass resolution for isotopes of hydrogen is achieved when ^4He ions are used as projectiles. Furthermore if mass resolution must be obtained only from differences in E_2 (recoil energy) for different recoil masses, Cu isotopes are preferably analyzed using lighter projectiles (^{35}Cl or ^{16}O). For light elements in the mass range of 12–20, a good separation is obtained for ^4He or ^{35}Cl projectiles, whereas the separation using ^{16}O projectiles is poor.

According to Eq. 4.52 optimal mass separation is obtained for $M_1 = 3.73\ M_2$. This condition can be used to choose the most suitable projectile for a given target isotope, for instance ^4He for ^1H. Note that when M_1 tends to M_2 mass resolution stems only from differences in the energies of the scattered ions given by $\delta E_1/\delta M_2$. These events can be identified by coincident detection. Mass resolution is also discussed in various sections of this book for different analytic methods for detecting recoil atoms (see for example Sections 4.2.5, 6.5.2, 8.3.1, and 9.4).

REFERENCES

1. Cohen, B. L., Fink, C. L., and Degman, J. H., Nondestructive analysis for trace amounts of hydrogen, *J. Appl. Phys.* **43**, 19 (1972).
2. L'Ecuyer, J., Brassard, C., Cardinal, C., Chabbal, J., Deschênes, L., Labrie, J. P., Terrault, B., Martel, J. G., and St.-Jaques, R., An accurate and sensitive method for the determination of the depth distribution of light elements in heavy materials, *J. Appl. Phys.* **47**, 381 (1976).
3. Terrault, B., Martel, J. G., St.-Jaques, R., and L'Ecuyer, J., Depth profiling of light elements in materials with high-energy ion beams, *J. Vac. Sci. Technol.* **14**, 492 (1977).
4. Doyle, B. L., and Peercy, P. S., Technique for profiling ^1H with 2.5-MeV Van de Graaff accelerators, *Appl. Phys. Lett* **34**, 811 (1979).
5. Turos, A., and Meyer, O., Depth profiling of hydrogen by detection of recoil protons, *Nucl. Instrum. Methods Phys. Res. Sect. B* **4**, 92 (1984).
6. Besenbacher, F., Stensgaard, I., and Vase, P., Absolute cross section for recoil detection of deuterium, *Nucl. Instrum. Methods Phys. Res. Sect. B* **15**, 459 (1986).
7. Pretorius, R., Peisach, M., and Mayer, J. W., Hydrogen and deuterium depth profiling by elastic recoil detection analysis, *Nucl. Instrum. Methods Phys. Res. Sect. B* **35**, 478 (1988).
8. Tirira, J., Trocellier, P., Frontier, J. P., and Trouslard, P., Theoretical and experimental study of low-energy ^4He-induced ^1H elastic recoil with application to hydrogen behaviour in solids, *Nucl. Instrum. Methods Phys. Res. Sect. B* **45**, 203 (1990).
9. Pászti, F., Kótai, E., Mezey, G., Manuaba, A., Pócs, L., Hildebrandt, D., and Strusny, H., Hydrogen and deuterium measurements by elastic recoil detection using alpha particles, *Nucl. Instrum. Methods Phys. Res. Sect. B* **15**, 486 (1986).
10. Pászti, F., Szilágyi, E., and Kótai, E., Optimization of the depth resolution in elastic recoil detection, *Nucl. Instrum. Methods Phys. Res. Sect. B* **54**, 507 (1991).
11. Klein, S., and Rijken, H. A., Pulse-shape discrimination in elastic recoil detection and nuclear reaction analysis, *Nucl. Instrum. Methods Phys. Res. Sect. B* **66**, 393 (1992).
12. Rijken, H. A., Klein, S., van IJzendoorn, L. J., and de Voigt, M. J., Elastic recoil detection analysis with high-energy alpha beams, *Nucl. Instrum. Methods Phys. Res. Sect. B* **79**, 532 (1993).

13. Thomas, J. P., Fallavier, M., Ramdane, D., Chevarier, N., and Chevarier, A., High-resolution depth-profiling elements in high atomic mass materials, *Nucl. Instrum. Methods Phys. Res.* **218**, 125 (1983).

14. Groleau, R., Gujrathi, S. C., and Martin, J. P., Time-of-flight system for profiling recoiled light elements, *Nucl. Instrum. Methods Phys. Res.* **218**, 11 (1983).

15. Houdayer, A., Hinrichsen, Gujrathi, S. C., Martin, J. P., Morano, S., Lessard, L., Oxorn, K., Janicki, C., Brebner, J., Belhadfa, A., and Yelon, A., Trace element and surface analysis at the university of Montreal, *Nucl. Instrum. Methods Phys. Res. Sect. B* **24/25**, 643 (1987).

16. Gujrathi, S. C., Aubry, P., Lemay, L., and Martin, J. P., Nondestructive surface analysis by nuclear-scattering techniques, *Can. J. Phys.* **65**, 950 (1987).

17. Gujrathi, S. C., and Bultena, S., Depth profiling of hydrogen using the high-efficiency ERD-TOF technique, *Nucl. Instrum. Methods Phys. Res. Sect. B* **64**, 789 (1992).

18. Goppelt, P., Gebauer, B., Fink, D., Wilpert, M., Wilpert, T., and Bohne, W., High-energy ERDA with very heavy ions using mass and energy-dispersive spectrometry, *Nucl. Instrum. Methods Phys. Res. Sect. B* **68**, 235 (1992).

19. Arai, E., Funaki, H., Katayama, M., and Shimizu, K., Stoichiometry and profiling of surface layers by means of TOF-E, ERDA, and RBS, *Nucl. Instrum. Methods Phys. Res. Sect. B* **68**, 202 (1992).

20. Hult, M., Elbuanani, M., Persson, L., Whitlow, H. J., Andersson, M., Zaring, C., Östling, M., Cohen, D. D., Dytlewski, N., Bubb, I. F., Johnston, P. N., and Walker, S. R., Empirical characterisation of mass distribution broadening in TOF-E recoil spectrometry, *Nucl. Instrum. Methods Phys. Res. Sect. B* **101**, 263 (1995).

21. Whitlow, H. J., Time of flight spectroscopy methods for analysis of materials with heavy ions, in *Proc. of High-Energy and Heavy Ion Beams in Materials Analysis* (J. R. Tesmer, ed.) (Materials Research Society, Albuquerque, New Mexico, 14–16 June 1989) pp. 243–56.

22. Kruse, O., and Carstanjen, H. D., High-depth resolution ERDA of H and D by means of an electrostatic spectrometer, *Nucl. Instrum. Methods. Phys. Res. Sect. B* **89**, 191 (1994).

23. Gossett, C., Use of a magnetic spectrometer to profile light elements by elastic recoil detection, *Nucl. Instrum. Methods Phys. Res. Sect. B* **15**, 481 (1986).

24. Dollinger, G., Elastic recoil detection analysis with atomic depth resolution, *Nucl. Instrum. Methods Phys. Res. Sect. B* **79**, 513 (1993).

25. Ross, G., Terreault, B., Gobeil, G., Abel, G., Boucher, C., and Veilleux, G., Inexpensive, quantitative hydrogen depth-profiling for surface probes, *J. Nucl. Mater.* **128/129**, 730 (1984).

26. Serruys, Y., and Tirira, J., ERDA analysis from hydrogen to oxygen using a modified ExB filter, submitted to *Nucl. Instrum. Methods Phys. Res. Sect. B*.

27. Petrascu, M., Berceanu, I., Brancus, I., Buta, A., Duma, M., Grama, C., Lazar, I., Mihai, I., Petrovici, M., Simon, V., Mihaila, M., and Ghita, I., A method for analysis and profiling of boron, carbon, and oxygen impurities in semiconductor wafers by recoil atoms in heavy ion beams, *Nucl. Instrum. Methods Phys. Res. Sect. B* **4**, 396 (1984).

28. Behrooz, A., Headrick, R., Seiberling, L., and Zurmühle, W., A UHV-compatible ΔE-E gas telescope for depth profiling and surface analysis of light elements, *Nucl. Instrum. Methods Phys. Res. Sect. B* **28**, 108 (1987).

29. Assmann, W., Huber, H., Steinhausen, Ch., Dobler, M., Glückler, H., and Weidinger, A., Elastic recoil detection analysis with heavy ions, *Nucl. Instrum. Methods Phys. Res. Sect. B* **89**, 131 (1994).

30. Arnoldbik, W. M., de Laat, C. T., and Habraken, F. H., On the use of a ΔE-E telescope in elastic recoil detection, *Nucl. Instrum. Methods Phys. Res. Sect. B* **64**, 832 (1992).

31. Arnoldbik, W. M., and Habraken, F. H., Elastic recoil detection, *Rep. Prog. Phys.* **56**, 859 (1993).

32. Prozesky, V., Churms, C., Pilcher, J., Springhorn, K., and Behrisch, R., ERDA measurements of hydrogen isotopes with a ΔE-E telescope, *Nucl. Instrum. Methods Phys. Res. Sect. B* **84**, 373 (1994).

33. Chu, W. K., and Wu, D., Scattering recoil coincidence spectrometry, *Nucl. Instrum. Methods Phys. Res. Sect. B* **35**, 518 (1988).

34. Hofsäss, H. C., Parikh, N. R., Swanson, M. L., and Chu, W. K., Depth profiling of light elements using elastic recoil coincidence spectrometry (ERCS), *Nucl. Instrum. Methods Phys. Res. Sect. B* **45**, 151 (1990).

35. Hofsäss, H. C., Parikh, N. R., Swanson, M. L., and Chu, W. K., Elastic recoil coincidence spectroscopy (ERCS), *Nucl. Instrum. Methods Phys. Res. Sect. B* **58**, 49 (1991).

36. Wahl, U., Restle, M., Ronning, C., Hofsäss, H., and Jahn, S. G., Li on bond center sites in Si, *Phys. Rev. B: Condens. Matter* **50**, 2176 (1994).

37. Nölscher, C., Brenner, K., Knauf, R., and Schidt, W., Elastic recoil detection analysis of light particles (^{1}H-^{16}O) using 30-MeV sulphur ions, *Nucl. Instrum. Methods Phys. Res. Sect.* **218**, 116 (1983).

38. Habraken, F. H., Light elements depth profiling using elastic recoil detection, *Nucl. Instrum. Methods Phys. Res. Sect. B* **68**, 181 (1992).

39. Möller, W., Hufschmidt, M., and Kamke, D., Large-depth profile measurements of D, ^{3}He, and ^{6}Li by deuteron-induced nuclear reaction, *Nucl. Instrum. Methods Phys. Res.* **140**, 157 (1977).

40. Williams, J. S., and Möller, W., On the determination of optimum depth-resolution conditions for Rutherford-backscattering analysis, *Nucl. Instrum. Methods Phys. Res.* **157**, 213 (1978).

41. Tirira, J., Frontier, J. P., Trocellier, P., and Trouslard, P., Development of a simulation algorithm for energy spectra of elastic recoil spectrometry, *Nucl. Instrum. Methods Phys. Res. Sect. B* **54**, 328 (1991).

42. Nagata, S., Yamagushi, S., Fujino, Y., Hori, Y., Sugiyama, N., and Kamada, K., Depth resolution and recoil cross section for analyzing hydrogen in solids using elastic recoil detection with ^{4}He beam, *Nucl. Instrum. Methods Phys. Res. Sect. B* **6**, 533 (1985).

43. Szilágyi, E., Pászti, F., and Amsel, G., Theoretical approximation for depth-resolution calculations in IBA methods, *Nucl. Instrum. Methods Phys. Res. Sect. B* **100**, 113 (1995).

44. Tirira, J., Trocellier, P., and Frontier, J. P., Analytical capabilities of ERDA in transmission geometry, *Nucl. Instrum. Methods Phys. Res. Sect. B* **45**, 147 (1990).

45. Wielunski, L. S., Benenson, R., and Lanford, A., Helium-induced hydrogen recoil analysis for metallurgical applications, *Nucl. Instrum. Methods Phys. Res. Sect. B* **218**, 124 (1983).

46. Wielunski, L. S., Benenson, R., Horn, K., and Lanford, A., High-sensitivity hydrogen analysis using elastic recoil, *Nucl. Instrum. Methods Phys. Res. Sect. B* **15**, 469 (1986).

47. Morimoto, S., Nagata, S., Yamaguchi, S., and Fujino, Y., Depth profiling of deuterium implanted in thin films by the elastic recoil detection technique of transmission geometry, *Nucl. Instrum. Methods Phys. Res. Sect. B* **48**, 478 (1990).

48. Tirira, J., Trocellier, P., Frontier, J. P., Massiot, Ph., Constantini, J. M., and Mori, V., 3D Hydrogen profiling by elastic recoil detection analysis in transmission geometry, *Nucl. Instrum. Methods Phys. Res. Sect. B* **50**, 135 (1990).

49. Forster, J. S., Leslie, J. R., and Laursen, T., Scattering recoil coincidence spectrometry: A new experimental technique for profiling hydrogen isotopes in low-Z thin films, *Nucl. Instrum. Methods Phys. Res. Sect. B* **45**, 176 (1990).

50. Forster, J. S., Leslie, J. R., and Laursen, T., Depth profiling of hydrogen isotopes in thin low-Z films, *Nucl. Instrum. Methods Phys. Res. Sect. B* **66**, 215 (1992).

6

Time of Flight ERDA

With Nick Dytlewski

6.1. INTRODUCTION

Determining mass separation between scattered and recoiled particles by measuring the TOF parameter of neutral or ionic species was first used at low energies; for example at the end of the seventies, Chen and coworkers studied gold targets using 8-keV Ar^+.[1] The MeV ERDA techniques were used for many years, but these were principally applied in materials research to determine depth profiles of a specific target matrix element, such as hydrogen or one of its isotopes. With the increasing development of multilayer thin-film microelectronic devices, metalized polymers and advanced ceramic materials, there is a need to depth profile a range of other light elements simultaneously, such as B, C, N, O, Al, Si, P, and some three-dimensional transition metals.

In the mid-1980s TOF ERDA was developed by various research groups, such as Groleau and coworkers,[2] Thomas and coworkers,[3] and Whitlow and coworkers[4] for application in such areas as surface modification of polymers for enhanced metal adhesion, optoelectronics, thin-film sensors, and interfacial diffusion or reactions in Si- and GaAs-based semiconductor devices. Low-energy scattering and recoiling were essentially used for surface structure determination,[5] whereas ERDA and RBS in the MeV energy range are basic techniques for bulk structural analysis.[2,3]

Currently problems involve characterizing multielemental stoichiometry in multilayer thin films, and hydrogen distribution is not generally investigated. Simultaneously and quantitatively depth profiling light elements (B, C, N, O, Al, Si, P, etc.), typically in the presence of heavier elements, presents an almost impossible task for techniques other than ERDA, except secondary ion mass spectrometry (SIMS). Nevertheless principles used to depth profile hydrogen, as described elsewhere in this book, are directly applicable to these more complex systems.

In conventional ERDA the easiest and most common method of recoil–projectile separation uses their differences in stopping power by placing an absorber foil over the entrance of the particle detector. Its thickness is just sufficient to stop the most energetically scattered particles. Only recoiled atoms with a mass lighter than the incident ion beam mass are transmitted into the detector. However since the energy of scattered ions can be quite high, a relatively thick absorber foil must be used, typically in the range of 6–12 μm of polyester for MeV ^{4}He$^+$. The penalty of using this form of discrimination is that recoiled particles may undergo significant energy loss in the absorber foil. They may also acquire so much energy straggling that the depth resolution of the method is degraded. By using other methods of particle discrimination (see Chapters 7–9), the requirement for a range foil can be by passed and better energy resolution and larger analyzable depth attained.

The absorber foil technique is naturally devoted to determining hydrogen and its isotopes using the MeV helium ion beams readily available from accelerators. Nevertheless it is possible to depth profile effectively target elements heavier than the incident helium beam using the *ExB* technique discussed in Chapter 7. Using a heavy ion beam, such as sulfur or chlorine, the ERDA technique can be extended to depth profile elements heavier than hydrogen[6]; however with a heavy-ion beam, a multitude of different recoil masses is detected, with the recorded energy spectrum the sum of the energy spectra of all recoiled particle types. Consequently as in Rutherford-backscattering analysis, it can be difficult to interpret unambiguously the resulting spectrum by associating recoil mass and stoichiometry with depth. To separate contributions from various elemental components, the recoil particle detector must employ some form of mass or atomic number identification in conjunction with energy detection. Time of flight ERDA is one such technique; it is typically applied with medium mass and heavy-ion beams from 10–30 MeV ^{32}S or ^{35}Cl to 200–400 MeV ^{127}I or ^{197}Au.

Chapter 6 describes the general principle of time of flight-ERDA. Section 6.4 presents the electrostatic mirror detector. In section 6.5 we discuss about efficiency and resolution of TOF-ERDA. In section 6.6 we analyze the data processing method.

6.2. GENERAL CONSIDERATIONS

The principle of TOF-ERDA is based on measuring the duration of a given particle flight path, using start and stop signals together with the measurement of the particle energy. Figure 6.1 displays a scheme of the experimental set-up used in TOF-ERDA. Mass separation is thus obtained through differences in the time of flight.[2–4,7]

A suitable choice of experimental geometry will not kinematically allow the heavy-ion beam to be scattered into the recoil detector, thus avoiding the recoil–projectile ambiguity. When $M_1 > M_2$ the use of a windowless detector is permitted for recoil angles ϕ larger than θ_{max}. The expression of θ_{max} in the laboratory frame is given by Eq. 2.7. With a recoil angle of 45°, ^{81}Br does not enter the recoil detector if scattered off elements lighter than Co, and for ^{127}I, elements lighter than Zr. Indeed the choice of experimental geometry must be a compromise among sample composition, incident ion, and the nature of the recoiled species to solve the recoil-scattering ambiguity.

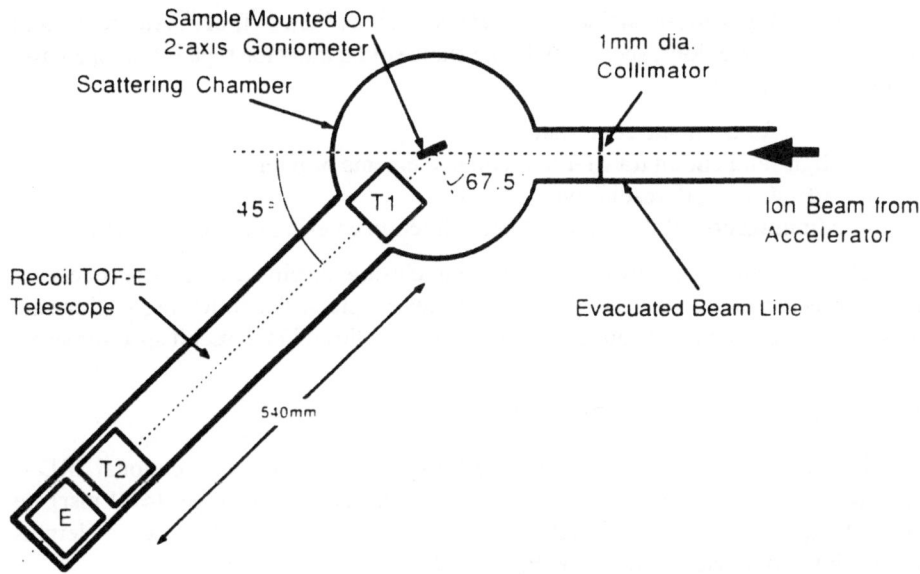

Figure 6.1. Experimental set-up used in TOF-ERDA. (From Ref. 4)

In samples containing very heavy elements, for example platinum, gold, or lead, some scattered beam does enter the detector. Nevertheless as the TOF system measures both recoiled particle energy and speed, thereby uniquely identifying its mass, the scattered beam component is easily identified. Remaining data are then sorted into separate energy spectra for all elemental constituents in the sample matrix. Consequently, by using a TOF system, both recoil-scattering and mass–depth ambiguities can be solved.

As with all analytic methods, there are limitations to mass and depth resolution. The TOF detector mass resolution is a function of many quantities, such as beam particle and energy, flight time and energy detector resolution, but it is best for light elements Li, Be, B, C, N, O, and F. Compared to heavier elements, in this mass region there is a relatively large difference in flight times between adjacent elements having equal recoil energies (coming from different depths). This region also exhibits no interferences for equal-mass isotopes of adjacent atomic number atoms, for example ^{40}Ar and ^{40}Ca (isobaric ambiguity). Sensitivity of elemental detection is mass-dependent, but it is around 0.1 at. %.

Since TOF-ERDA measures all elements in the sample, to a certain extent it is self-calibrating. Stoichiometric ratios can be referenced to a major sample component, and since the total elemental composition must be 100%, there is no need to measure the beam charge or to calibrate the detector solid angle.

In most applications incident ion beams, such as ^{81}Br, ^{127}I, or ^{197}Au are used in preference to lighter ions, such as ^{35}Cl. The choice of incident ion type is a compromise among four factors:

- Need for adequate count rates
- High-momentum transfer to recoiled elements of interest
- Adequate depth resolution
- Minimization of the scattered beam fraction entering the TOF detector

The TOF-ERDA is only a detection method using the physical concepts of elastic collision as thoroughly discussed in the previous chapters. The following sections are thus strictly devoted to outlining specific characteristics of TOF recoil spectrometry.

6.3. TIME OF FLIGHT DETECTOR

The experimental geometry of TOF-ERDA is very similar to conventional ERDA. To have good mass and energy resolutions, small telescope solid angles (≈0.1 msr) are used. A representation of the high-resolution timing set-up used in Lucas Heights Research Laboratories is shown on Figure 6.2.

In this configuration the TOF detector is located at 45° to prevent the elastically scattered incident beam from entering the detector. Recoiled particles traveling to the TOF detector telescope first pass through two electrostatic mirror time detectors, which measure flight time, then they are stopped in an ion-implanted silicon detector that provides energy information. Time detectors produce fast timing pulses, generated by the secondary electrons produced as the recoil particle travels through a thin carbon foil. Flight time and energy measurements provide sufficient information to uniquely identify the mass of the recoiled particle. In lower timing resolution configurations, the second time detector is omitted and instead the timing pulse is obtained from the

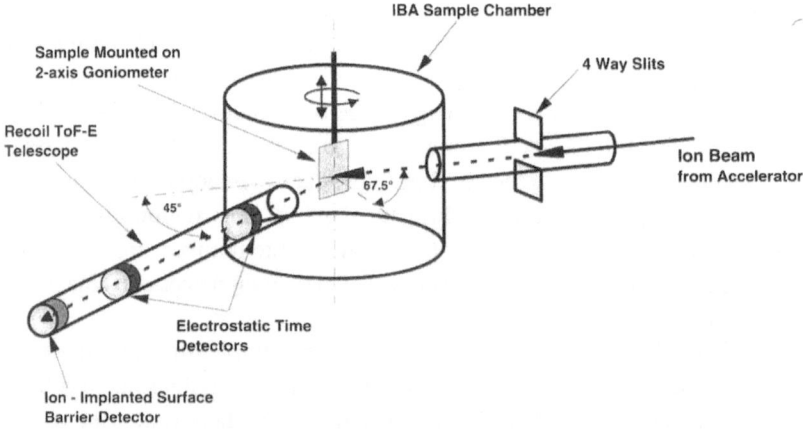

Figure 6.2. Typical experimental geometry for TOF-ERDA measurements.

energy detector preamplifier. The timing resolution obtainable from silicon detectors is inferior to those from electron mirror systems, and consequently it may not be possible to have adequate mass resolution over a wide range of recoil masses. The effect of timing resolution on mass resolution is further discussed in Section 6.5.

On the basis of depth-resolution considerations, a recoil angle of 30° can also be used (with a similar ERDA cross section). However this situation is applicable only to small analyzed depths, and it strongly depends on surface topography. Nevertheless various parameters, such as the nature and energy of the incident beam target composition, and target roughness dictate angle choice. If 30° is used, the incident beam scatters into the TOF detector when $M_2 > M_1/2$, which corresponds to Cu for an ^{127}I beam and Mo for ^{197}Au. Since a large class of microelectronic sample materials involves Si_xGe_{1-x} and GaAs, it is desirable to have a minimal background component due to the scattered incident beam. Selecting the angle thus depends on which incident beam types are available. An additional consideration is the available target beam current. On low-energy accelerators using $^4He^+$-induced ERDA, there is enough primary beam current to obtain (after collimation and focusing) a small target beam spot with useful target current. However this may not be the case with very energetic heavy ions. In such a case a compromise between acquisition time and the target current must be found, taking into account the behavior of the target sample under irradiation (see Chapter 14).

In the double-time detector system, the flight time t_f of a particle with mass M_2 and energy E_r is given by:

$$t_f = \ell \left(\frac{M_2}{2E} \right)^{1/2} \tag{6.1a}$$

More accurately taking into account energy loss in the entrance window:

$$t_f = \ell \left\{ \frac{M_2}{2[E_r - S(E_r)]x} \right\}^{1/2} \tag{6.1b}$$

where $S(E_r)$ x is the energy loss of the recoil particle traveling through the carbon foil in the time detector and ℓ is the timed flight length. Typical foil thicknesses are 5–20 $\mu g/cm^2$, and flight paths are 50 cm. Flight times are thus in the range of 30–100 ns. Equation 6.1 assumes that the burst of secondary electrons providing the timing pulse is generated when the recoil particle just penetrates the surface of the carbon foil. This energy correction term is usually small, a few hundreds of keV, small enough to be neglected in determining t_f, but it can be significant for very thick foils or very heavy-mass recoils.

The energy detector sees a recoil particle of energy:

$$E = E_r - 2 S(E_r) x \tag{6.2}$$

For heavy recoils from compounds, such as GaAs, the total energy loss in traversing two 20 $\mu g/cm^2$ carbon foils is greater than 1 MeV, which must be considered in data analysis. An additional complication arises in data analysis because pulse height defects for heavy ions in the energy detector result in mass-dependent energy calibration. The advantage of using a double time detector system is that once recoil masses are identified, energy information can be derived from the flight time rather than the residual energy, thus avoiding a potentially complex mass-dependent energy calibration procedure.

Since the first time detector subtends a larger solid angle than the second and considering the angular spread due to travel through the first two carbon foils, a large number of recoils do not enter the second time detector or energy detector. Thus if an electronic start trigger is generated by the first time detector, there can be a large number of cases when no corresponding stop trigger is generated. Therefore we must wait for a system reset to occur to trigger off another recoil event, typically when the time–analogic converter (TAC) has run past its selected conversion time range. To avoid long system dead times caused by undetected particles and also to discriminate better random coincidences, it is customary to use the timing pulse from the second time detector as a start signal and to use the pulse from the first detector delayed by t_0 as the stop signal. In addition a triple coincidence between both time detectors and residual energy detector is used to discriminate further against undetected recoils and false triggers. A biparametric coincidence display is thus generated, as shown in Figure 6.3, where the x-axis is the analogic digital converter (ADC) pulse height from the energy detector, and the y-axis is the $(t_0 - t_f)$ ADC pulse height from the timing TAC.

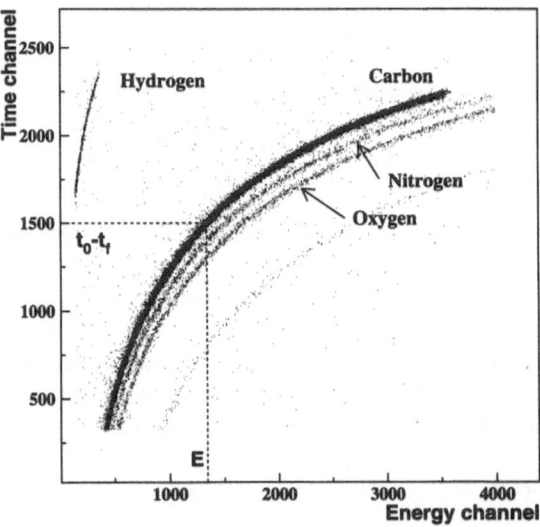

Figure 6.3.   The TOF-ERDA coincidence spectrum for a polyimide sample ($C_{22}H_{10}N_2O_6$) measured with a 84-MeV [127]I beam. Each recorded event is a data point plotted at the intersection of the measured energy (*abscissa*) and delayed flight time (*ordinate*).

Figure 6.3 shows that a family of curves is generated, with each representing a different recoil mass and increasing recoil masses arranged in descending order. If there are no energy losses in the TOF detector system, the curves generated have the simple analytic form:

$$t = t_0 - t_f = t_0 - \ell \left(\frac{M_2}{2E_r} \right)^{1/2} \tag{6.3}$$

The maximum abscissa value of each curve in Figure 6.3 reflects the different surface recoil energies. As we travel along each track from right to left, the recorded event density is a spectrum of the concentration versus depth profile for that recoil mass. This is more clearly illustrated in the corresponding isometric projection shown in Figure 6.4.

From Eqs. 6.1 and 6.2, horizontal curves of mass versus energy can be constructed by using Eq. 6.3 to transform variables. As an example Whitlow and coworkers[8] used TOF-ERDA to study the motion of nitrogen occurring in a silicon metal oxide (SiMOX)

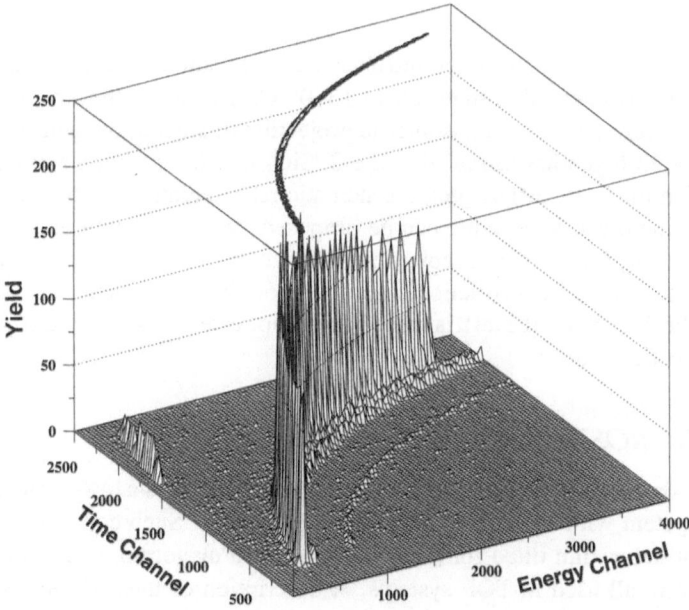

Figure 6.4. Isometric projection of the event density for the polyimide sample in Figure 6.3, illustrating the relative contribution from different sample constituents. The largest profile track is carbon, with the smaller nitrogen and oxygen components visible. The track of lowest intensity is aluminium from the target holder. Detection efficiency for hydrogen is comparatively lower, resulting in a track height much less than expected from elemental stoichiometry.

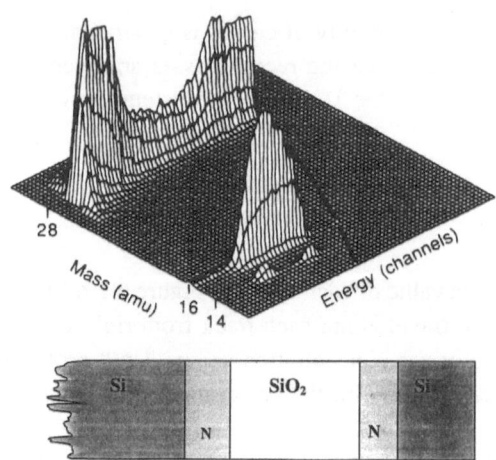

Figure 6.5. (a) Isometric plot of the TOF-ERDA recoil spectrum of a nitrogen- and oxygen-implanted Si sample, subsequently annealed to form a SIMOX structure. The solid line in the isometric plot represents the location of surface recoils. (b) Schematic description of the layered structure deduced from the isometric plot.

synthesized structure. Oxygen and nitrogen were implanted in a heated Si substrate and subsequently thermally annealed at 1300° C, producing the layered structure $Si/SiO_2/Si$. Figure 6.5 shows an isometric projection of silicon, oxygen, and nitrogen recoil yields; nitrogen has migrated to the Si/SiO_2 interfaces giving a bimodal depth profile. From the TOF-ERDA plot, a schematic representation of the location of all elements in the SiMOX structure can be proposed. By subdividing each mass curve into small energy axis bins, recoil-yield histograms for different masses can be obtained. The matrix reconstruction algorithms, outlined in Chapter 11, can thus be applied to the data. Since the task is somewhat complicated, this procedure is described further in Section 6.6.

6.4. ELECTROSTATIC MIRROR DETECTOR

The essential requirement of the time pick-off detector is a low-noise, high-electron gain system with subnanosecond timing resolution. Such configurations as the plastic scintillator, thin tilted foil, magnetic half-turn cyclotron, and electron mirror detectors were all used in TOF systems. A description of these detectors and their respective merits and disadvantages is given in Ref. 9.

The timing detector described here is an electron mirror design, based on that developed by Busch and coworkers.[10] This type of detector (see Figure 6.6) has good timing characteristics.

Operation is based on secondary electrons produced by the passage of a recoil particle through a thin self-supporting carbon foil traveling in isochronous paths to a

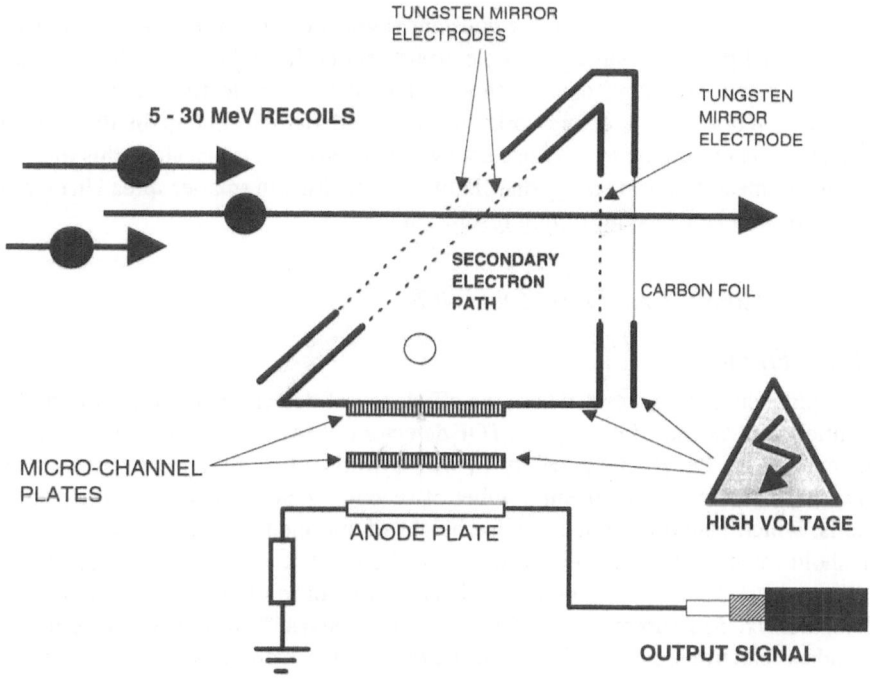

Figure 6.6. Secondary electron trajectory inside the electrostatic time detector.

pair of microchannel plates (MCPs) stacked in a chevron configuration that amplifies the secondary electron burst onto an anode plate. Backward-emitted secondary electrons are collected by an acceleration grid maintained at a positive voltage of 2 kV with respect to the carbon foil. This grid and the other two grids constituting the mirror electrodes consist of a fine wire mesh with a high (95 %) open area. After being accelerated by the 2-kV potential, electrons pass through a field-free region to an electrostatic mirror field inclined at 45°, with inner and the outer grids, respectively, maintained at a negative voltage of 2 and 6 kV. On entering this field electrons travel in a parabolic trajectory, and they are deflected 90° downward onto the stacked pair of MCPs. The two MCPs are separated by a 100-μm spacer to allow the electron cloud exiting from the first MCP to spread over channels in the second MCP to maintain a high gain[11,12] and avoid saturating MCP channels when a large number of secondary electrons are produced by a very heavy recoil particle. Electron gains are in the range of 10^6–10^7, with a pulse rise time of a few hundred picoseconds and total width around 1 ns. The anode electron pulse is fed into a fast constant-fraction discriminator that provides the timing pick-off pulse.

The isochronous condition (or constant flight times from the carbon foil to the microchannel plates) is valid only if electrons emitted from the carbon foil have zero energy and the deflecting electric field is inclined at 45° to the incident electron trajectory. While the detector works if the recoil particle enters in the opposite direction, it is preferable to only use backward-emitted electrons, since this direction has a lower mean electron energy distribution,[13] resulting in smaller spread in electron flight time and hence better timing resolution.

6.5. EFFICIENCY AND RESOLUTION

6.5.1. Efficiency

Unlike silicon detectors where the efficiency of detecting an incident particle is essentially 100%, the efficiency of a TOF detector is mass-dependent for the very light elements H, Li, Be, B, and possibly C, with hydrogen being the least efficient. This situation is due to the low mean number of secondary electrons produced by light Z recoils, which cannot produce a pulse height of sufficient magnitude to exceed the threshold value of the timing electronics. The theory of secondary electron production by Sternglass[14] states that electron yield should be proportional to the ion-stopping power dE/dx. Measurements by Clerc[15] and Clouvas[16] on the mean number of secondary electrons emitted from thin carbon foils for different ions and different energies support such a dependence, with maximum secondary yield occurring at an energy near the Bragg maximum. For a 6-μg/cm^2 carbon foil, Clerc and coworkers[15] measured a mean secondary electron yield of approximately 6 for a 4-MeV ^{4}He particle and 95 for a 20-MeV ^{32}S.

It thus follows that a recoil particle like hydrogen may not produce enough secondary electrons to produce a signal detectable above the system noise. However some hydrogen recoils can be detected because of the statistical distribution of the number of emitted secondary electrons. Accordingly the efficiency of particle detection increases with increasing recoil mass and eventually reaches a constant value. The mass value beyond which efficiency can be regarded as constant depends on the number and time distribution of electrons reaching the anode plate, system noise, and how low the threshold on the constant-fraction timing discriminators can be set. A comparative measurement of recoil yields for carbon and silicon from a silicon carbide target, or oxygen and silicon from a silicon dioxide target, can establish the validity of measured stoichiometry for these light mass recoils. Moreover as the sensitivity for hydrogen isotopes is poor, it is not very easy to select a hydrogen standard with a known stoichiometry (see Chapter 14).

Detection efficiency for light elements is also affected by the thickness and nature of the secondary electron-producing foil. Clerc and coworkers[15] found that electron emissivity from two 6-μg/cm^2 carbon foils is greater than twice that of one foil; this is most likely due to high-energy d-electrons from the first foil causing additional electron yield in the second foil. Girard and coworkers[17] measured electron yield as a function of foil thickness from 10–100 µg/cm^2. They found that as foil thickness initially increases, so does electron yield, but it attains a nearly constant value above 40 µg/cm^2. Electron yields can be enhanced by evaporating a thin coating of LiF or MgO onto the carbon foil.[17-19]

6.5.2. Mass Resolution

The ability to resolve nearly equal masses depends on two quantities: the separation between centroids of TOF mass tracks and the dispersion of data about the centroids. Both quantities depend on incident ion beam type and energy, with both centroid separation and dispersion becoming smaller as incident beam energy increases. For two recoil particles of mass M_{2a} and M_{2b} with equal energies, difference in flight times Δt_f (or track separation) is obtained from Eq. 6.4:

$$\Delta t_f = \ell \left(\frac{M_{2b}}{2E_r} \right)^{1/2} \left[1 - \left(\frac{M_{2a}}{M_{2b}} \right)^{1/2} \right] \tag{6.4}$$

From Eq. 6.4, we see that it is more difficult to separate adjacent heavy elements than adjacent light elements. This effect is illustrated in the TOF-ERDA spectrum in Figure 6.7 for a glass ceramic composed of approximately 55% fluorophlogopite mica and 45% borosilicate glass, where for the same separation of 1 amu, ^{10}B and ^{11}B are resolved, but not ^{27}Al and ^{28}Si.

With increasing beam energy, flight time difference decreases further, with neighboring mass tracks becoming less resolvable. To compensate for this effect, we simply increase the length of the timed flight path. For a fourfold increase in beam energy, doubling the electrostatic mirror separation restores track separations. For a point-source target beam spot, dispersion of data about the track centroid is due to statistical fluctuations in the total charge accumulated in the energy detector (energy resolution), variations in flight path due to a finite solid angle, and the resolution of the time pick-off and flight time measurement electronics. Geometrically the intersection of the two lines defined in Figure 6.3 varies horizontally due to energy resolution effects and vertically from timing effects. The resultant dispersion of the detected mass can be approximated by combining the different terms and using Eq. 6.3:

$$\frac{\Delta M_2}{M_2} = \left[\left(\frac{\Delta E}{E} \right)^2 + \left(\frac{2\Delta \ell}{\ell} \right)^2 + \left(\frac{2\Delta t}{t} \right)^2 \right]^{1/2} \tag{6.5}$$

Alternatively the dispersion of data about the centroid of each mass track can be described by a Gaussian distribution with a standard deviation σ_{M_2}, given by Eq. 6.6:

$$\sigma_{M_2} = \left[\left(\frac{m}{E} \right)^2 (\Delta E)^2 + \left(\frac{2m}{\ell} \right)^2 (\Delta \ell)^2 + \left(\frac{8mE}{\ell^2} \right) (\Delta t)^2 \right]^{1/2} \tag{6.6}$$

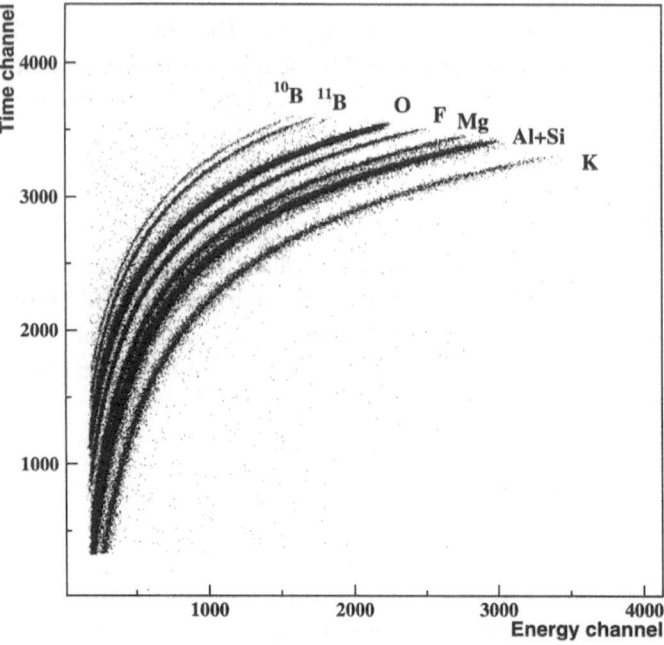

Figure 6.7. The TOF-ERDA spectrum of a glass ceramic composed of SiO_2 (46 %), Al_2O_3 (16 %), MgO (17 %), K_2O (10 %), B_2O_3 (7 %), and F (4 %) by weight. The measurement was made with a 84-MeV ^{127}I beam. Isotopes of boron are resolved, but not Al and Si, although the mass separation is 1 amu in both cases.

Typically the variation in flight path $\Delta\ell$ is 3×10^{-5} m, and timing resolution Δt is 0.2 ns (FWHM). For 30-MeV ^{59}Co recoils timed over a flight path of 0.5 m, Eq. 6.6 predicts the standard deviation to be 0.85 for an energy resolution of 1-MeV FWHM. This value is only an estimate, since the detector energy resolution is not known a priori for heavy ions. It is also an underestimate of what is actually observed, since such effects as a finite beam spot size, detector acceptance angle, energy straggling, and variations in accelerator energy causing beam spot walk across the sample material all contribute to an increase in mass variance. The dominant term in Eq. 6.6 is the energy resolution of the charged particle detector. Ghetti and coworkers[20] measured the response function of silicon p-i-n diode detectors for heavy ions with energies of 0.05–0.5 MeV/amu. While silicon detectors may show markedly different response functions, their energy resolution can be described reasonably well with the a + b$E^{1/3}$ law of Amsel,[21] where values a and b are both mass-dependent. Some a and b values were also tabulated by O'Connor and coworkers.[22] Mass resolution should thus approximately follow changes in detector energy resolution, with heavy-mass recoils having a larger variance than light-mass recoils. In addition mass resolution should become better as recoil energy increases.

For example Hult and coworkers[23] proposed a simple approach to study the dispersion of data about the track centroid. Substituting the one-third power law for the charged particle detector into Eq. 6.6 gives very approximately a semiempirical equation for the standard deviation:

$$\sigma_{M_2} = C_1 + C_2 m^{3/2} E^{-1} + C_3 m^2 E^{-2/3} + C_4 m E^{1/2} \tag{6.7}$$

where the second and third terms are associated with detector energy resolution and the first and last terms with TOF variations. Figure 6.8 shows measured values for the standard deviation and the globally fitted equation (6.7). Fitted constants C_1, C_2, C_3, and C_4 are 5.01×10^{-2}, 7.63×10^{-3}, 1.69×10^{-3}, and 9.65×10^{-4}, respectively, for the experimental configuration used on the ANTARES facility at ANSTO Lucas Heights Research Laboratories. Plotted data points were obtained by subdividing mass tracks into small energy intervals and fitting a Gaussian function to the data within each interval. As recoil energy increases, dispersion of data about the centroid of the mass track decreases. This effect is illustrated in measurements by Goppelt and his group.[24] Using a TOF system similar to that previously described, with two time detectors separated by 134 cm and an ion-implanted silicon energy detector, they were able to achieve a complete mass separation of nickel isotopes (masses 59, 60, 61, 62, and 64) using a 340-MeV ^{129}Xe beam.

Whitlow and coworkers discuss in detail the mass resolution of TOF-ERDA devices constituted of carbon foil time-zero detectors and ion-implanted silicon

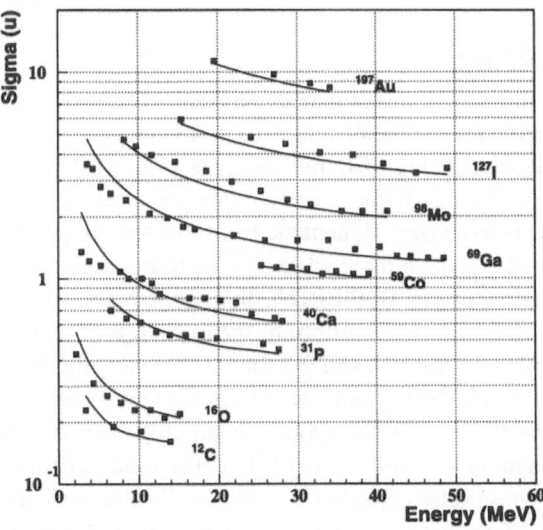

Figure 6.8. Standard deviation of a Gaussian function representing the dispersion of data about the mass track centroid compared with the semiempirical curve of Eq. 6.9. (Data from Ref. 23)

detector.[25] Comparing theoretical estimates based on literature survey with experimentally determined second moments of mass distribution, they outline several important conclusions:

- Mass resolution improves with increasing recoil energy from 0.05–0.5 MeV/amu.
- Mass resolution degrades with increasing recoil mass number.
- Mass resolution for low-energy (< 0.3 MeV/amu) recoils is dominated by the contribution from the silicon detector, and thus it is practically independent of TOF length.
- Time detector pair resolution of less than 200 ps is compatible with TOF length less than 20 cm at 0.1 MeV/amu.
- Mass resolution is not significantly degraded when carbon foil thickness variation is around 10 %.

For incident ions, such as He, N, Ne, and Ar, with energies up to 10 MeV, the mass separation of the two isotopes of Ga (^{69}Ga and ^{71}Ga) in GaAs samples can easily be achieved, and a depth resolution of 3 nm can be reached, as reported by Stanescu and coworkers, by using an RBS geometry.[26] Martin and coworkers, investigating superconductor multilayers by 40–110 MeV chlorine and iodine ions, reported a mass resolution of 4–5 amu in the region of Ga and a depth resolution of 15 nm at the surface.[27] As mentioned here above, using very high energy heavy ions, for example 340 MeV ^{129}Xe, Goppelt and coworkers have obtained a mass resolution of 1 amu in the region of Ni.[24]

6.5.3. Depth Resolution

The best depth resolution is obtained at a beam energy near the maximum of the stopping power curve for the incident ion. To profile oxygen in thin-film $YBa_2Cu_3O_7$ high T_C superconductors using ^{35}Cl, this energy is around 30 MeV, whereas for ^{127}I it is around 400 MeV. On most tandem accelerators, beam energies of ^{127}I are limited to about 100 MeV. Due to its larger kinematic factor for recoiling light elements, the use of ^{35}Cl would initially seem a reasonable choice, however the recoil cross section is too small for practical use in TOF-ERDA. In a tandem accelerator operating with a terminal voltage of 7 MV, the equilibrium mean charge of ions emerging from the carbon stripper foil is 7.3 for ^{35}Cl and 9.0 for ^{127}I.[28] Selecting the charge states ^{35}Cl^{8+} (63 MeV) and ^{127}I^{10+} (77 MeV) produces ^{16}O surface recoil energies at 45° of 27 MeV and 15 MeV, respectively; however typical recoil cross-section values are about 694 mb/sr using ^{35}Cl and 35,480 mb/sr using ^{127}I, a 50:1 ratio in counting rate in favor of the iodine beam. These values coupled with the small solid angle used in TOF-ERDA, around 0.1 msr, predicts that a very low recoil counting rate occurs using Cl. Doubling ^{127}I beam energy to increase recoil energy and obtain better depth resolution is always more profitable than halving the ^{35}Cl beam energy to obtain a higher count rate. So in

TOF-ERDA it is generally desirable to use a very heavy incident ion beam at a high energy. To use the TOF system with much larger solid angles, position-sensitive detectors are required to correct for differences in flight paths. These devices are currently being developed in some laboratories[29] (see also Chapter 9).

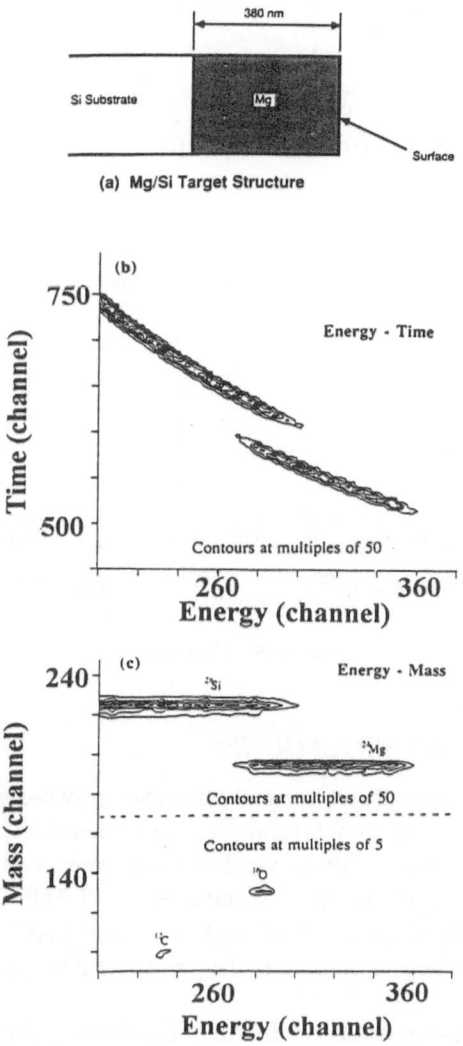

Figure 6.9. Analysis of data from a 380-nm Mg overlayer on a silicon substrate investigated by a 48-MeV ^{79}Br^{8+} ion beam. (a) Scheme of the target structure, (b) energy–time spectrum, (c) energy–mass spectrum, (d) mass spectrum, (e) ^{16}O recoil energy spectrum, (f) ^{24}Mg recoil energy spectrum, (g) ^{28}Si recoil energy spectrum. (Data from Ref. 4)

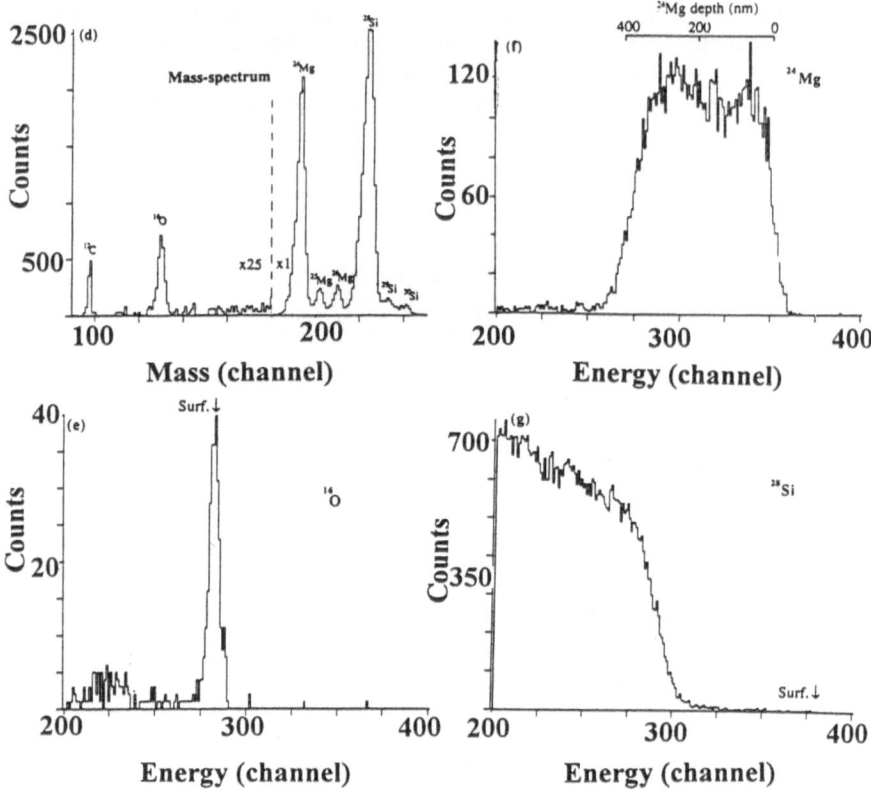

Figure 6.9. Continued.

6.6. DATA ANALYSIS PROCEDURE

The TOF-ERDA raw data are currently obtained as two-dimensional plots with time versus energy resulting from biparametric coincidence display (see Figure 6.3) or as three-dimensional plots with the yield of recoil events on the vertical axis (see Figure 6.4). Off-line graphic production and analysis of TOF-ERDA are generally conducted using the CERN Physics Analysis Workstation (PAW) software. Figure 6.9 shows data-processing steps according to the procedure described by Whitlow[4] and Martin.[27]

After defining mass contours on the flight time/energy diagrams and using Eq. 6.3, projected curves of mass versus energy are constructed, as shown in Figure 6.5. Then two-dimensional histograms with recoil yield versus mass for the whole mass range, or recoil yield versus energy for a given mass, corresponding to elemental depth profile, can be deduced from previous curves.

To extract depth profiles from recoil energy spectra (for example, Figure 6.9f and g), the corresponding recoil cross section must be considered depending on the incident energy used, so that both classical formalism and eventually non-Rutherford cross sections have to be applied, as discussed in Section 3.2 (e.g., recent papers from Räisänen and coworkers[30,31]). The last step consists in applying the general algorithms described in Chapter 11, which are available for any type of recoil energy spectra.

6.7. CONCLUSION

The TOF-ERDA is an efficient technique for readily obtaining depth profiles of elements in complex structures. In many practical situations all that is required is a knowledge of the location of elemental species and relative changes caused by different sample-processing conditions. Raw data obtained using TOF-ERDA depicts this qualitative information.

It is possible to obtain an actual concentration from measured data using matrix reconstruction algorithms described in this book. However a general matrix reconstruction procedure does not as yet exist, due to the complexity in programming the many dependent mathematical equations, which involve sample descriptions where there are large variations in matrix stoichiometry to consider.

The TOF technique is best suited to the light-elements range, where a good mass separation is obtained, but it can also be applied to medium- and heavy-mass elements, provided there are sufficient differences in the recoil masses. Moreover this kind of investigations requires a high-energy heavy-ion accelerator; thus it is not as easily accessible as other types of recoil spectrometry: conventional ERDA (Chapter 5), ExB filter (Chapter 7), or coincidence spectrometry (Chapter 9).

REFERENCES

1. Chen, Y. S., Miller, G. L., Robinson, D. A. H., Wheatley, G. H., and Buck, T. M., Energy and mass spectra of neutral and charged particles scattered and desorbed from gold surfaces, *Surf. Sci.* **62**, 133 (1977).
2. Groleau, R., Gujrathi, S. C., and Martin, J. P., Time-of-flight system for profiling recoiled light elements, *Nucl. Instrum. Methods Phys. Res.* **218**, 11 (1983).
3. Thomas, J. P., Fallavier, M., Ramdane, D., Chevarier, N., and Chevarier, A., High-resolution depth profiling of light elements in high atomic mass materials, *Nucl. Instrum. Methods Phys. Res.* **218**, 125 (1983).
4. Whitlow, H. J., Possnert, G., and Petersson, C. S., Quantitative mass and energy-dispersive elastic recoil spectrometry: Resolution and efficiency considerations, *Nucl. Instrum. Methods Phys. Res. Sect.* B **27**, 448 (1987).
5. Rabalais, J. W., Schultz, J. A., and Kumar, R., Surface analysis using scattered primary and recoiled secondary neutrals and ions by TOF and ESA techniques, *Nucl. Instrum. Methods Phys. Res.* **218**, 719 (1983).
6. Nölscher, C., Brenner, K., Knauf, R., and Schmidt, W., Elastic recoil detection analysis of light particles (^{1}H - ^{16}O) using 30-MeV sulphur ions, *Nucl. Instrum. Methods Phys. Res.* **218**, 116 (1983).

7. Gujrathi, S. C., in *Metallization of Polymers* (Sacher, E., Pireaux, J.-P., and Kowalczyk, S. P., eds.) (ACS Symposium Series 440. American Chemical Society, Washington, D.C. 1990), pp. 88–109.

8. Whitlow, H. J., Petersson, C. S., Reeson, K. J., and Hemment, L. F., Mass-dispersive recoil spectrometry studies of oxygen and nitrogen redistribution in ion-beam-synthesized buried oxynitride layers in silicon, *Appl. Phys. Lett.* **52**, 1871 (1988).

9. Whitlow, H. J., Time of flight spectroscopy methods for analysis of materials with heavy ions: a tutorial, in *Proc. of High-Energy and Heavy Ion-Beams in Materials Analysis* (J. R. Tesmer, ed.) (Materials Research Society, Albuquerque, 1990), pp. 243–256.

10. Busch, F., Pfeffer, W., Kohlmeyer, B., Schüll, D., and Pülhoffer, F., A position-sensitive transmission time detector, *Nucl. Instr. Meth.* **171**, 71 (1980).

11. Smith, A. D., and Allington-Smith, J. R., A study of microchannel plate intensifiers, *IEEE Trans. Nucl. Sci.* **33**, 295 (1986).

12. Hammamatsu Technical Manual RES-0795, Characteristics and applications of microchannel plates (1989).

13. Pferdekämper, K. E., and Clerc, H. G., Energy distribution of electrons ejected from a thin carbon foil by alpha particles and fission products, *Z. Phys.* **A275**, 223 (1975).

14. Sternglass, E. J., Theory of secondary electron emission by high-speed ions, *Phys. Rev.* **108**, 1 (1957).

15. Clerc, H. G., Gehrhardt, H. J., Richter, L., and Schmidt, K. H., Heavy-ion-induced secondary electron emission. A possible method for Z-identification, *Nucl. Instrum. Methods Phys. Res.* **113**, 325 (1973).

16. Clouvas, A., and Katsanos, A., Heavy-ion-induced electron emission from thin carbon foils, *Phys. Rev. B: Condens. Matter* **43**, 2496 (1991).

17. Girard, J., and Bolore, M., Heavy-ion timing with channel plates, *Nucl. Instrum. Methods Phys. Res.* **140**, 279 (1977).

18. Kavalov, R. L., Margaryan, Yu. L., Panyan, M. G., and Papyan, G. A., A zero-time detector of charged particles based on secondary electron emission from low-density dielectrics, *Nucl. Instrum. Methods Phys. Res. Sect. A* **237**, 543 (1985).

19. Starzecki, W., Stefanini, A. M., Lunardi, S., and Signorini, C., A compact time-zero detector for mass identification of heavy ions, *Nucl. Instrum. Methods Phys. Res. Sect. B* **193**, 499 (1982).

20. Ghetti, R., Jakobsson, B., and Whitlow, H. J., Measurements of the response function of silicon diode detectors for heavy ions using a time of flight technique, *Nucl. Instrum. Methods Phys. Res. Sect. A* **317**, 235 (1992).

21. Amsel, G., Cohen, C., and L'Hoir, A., Experimental measurements, mathematical analysis, and partial deconvolution of the asymmetrical response of surface-barrier detectors to MeV 4He, 12C, 14N, and 16O ions, in *Ion Beam Surface Layer Analysis*, Vol. 2 (O. Meyer, G. Linker, and F. Kappeler, eds.) (Plenum, New York, 1976), pp. 953–64.

22. O'Connor, D. J., and Tan, C., Application of heavy ions to high-depth resolution RBS, *Nucl. Instrum. Methods Phys. Res. Sect. B* **36**, 178 (1989).

23. Hult, M., El Bouanani, M., Persson, L., Whitlow, H. J., Andersson, M., Zaring, C., Östling, M., Cohen, D. D., Dytlewsli, N., Bubb, I. F., Johnston, P. N., and Walker, S. R., Empirical characterisation of mass distribution broadening in ToF-E recoil spectrometry, *Nucl. Instrum. Methods Phys. Res. Sect. B* **101**, 263 (1995).

24. Goppelt, P., Gebauer, B., Fink, D., Wilpert, M., Wilpert, Th., and Bohne, W., High-energy ERDA with very heavy ions using mass- and energy-dispersive spectrometry, *Nucl. Instrum. Methods Phys. Res. Sect. B* **68**, 235 (1992).

25. Whitlow, H. J., Jakobsson, B., and Westerberg, D. L., Mass resolution of recoil fragment detector telescopes for 0.05–0.5 A MeV heavy recoiling fragments, *Nucl. Instrum. Methods Phys. Res. Sect. A* **310**, 636 (1991).

26. Stanescu, T. M., Meyer, J. D., Baumann, H., and Bethge, K., Time-of-flight spectrometry for materials analysis, *Nucl. Instrum. Methods Phys. Res. Sect. B* **50**, 167 (1990).

27. Martin, J. W., Cohen, D. D., Dytlewski, N., Garton, D. B., Whitlow, H. J., and Russell, G. J., Materials characterisation using heavy-ion elastic recoil time-of-flight spectrometry, *Nucl. Instrum. Methods Phys. Res. Sect. B* **94**, 277 (1994).
28. Shima, K., Kuno, N., Yamanouchi, M., and Tawara, H., Equilibrium charge fractions of ions of $Z =$ 4–92 emerging from a carbon foil, *Atom. Data Nucl. Data Tables* **51**, 173 (1992).
29. Laegsgaard, E., Position-sensitive semiconductor detectors, *Nucl. Instrum. Methods Phys. Res.* **162**, 93 (1979).
30. Räisänen, J., Rauhala, E., Knox, J. M., and Harmon, J. F., Non-Rutherford cross sections in heavy-ion elastic recoil spectrometry : 40–70 MeV ^{32}S ions on carbon, nitrogen, and oxygen, *J. Appl. Phys.* **75**, 3273 (1994).
31. Räisänen, J., and Rauhala, E., Angular distributions of ^{12}C, ^{14}N and ^{16}O ion elastic scattering by sulfur near the Coulomb barrier and the high-energy limits of heavy-ion Rutherford scattering, *J. Appl. Phys.* **77**, 1762 (1995).

Depth Profiling by Means of the ERDA ExB Technique

With Guy Ross

7.1. INTRODUCTION

The power of the ERDA method is essentially limited by the technique used to remove the scattered particle beam. In the most used method,[1] an absorber prevents the scattered particles, as well as undesired recoils from different depths, from reaching the detector. However the absorber introduces straggle into the energy of the detected particles, thereby degrading the depth resolution. An absorber can also not be used to detect particles heavier than incident ions.

Other approaches can be considered: for example when a heavy-ion beam is incident on a relatively light sample, there is a maximum scattering angle (θ_{max}), and when the detector is at an angle larger than θ_{max}, only recoils are detected.[2] This technique is limited and not applicable with a helium beam. Another example is using coincidence detection, which is the most efficient method of background elimination; unfortunately it is usable only with thin samples (in transmission geometry), and depth resolution is limited. A powerful method of distinguishing recoils of interest from all other particles uses a TOF technique; however, this method is not straightforward and needs expensive equipment. For special conditions (such as increased depth resolution or sensitivity near the surface, an inexpensive apparatus requirement, etc.), a low-energy (< 1 MeV) ion beam may be required. Since many of the preceding approaches are not appropriate because transmission geometry cannot be used, an absorber may block recoils as well as scattered particles, and telescope detectors are unusable due to the thickness of the entrance window.

An *ExB* filter is really appropriate for removing undesired particles.[3-6] For this purpose it is not used as a Wien filter (which is a velocity selector) but as an achromatic mass and charge selector. Figure 7.1 shows a set-up based on a 400-kV Van de Graaff accelerator.[4] A 350-keV $^4He^+$ or $^3He^+$ beam is incident at an angle of 65° with respect to the normal to the surface. The recoils R+ (H^+, D^+, T^+, $^3He^+$, or $^4He^+$) are energy-

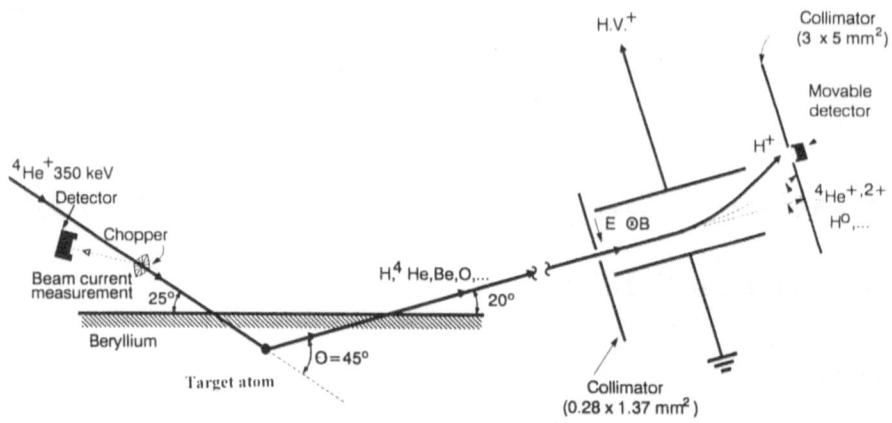

Figure 7.1. The ERDA *ExB* method showing the interaction geometry, *ExB* filter, collimator positioning, and detector.

analyzed at an angle of 45° to the beam. Scattered projectiles and other elastic recoils emerging in various charge states are eliminated with the *ExB* filter.

Chapter 7 describes the properties of such a filter (Section 7.2) and specifies some practical considerations (Section 7.3). The procedure for using the filter and the ERDA technique with an ion beam of different energies (0.35, 1, and 2.5 MeV) are described in Sections 7.4 and 7.5, then discussed using experimental examples in terms of advantages and weaknesses of the ERDA *ExB* method. In Section 7.6 the theory of a novel variant of the *ExB* filter devised for improving selectivity and extending applications to higher masses is presented, and simulation calculations are discussed.

7.2. PHYSICS AND PROPERTIES OF THE ExB FILTER

When a particle is going through crossed (perpendicular and superimposed) magnetic and electric fields (*ExB*), the exact movement equations are

$$m \frac{dv_y}{dt} = q(E - v_x B)$$

$$m \frac{dv_x}{dt} = q v_y B \tag{7.1}$$

where m refers to the mass of the entering particle (scattered projectile or ion) and v to its velocity. Assuming that its trajectory is normal to both fields, i.e., $v_x = v_0$, $v_y = 0$, they reduce to:

$$m \frac{dv_y}{dt} = q(E - v_0 B) \tag{7.2}$$

Then at the exit of the fields, the particle is deviated by an angle θ given by Eq. 7.3:

$$\theta \cong \sin\theta = \frac{qL(E - v_0B)}{mv_0^2} \tag{7.3}$$

where L is the total length of the *ExB* fields. For our purpose we are interested in adjusting the fields so that they act as a mass and charge selector. Because detecting particles of different energies is required, the *ExB* fields should deviate particles of all energies E with the same angle, this happens when:

$$\frac{d\theta}{dE} = 0 \Rightarrow E = \frac{Bv_0}{2} \tag{7.4}$$

that is when deviation due to the electric field is opposite and equal to half of the deviation due to the magnetic field. Note that the velocity appears in Eq. (7.4), which means that for this specific adjustment, the *ExB* fields are nearly energy-independent. Inside the *ExB* fields the particle trajectory is given by Eq. 7.5:

$$y = \frac{q(E - v_0B)\, x^2}{2mv_0^2} \tag{7.5}$$

These *ExB* field properties have been used to remove undesired particles for ERDA.[3-6] This technique is called ERDA *ExB*. The best geometry for this purpose is adjusting crossed magnetic and electric fields so that deviation due to an electric field is opposite to the deviation due to a magnetic field. The *ExB* filter should not be employed in the usual velocity selector mode characterized by null deflection and maximum energy dispersion (Wien filter), but in a nearly achromatic mass and charge selector mode. By varying the E/B ratio, this filter offers interesting properties, one of which concerns its capability of isolating a given ion while keeping ion deflection nearly energy-independent. This fact allows us to use a narrow collimator and to detect particles of different energies simultaneously.

Figure 7.2 shows trajectories of H^+, $^4He^+$, and $^4He^{2+}$ in an *ExB* filter (calculated with Eq. 7.5.) with an electric field of 15 kV/cm and a magnetic field of 2.9 kG.[6] At the exit of the filter, the proton recoil energy from 0.5–1.2 MeV is focused on a deflection length of less than 3 mm, and it is clearly separated from $^4He^+$ and $^4He^{2+}$ ions.

Figure 7.3 shows results of simple calculations (neglecting fringe effects) of the deflection angle θ for filter length L = 76 mm and B field of 1 kG as a function of the electric field E and recoil energies for hydrogen isotopes. At an electric field of 1.7 kV/cm, the H^+ deflection is nearly energy-independent for recoil energies between 112 keV (surface with a 350-keV ^4He beam) and 40 keV (depth of $\approx$ 100 nm), and it is isolated from other recoils and scattered particles. This fact allows us to use a narrow

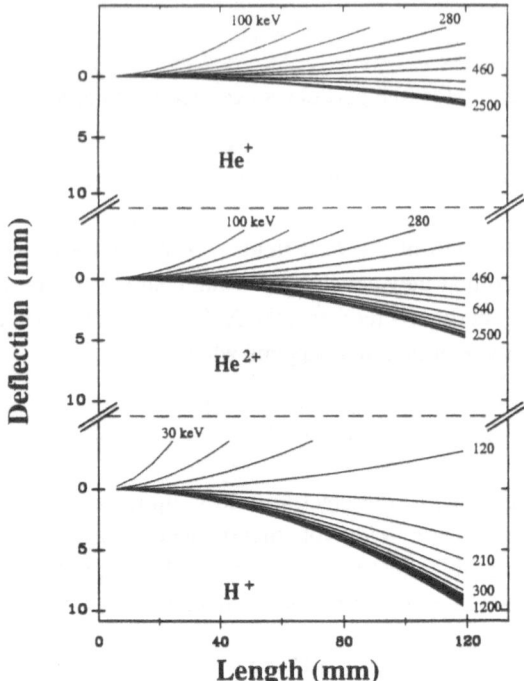

Figure 7.2. Proton and helium trajectories over the length (115 mm) of the *ExB* filter for E = 15 kV/cm and B = 2.9 kG. The ⁴He particle energies from 100 to 2500 keV are represented, while in case of hydrogen recoils, the energy range is 30–1200 keV [6].

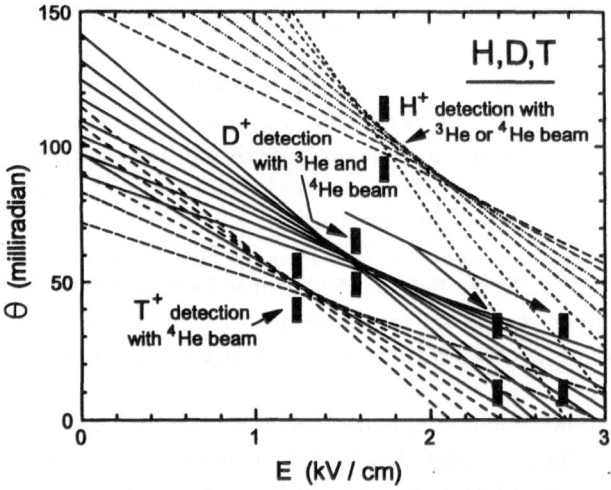

Figure 7.3. Deflection angle as a function of the electric field in the *ExB* filter of 76-mm length and 1-kG magnetic field for H⁺ (-•-•-), D⁺ (—), and T⁺ (- - -). Shaded bars indicate recommended angular positions for collimators and *E* field for H, D, and T profiling, respectively. Incident ion energy is 350 keV, and recoil energies (in keV) for various hydrogen isotopes are indicated on each line.

collimator, which affords an excellent geometry, i.e., an easy rejection of various backgrounds.

7.3. PRACTICAL CONSIDERATIONS

7.3.1. Correction for Undetected Charge Fractions

Scattered projectiles and other elastic recoils emerge from the sample in various charge states. Figure 7.4 shows an example of H^-, H^0, H^+, D^-, D^0, D^+, He^0, He^+, and He^{++} fractions as a function of particle velocity (corresponding to an energy up to 350 keV) when H, D, or He is scattered onto an oxidated aluminum sample.[7] It was reported that the charge fractions are similar for metals and semiconductors.

Because of its properties the *ExB* filter selects only a specific charge state (e.g., only R^+). Moreover all neutral particles of different masses are not deflected in the *ExB* filter. Consequently it is not possible to separate R^0 from other neutrals, so the R^+ spectrum must be corrected for undetected R^0, R^-, R^{2+}, etc., fractions by using charge fractions accurately measured previously or available in the literature.[7–9] This fact is not really restrictive for hydrogen isotopes but this is not sure for heavier elements for which many charge states are possible. Fortunately only two or three charge states have to be considered in most cases. It is recommended to take special care when investigating a new material (especially an insulator).

7.3.2. Depth Resolution

The principal factors affecting depth resolution are described in Section 4.2.3. With the ERDA *ExB* technique, the number of factors can usually be reduced to three: detector energy resolution, energy straggling, and MS. Depth resolution at the surface depends on detector resolution (3 keV in the example for H detection), and it is

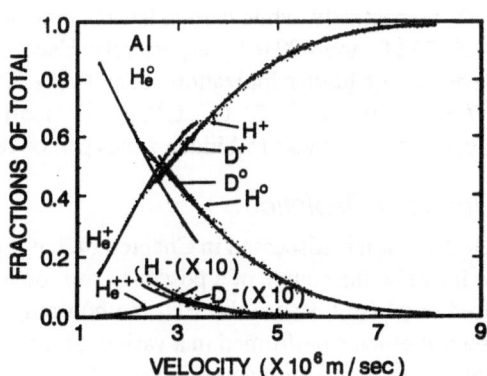

Figure 7.4. The H, D, and He charge fractions as a function of the velocity of scattered particles on an Al sample.

estimated to be ≈ 4 nm. Unfortunately deeper in, energy straggling and MS degrade depth resolution. Energy straggling can be evaluated by means of the Bohr formula[10] (Eq. 2.18):

$$\delta E_{st} = 2.355(4\pi e^4 Z_1^2 Z_2 Nx)^{1/2} \tag{7.6}$$

where N is the sample atomic density and x is the distance through the sample. Multiple scattering is particularly important, and it cannot be calculated analytically because small-angle collisions are strongly screened. It can be evaluated[11] approximately by means of the LSS expression[12] for the screened Coulomb cross section.

From these evaluations we deduce a practical equation (which is a fit to the total depth resolution) for evaluating the depth resolution δx at a depth x:

$$\delta^2 x = \delta^2 x_d + \delta x_{c1} x + \delta_{c2} x^2 \tag{7.7}$$

where $\delta^2 x_d$ is the detector contribution, δx_{c1} and δ_{c2} are both for the straggling and MS contributions, respectively. Good agreement has been found between evaluations using this equation and results of recent detailed calculations by Szilágyi and Pászti.[13] Depth resolution depends on the material and detected recoil; it is given in Section 7.4 for each combination of recoil and material considered.

7.3.3. Depth Profiling in Low-Z Materials

The energy of recoiling target atoms can be comparable to the recoil energy of implants when depth profiling is performed in very low-Z target materials.[4] Thus the capability of the *ExB* filter to discriminate implants from target material must be verified by calculating the low-Z ion deflection in this filter. Figure 7.5 shows the deflection angle as a function of the electric field for different recoil energies (from 40–160 keV) and different ionization states of beryllium and carbon, which are the most common very low-Z materials, Figure 7.5 defines forbidden zones for those materials. Beryllium has four ionization states with ionization potentials (V_i) of 9.3, 18.2, 153.9, and 217.7 V, respectively, while carbon has six ionizations states with V_i of 11.3, 24.4, 47.9, 64.5, 392.0, and 490.0 V, respectively. Usually in the low recoil energy range, the percentage of higher ionization states ($V_i > 50$ V) is lower than 0.1%.[8,9] So in Figure 7.5, only Be^+, Be^{2+}, C^+, C^{2+}, C^{3+}, and C^{4+} really define forbidden zones. However for given analyses, these forbidden zones can be restrictive.

7.3.4. Ion-Beam-Induced Depletion

Ion-beam-induced depletion is discussed in Chapter 14, but it requires particular attention here because it can be the source of a possible error, owing to the relatively large ion doses required and also because the *ExB* method may be suitable for its measurement. Many measurements performed in a variety of materials showed that a large quantity (up to 50%) of hydrogen may be released by the analysis beam.[14–18] Figure 7.6 shows an example of hydrogen depletion in a deposited carbon layer (saturated with hydrogen) and beryllium implanted with hydrogen. In the former a

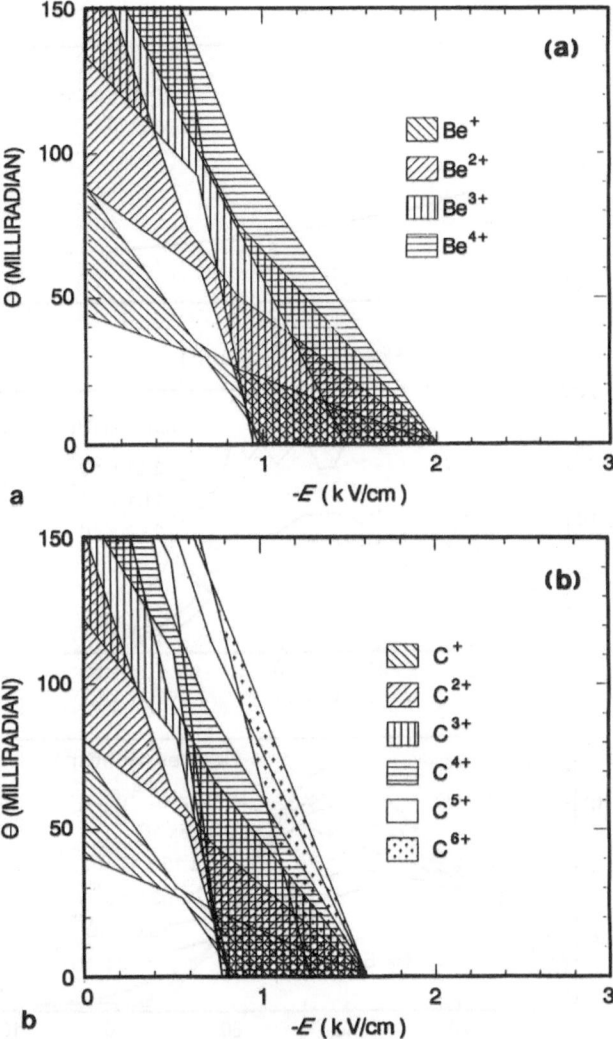

Figure 7.5. Deflection angle as a function of the *E* field in the *ExB* filter for recoils (a) Be$^{+,2+,3+,4+}$, (b) C$^{+,2+,3+,4+,5+,6+}$ defining the zones where detection is not recommended.

large quantity of hydrogen is released at the beginning of the irradiation by the analysis beam (Figure 7.6a), but the shape of the profile (still convoluted with the depth resolution) was not changed (Figure 7.6b), suggesting that depletion occurred simultaneously throughout the whole analysis depth. However when hydrogen is implanted, there is a nonuniform depth distribution. In such a case the shape of the depth profile may be modified by the analysis beam, as shown in Figure 7.6c, where depth resolution was deconvoluted by means of superpositioning a delta function and cubic splines.[19]

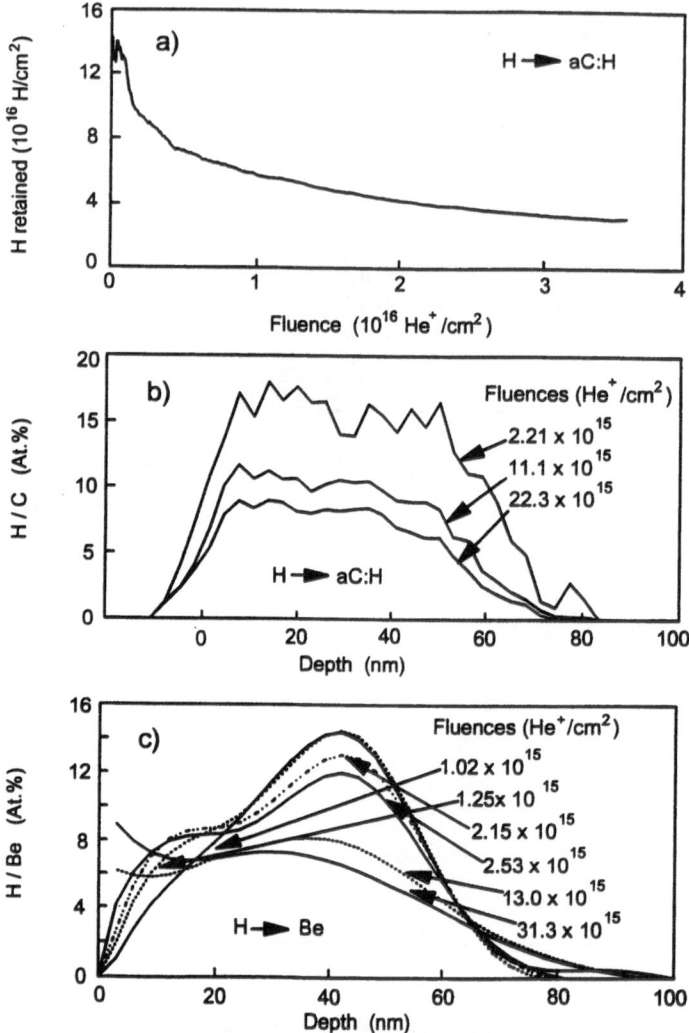

Figure 7.6. Ion-induced depletion of H in a deposited carbon layer saturated with H. (a) Behavior of the total quantity of H as a function of the fluence of the incident ions and (b) modification of the H-depth profile in the carbon layer. (c) Depth profiles of H implanted in a Be sample and analyzed by means of 350-keV [4]He of different fluences.

Results of published work can be used to estimate the effects of this depletion. Briefly nonnegligible hydrogen and deuterium depletions were observed in Be, C/B deposits, pyrolytic C, vitreous C, C deposits, Si_3N_4 deposits, TiC deposits and samples exposed to plasma discharges in tokamaks. However only a negligible amount of hydrogen or deuterium depletion was detected in Si and SiC; no tritium removal was observed in Sc, and no helium depletion was seen in Be. Consequently we must be very careful in depth profiling hydrogen isotopes in the first list of materials. It is recommended to use models or correction curves given in the literature.[14-18] When the depletion rate in a given material is not known, it is suggested to use a low-current beam ($\approx$2 nA) and to perform many analyses at different lateral positions (if there is no spatial irregularity), limiting each analysis to a beam charge of $\approx$0.5 μC.

7.4. ADJUSTMENTS FOR A 350-keV HELIUM BEAM

The ERDA *ExB* method was first developed for a 350-keV helium beam.[3] In this set-up (see Figure 7.1a) recoils R^+ (H^+, D^+, T^+, $^3He^+$, or $^4He^+$) are energy-analyzed at an angle of 45° to the beam in an ion-implanted detector with a resolution (standard deviation) of 3 keV. Therefore adjustments are first detailed for this low energy (350 keV) and this geometry (Figure 7.1a).

7.4.1. Hydrogen Depth-Profiling

Either a ^3He or a ^4He beam with an energy of 350 keV can be used to depth profile H.[4] Figures 7.7 and 7.8 show the deviation of ^3He ions and ^4He ions in the *ExB* filter, respectively. Figure 7.3 shows that H^+ can easily be detected at a magnetic field of 1 kG and an electric field of 1.7 kV/cm; the deviation of H^+ in the filter is then 110 mrad.

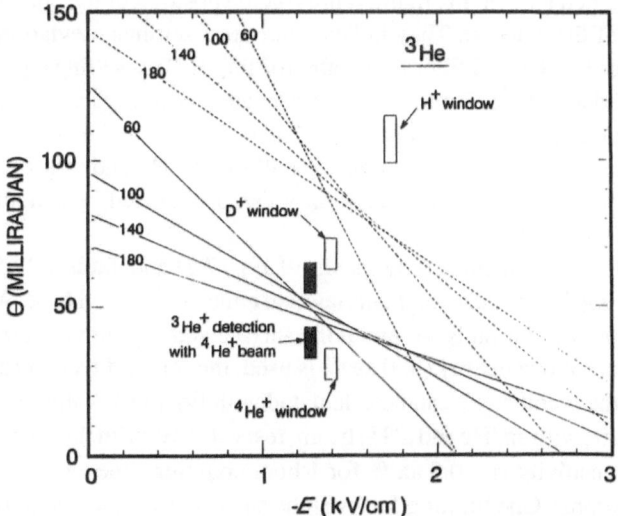

Figure 7.7. Deflection angle as a function of the *E* field in the *ExB* filter for $^3He^+$ (——) and $^3He^{2+}$ (- - -) scattered ions. Shaded bars indicate recommended settings for ^3He profiling, while rectangles indicate settings for H, D, and ^4He profiling by means of an ^3He beam (taken from Figures 7.2 and 7.5). Scattered ^3He ion energies are indicated on each line in keV units.

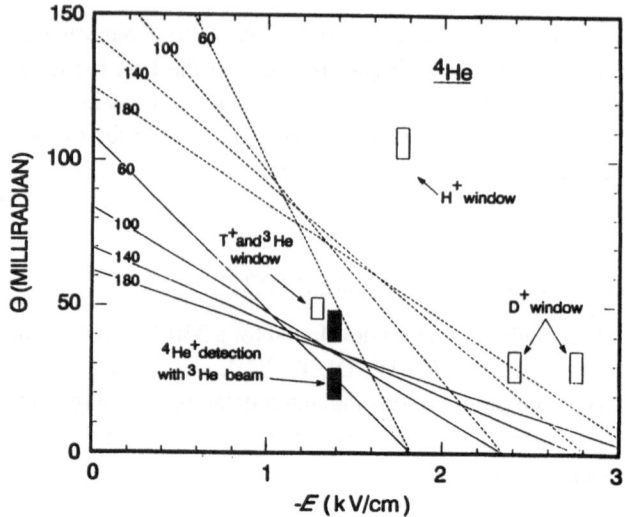

Figure 7.8. Deflection angle as a function of the E field in the ExB filter for $^4He^+$ (—) and $^4He^{2+}$ (- - -) scattered ions. Shaded bars indicate recommended settings for ^4He profiling, while rectangles indicate settings for H, D, T, and ^3He profiling, respectively, by means of an ^4He beam (taken from Figures 7.2 and 7.4). Scattered ^4He ion energies are indicated on each line in keV units.

Two series of hatched bars in this figure represent the position of the collimator edges located in front of the detector. They indicate the exact settings (deviation angle and electric field value) of the ExB filter for depth profiling H. This setting (space between hatched bars), deduced from Figure 7.3, was carried over in Figures 7.7 and 7.8 and shown as a rectangle or a window. Figures 7.7 and 7.8 show that neither ^3He nor ^4He scattered ions cross this H^+ window, consequently both ^3He and ^4He beams can be used. Also H^+ with an energy as low as 40 keV can be detected, which allows depth profiling to ≈ 100-nm depth.

Depth resolution is calculated by means of Eq. (7.7) and Table 7.1 gives values of $\delta^2 x_d$, and δx_{c1}, and δ_{c2} for different materials and either an ^3He or ^4He beam; typically in carbon the depth resolution is 4.0 nm at the surface and 12.5 nm at 100 nm depth.

When a beam of low energy (350 keV) is used, the forward recoil cross sections as a function of the recoil angle can be calculated with Eq. (3.6). Values for hydrogen are 7.7 and 12 barns with an ^3He and a ^4He beam, respectively. With the set-up described previously, the sensitivity is ≈ 0.5 at. % for 1 hour counting time and a beam current density of 25 nA/mm². One limiting factor is the small solid angle due to the distance between sample and collimator. This distance was recently shortened by a factor of three, reducing the counting time and the beam fluence. However the relative background due to scattering on the electrode plates of the electric field remained the same, and the limit of detection has not really been improved. Fortunately this background

Table 7.1. Values of $\delta^2 x_d$, δx_{c1}, and δ_{c2} for the Depth Resolution of H Calculated by
$$\delta^2 x = \delta^2 x_d + \delta x_{c1} x + \delta_{c2} x^2$$

Beam	Material	$\delta^2 x_d$ (nm^2)	δx_{c1} (nm)	δ_{c2} ($\times 10^{-3}$)	Resolution (nm) at 50-nm depth
^{3}He	Be	21.62	0.665	3.22	7.9
^{3}He	C	12.07	0.628	3.96	7.3
^{3}He	Si	20.06	0.741	17.62	10.1
^{3}He	SiC	8.75	0.683	8.05	7.9
^{3}He	TiC	6.52	0.861	11.75	8.9
^{4}He	Be	24.47	0.794	3.96	8.6
^{4}He	C	14.24	0.729	6.06	8.1
^{4}He	Si	24.44	0.946	23.35	11.4
^{4}He	SiC	10.19	0.845	10.20	8.8
^{4}He	TiC	8.26	1.164	16.19	10.3

can be reduced with high-voltage plates of low-Z materials, transparent electrodes (grids), or a large gap between particles and plates.

An example of measured depth profiles for hydrogen implanted in silicon at different energies is shown in Figure 7.9; depth profiles are still convoluted with the depth resolution. The vertical scale should be read as concentration in H at.%, but for the sake of visual clarity, profiles were adjusted to different heights. Obviously the ERDA *ExB* technique allows us to distinguish easily profiles of hydrogen implanted at very low energies (0.2–1 keV), even when the energy discrepancy (200 eV in this example, but 50 eV is easily obtainable) is small.

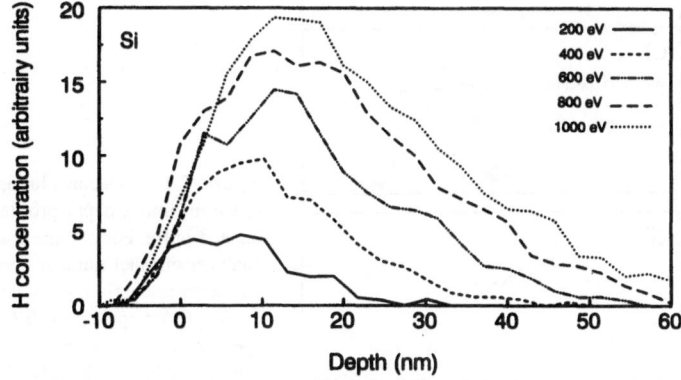

Figure 7.9. Depth profiles of H implanted at different low energies in Si as measured by the ERDA-*ExB* method with a 350-keV ^{4}He beam.

7.4.2. Deuterium Depth Profiling

Deuterium can be depth profiled by means of either an ^{3}He or ^{4}He beam; however the settings of the *ExB* filter are different. When an ^{3}He beam is used, the electric field of the *ExB* filter is set at a value for which ion deflection is nearly energy-independent. This setting is shown in Figures 7.3 and 7.7 for an electric field of 1.4 kV/cm (magnetic field of 1 kG). Average ion deflection is then 70 milliradians, and we can detect D$^+$ with an energy range of 70–175 keV (D$^+$ at lower energy is superimposed with He^{++}), which allows depth profiling to ≈100-nm depth.

When an ^{4}He beam is used, the *ExB* filter is adjusted differently. The electric field cannot be set at a minimum ion dispersion in energy because for this setting, we cannot discriminate the D$^+$ from the ^{4}He^{2+}; therefore a higher value of the electric field must be used. An electric field of 2.7 kV/cm is used to detect D$^+$ from 115 to 160 keV, which allows us to depth profile D$^+$ in the first 45 nm only. To measure more deeply (up to 80 nm), the electric field must be set at a lower value (2.4 kV/cm) to detect D$^+$ with an energy from 70 to 125 keV. The reason for this procedure is that from 70 to 115 keV

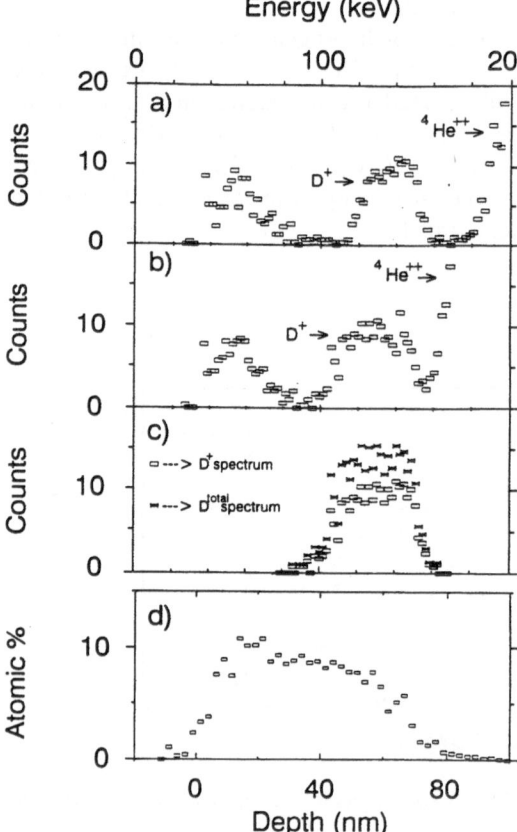

Figure 7.10. Deconvolution of a raw spectrum into a depth profile for D implanted in Be. For a–c the abscissa is the analyzer channel number converted into an energy scale; for d it is the corresponding depth computed by ERDA. (a) D$^+$ spectrum from 115–160 keV, (b) D$^+$ spectrum from 70–125 keV, (c) D$^+$ spectrum (□) and D^{total} spectrum (*) corrected for the undetected charge fraction, (d) D depth profile.

D^+ follows the same trajectory as $^4He^{2+}$ from 140 to 230 keV in the *ExB* filter; consequently we must proceed in two steps.

Figure 7.10 shows a typical D^+ spectrum, measured by means of an ^4He beam, and its deconvolution. This spectrum is based on an analysis of deuterium implanted at 1.75 keV in Be. Plot a is the D^+ spectrum from 115 to 160 keV; Plot b is the D^+ spectrum from 70 to 125 keV. Some D^+ is still detected at an energy higher than 125 keV, but this signal may be superimposed on the scattered ^4He signal. Plots a and b are combined to obtain the D^+ spectrum from 70 to 160 keV shown in Plot c ($\square$). The measured charge fraction of deuterium diffused from the beryllium is used to correct Plot c, channel-by-channel, to obtain the total energy spectrum of D^{total} shown in Plot c (*). Plot d is the final D depth profile obtained by means of a deconvolution using the computer code ERDA.

Table 7.2 gives the values of δ^2x_d, δx_{c1}, and δ_{c2} required to calculate (with Eq. 7.7) the depth resolution for either an ^3He or ^4He beam. Typically depth resolution is not very different from that of H. Cross sections are 3.0 and 4.3 b for ^3He and ^4He beams, respectively. A relatively large background signal, mainly due to particles scattering on electrodes in the electric field and the collimator edge at the entrance of the *ExB* filter, limits the sensitivity to $\approx 1\%$. The presence of neutral scattered particles are a consequence of weak ion deflection. Thus, the background is negligible when hydrogen is depth profiled (110 mrad with an ^4He beam), but it is larger when deuterium is detected (25 mrad). Using a larger magnetic field would help increase deflection and thereby reduce the background.

Figure 7.11 shows measured depth profiles (convoluted with the depth resolution) for deuterium implanted at different energies in carbon; here also the concentration scale was adjusted for the sake of visual clarity. The excellent depth resolution of the ERDA *ExB* technique allows us to distinguish depth profiles of deuterium implanted with a small difference in energy.

Table 7.2. Values of δ^2x_d, δx_{c1}, and δ_{c2} for the Depth Resolution of D Calculated by $\delta^2x = \delta^2x_d + \delta x_{c1}x + \delta_{c2}x^2$

Beam	Material	δ^2x_d (nm^2)	δx_{c1} (nm)	δ_{c2} ($\times 10^{-3}$)	Resolution (nm) at 50-nm depth
^3He	Be	17.14	0.775	3.92	8.1
^3He	C	10.63	0.790	5.14	7.9
^3He	Si	17.84	0.870	21.27	10.7
^3He	SiC	7.54	0.802	9.65	8.5
^3He	TiC	5.73	1.009	13.96	9.5
^4He	Be	18.79	0.918	4.75	8.8
^4He	C	12.01	0.967	6.42	8.7
^4He	Si	19.79	1.035	26.13	11.7
^4He	SiC	8.21	0.948	11.65	9.2
^4He	TiC	6.46	1.235	17.41	10.6

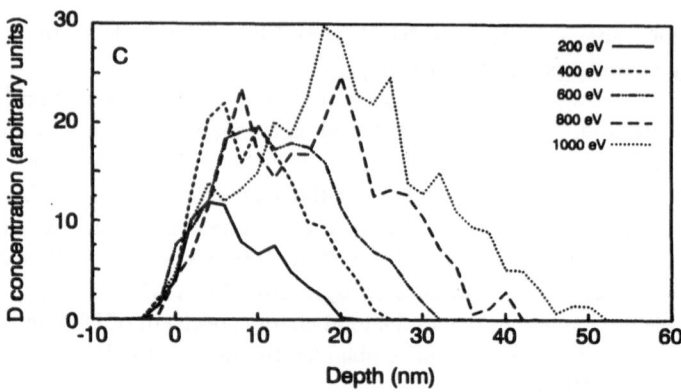

Figure 7.11. Depth profiles of D implanted at different low energies in C as measured by the ERDA-*ExB* method with a 350-keV ^{4}He beam.

7.4.3. Tritium Depth Profiling

Because ^{3}He and T follow the same trajectory in the *ExB* filter, tritium can be depth profiled only by means of an ^{4}He beam. Figures 7.3 and 7.8 show that T^+ can be detected at an electric field of 1.25 kV/cm, for which its deviation is $\approx$50 mrad. We must be very careful in adjusting the collimator aperture because T^+ is detected just between the scattered ^{4}He$^+$ and ^{4}He^{2+}, and the gap is narrow. Also some recoils from very low-Z material (e.g., Be^{3+} and Be^{4+}, which should be negligible) may perturb measurements. Nevertheless T^+ with an energy from 60 to 180 keV can be detected, which is enough to reach a depth of $\approx$100 nm in light materials.

Depth resolution is calculated by means of Eq. (7.7) and Table 7.3 gives the values of $\delta^2 x_d$, δx_{c1}, and δ_{c2}. Typically in carbon depth resolution is 3.8 nm at the surface, decreasing to 14.9 at a 100-nm depth. The cross section, calculated with Eq. (3.8), is 2.6 b.

Figure 7.12 shows a tritium depth profile (convoluted with the depth resolution) from a ScT$_2$ layer ($\approx$150 nm thick) deposited on a Si substrate. As mentioned previously

Table 7.3. Values of $\delta^2 x_d$, δx_{c1}, and δ_{c2} for the Depth Resolution of T Calculated by
$$\delta^2 x = \delta^2 x_d + \delta x_{c1} x + \delta_{c2} x^2$$

Beam	Material	$\delta^2 x_d$ (nm^2)	δx_{c1} (nm)	δ_{c2} ($\times 10^{-3}$)	Resolution (nm) at 50-nm depth
^{4}He	Be	18.28	0.995	5.17	9.0
^{4}He	C	11.76	1.023	6.79	8.9
^{4}He	Si	18.56	0.989	24.60	11.4
^{4}He	Sc	19.34	1.055	39.16	13.0
^{4}He	SiC	8.02	9.58	11.70	9.2
^{4}He	TiC	6.44	1.243	13.37	10.1

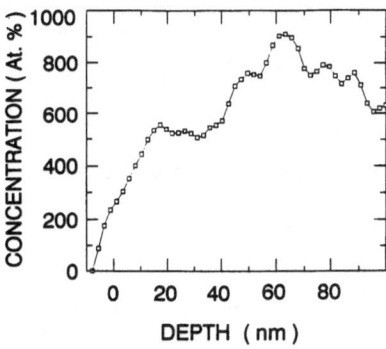

Figure 7.12. The T depth profile in a ScT_2 deposit by means of a 350-keV ^{4}He beam.

the absolute quantity of tritium in this profile does not seem to be realistic. We expected a T concentration (T/Sc) of ≈200% instead of ≈700% as measured by the ERDA *ExB* technique. This result was checked carefully (with the ERDA computer code, hand calculations, and another simplified computer code); all checks gave this high concentration. We also analyzed a ScD_2 layer (which should be similar to the ScT_2 layer), but no artifact has been observed. We assume that the cross section is far from the value for the Rutherford cross section even in this low energy range (≈350 keV).

7.4.4. ^{3}He Depth Profiling

An ^{4}He beam is used to depth profile ^{3}He. As shown in Figures 7.7 and 7.8, because they have the same charge over mass ratio, ^{3}He$^+$ is detected at the same electric field as T (1.25 kV/cm), and its deviation in the *ExB* filter is also ≈50 mrad. Thus ^{3}He can be depth profiled only when there is no tritium in the sample and vice versa. Likewise ^{3}He$^+$ with an energy from 80 to 180 keV can be detected, which allows a depth profile up to 50 nm under the surface. Some recoils from very low-Z materials may follow the same trajectory as the ^{3}He$^+$ in the *ExB* filter.

Table 7.4 gives the values of $\delta^2 x_d$, δx_{c1}, and δ_{c2} used to compute depth resolution. Typically at the surface of beryllium, the depth resolution is 4.5 nm while it decreases

Table 7.4. Values of $\delta^2 x_d$, δx_{c1}, and δ_{c2} for the Depth Resolution of ^{3}He Calculated by
$$\delta^2 x = \delta^2 x_d + \delta x_{c1} x + \delta_{c2} x^2$$

Beam	Material	$\delta^2 x_d$ (nm^2)	δx_{c1} (nm)	δ_{c2} ($\times 10^{-3}$)	Resolution (nm) at 50-nm depth
^{4}He	Be	19.53	1.364	8.42	10.4
^{4}He	Al	21.25	1.797	44.95	14.9
^{4}He	Si	23.19	1.569	42.81	14.4
^{4}He	SiC	9.67	1.492	20.34	11.6
^{4}He	TiC	7.89	1.993	30.53	13.6

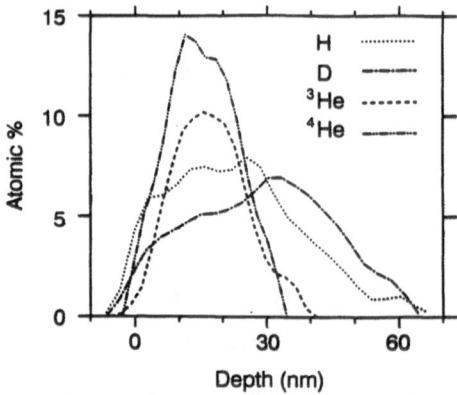

Figure 7.13. Depth profiles of H, D, ^{3}He, and ^{4}He implanted at 1 keV in Be. The H, D, and ^{4}He were profiled by means of a 350-keV ^{3}He beam.

to 10.5 nm at a 50-nm depth; the cross section is 10.4 b. The sensitivity is 1%, limited by the low deflection angle. An example of an ^{3}He depth profile is shown in Figure 7.13 (Figure 7.13 is discussed in the next section).

7.4.5. ^{4}He Depth Profiling

An ^{4}He is depth profiled in light materials by means of an ^{3}He beam. Figures 7.7 and 7.8 show that at an electric field of 1.4 kV/cm, ^{4}He$^+$ ions are detected in an energy range of 80–180 keV (maximum depth probed $\approx$50 nm); the deviation of ^{4}He$^+$ ions is only 35 mrad. Here also recoils from very low-Z materials may perturb measurements, especially when a lower plate voltage is used. The cross section is 5.9 b. Neutral background due to low deviation reduces sensitivity to $\approx$2%.

The values of $\delta^2 x_d$, δx_{c1}, and δ_{c2} used to compute depth resolution are given in Table 7.5. At the surface in beryllium, depth resolution is 5.3 nm, decreasing to 19 nm at 50 nm depth.

Figure 7.13 shows a depth profile of ^{3}He and ^{4}He, as well as those of H and D (implanted at 1 keV in Be but at different doses), obtained by means of a 350-keV ^{3}He

Table 7.5. Values of $\delta^2 x_d$, δx_{c1}, and δ_{c2} for the Depth Resolution of ^{4}He Calculated by
$$\delta^2 x = \delta^2 x_d + \delta x_{c1} x + \delta_{c2} x^2$$

Beam	Material	$\delta^2 x_d$ (nm^2)	δx_{c1} (nm)	δ_{c2} ($\times 10^{-3}$)	Resolution (nm) at 50-nm depth
^{3}He	Be	22.36	2.275	14.27	13.1
^{3}He	Al	22.75	2.037	51.11	15.9
^{3}He	Si	24.94	1.550	54.63	15.5
^{3}He	SiC	10.89	1.816	24.94	12.8
^{3}He	TiC	8.14	2.174	33.46	14.1

(^{4}He) beam. As expected mean ranges as well as profile widths are higher in H and D than in ^{3}He and ^{4}He. The surface H peak due to absorbed water vapor is hardly detected because of the use of a nitrogen-cooled trap (not because of a lack of resolution or sensitivity).

7.5. DEPTH PROFILING WITH A HIGH-ENERGY (MeV) BEAM

7.5.1. ^{4}He 1-MeV Beam

A higher beam energy can be employed with the ERDA *ExB* technique. According to simple calculations,[4] depth probed up to ≈250 nm can be reached with a 1-MeV ^{4}He beam. In that case the magnetic field must be increased; in fact a higher magnetic field increases the performance of the *ExB* filter even at 350 keV because the deviation is larger and the background coming from particles scattered on the collimator edge decreases.

Figure 7.14 shows *ExB* filter settings when a 1-MeV He beam and a 2-kG magnetic field are used. An electric field of 5.2 kV/cm is used to isolate the H$^+$. Once again two steps have to be followed to detect the D$^+$ isotope when an ^{4}He beam is used [the cross section of d(^{3}He,^{4}He)p nuclear reaction is important at 1 MeV, so a ^{3}He beam is not recommended with the ERDA *ExB* technique]. An electric field of 8 kV/cm is used to detect deuterium up to ≈350 keV, while an electric field of 7 kV/cm is indicated to depth profile deuterium more deeply. Depth profiles of tritium and ^{3}He are still obtainable; the suggested electric field is 3.8 kV/cm (not shown in Figure 7.14). An electric field of 4.5 kV/cm is recommended to depth profile ^{4}He by means of an ^{3}He beam.

Equation (7.7) and the values of $\delta^2 x_d$, δx_{c1}, and δ_{c2} given in Table 7.6 can be used to estimate depth resolution for each isotope in Be and Si. A glance at Eq. 3.8 convinces us that sensitivity is reduced when a higher energy beam is used.

7.5.2. ^{4}He 2.5-MeV Beam

B. Roux *et al.*[6] demonstrated that H can be depth profiled by means of the *ExB* technique with a 2.5-MeV ^{4}He beam. Researchers used a geometry similar to that in Figure 7.1a with a detection angle of 30° and a beam incident at 75° with respect to the normal of the target surface. The *ExB* filter of 115-mm length was constructed from a permanent magnetic field of 2.9 kG and a variable (0–25 kV) electric field. A 0.37 × 4 mm^2 collimator at the entrance of the *ExB* filter and located 150 mm from the sample defines a solid angle of 6.6 × 10^{-5} sr. Note that at these energies, Rutherford cross sections are no longer suitable (cf. Chapter 3).

Figure 7.2 shows the nearly energy-independent ion deflection for H$^+$, ^{4}He$^+$, and ^{4}He^{++} obtained with an electric field of 15 kV/cm. The proton recoil energy from 0.5 to 1.2 MeV is focused on a deflection length of less than 3 mm, and it is clearly separated from ^{4}He$^+$ and ^{4}He^{++} ions. Since the ^{4}He scatter on the high-voltage plates induces a continuous background in the detector, this effect is important for samples of atomic number larger than 25. It was also reported that H was depth profiled in Nb

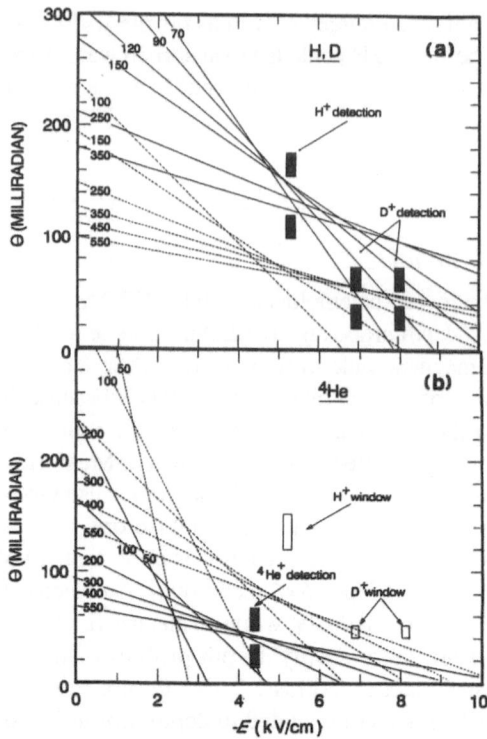

Figure 7.14. Deflection angle as a function of the E field in the ExB filter of 76-mm length and 2-kG B field. (a) H^+ (—) and D^+ (- - -); Shaded bars indicate recommended settings for H and D profiling, respectively. (b) The $^4He^+$ (—) and $^4He^{2+}$ (- - -); shaded bars indicate recommended settings for ^4He profiling, while rectangles indicate settings for H and D profiling, respectively, by means of an ^4He beam. Incident ion energy is 1 MeV, and recoil energies (in keV) for different ions are indicated on each line.

Table 7.6. Values of δ^2x_d, δx_{c1}, and δ_{c2} for the Depth Resolution of H, D, ^4He (With a 1-MeV Beam) Calculated by $\delta^2x = \delta^2x_d + \delta x_{c1}x + \delta_{c2}x^2$

Beam	Implant	Material	δ^2x_d (nm^2)	δx_d (nm)	δx_{c1} ($\times 10^{-3}$)	Resolution (nm) at 50-nm depth
^3He	H	Be	85.21	4.273	0.0	17.1
^3He	H	Si	76.11	9.987	3.707	24.2
^3He	^4He	Be	20.81	1.134	4.245	9.4
^3He	^4He	Si	16.36	0.934	15.568	10.0
^4He	H	Be	53.67	2.100	0.0	12.6
^4He	H	Si	41.37	4.900	3.839	17.2
^4He	D	Be	28.58	1.450	0.0	10.1
^4He	D	Si	25.58	2.669	2.579	12.9

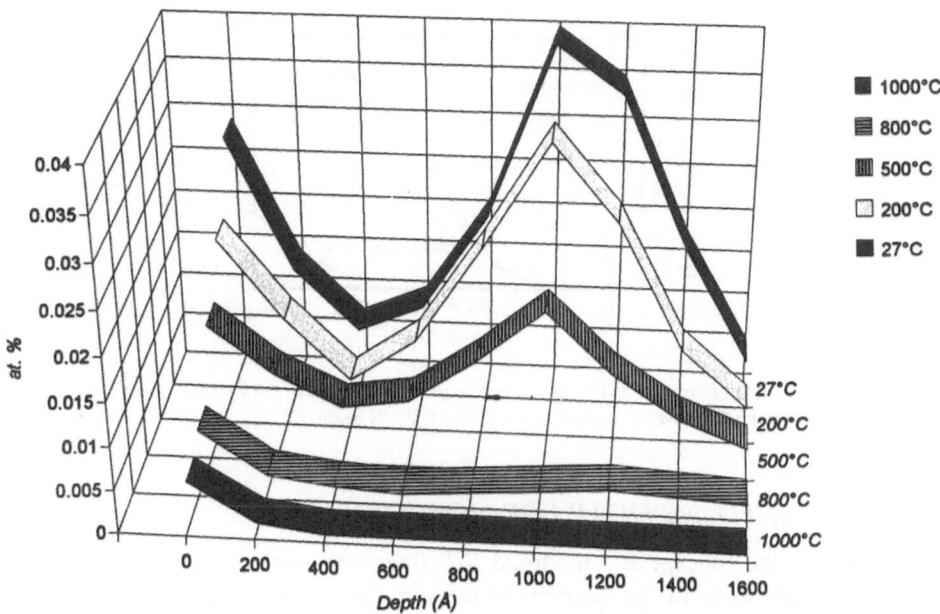

Figure 7.15. H depth profile by means of 2.5-MeV ^{4}He beam.[20] H was implanted at an energy of 10 keV in $B_{0.2}C_{0.8}$.

to a depth of 200 nm with a depth resolution of 7.3 nm at the surface and a sensitivity of 0.5 at. %.

An example of an H depth profile obtained by means of a 2.5-MeV ^{4}He beam is shown in Figure 7.15.[20] Hydrogen was implanted at an energy of 10 keV in a compound of boron (20%) and carbon (80%). Different depth profiles show the hydrogen evolution as a function of annealing temperature. There was no notable background associated with this measurement.

7.5.3. ^{15}N 2.65-MeV Beam

Recently F. Schiettekatte and the Lyon group[21] demonstrated that a 2.65-MeV ^{15}N beam is appropriate for depth profiling hydrogen and helium by means of the ERDA *ExB* technique. Researchers used the same set-up as described by B. Roux et al.[6] The forward recoil cross section, as calculated with Eq. (3.6) is 14.3 b for H and 4.5 b for ^{4}He, respectively. A depth probe up to 250 nm and a sensitivity of ≈0.1 % were reported.

Figure 7.16 shows H and He depth profiles obtained by means of a 2.65-MeV ^{15}N beam. Hydrogen and helium were implanted in beryllium with energies of 0.8 keV for H and 10 keV for He. A depth resolution (standard deviation) of ≈4.5 nm at the surface was reported for H. The background was subtracted from the He profile but not from the H profile. Its level is indicated as ≈0.1 at. %.

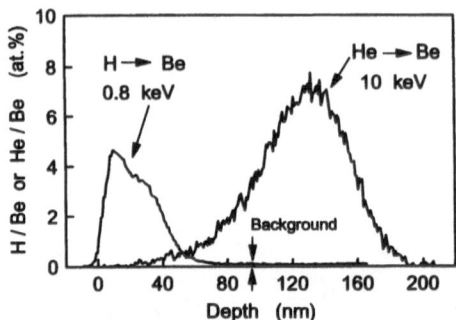

Figure 7.16. Depth profiles of H and He implanted in Be with energies of 0.8 keV for H and 10 keV for He, as measured by the ERDA-*ExB* method with a 2.65-MeV ^{15}N beam.[17]

7.6. MODIFIED ExB FILTER FOR HEAVIER ELEMENTS

Recently the concept of a modified *ExB* filter was proposed to improve mass selectivity and extend the applications to the very frequently encountered light elements, i.e., up to oxygen.[21,22] It is designed under the hypothesis of a 1-MeV He⁺ beam and 30° scattering angle.

In principle the *ExB* filter enables us to separate ions of different mass and charge state and thereby eliminate both scattered incident ions (i.e., to solve the recoil/primary ambiguity) and to analyze separately each recoil type (i.e., to solve the mass–depth ambiguity). However since, for $E = v_0 B / 2$,

$$\left. \frac{\partial Y}{\partial m} \right|_{E=Cte} = \frac{-q\,BL^2}{8\,(2E)^{1/2}}\, m^{-3/2} \tag{7.8}$$

where E is the kinetic energy, mass separation decreases rapidly and the trend of neighboring ion species to overlap becomes more and more important with increasing mass. Meanwhile the deflection of ions of the same mass and charge state is only approximately independent of their energy, since Eq. 7.4 depends on initial velocity. Dependence on energy of the deflection due to the electric force is much stronger than dependence of the magnetic deflection. This chromatism is more pronounced at low energies and favors overlap of the windows for neighboring elements, as seen in Figure 7.2. This overlap limits the energy range for a really independent analysis of each ion species. Moreover since only the magnetic deflection depends on mass, the high *B* fields necessary for separating large masses require unrealistic *E* fields to satisfy Eq. 7.4, which would themselves cause considerable divergence of low-energy trajectories.

Increasing the filter length is not sufficient to improve mass separation, since divergence of trajectories for different energies increases simultaneously with mass separation. Indeed no realistic field strengths nor filter dimensions can be found for separating carbon, nitrogen, and oxygen in the 100–500 keV range. It is still possible to find a modified spatial repartition of the fields that reduces the divergence of trajectories.

7.6.1. The B-ExB-B Filter

The chromatism of the classical *ExB* filter is characterized by a stronger curvature of low-energy trajectories. Since both fields act in opposite directions, it should be possible to have a more equilibrated compensation if trajectories changed curvature, thereby allowing this overcompensation to act in an opposite direction on both sides of the inflexion point.

Practically E ranges between limits E_{min} and E_{max}. The E_{max} is determined by the energy of primary ions and the scattering geometry, and E_{min} is some practical limit at which straggling becomes too important or the detected signal merges into the electronic noise of the detector. Ion divergence of a given mass is characterized by:

$$\frac{\partial^2 y}{\partial x \partial E} = \frac{qx}{2E^2}\left[\left(\frac{3E}{2m}\right)^{1/2} B - E\right] \tag{7.9}$$

which is positive for $E < B\sqrt{3E_{min}/2m}$ and negative for $E > B\sqrt{3E_{max}/2m}$. This suggests that a suitable change in the E/B ratio offers some compensation between a divergent and a convergent part of the trajectories.

A simple arrangement with $E = 0$ in the first part of the filter and $E > 0$ along the y-axis in the second part realizes the desired curvature inversion, but a better reduction in energy divergence is obtained with a third part, where again $E = 0$, to bend back the low-energy trajectories that are too strongly bent by the electric field. On the whole we obtain a sort of pseudofocusing that groups ions with given mass and charge within a limited deflection range.

This *B-ExB-B* filter is thus identical to the classical one, except that electrodes are shorter than the magnet length and the entrance slit is shifted away from the filter axis to accommodate deflections as large as possible. However since large B fields are necessary for heavy ions and unsuitable for light ones, the magnet must be replaced by an electromagnet to analyze ions heavier than helium.

Let x_1 be the length of the first part ($E = 0$), ℓ the length of the second part ($E > 0$), x_2 the length of the third part ($E = 0$), and

$$L = x_1 + \ell + x_2 \tag{7.10}$$

total length of the filter.

In the usual approximation of Eq. (7.2), the deflection at the filter exit is

$$Y = \frac{q}{2mv_0^2}[E\ell\,(\ell + 2x_2) - Bv_0L^2] \tag{7.11}$$

We can expect a good concentration of ions of a given mass at the filter exit if equal deflections of extremal energies are obtained, i.e., if:

$$Y(E_{max}) = Y(E_{min}) \qquad (7.12)$$

where E_{max} is the largest energy—admittedly 1 MeV for scattered primary ions and the maximum recoil energy at 30° recoil angle under 1-MeV bombardment for other ions—and E_{min} the lowest usefully detectable energy—admittedly 100 keV. This condition weakly depends on the ion mass.

Deflection should not exceed filter dimensions; this determines x_1. Then the relation between ℓ and x_2 is given by Eq. (7.12) applied to the heaviest ion species with the maximum allowable B and E fields (respectively B_{lim} and E_{lim}):

$$\ell(\ell + 2x_2) = \frac{B_{lim}L^2}{E_{lim}}\left(\frac{2}{m}\right)^{1/2} R \ \text{ with } \ R = \frac{[(E_{max})^{1/2} - (E_{min})^{1/2}](E_{min}E_{max})^{1/2}}{E_{max} - E_{min}} \quad (7.13)$$

Under these conditions, the ratio of E to B for other ion species is fixed by Eq. (7.12).

7.6.2. Filter Design

A maximum value of 5 kG is assumed for the B field, which is realizable in the 25-mm air gap of an external electromagnet. This requires a length L = 200 mm to separate oxygen from nitrogen. An electrode spacing of 25 mm was chosen to reach an electric field E_{lim} = 16 kV/cm with a ±20 kV electrode potential.

The entrance slit is shifted 8 mm from the median plane toward the negative electrode to accommodate large deflections. Its dimension must be optimized experimentally but the choice of 0.4 × 4 mm, i.e. a collection solid angle of 7×10^{-5} sr and an incertitude of 0°4' on the scattering angle, has been proved reasonable for light ions up to helium.[6] Since scattering cross-sections increase with ion mass, it should be even more suitable for heavier ions.

The range of deflection of different elements has a variable width, and in some cases the gap between them is narrow. Hence the selection collimator in front of the detector must be composed of two separately movable slits.

A simplified set-up was also studied for the limited purpose of hydrogen analysis with helium beams with energy up to 3 MeV. In this case the magnetic field does not have to be varied, so a permanent magnet can be used as in classical ExB filters. The magnetic field B = 2.9 kG and length L = 115 mm are the same as in Ref. 6 for the sake of comparison.

Since Eq. 7.13 is derived using Eq. 7.2, and Eq. 7.12 is only a tentative convergence condition, we must still compute the trajectories completely and improve the choice of parameters. This is done by using the exact movement equations (7.1), because separating the largest masses requires the best possible values of the parameters. Optimized dimensions of the 200-mm filter are $x_1 = x_2 = 70$ mm, $\ell = 60$ mm ($x_1 = x_2$ is merely fortuitous). For the 115-mm simplified filter with fixed B field, $x_1 = 18$

mm, $\ell = 66$ mm, and $x_2 = 31$ mm. If the primary energy is limited to 1 MeV, these values become $x_1 = 17$ mm, $\ell = 23$ mm, and $x_2 = 75$ mm with a B field of only 0.2 T.

Equation 7.13 is mass-dependent, so we must resort to different arguments depending on whether or not the B field can be varied. When selectivity for high masses is essential, Eq. 7.13 must be applied to the heaviest element, i.e., oxygen, for which the maximum E and B fields are required. Convenient deflections for light elements are then found by reducing the magnetic field. On the other hand, for the simplified set-up, since the B field cannot be reduced, the important requirement is no longer to separate as much as possible the heaviest elements but to have enough deflection by the electric force to keep the lightest element inside the filter limits. In this case Eq. 7.13 must be applied to hydrogen.

In the filter with the electromagnet, the electric field and the magnetic field must be optimized for each ion species using Eq. 7.12 and improving the solution by simulation according to Eq. 7.1. The best values for each ion species according to the two criteria of minimum spread for trajectories and maximum gap between neighboring species are tabulated in Ref. 22. For the variant with fixed B field, Eq. 7.13 is no longer the essential requirement, and the optimal E field is determined by the necessity of keeping hydrogen deflection inside the filter limits.

7.6.3. Performance Evaluation

Trajectories simulated according to Eq. 7.1 are shown with a strongly expanded y-axis in Figures 7.17 (200-mm filter with electromagnet, 1-MeV He$^+$ beam) and 7.18 (115-mm filter with fixed B field, 3-MeV He$^+$ beam) for most usual ion species. It was verified in Ref. 23 using the trajectory tracing code ZGOUBI[24] that edge effects of the electric field are negligible. Edge effects are much more difficult to predict for the magnetic field, since they strongly depend on the detailed design of the magnet's polar pieces. However these effects are not expected to be important, since trajectories remain almost parallel to the filter axis. The selectivity of the modified filter essentially results from the pseudofocusing effect of curvature inversions in low-energy trajectories.

Gaps between carbon, nitrogen, and oxygen when these elements are focused (Figure 7.17) are just large enough, taking into account the width and divergence of the beam impinging on the filter. Thus it is clear that no extension of the energy range is possible for large masses except by increasing the filter length. But this would probably lead to increased divergence of low energies.

Comparing Figures 7.2 and 7.18, we see that selectivity for low masses (helium and below) is considerably improved in the modified filter, which allows independent detection at much lower energies than the original one for hydrogen and helium even with the simplified set-up with the fixed B field.

Concerning selectivity for higher masses, the filter with electromagnet offers the possibility of separating carbon, nitrogen, and oxygen in a large energy range and hydrogen and helium in the whole energy range. Thus all elements of usual polymers are accessible. Since large magnetic separation is required, and therefore a large electric field, divergence of low energies due to overcompensation by the electric field becomes unavoidable for large masses and causes some overlapping. This overlap is less disturbing at the low energy limit, where it results only in a slight reduction of the analyzable depth. As a consequence, a larger value of E_{min} is taken in Eqs. 7.12 and 7.13, and thus overlap at high energy is avoided.

A small energy change is induced by the filter because the electric field modifies ion velocity as soon as it is no longer oriented along the filter axis. This loss is less than 1% at high energy and 3% for carbon and heavier elements around 220–240 keV, due to the large deflections required to separate these elements. A correction is possible using final energy values provided by the simulations. Energy loss is not larger than in a classical *ExB* filter for equal mass and energy. Hydrogen is separated and well-focused over the whole energy range. Since the deviation varies slowly at high energies, separation is still possible with a larger value of E_{max}. Monocharged helium is well-separated as long as tritium is not involved; other less focusing conditions allow separation from tritium.

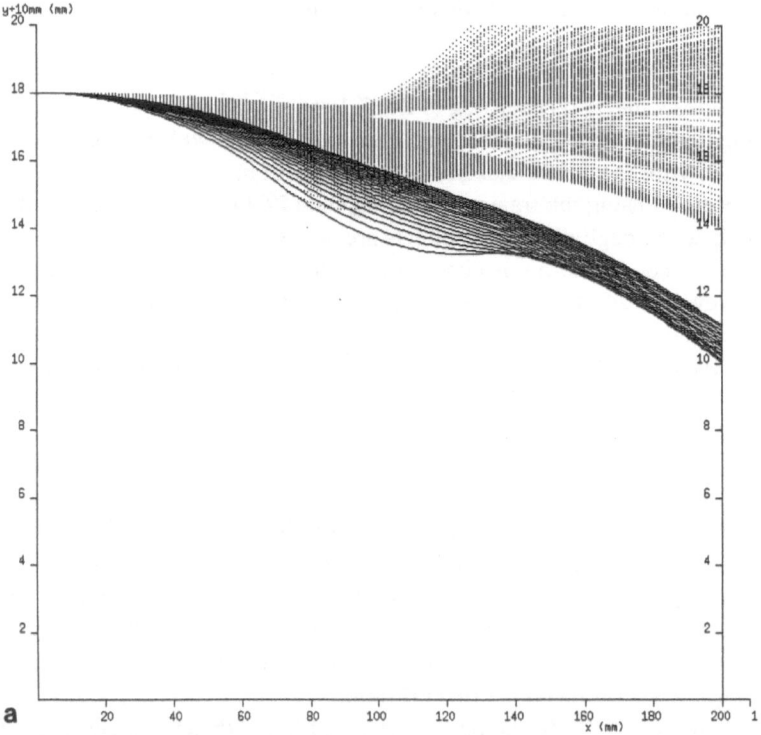

Figure 7.17. Ion trajectories in the 200-mm *B-ExB-B* filter with variable B field for 1-MeV He⁺ primary beam. Focused species in black, neighboring species in grey. (a) H⁺, (b) He⁺ (T⁺ interference), (c) He⁺⁺ (D⁺ interference), (d) C⁺ (B⁺ interference), (e) N⁺, (f) O⁺; trajectories spacing: 20 keV.

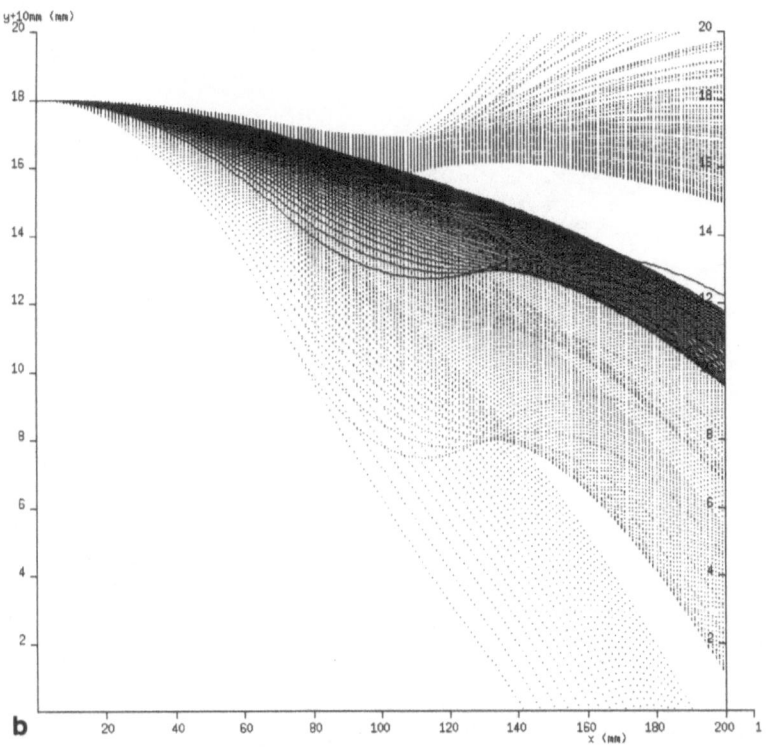

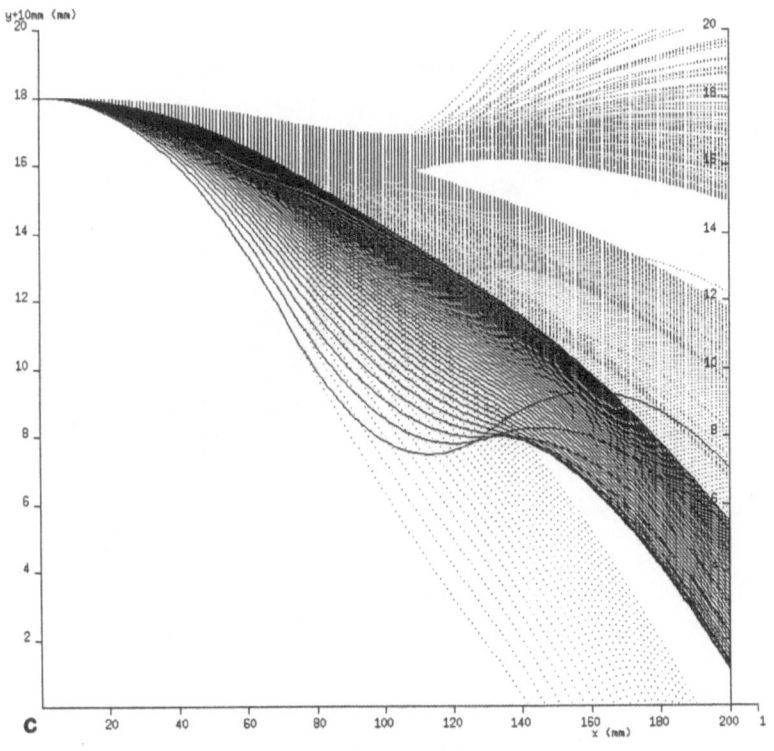

Figure 7.17. Continued.

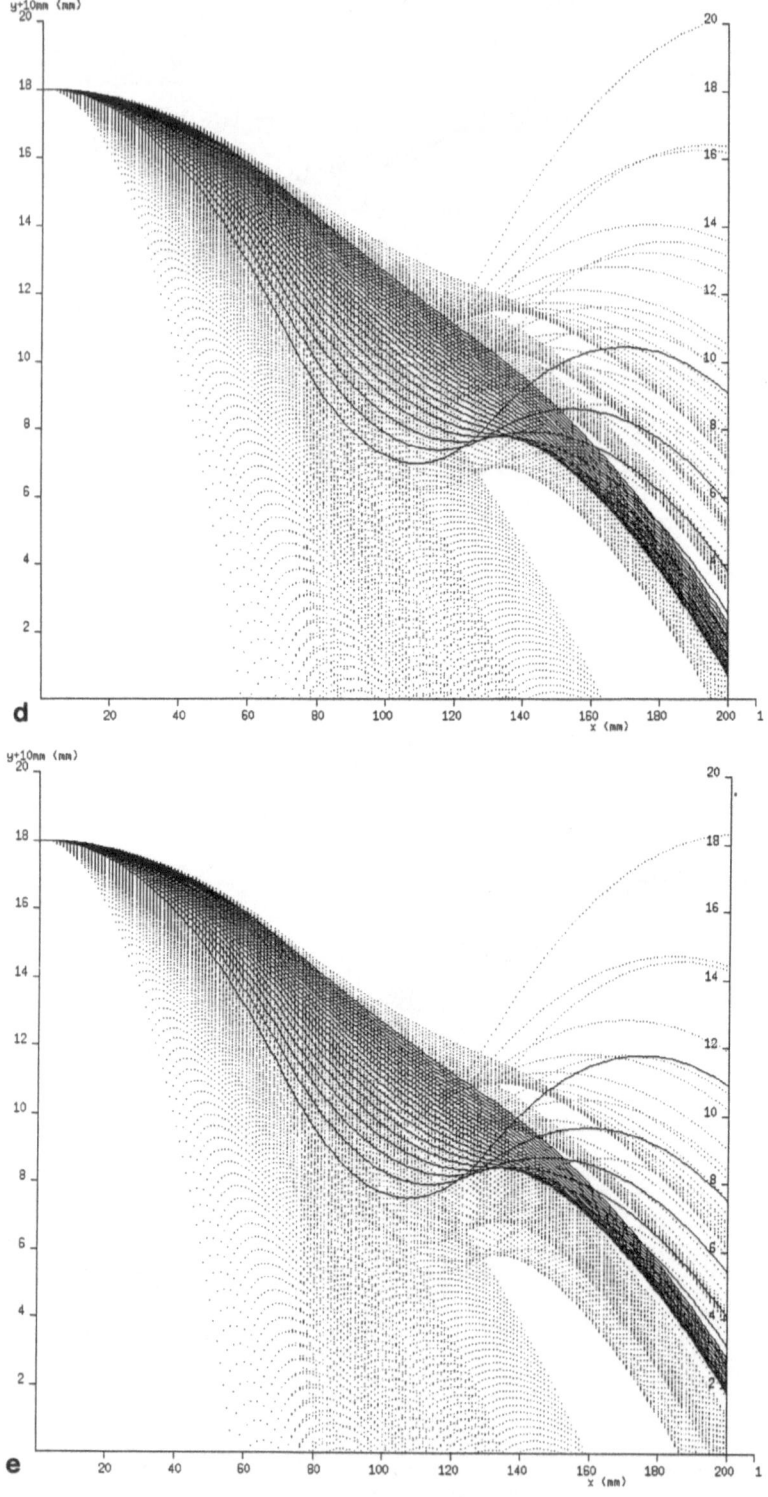

Figure 7.17. Continued.

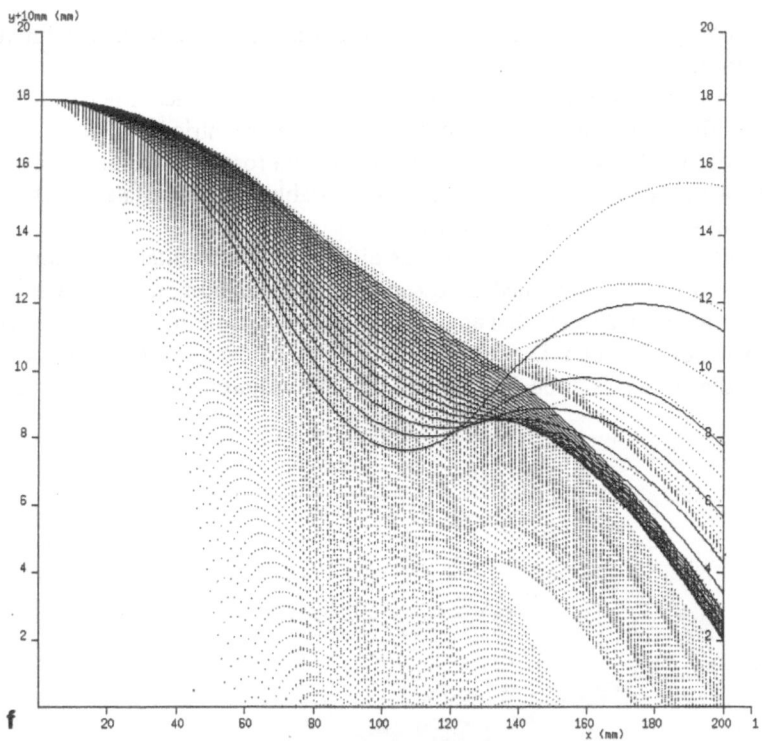

Figure 7.17. Continued.

As in classical *ExB* filters, since the magnetic force is the same for deuterium and He^{++}, these elements strongly overlap. Only deuterium above 500 keV (1700 keV in the variant with a fixed *B* field and 3-MeV He$^+$ beam) is outside the He^{++} window. Beam divergence may even reduce the accessible range. Deuterium detection is possible with ^{3}He projectiles in the whole energy range. Both tritium and ^{3}He follow identical trajectories, and they are well-separated from other elements.

Good focusing of carbon is obtained above 180 keV, but energies below 170 keV fall in the nitrogen window. Boron strongly overlaps with carbon, but both elements are not frequently encountered in the same target. Nitrogen below 160 keV falls in the oxygen window, and carbon below 170 keV interfers in the nitrogen window. Independent detection is also possible for oxygen except for interference of nitrogen below 160 keV and carbon below 120 keV.

Except for He^{++}, which behaves almost like deuterium, multicharged ions are strongly deflected and cannot interfere with monocharged ions. It is clear that dividing both fields by the ion charge brings multicharged ions on the same trajectories as their monocharged species. This allows similar focusing, but interference with mono-charged light ions occurs.

Determining fractions of different charge states is thus no less important than for the classical filter (cf. Section 7.3.1), but it becomes more intricate. Careful measurements of these fractions are an important prerequisite for experimental applications. Standards, which should be chosen among very stable and pure materials (BN, SiC, SiO_2, etc.) may be used for this purpose.

Since each element is detected under different experimental conditions (field, detector position), the cumulated dose certainly exceeds acceptable values in most materials during such a sequential analysis if we do not resort to rastering or periodic displacements of the target. Due to the vulnerability of many materials, in particular polymers, sensitivity is probably limited by the requirement to minimize beam damage.

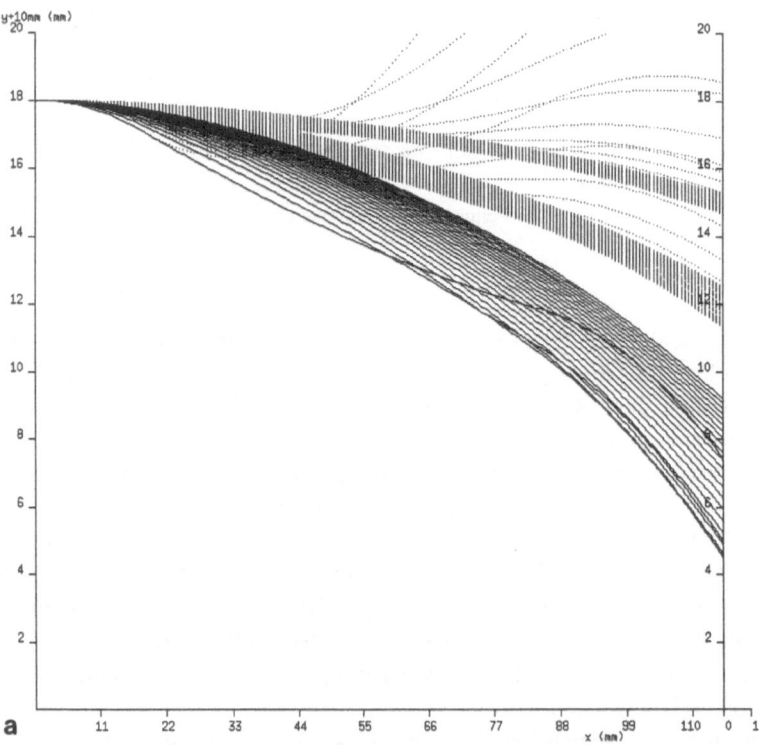

Figure 7.18.   Ion trajectories in the 115-mm *B-ExB-B* filter with B = 2.9 kG for 3-MeV He$^+$ primary beam. Focused species in black, neighboring species in grey. (a) H$^+$, (b) He$^+$, (c) He^{++}, (d) T$^+$; trajectories spacing: 60 keV.

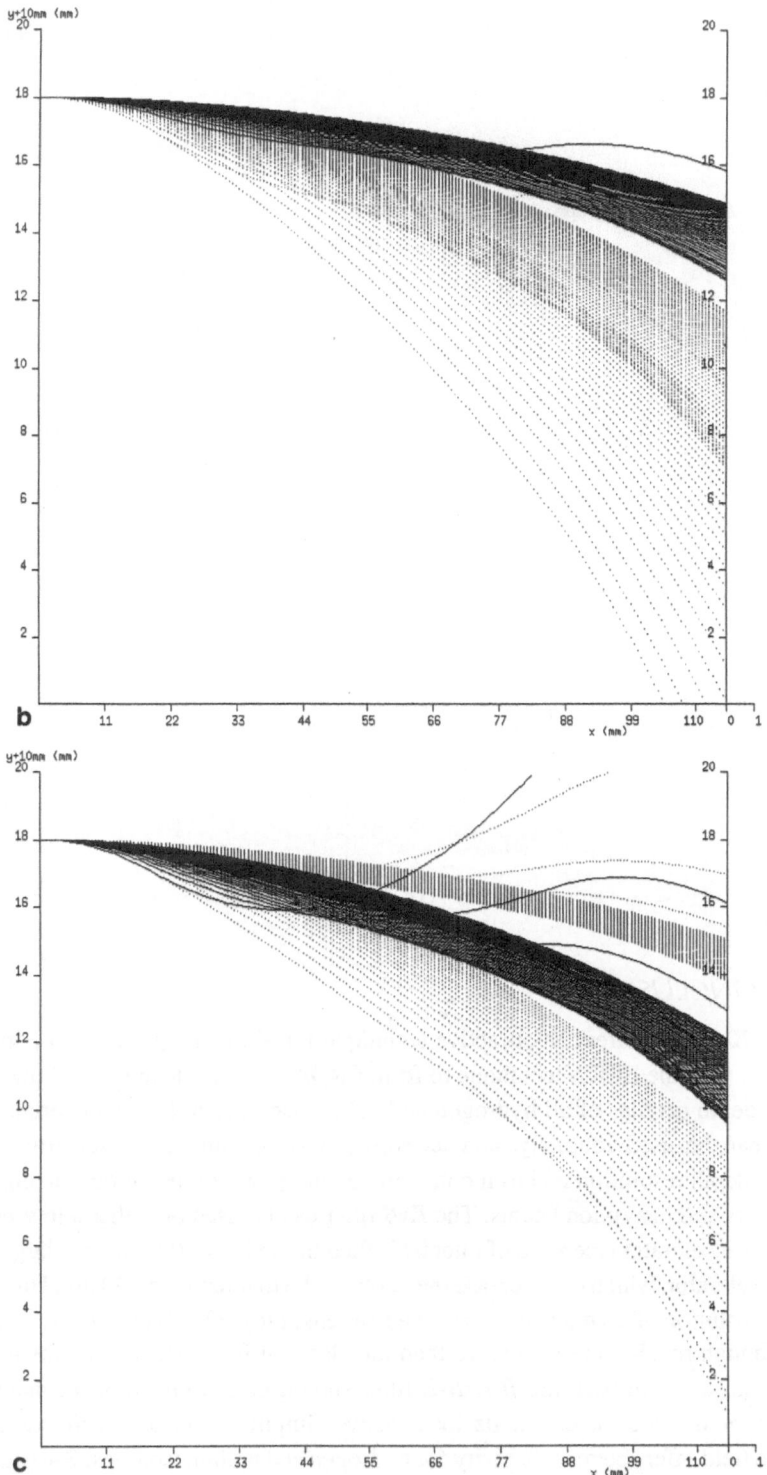

Figure 7.18. Continued.

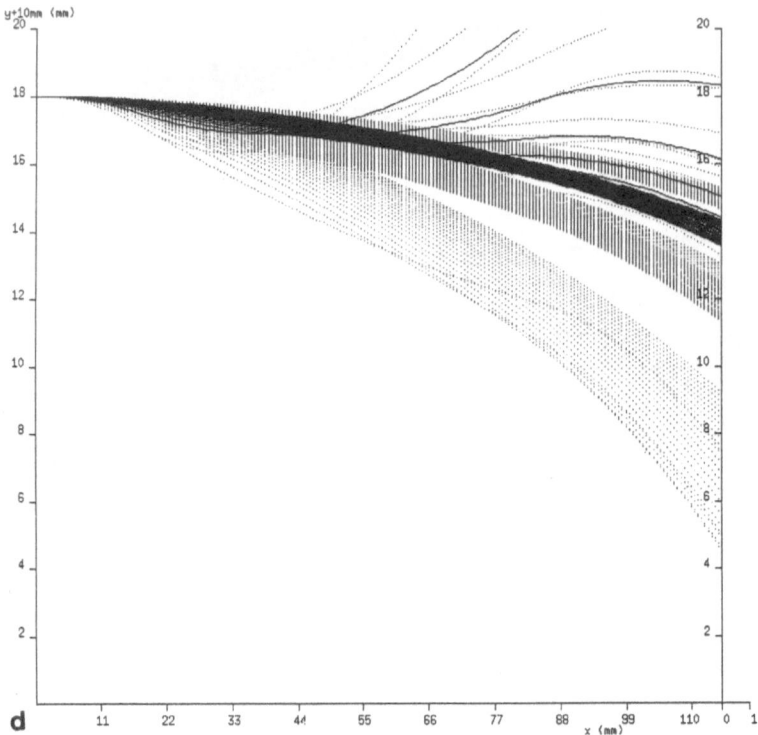

Figure 7.18. Continued.

7.7. CONCLUSION

The *ExB* filter offers an attractive technique for eliminating scattered particles. Combined with the ERDA technique to form ERDA *ExB*, this technique allows us to measure depth profiles of all hydrogen and helium isotopes in low-Z materials. This method can be used routinely, and its combination of simplicity, sensitivity, and excellent depth resolution makes it competitive with powerful techniques using large accelerators and heavy-ion beams. The *ExB* filter can be used on either a low-energy (400 keV) accelerator (the zone of interest is then limited to ≈100 nm) or a larger (2.5 MeV) accelerator, which can increase the depth probe to more than 250 nm. Until now a depth resolution of ≈5 nm at the surface and a sensitivity ≈0.1 % have been reported.

Although much more expensive than the classical filter when elements heavier than helium are analyzed, the *B-ExB-B* filter should be a useful tool for analyzing polymers or organic materials. Its inexpensive simplified version with unvariable magnetic field offers better selectivity for hydrogen and helium isotopes. But practical application to heavier elements requires further work to determine charge fractions in recoils.

The principal limitations of *ExB* filters stem from scattering of detected particles on the collimator edges and electrodes of the electric field, which induces a large background; the small solid angle, which imposes large current densities or longer analysis time; the correction for undetected charge fractions (especially neutrals). In fact no real optimization of this technique has been made yet; for example using beryllium or transparent electrode plates as well as increasing the distance between them would reduce scattering; there is no fundamental reason why a small solid angle is required; a heavier ion beam increases both depth resolution and sensitivity; however the depth probe is reduced; using a variable magnetic field (electromagnet) improves the versatility of the *ExB* filter. It seems obvious that optimization of the ERDA *ExB* method improves the sensitivity to less than 0.01 % and reduces damage induced by the ion beam (by reducing the current density of the beam). The ERDA *ExB* technique has not yet achieved its maximum potential.

REFERENCES

1. L'Ecuyer, J., Brassard, C., Cardinal, C., Chabbal, J., Deschenes, L., Labrie, J. P., Terreault, B., Martel, J. C., and Saint-Jacques, R. J., An accurate and sensitive method for the determination of the depth distribution of light elements in heavy materials, *J. Appl. Phys.* **47**, 381 (1976).
2. Ross, G. G., and Terreault, B., High-precision depth profiling of light isotopes in low-atomic-mass solids, *J. Appl. Phys.* **51**, 1259 (1980).
3. Ross, G. G., Terreault, B., Gobeil, G., Abel, G., Boucher, C., and Veilleux, G., Inexpensive, quantitative hydrogen depth profiling for surface probes, *J. Nucl. Mater.* **128/129**, 730 (1984).
4. Ross, G. G., and Leblanc, L., Depth profiling of hydrogen and helium isotopes by means of the ERD-ExB technique, *Nucl. Instrum. Methods Phys. Res. Sect. B* **62**, 484 (1992).
5. Ross, G. G., Leblanc, L., Terreault, B., Pageau, J. F., and Gollier, P. A., Nuclear microanalysis by means of 350-keV Van de Graaff accelerator, *Nucl. Instrum. Methods Phys. Res. Sect. B* **66**, 17 (1992).
6. Roux, B., Chevarier, A., Chevarier, N., Wybourn, B., Antoine, C., Bonin, B., Bosland, P., and Cantacuzene, S., High-resolution hydrogen profiling in superconducting materials by ion beam analysis (ERD-ExB), *Vacuum* **47**, 629 (1995).
7. Ross, G. G., and Terreault, B., H^-, H^0, H^+, He^0, He^+, and He^{2+} fractions of projectiles scattered from 14 different materials at 30–340 keV, *Nucl. Instrum. Methods Phys. Res. Sect. B* **15**, 146 (1986). Ross, G. G., and Leblanc, L., Charge fractions of deuterium scattered from 17 different materials at 50–340 keV, *Nucl. Instrum. Methods Phys. Res. Sect. B* **48**, 134 (1990).
8. Wittkower, A. B., and Betz, H. D., Equilibrium-charge-state distributions of energetic ions ($Z > 2$) in gazeous and solid media, *Atomic Data* **5**, 113 (1973).
9. Lennard, W. N., Phillips, D., and Walker, D. A. S., Equilibrium charge distributions of the ion beams exiting carbon foils, *Nucl. Instrum. Methods Phys. Res.* **179**, 413 (1981). Lennard, W. N., Jackman, T. E., and Phillips, D., Mean charge of ions ($5 \leq Z_1 \leq 25$) emerging from aluminium foils, *Phys. Rev. A:Gen. Phys.* **24**, 2809 (1981).
10. Bohr, N., The penetration of atomic particles through matter, *Mater. Fys. Medd. Dan. Vid. Selsk.* **18** (8), 1 (1948).
11. Leblanc, L., Ross, G. G., and Terreault, B., (1986). INRS-Energie, internal report, NRG-609 (in French), summarized in Ross, G. G., and Terreault, B., Ranges of 0.7–2.1 keV hydrogen ions in Be, C, and Si, *Nucl. Instrum. Methods Phys. Res. Sect. B* **15**, 61 (1986).

12. Linhard, J., Nielsen, V., and Scharff, M., Approximation method in classical scattering by screened Coulomb fields, *K. Dan. Vidensk. Selsk. Mat. Fys. Medd.* **36** (10), 1 (1968). Lindhard, J., Scharff, M., and Schiott, H. E., Range concepts and heavy ion ranges, *K. Dan. Vidensk. Selsk. Mat. Fys. Medd* **33** (14), 1 (1963).

13. Szilágyi, E., and Pászti, F., Theoretical calculation of the depth resolution of IBA methods, *Nucl. Instrum. Methods Phys. Res. Sect. B* **85**, 616 (1994).

14. Roth, J., Scherzer, B. M. U., Blewer, R. S., Brice, D. K., Picraux, S. T., and Wampler, W. R., Trapping, detrapping and replacement of keV hydrogen implanted into graphite, *J. Nucl. Mater.* **93–94**, 601 (1980).

15. Nagata, S., Yamaguchi, S., Bersåker, H., and Emmoth, B., Ion-induced release of H and D implanted in Be, C, Si, and SiC, *Nucl. Instrum. Methods Phys. Res. Sect. B* **33**, 739 (1988).

16. Adel, M. E., Amir, O., Kalish, R., and Feldman, L. C., Electron spin resonance investigation of ion-beam-modified amorphous hydrogenated (diamondlike) carbon, *J. Appl. Phys.* **66**, 3248 (1989).

17. Ross, G. G., and Richard, I., Influence of the ion-beam-induced desorption on the quantitative depth profiling of hydrogen in a variety of materials, *Nucl. Instrum. Methods Phys. Res. Sect. B* **64**, 603 (1992).

18. Quillet, V., Interaction de faisceaux d'ions avec des polymères, thesis, Université Paris-VII. (1992).

19. Schiettekatte, F., Marchand, R., and Ross, G. G., Deconvolution of noisy data with strong discontinuity and uncertainty evaluation, *Nucl. Instrum. Methods Phys. Res. Sect. B* **93**, 334 (1994).

20. Vidovic, Z., private communication.

21. Schiettekatte, F., Chevarier, A., Chevarier, N., Plantier, A., and Ross, G. G., Quantitative depth profiling of light elements by means of the ERD-*ExB* technique (Twelfth International Conference on Ion Beam Analysis, 22–26 May 1995, Tempe, AZ, to be submitted).

22. Serruys, Y., and Tirira, J., A modified ExB filter for analysing higher masses with ERDA. (Second French-Australian Workshop on the Applications of Ion Beam Analysis, Lucas Heights, Australia, 1–3 Feb. 1995); Serruys, Y., and Tirira, J., ERDA analysis from hydrogen to oxygen using a modified ExB filter, submitted to *Nucl. Instrum. Methods Phys. Res. Sect. B*.

23. Delferriere, O., private communication.

24. Méot, F., The ray-tracing code ZGOUBI (Third International Workshop on Optimization and Inverse Problems in Electromagnetism, CERN, Geneva, 19–21 Sept. 1994).

Recoil Spectrometry with a ΔE-E Telescope

8.1. INTRODUCTION*

In ERDA spectrometry an absorber limits the method to determining target elements lighter than the incident ions, as for example determining hydrogen isotopes when applying MeV helium-4 ions. In both MeV ^{4}He$^+$-induced elastic recoil spectrometry and in HI-ERDA analysis, mass separation between scattered ions and recoiled nuclei can be improved by measuring their difference in stopping power. This can be done in either a gas-filled ionization chamber[1] or a solid-state transmission detector[2]; the residual energy in this case is measured by a thick silicon surface barrier detector.

The association of a thin ΔE detector with a thick E detector, called telescope, is derived from heavy-ion nuclear physics,[3] where particle identification telescopes have been used for about 25 years. A telescope can also be constituted by an ionization chamber coupled with a thick silicon surface barrier detector. For such combined systems the silicon detector is generally located inside the gas chamber.[4] In the case of a solid-state telescope, the thin silicon detector typically has a thickness in the range of 10–20 μm for energy detection[5] less than 1 MeV/amu. At higher energies it can be up to 200 μm, depending on the atomic number of the incident ion. Thickness of the depleted layer of the residual energy detector is typically in the range of 500–1500 μm.

ΔE-E telescopes give information on both the atomic number and energy. For the lightest atoms (Z < 4), isotopes can be separated. A limitation may arise in terms of mass resolution and Z separation in their use for further application due to both energy resolution and energy straggling in the ΔE detector.[6] A TOF section may thus be combined in a TOF telescope to improve both mass and energy resolution.[6,7]

---

*The authors wish to acknowledge Bernard Berthier for his contribution and Wim Arnoldbik for his critical reviewing of this chapter.

8.2. EXPERIMENTAL CONSIDERATIONS

Figure 8.1 shows the experimental arrangement obtained in the case of a solid-state telescope; Figure 8.2 shows the general outline of a combined gas and solid-state telescope.

Theoretically the transition from conventional ERDA measurement to ΔE-E ERDA involves no more than replacing the absorber foil by a thin transmission detector, thus basic considerations discussed in previous chapters are still relevant. A new consideration is added by the enormous amount of scattered particles stopped in the absorber in most conventional ERDA experiments. These particles could damage the ΔE detector and induce pile-up effects. This problem can be solved by using incident ions so that $M_1 > M_2$, then the maximum scattering angle [$\arcsin(M_2/M_1)$] obtains a value smaller than the detection angle.

The most classic way of displaying telescope recoil data is a two-dimensional plot, where the x and the y-axes correspond to residual energy E_R measured in the thick silicon detector and to the energy loss ΔE measured in the thin detector, respectively (see for example Figures 8.5 and 8.7 and also Chapters 6 and 13). Each particle type is thus represented by a cloud of points. The horizontal size of the clouds is proportional to the particle energy interval explored, and the density of points is directly connected to the recoil yield. The cloud at the lowest ΔE value denotes the lightest particles, the protons, and the other clouds are displayed by order of increasing mass.

In classic IBA, as for example conducting conventional NRA, RBS, and ERDA investigation, the mass number of the incident particles involved does not usually exceed 4. The use of a ΔE-E telescope is particularly well-adapted because protons, deuterons, and alpha particles issued from elastic scattering and nuclear reaction events can be well-discriminated. Although Figure 8.3 does not correspond to an ERDA experiment, it illustrates the mass separation ability of a thin ΔE detector (17 μm) alone in the case of determining light-element distributions in the near surface region of a UO_2 sample by combining NRA and RBS detection. This representation is equivalent to a projection of E-ΔE data onto the ΔE axis. Applying a 2.2-MeV deuteron beam[8] and placing the telescope at 40° with respect to the incident beam direction, three different regions can be identified in the energy loss spectrum:

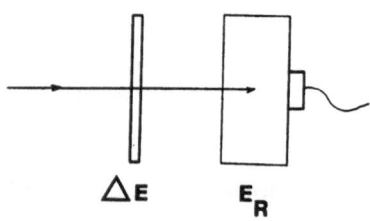

Figure 8.1. Schematic view of a solid-state telescope.

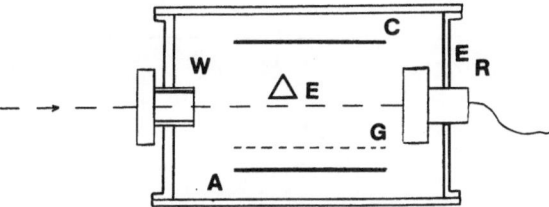

Figure 8.2. Outline of a ΔE-E telescope composed of a gas ionization chamber associated with a thick silicon surface barrier detector (C, A, and G denote, respectively, the cathode, the anode, and the grid of the ionization chamber; W denotes the thin plastic entrance window).

- Protons coming from ^{12}C(d, p) and ^{16}O(d, p) nuclear reaction events
- Scattered deuterons essentially from elastic collisions with uranium nuclei
- Alpha particles issued from ^{16}O(d, α) nuclear reaction events

For complete mass discrimination, such a projection is not sufficient, so we must define regions in the E-ΔE plane. As for TOF-ERDA data in each region can be converted into an energy spectrum for a single element and interpreted with classical methods (Sections 6.6 and 11.6.3).

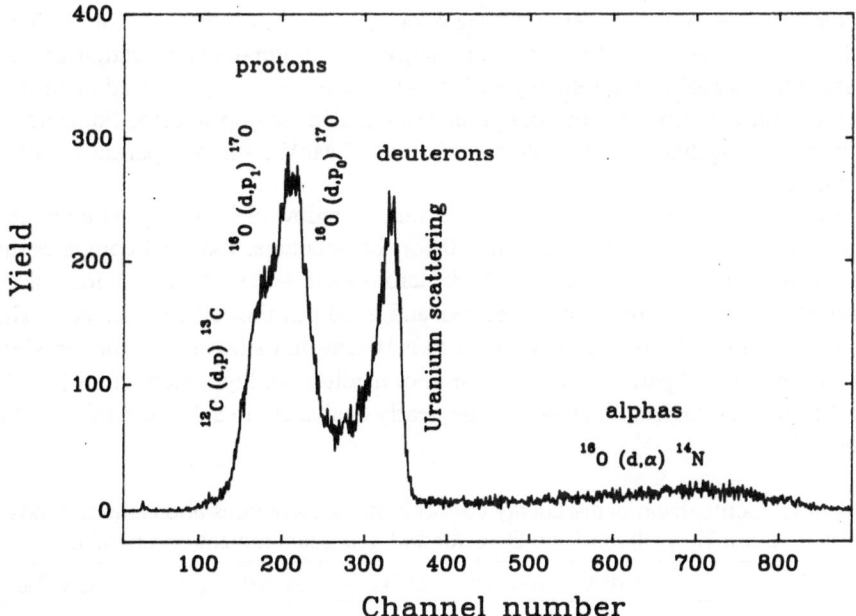

Figure 8.3. Energy loss spectrum from a 17-μm ΔE silicon detector obtained for 2.2-MeV deuteron beam irradiation of a thick UO_2 target leached in carbonated groundwater. (Data from Ref. 8)

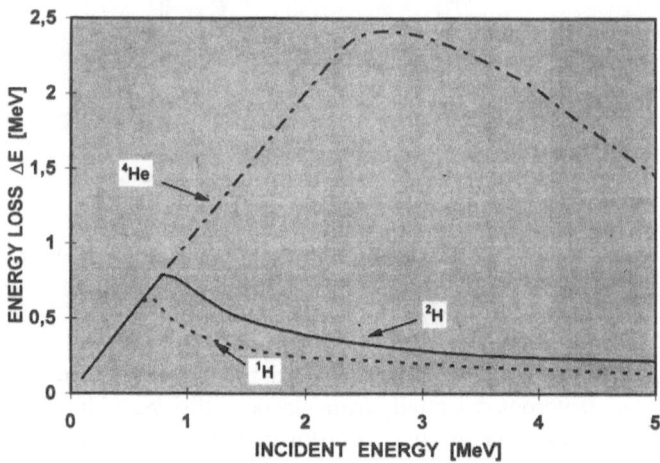

Figure 8.4. Variation of the energy loss in a 10-μm thin solid detector for protons, deuterons, and alpha particles up to 5 MeV.

Figure 8.4 shows the energy loss in a thin ΔE detector (10 μm) versus the incident energy for protons, deuterons, and alpha particles up to 5 MeV. From Figure 8.4 we can determine that the minimum incident energy for traversing the 10-μm-thick detector is 0.73 MeV for protons, 0.87 MeV for deuterons and 2.56 MeV for ^{4}He ions. As shown in Figure 8.4, the threshold energy for helium-4 identification can be decreased to 0.8 MeV. At this energy helium-4 ions are completely stopped in the thin detector, while deuterons and protons produce a signal in the stop detector. For incident energy values less than the threshold for protons (0.7 MeV), different particles cannot be distinguished.

A classic ionization chamber has an entrance window of 0.3-μm plastic foil and a length of about 10 cm; it is filled with 50 Torr of isobutane. Table 8.1 compares the energy loss of 3-MeV protons, 2-MeV deuterons, and 4-MeV helium-4 ions in an ionization chamber as previously defined and a 10-μm-thin silicon detector. The ionization chamber is roughly equivalent to a 10-μm-thin silicon detector for MeV light ions in terms of particle energy losses. For medium- or high-energy ions (A > 5, a few MeV/amu), ionization chambers are nearly equivalent to a 2–3 μm thin silicon detector.[9]

Table 8.1. Comparison of the Energy Loss of Protons, Deuterons and Helium-4 Ions in a 10-μm Thin Silicon Solid Detector and in a Usual Ionization Chamber.

	ΔE (keV) 3-MeV ^{1}H$^+$	ΔE (keV) 2-MeV ^{2}H$^+$	ΔE (keV) 4-MeV ^{4}He$^+$
0.3-μm Plastic foil	5	11	39
10-μm Si	205	443	1930
500-Torr cm isobutane	239	581	2464

8.3. PERFORMANCES

8.3.1. Mass and Charge Separation

The mass and charge separation power of a telescope can roughly be described by Eq. 8.1 or 8.2:

$$\frac{\Delta M}{M} = \left(\left(\frac{\Delta E_R}{E_R} \right)^2 + \left(\frac{\Delta dE}{dE} \right)^2 + \left(\frac{\Delta dx}{dx} \right)^2 \right)^{1/2} \tag{8.1}$$

$$\frac{\Delta Z}{Z} = \frac{1}{2} \left(\left(\frac{\Delta E_R}{E_R} \right)^2 + \left(\frac{\Delta dE}{dE} \right)^2 + \left(\frac{\Delta dx}{dx} \right)^2 \right)^{1/2} \tag{8.2}$$

where the last two terms are derived from the stopping power measured through the energy loss dE in a thin detector of thickness dx. At high energy the Bethe–Bloch relation giving the energy loss dE of an incident ion (energy E) passing through a thin target (dx) of a material with mass M and charge Z results in dE approximately proportional to $M Z^2 dx/E$. Assuming a precision of about 1 % for energy loss and residual energy measurements and a precision of about 10 % on the ΔE detector thickness evaluation, mass separation for a given atomic number Z can be optimized down to 1 amu up to M = 8. For a given mass M, the charge separation can be optimized down to $\Delta Z = 1$ up to Z = 16.

Figure 8.5 illustrates the mass and charge separation capability of a ΔE-E solid-state telescope (17 μm, 1500 μm) for a classic NRA measurement used in investigating carbon, nitrogen, and oxygen in a zirconium alloy sample by applying a 2.2-MeV deuteron microbeam.[10] Each particle contribution appears as a parabolic cloud. These clouds are vertically displayed by increasing order of atomic mass and horizontally separated by increasing order of emitted energy. Six individual groups of particles are discriminated: ^{16}O(d, p), ^{12}C(d, p_0), ^{14}N(d, p_0), scattered deuterons, ^{14}N(d, α_1), and ^{14}N(d, α_0). Figure 8.5b shows the projection on the residual energy axis of the ^{14}N(d, α_1) and ^{14}N(d, α_0) groups as yield versus energy spectra.

The analog-digital converter that drives the signals of both detectors is generally activated by pulses from the E detector because in the case of hydrogen investigations, recoil atoms do not lose enough energy in the ΔE detector to generate measurable pulses. Moreover numerous particles having an energy too small to enter the E detector do not contribute to system dead time.[11]

8.3.2. Depth Resolution

In ERDA the presence of an absorber foil in front of the silicon surface barrier detector induces a systematic energy loss for each recoil atom. Moreover the recoil energy distribution is broadened by energy straggling through the absorber foil. In comparison the presence of a thin ΔE detector instead of the absorber foil and its association with a thick detector allow us to obtain simultaneously the energy loss of

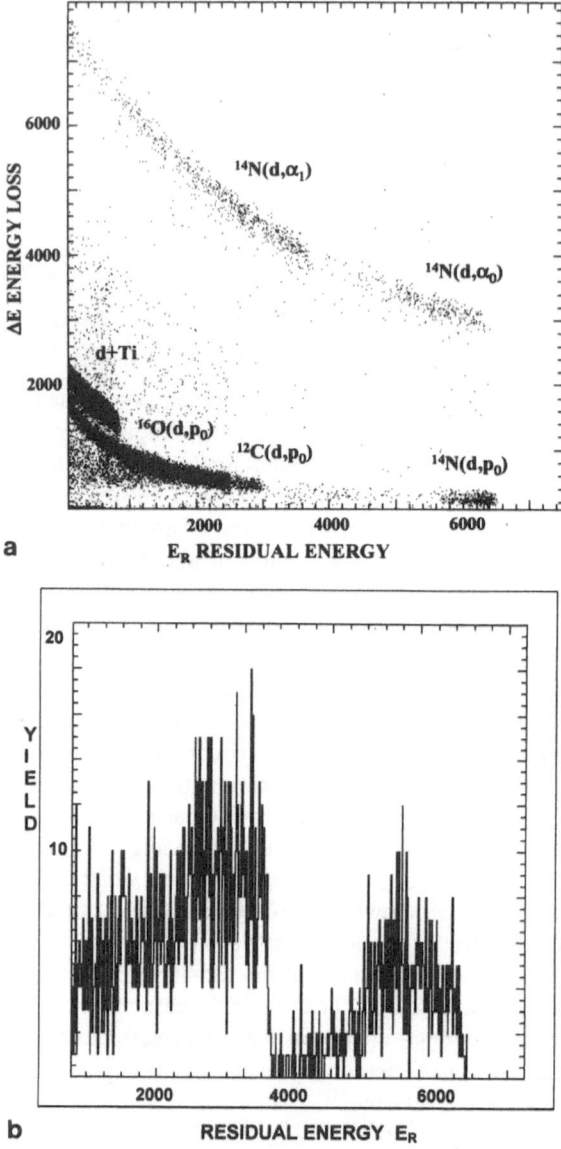

Figure 8.5. (a) Two-dimensional spectrum ($\Delta E = f(E_R)$) for a Zr alloy sample investigated by a 2.2-MeV deuteron microbeam: protons and alpha particles emitted consecutively to (d,p) and (d,α) nuclear reactions on C, N, and O; scattered deuterons are easily discriminated. (b) Alpha particle yield obtained by projection on the residual energy axis in the interval 1000–7000 keV. (Data from Ref. 10)

the recoil atom (its atomic number) and its residual energy E_R. These values can be summed to deduce total energy in the recoil atom. This summation $\Delta E + E_R$ permits us to eliminate the contribution from straggling in the ΔE detector (except for straggling occurring in the dead layers of both detectors). Eliminating straggling allows us to improve depth resolution. Additional effects of spectral overlap between neighboring elements or the background signal ratio have not been reported.

Much attention must be paid to the choice of the thick detector. The active area of the thick detector must be adjusted to take into account angular dispersion caused by the thin detector and to avoid counting losses. The presence of nonlinear response zones near the edge of the detector may also play a role in this particular configuration. Another parameter must be considered: the thickness of its dead layer. In the case of heavy-ion recoil spectrometry, if recoil atoms lose the major part of their energy in the thin detector, they may be stopped near the end of the dead layer of the thick detector, so straggling effects may strongly disturb data collection in this case. Finally detectors should be placed as close to each other as possible, and their respective dead layers should be as thin as possible.

Depth resolution in telescope ERDA measurement can be experimentally assessed by investigating a thin oxide layer coated on a thick substrate, as described by Arnoldbik and Habraken.[12] For a 50-nm Si_3N_4 layer on c-Si bombarded with a 43-MeV ^{63}Cu beam (recoil angle = 37°, angle of incidence = 33°), researchers showed that it is possible to discriminate oxygen contributions from native oxide layers on the surface and at the interface (see also Figure 8.8). This separation reveals a depth resolution better than 25 nm. Depth resolution can be improved considerably by reducing the acceptance angle of the telescope to a few tenths of a degree by means of a suitable diaphragm, of course such an improvement implies a loss of sensitivity.

8.3.3. Sensitivity

The detection limit of the method is illustrated by the oxygen contamination level of a freshly prepared thin layer of amorphous silicon, described as in Ref. 11. Using the same experimental set-up as previously described, the researcher found an oxygen limit of about 50 at. ppm. This value is probably limited by a background yield due to the oxygen content present in the slits and sample holder. Detection limits for other elements, such as boron, carbon, and nitrogen, appear at least an order of magnitude lower. These detection limits are probably governed exclusively by the degradation of the sample during ion bombardment.

In opposition to conventional ERDA spectrometry, where elemental contents can be deduced only by subtracting the unknown background yield from each contribution, the influence of the background on the minimum detectable content in telescope measurement is minimized by selecting suitable ΔE window on the two-dimensional elemental spectrum.[11] The final limitation factor in lowering the detection limit is probably the degradation of the sample during analysis, which limits the total allowable ion dose (see Chapter 14).

The general method used for telescope data processing, essentially interpreting (ΔE,E) matrices, presents the same characteristics as those discussed for TOF spectrometry in Chapter 6.

8.4. EXAMPLES

In this section we discuss only a small number of examples of using a telescope for ERDA measurements in MeV ^{4}He$^+$ recoil spectrometry and HI-ERDA. Chapters 6 and 13 contain several other examples of the analytical capabilities of ΔE-E telescopes.

First of all a ΔE-E telescope can be used for hydrogen determination by ERDA. Recently Ref. 13 reported data on the bombardment of titanium hydride samples with 4 MeV ^{4}He$^+$ ions, at the National Accelerator Centre in Faure (South Africa). Using a 13.6-μm silicon transmission detector, these authors showed that it is possible to separate hydrogen, deuterium, and tritium contributions from the total ERDA spectrum. Figure 8.6 illustrates this separation for an accumulated charge of 30 μC.

The ΔE-E telescope devices have been used essentially for heavy-ion ERDA, since the beginning of the nineties.[5–7,11,12,14–16] In a recent review paper, Habraken emphasized the use of telescope detectors for light-element depth profiling in solids using heavy-ion beams.[14] He also noticed some perturbing effects when applying silicon detectors in telescope devices to detect heavy ions and particularly the energy resolution degradation and the increase in pulse height defects for increasing Z in particles to be measured. In the following example heavy-ion beams (^{28}Si, ^{63}Cu, and ^{107}Ag) are used to characterize silicon nitride layers deposited on silicon substrates.[12] Figure 8.7 compares the ERDA spectra for a 180-nm-thick layer by using conventional recoil spectrometry with a Mylar absorber foil (30-MeV ^{28}Si beam) and telescope recoil spectrometry (78-MeV ^{107}Ag beam). Figure 8.8 illustrates the analysis of a 50-nm-thick silicon nitride layer. These results clearly show the ability of the method to discriminate successive oxide layers formed at the nitride surface and the silicon interface, respectively.

The RBS and recoil spectrometry were used to examine the depth distribution of metallization contacts on GaAs in Ref. 15. A recoil detector telescope combined with a TOF spectrometer was attached to a beam line of the tandem accelerator ANTARES at ANSTO Lucas Heights Research Laboratories in Australia. A 77-MeV ^{127}I^{10+} beam was employed to analyze GaAs samples with thin film overlayers: Si (220 nm)/Co (50 nm)/<100>GaAs. Data were collected until approximately 10^5 events were recorded, which was sufficient for quantitative evaluation. The analysis duration varied from 15 to 100 min, depending on the beam current (typically around 10 nA). The effect of annealing in the range 300–600 °C was investigated. Data showed that CoSi$_2$ formed during annealing at and above 500 °C with no detectable reaction between the GaAs substrate and the CoSi$_2$ overlayer.[15]

Using a combined gas–solid telescope, Assmann found that it was possible to determine elemental distributions in TiN$_x$O$_y$ films (50–100 nm) evaporated on a

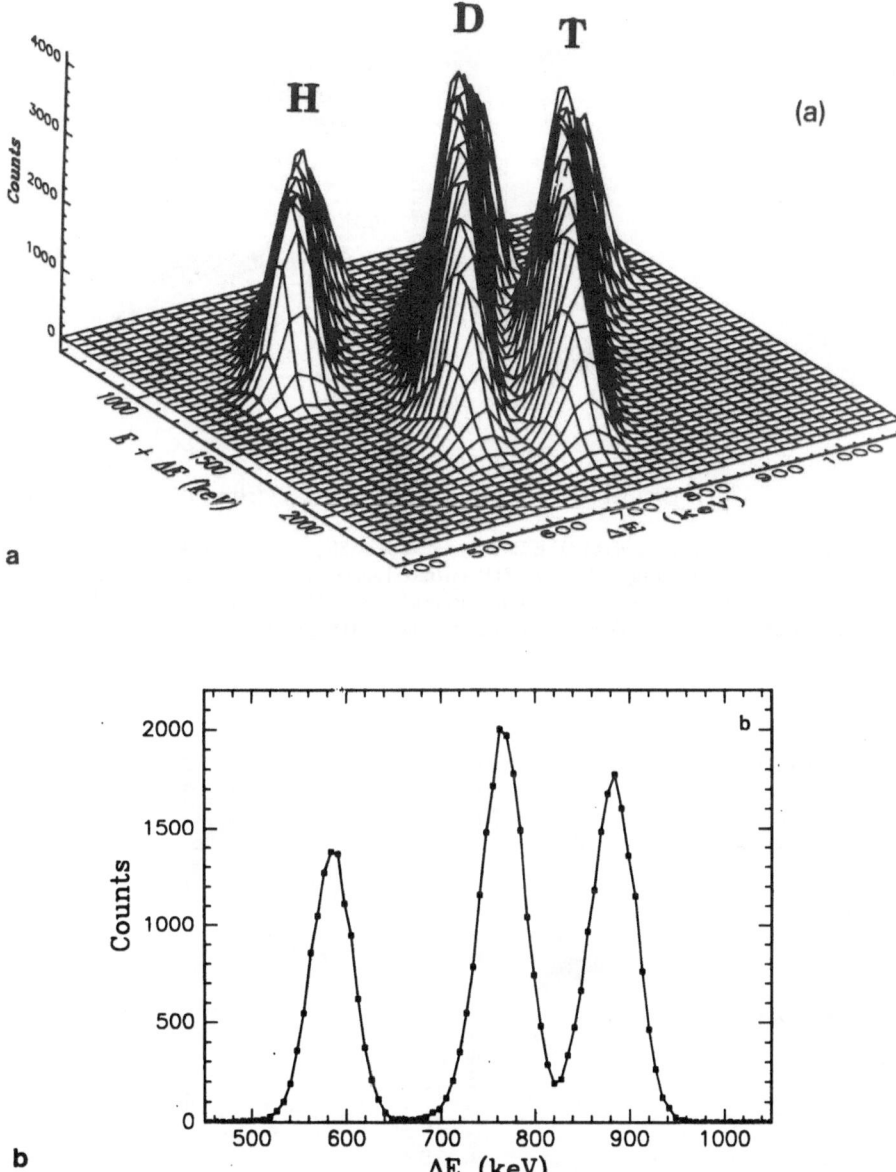

Figure 8.6. (a) Three-dimensional plot of hydrogen, deuterium, and tritium distributions in a titanium hydride sample bombarded with 4-MeV helium ions. The *x* and *y* axes represent total energies (sum of ΔE and E) of recoils and the ΔE signals, respectively. (b) Two-dimensional plot obtained by cutting the three-dimensional diagram in the interval of 470–510 keV. (Data from Ref. 13)

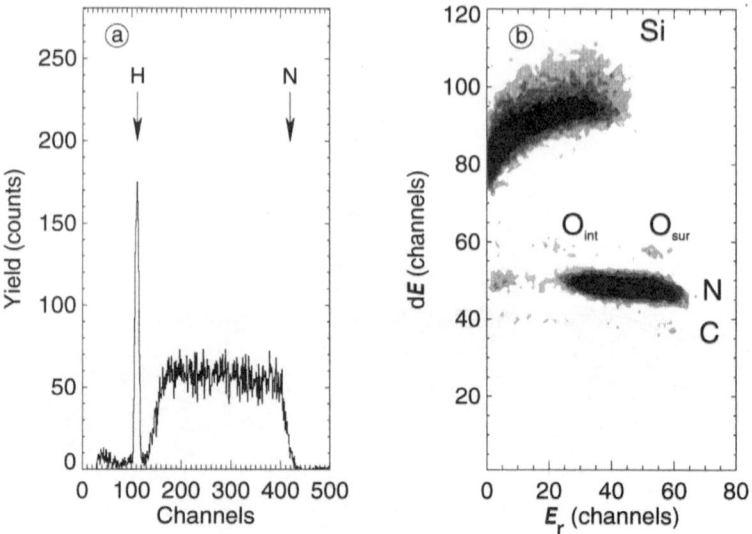

Figure 8.7. (a) Conventional and (b) ΔE-E ERDA spectra of a 180-nm Si_3N_4 sample. The conventional ERDA spectrum is measured using a 30-MeV ^{28}Si beam (recoil angle = 45°, angle of incidence = 28°, Mylar thickness = 9 μm). The ΔE-E_R spectrum is measured using a 78-MeV ^{107}Ag beam (recoil angle = 37°, angle of incidence = 25°). O_{sur} and O_{int} denote two native oxide layers. (Data from Ref. 12)

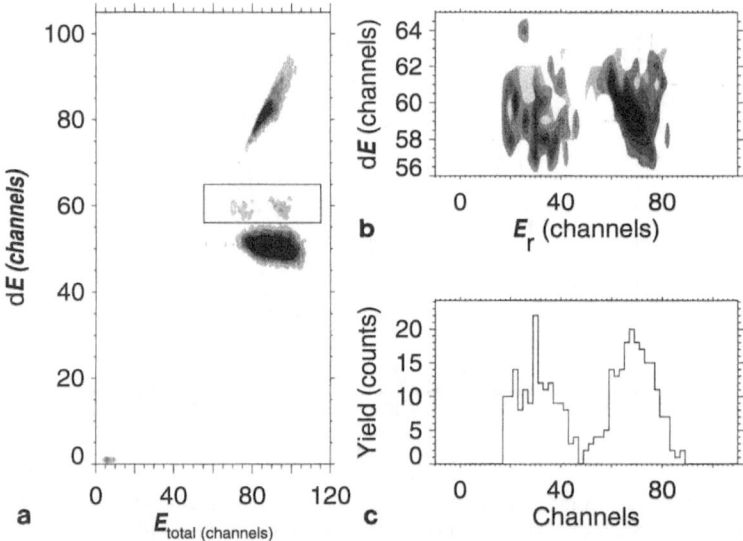

Figure 8.8. (a) A ΔE-E ERDA spectrum of a 50-nm Si_3N_4 sample. This spectrum is measured with a 43-MeV ^{63}Cu beam (recoil angle = 37°, angle of incidence = 33°). (b) Plots show the region that can be ascribed to oxygen and (c) the corresponding projection on the x-axis. (Data from Ref. 11)

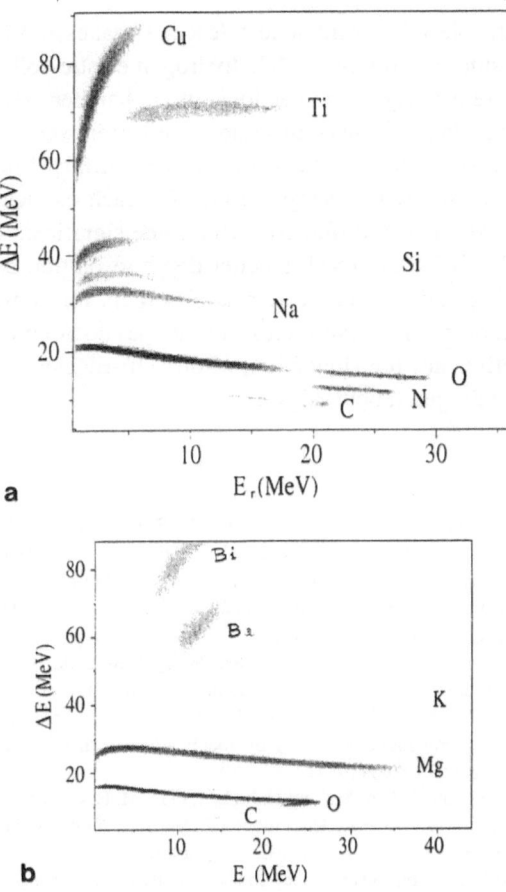

Figure 8.9. The (ΔE, E) matrix of ERDA with a 170-MeV ^{127}I beam using a large-area ionization chamber: (a) on a TiN$_x$O$_y$ film deposited on a Cu coated glass, (b) on a BaBiKO film coated on MgO. (Data from Ref. 17)

Cu-coated glass substrate by using 170-MeV ^{127}I ions. The Z separation power of this system was adapted to discriminate C, N, and O from the thin film; Na, Si, and Ca from the glass; Ti and Cu as shown in Figure 8.9a.[17] Another example was given in the same work by Assmann and coworkers concerning the analysis of a BaBiKO film coated on a MgO substrate (Figure 8.9b).[17]

8.5. CONCLUSION

The use of a ΔE-E transmission telescope for ERDA appears to be a very elegant way of partially resolving the mass–depth and recoil–projectile ambiguities. This type of detection is suitable for low- or high-energy helium beams or HI-ERDA.

The basic principle of a solid-state telescope makes it very easy to use for hydrogen isotopes studies through the ^{4}He/hydrogen elastic collision. In the case of HI-ERDA, the low sensitivity of the gas ionization chamber associated with a thick silicon surface barrier detector, make this telescope system very suitable.

Telescopes and TOF spectrometers provide very similar biparametric data that can be converted into separated energy spectra for each element in a similar way, although measured parameters differ. However some significant differences have to be considered: TOF-ERDA can reach a better depth resolution because carbon foils induce less straggling and angular scatter as absorbers for conventional ERDA or transmission silicon detectors. Nevertheless, telescopes do not present the problem of variable detection efficiency for a low Z ion, while TOF-ERDA in particular has a very low efficiency for hydrogen (about 30 %).

REFERENCES

1. Stoquert, J. P., Guillaume, G., Hage-Ali, M., Grob, J. J., Ganter, C., and Siffert, P., Determination of concentration profiles by elastic recoil detection with a ΔE-E gas telescope and high-energy incident heavy ions, *Nucl. Instrum. Methods Phys. Res. Sect. B* **44**, 184 (1989).
2. Yu, R., and Gustafsson, T., Determination of the abundance of adsorbed light atoms on a surface using recoil scattering, *Surf. Sci.* **177**, L987 (1986).
3. Knoll, G. F., *Radiation Detection and Measurement* (Wiley, New York, 1979), pp. 406–9.
4. Petrascu, M., Berceanu, I., Brancus, I., Buta, A., Duma, M., Grama, C., Lazar, I., Mihai, I., Petrovici, M., Simion, V., Mihaila, M., and Ghita, I., A method for analysis and profiling of boron, carbon, and oxygen impurities in semiconductor wafers by recoil atoms in heavy-ion beams, *Nucl. Instrum. Methods Phys. Res. Sect. B* **4**, 396 (1984).
5. Arnoldbik, W. M., de Laat, C. T. A. M., and Habraken, F. H. M. P., On the use of a dE-E telescope in elastic recoil detection, *Nucl. Instrum. Methods Phys. Res. Sect. B* **64**, 832 (1992).
6. Whitlow, H. J., Possnert, G., and Petersson, C. S., Quantitative mass- and energy-dispersive elastic recoil spectrometry: Resolution and efficiency considerations, *Nucl. Instrum. Methods Phys. Res. Sect. B* **27**, 448 (1987).
7. Arai, E., Funaki, H., Katayama, M., and Shimizu, K., TOF-ERD experiments using a 10-MeV ^{35}Cl beam, *Nucl. Instrum. Methods Phys. Res. Sect. B* **68**, 202 (1992).
8. Trocellier, P., Gallien, J. P., and Cachoir, C., Alteration mechanisms of uranium dioxide by granitic groundwater, communication to Migration'95 Conference, (Saint-Malo, France, 10–15 Sept. 1995).
9. Ziegler, J. F., *Stopping Powers and Ranges in all Elements* (Pergamon, New York, 1977).
10. Berthier, B., personal communication, 1995.
11. Arnoldbik, W. M., Elastic recoil detection and hydrogen chemistry in silicon oxynitrides, thesis, University of Utrecht, 1992.
12. Arnoldbik, W. M., and Habraken, F. H. M. P., Elastic recoil detection, *Rep. Prog. Phys.* **56**, 859 (1993).
13. Prozesky, V. M., Churms, C. L., Pilcher, J. V., Springhorn, K. A., and Behrisch, R., ERDA measurement of hydrogen isotopes with a ΔE-E telescope, *Nucl. Instrum. Methods Phys. Res. Sect. B* **84**, 373 (1994).
14. Habraken, F. H. M. P., Light-element depth profiling using elastic recoil detection, *Nucl. Instrum. Methods Phys. Res. Sect. B* **68**, 181 (1992).
15. Hult, M., Whitlow, H. J., Östling, M., Lundberg, N., Zaring, C., Cohen, D. D., Dytlewski, N., Johnston, P. N., and Walker, S. C., RBS and recoil spectrometry analysis of CoSi$_2$ formation on GaAs, *Nucl. Instrum. Methods Phys. Res. Sect. B* **85**, 916 (1994).

16. Assmann, W., Ionization chambers for materials analysis with heavy-ion beams, *Nucl. Instrum. Methods Phys. Res. Sect. B* **64**, 267 (1992).

17. Assmann, W., Huber, H., Steinhausen, Ch., Dobler, M., Glückler, H., and Weidinger, A., Elastic recoil detection analysis with heavy ions, *Nucl. Instr. Meth. Phys. Res. Sect. B* **89**, 131 (1994).

18. ...

19. ...

9

Coincidence Techniques

With Hans Hofsäss

9.1. INTRODUCTION

Conventional ERDA techniques provide the possibility of multiple-element analysis, in particular for profiling different hydrogen isotopes,[1] and a rather good depth resolution down to 10 nm [2] (see Chapter 4). Problems may arise mainly because of a glancing angle of incidence: Probing depth is limited to less than 1 μm; depth resolution is influenced by surface roughness; and a slight error in the recoil angle can lead to a relatively large error in identifying the depth. The latter problem can be solved by a double-detector method for a more precise determination of the recoil angle.[3] Because only recoiled particles are usually detected, intrinsic difficulties arise in conventional ERDA:

- A reasonable depth resolution is obtained only for small-detector solid angles, thereby limiting the sensitivity of the technique.
- The recoil mass and depth where a scattering event took place cannot be determined unmistakably; we called this the *mass–depth ambiguity* (Section 5.2).
- In some cases it is difficult to determine whether a recoiled target atom or a scattered projectile were detected; we *called this the projectile–recoil ambiguity*.

For free-standing thin-film samples and an experimental set-up using transmission geometry, these intrinsic difficulties can be solved by coincidence techniques (Figure. 9.1).

In the first ERDA experiments performed by Cohen *et al.* in 1972,[4] Smidt and Pieper in 1974,[5] Moore *et al.* in 1975,[6] and L'Ecuyer *et al.* in 1976,[7] transmission geometry (Section 5.4) was used to measure the depth distribution of light elements in thin-film samples (and coincident detection of scattered and recoiled particles was already applied to achieve mass selectivity, i.e., to avoid the mass–depth ambiguity, reduce background counts, and improve the detection limit). Coincidence techniques

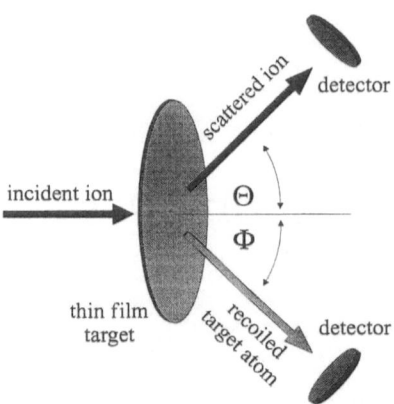

Figure 9.1. Experimental set-up for ERDA using the transmission geometry and coincident detection of scattered and recoiled particles.

were also applied to deuterium depth profiling by NRA using the $D(^3He,\alpha)H$ reaction.[8] In this case measuring alpha particles in coincidence with high-energy protons allows a reduction of the background arising from the Rutherford scattering of ^3He and other reaction products.

The problem of both mass–depth and projectile–recoil ambiguities discussed in Section 5.2 can be overcome by coincident detection of scattered and recoiled particles. Other limitations of conventional ERDA, such as sensitivity, can also be improved by using coincidence techniques. By detecting both the recoiled target atom as well as the scattered projectile, complete information about a scattering event, (i.e., E_1, E_2, θ, and ϕ) is obtained, so that the recoil mass M_2 and the depth can easily be determined. To assign a recoiled target atom and a scattered ion to the same scattering event, it is necessary to detect both particles in time coincidence. This is possible only if at least one of the particles involved in a scattering event is transmitted through the sample without significant energy loss, energy straggling, or angular straggling. *Coincidence techniques in ERDA therefore require a transmission geometry, and they can be applied only to free-standing, sufficiently thin films typically of a few µm thickness.* Analytic capabilities of ERDA in transmission geometry were discussed in Section 5.4, additional aspects of transmission ERDA that must be considered for coincidence techniques are treated in Section 9.2.

Several modifications of the ERDA technique, which make use of time coincidence methods, were developed, and these are referred to in the literature as proton–proton scattering,[4,9] coincident elastic recoil detection analysis (CERDA),[10–12] scattering recoil coincidence spectroscopy (SRCS),[13–16] or elastic recoil coincidence spectroscopy (ERCS).[17,18] These modifications of ERDA can be divided into two groups of coincidence techniques.

In the first group, which we denote as CERDA, are all those coincidence techniques using Eq. 2.6b, by *properly adjusting* θ *and* φ, to achieve mass selectivity and reduce background signals.[6,10,11,19,20] Depth information is then derived from the total energy $E_{total} = E_2 + E_1$. Since for elastic scattering the sum of the energies of scattered and recoiled particles equals projectile energy before the scattering event, a variation of E_{total} is due only to the energy loss particles experience while passing through the target material.

If $M_1 = M_2$, i.e., for proton–proton[4,9] or alpha–alpha scattering,[5] it follows from Eq. 2.6b that the sum of the scattering and recoil angle is exactly θ + φ = 90°. For incident projectiles deflected by 45°, recoiled target atoms must emerge at 45° on the opposite side of the beam and in the scattering plane. Positioning the two detectors in such a way, we end up with a technique absolutely specific to one single element, which is identical to incoming projectiles. The few CERDA experiments of this type are discussed in Section 9.3.

For the general case $M_1 \neq M_2$, the sum of the scattering and recoil angle deviates from 90°. For a given projectile mass M_1, detector positions can be adjusted so that one is selective to a specific recoil mass M_2. This is the basis of multiple-element detection using CERDA; it is described in Section 9.4.

Coincidence ERDA techniques of the second group (Sections 9.5–9.6), denoted as SRCS or ERCS, were first proposed by Chu and Wu.[15,16] The SRCS was initially proposed as a technique for efficiently profiling hydrogen using an incident ^{4}He ion beam.[13,14] The ERCS is an extension of the SRCS technique that can be applied to simultaneously depth profiling heavier light elements, such as carbon, oxygen, or nitrogen.[18] The key idea of SRCS and ERCS is *the elimination of* θ *and* φ *measurements* by measuring E_1 and E_2 in coincidence; its major advantage lies in using large solid angle detectors without sacrificing depth resolution.

Improvements in the sensitivity, depth resolution, and mass selectivity of coincidence ERDA techniques can be accomplished with position-sensitive detectors. In Section 9.7 we briefly describe state-of-the art position-sensitive detectors and their possible application for coincidence ERDA measurements.

9.2. TRANSMISSION GEOMETRY AND COINCIDENCE TECHNIQUES

Coincidence ERDA techniques require transmission geometry, which means that at least one of the particles involved in the scattering process, either the scattered projectile or the recoiled target atom, must be detected in the forward direction. Therefore we must consider scattering kinematics for the case when either scattering angles or recoil angles or both are in the range of 0°–90°. To complement to Section 5.4, we add a few points that may be relevant for the coincident detection of scattered and recoiled particles.

First let us distinguish between the two cases $M_1 < M_2$ and $M_1 > M_2$. For scattering heavier projectiles to lighter target atoms, there exists a maximum forward-scattering angle given by Eq. 2.7. For example when profiling hydrogen or deuterium with incident ^{4}He projectiles using the SRCS technique (see Section 9.5), the maximum scattering angle is $\theta_{max} = 14.47°$ for H and $\theta_{max} = 30°$ for D. This requires a detector positioned almost in the forward direction, which can easily be flooded by projectiles scattered at heavier target atoms because of a large scattering cross-section for small θ. On the other hand, a detector positioned at an angle $\theta > \theta_{max}$ can detect only light recoils, which enables discrimination between scattered and recoiled particles. If projectiles of mass M_1 are scattered at angles $\theta < \theta_{max}$, then recoils appear at two different recoil angles and with two different recoil energies. The projectile scattering cross section (Eq. 3.9) for angles approaching θ_{max} goes to infinity within a narrow range of $\Delta\theta < 0.01°$. This very sharp singularity is a result of the transformation (see Eq. 3.6) of differential cross section from the center of mass system to the laboratory system.[21] In practice this singularity is not important and has not yet been measured. However we may think of using it as a precision alignment of a detector in the forward direction with respect to the beam or to measure beam divergence. In most of coincidence ERDA experiments carried out so far, projectiles with $M_1 \leq M_2$ were used for several reasons.[4-10,19] In this case a maximum scattering angle θ_{max} does not exist, which gives greater flexibility in choosing the scattering recoil geometry.

Secondly, for $M_1 = M_2$, we find that the mass resolution $\delta E_2/\delta M_2$ (or $\delta K'/\delta M_2$) is zero. This situation is indicated in Figure 5.8. For small masses, i.e., for proton–proton or ^{4}He-^{4}He-scattering, this is not a severe problem, since M_2 changes in discrete steps of $\delta M_2 = 1$ and recoil energies of neighboring masses are therefore still well-separated. Identifying these scattering events is however preferably done by coincident detection of both particles, taking advantage of the fact that the sum of scattering and recoil angles is exactly $\theta + \phi = 90°$[4,5,9] (see Section 9.3).

As a third point we consider the Rutherford recoil cross section (Eq. 3.8) in the laboratory system. For increasing recoil angles, $(d\sigma/d\Omega)_\phi$ increases rapidly. Recoil angles of practical interest for most coincidence techniques are in the range of $30°–70°$, for which $(d\sigma/d\Omega)_\phi$ varies by more than an order of magnitude. The same problem arises when forward-scattered and forward-recoiled particles are detected in coincidence, which is the case for ERCS and SRCS, since the scattering cross section $(d\sigma/d\Omega)_\theta$ also varies strongly with the scattering angle. As a consequence the quantitative analysis of coincidence ERDA measurements requires a precise knowledge of detector positions and solid angles and a careful scattering yield calculation[18] (see also Section 9.6.3).

9.3. SINGLE-ELEMENT ANALYSIS WITH CERDA

As one of the first ERDA measurements described in the literature, Cohen *et al.* present proton–proton scattering as a nuclear-scattering technique that is absolutely specific to hydrogen, and it has an extremely low detection limit of about 1 at. ppm.[4] In this work a 17-MeV proton beam was incident perpendicular to the surface of a free-standing hydrogen-containing thin film. Scattered protons and recoiled target hydrogen atoms were detected in time coincidence by two detectors behind the sample, covering solid angles of 3.3×10^{-3} sr, positioned in a plane at angles of +45° and –45° with respect to the direction of the beam. Different metal foils as well as Teflon and a thin carbon sheet were analyzed. The coincidence ERDA spectrum of an iron foil is shown in Figure 9.2. Two peaks around channels 200 and 240 arise from hydrogen on the surfaces with a concentration of 10^{18} cm^{-2} on Side A and 2.5×10^{17} cm^{-2} on Side

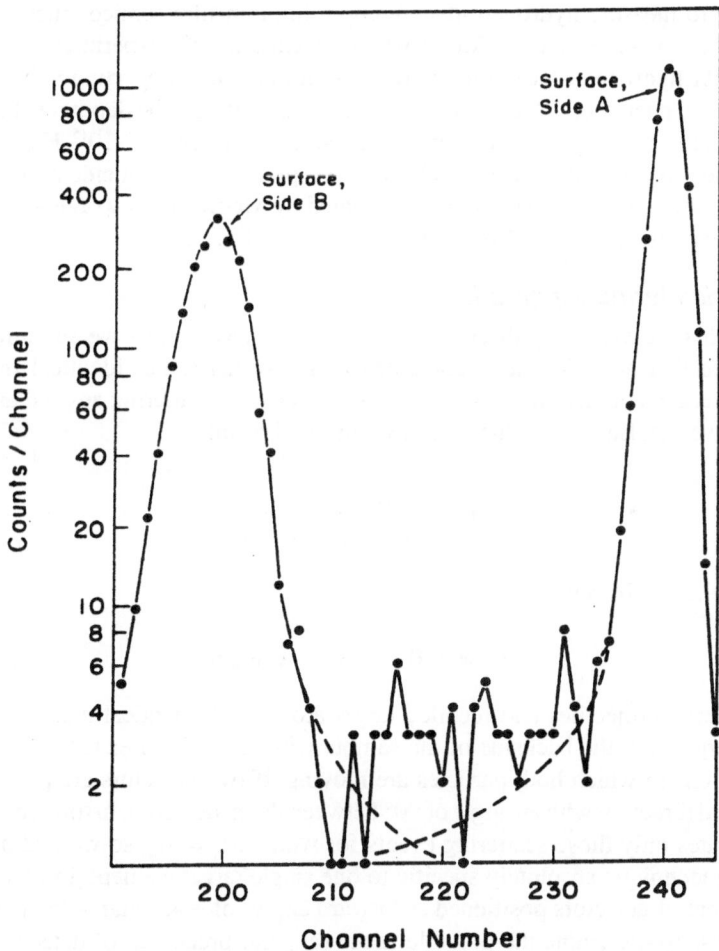

Figure 9.2.   Total energy CERDA spectrum of hydrogen in a 50.8-μm (2-mil) iron foil. Side A is the back surface of the sample and Side B the front surface where the beam is incident. Channels correspond to the total energy of scattered and recoiled hydrogen. (Data from Ref. 4)

B. The broader hydrogen peak from Surface B is caused by MS events (see Section 9.3.3). From the scattering events between the two surface hydrogen peaks, the residual bulk hydrogen concentration of the iron foil can be determined. The estimated concentration of 10 ppm proves the extreme sensitivity of the technique.

Only a few more attempts were made to use this technique for measuring light-element concentration profiles with high sensitivity. Smidt and Pieper applied α-α scattering to measure ^4He concentration profiles in metal thin films.[5] From scattering 48.5-MeV α particles at a 33-μm-thick ^4He-implanted vanadium foil, the peak concentration of 64 ppm, determined from the implanted He dose and energy, could easily be detected. Willemsen et al.[9] explores the applicability of proton-proton scattering to measure hydrogen in self-supporting thin-film semiconductor samples using rather low-energy (1–2 MeV) proton beams. In this experiment researchers analyzed the hydrogen concentration in a 200-nm-thick self-supporting Si_3N_4 film. For a primary ion energy of 2 MeV and a scattering angle of 45°, the depth resolution was about 70 nm, and the detection limit was estimated to be 3×10^{15} H/cm^2, which corresponds to about 0.1 at.%. Finally oxygen profiling by coincident detection in ^{16}O-^{16}O-scattering experiments was investigated by Moore and coworkers as part of a series of experiments for multielement analysis with CERDA.[6]

9.3.1. Scattering Kinematics

Kinematics were fully discussed in Chapter 2, nevertheless we add a few comments here that are relevant to coincidence spectrometry. For identical masses of projectile and recoil, i.e., $M_1 = M_2$, the relation between scattering angle θ and recoil angle ϕ given by Eq. 2.6b reduces to the simplified form:

$$\tan \theta = \frac{\sin (2\phi)}{1 - \cos (2\phi)} = \frac{1}{\tan \phi} \tag{9.1}$$

Equation 9.1 can be rewritten as:

$$\cos (\theta + \phi) = 0 \qquad \text{or} \qquad \theta + \phi = 90° \tag{9.2}$$

Scattered projectiles and recoiled target atoms with identical masses $M_1 = M_2$ must emerge from the backside of the sample with an angle of exactly 90° between the directions in which both particles are moving. If two detectors are positioned in the forward direction with an angle of 90° between them, detection in time coincidence discriminates only those scattering events for which $M_1 = M_2$, so we end up with a scattering technique absolutely specific to one single target element. In particular an arrangement of detectors positioned at forward angles of + 45° and – 45° has several advantages. To determine the possible range (angular precision) of detection angles for which only one single element is detected, we must use Eq. 2.6b, this is shown in Figure 9.3 for four different projectiles. We see that for proton-proton scattering or alpha-alpha scattering, rather large solid angle detectors can be used. For example a

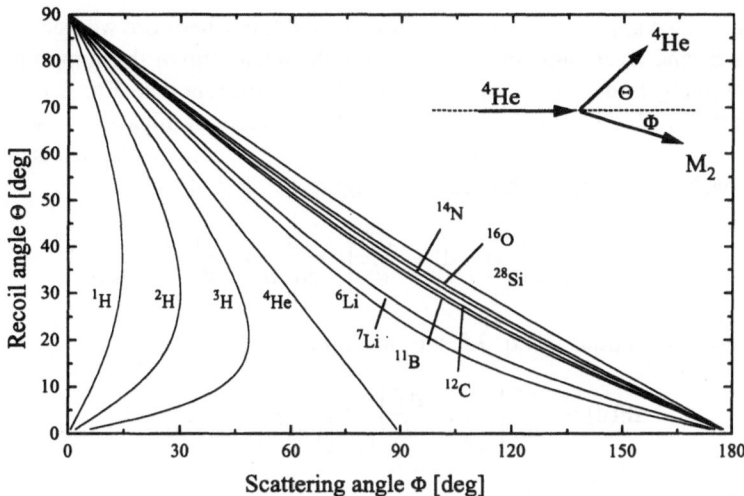

Figure 9.3. A complement to Figure 2.3, this figure shows the relation between scattering and recoil angles θ and φ, calculated for ^{1}H, ^{4}He, ^{7}Li, ^{12}C projectiles. Arrows indicate the angular precision needed to detect only recoil atoms having the same mass as the projectiles if detectors are positioned at forward angles of ± 45°.

circular-shaped detector covering an angular range 45 ± 5° provides a solid angle of 24 msr.

Energies of scattered and recoiled particles are derived from kinematic factors K and K′ given by Eqs. 2.10 and 2.12. For $M_1 = M_2$ we find the simple results:

$$K = \cos^2 \theta$$

$$K' = \cos^2 \phi = \sin^2 \theta \qquad (9.3)$$

Since $\theta + \phi = 90°$ we can replace φ by θ using $\cos \phi = \sin \theta$. For the special case of $\theta = \phi = 45°$, we obtain $K = K' = 0.5$, so that after the scattering event both the scattered and the recoiled particles have an energy:

$$E_1 = E_2 = \frac{E_0}{2} \qquad (9.4)$$

If energy loss is ignored, the detected total energy $E_{total} = E_1 + E_2$ is constant and equal to the energy E_0 of the incident projectile. The total energy that is actually measured is less than E_0 by the energy loss ΔE of the particles as they traverse the thin film.

To estimate ΔE as a function of the position in the target where the scattering event takes place, we assume that the total energy loss is small compared to the ion energies, so we can use the approximation $dE/dx|_E \approx$ constant. Furthermore we assume that E_0 as well as E_1 and E_2 are high enough (far from the maximum of the stopping power), so that stopping is described by the Bethe–Bloch formula (see Eq. 2.17) and the approximation $dE/dx|_E \propto E^{-1}$ can be applied. For a target of thickness t, we can now derive an expression for the position dependence $\Delta E(d)$ where the depth d is measured from the back surface of the thin film (see Figure 1 in Ref. 4 for details).

$$\Delta E(d) \approx \left.\frac{dE}{dx}\right|_{E_0} (t-d) + \left.\frac{dE}{dx}\right|_{E_1} \frac{d}{\cos\theta} + \left.\frac{dE}{dx}\right|_{E_2} \frac{d}{\cos\phi} \qquad (9.5)$$

With the approximations in Eq. 9.5, we can express ΔE as:

$$\Delta E(d) \approx \left.\frac{dE}{dx}\right|_{E_0} t + \left.\frac{dE}{dx}\right|_{E_0} d \left(\frac{1}{\cos^3\theta} + \frac{1}{\sin^3\theta} - 1\right) \qquad (9.6)$$

For $\theta = 45°$ we find the result:

$$\Delta E(d) \approx \left.\frac{dE}{dx}\right|_{E_0} \cdot (t + 4.6d) \qquad (9.7)$$

which was derived by Cohen et al.[4] The bracket on the right-hand side of Eq. 9.6 contains the angular dependence, and it is plotted in Figure 9.4 as a function of the scattering angle θ. The variation of $\Delta E(d)$ for scattering angles in the range 40°–50°

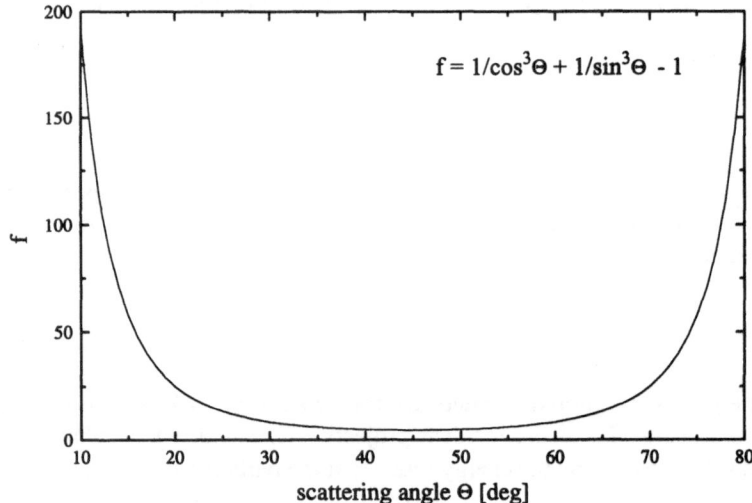

$$f = 1/\cos^3\Theta + 1/\sin^3\Theta - 1$$

scattering angle Θ [deg]

Figure 9.4. Angular dependence of $\Delta E(d)$ calculated from Eq. 9.6. For scattering angles in the vicinity of 45°, $\Delta E(d)$ is nearly independent from θ.

is only 6 % of the value corresponding to $\theta = 45°$. Therefore the depth resolution (Eq. 4.33) is not seriously affected if the detector solid angles are increased to about 0.02 sr for circular-shaped detectors. For example in the experiments performed by Willemsen et al.[9] two detectors of different size were used, a small-area detector because of its superior energy resolution and a large-area detector of about 2 msr to determine the occurrence of a coincident event.

9.3.2. Experimental Set-Up and Coincidence Measuring Systems

The time acceptance window τ needed to observe a coincidence event can be estimated from the TOF differences Δt of scattered and recoiled particles by:

$$\Delta t = D \left[\left(\frac{M_1}{2E_1} \right)^{1/2} - \left(\frac{M_2}{2E_2} \right)^{1/2} \right] \tag{9.8}$$

with D the target detector distance. In particular if the energy of recoil is $E_2 = E_1 + \delta E$ with $\delta E \to 0$, we can write:

$$\Delta t = D \left[\left(\frac{M_1}{2E_1} \right)^{1/2} - \left(\frac{M_2}{2E_2} \right)^{1/2} + \left(\frac{M_2}{2E_1} \right)^{1/2} \frac{\delta E}{2E_1} \right] \tag{9.9}$$

Indeed for $M_2 = M_1$ and $\theta = 45°$, both particles should arrive at the same time, so that $\tau \to 0$. Because of finite-detector solid angles, energy straggling occurring in thicker targets, and a limited time resolution of the electronics, a finite-time window is required. Assuming detectors accepting particles in an angular range of $45 \pm 3°$, detected energies are about $0.5E_0$, with a variation of about $\pm 0.05\ E_0$. For typical target detector distances of 10 cm, we find a TOF difference of:

$$\Delta t \approx 0.5 \left(\frac{M}{E_0} \right)^{1/2} \tag{9.10}$$

with E measured in MeV, M in amu and Δt in nsec.

For most masses and energies the required coincidence time window $\tau \approx \Delta t$ can be of the order of 1 nsec. A larger coincidence time window increases the random coincidence rate and thus deteriorates the detection limit.

The main contribution limiting the sensitivity of *identical-particle* CERDA are random coincidence events. The problem of avoiding random coincidences was discussed in detail by Cohen et al.[4] and Moore et al.[6] Cohen points out several ways of reducing the random coincidence rate: choosing τ as small as possible, reducing the beam current, and making an energy selection more carefully. In the case of high-energy protons, Cohen et al. estimated a detection limit for hydrogen of 1 ppm (atomic). Experimentally they chose a time acceptance window of $\tau = 2$ nsec and adjusted the beam current, so that count rates in each detector did not exceed 10 kHz. In this way a detection limit of 10 ppm was achieved. Willemsen et al.[9] applied a coincidence ERDA set-up consisting of a small-area and a large-area detector, only the energy signal of the small-area detector was analyzed. The measuring system is therefore rather simple with only one main amplifier and one ADC, gated by the coincidence unit. The rather poor detection limit of 0.1 at. % for hydrogen is a consequence of the limited detector solid angle of the small-area detector and the large time acceptance window of $\tau = 100$ ns. The latter produces a large random coincidence yield.

Moore *et al.*[6] have used a more powerful measuring system (Figure 9.5), allowing the discrimination of start and stop signals, as well as a precise selection of the time acceptance window. Coincidence events are recorded by detectors D_1 positioned at angle θ_1 and D_2 at position θ_2. Detector D_2 was chosen to deliver the START signals, and D1 the STOP signals. Elastic scattering of the incident projectiles from target atoms is characterized by a well-defined time difference between the corresponding START and STOP pulses. This time difference appears as a distinct peak in the time spectrum representing real coincidences. The real coincidence signal is used to gate the multichannel analyzers MCA1 and MCA2 that record the energy spectra of detected scattered and recoiled particles. The time spectrum was recorded with an additional multichannel analyzer MCA3. A similar electronic system can be used in connection with state-of-the-art multiparameter data acquisition systems. Instead of gating MCA1 and MCA2 individually, we would then store the complete time spectrum together with both energy spectra in a three-dimensional parameter array (see also Sections 9.6 and 9.7).

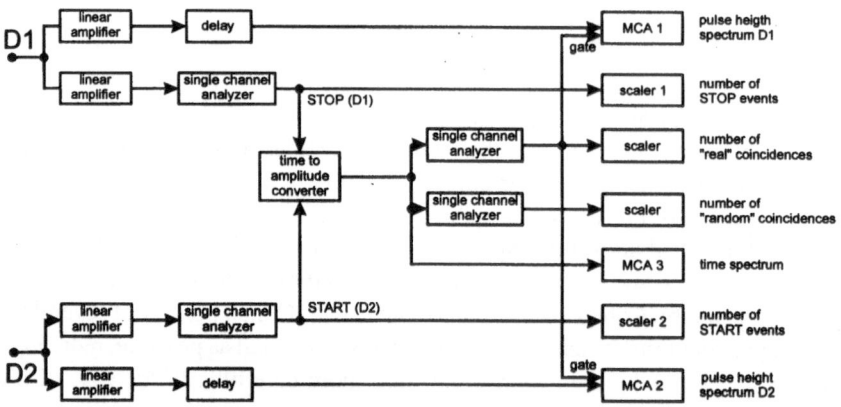

Figure 9.5. Functional outline of the electronic system used by Moore *et al.* for elastic-scattering coincidence detection. Real coincidence events recorded by detectors D_1 and D_2 produce a well-defined peak in the time spectrum. These real coincident events are used to gate multichannel analyzers that record pulse height spectra from each detector. (Data from Ref. 6)

9.3.3. Multiple Scattering

The MS of particles is discussed in detail in Section 3.3. We add several comments on MS effects that are relevant for coincidence techniques (see also Section 9.6.3). The MS of incident projectiles before the scattering event changes the direction of the incident projectiles by a small angle. After the scattering event, MS changes the trajectory of recoil atoms, for example projectile and recoil having identical masses, the angle enclosed by the scattered and recoiled particles is still $\theta + \phi = 90°$. However the angles with respect to the incident beam direction deviate slightly from the original value, for example 45°. This causes little difficulty because ion energy is still high enough, so the angular spread is usually small. Such a scattering event is detected if detector solid angles are sufficiently large. Multiple scattering of either the scattered or recoiled particle, which now possesses approximately half the incident energy leads

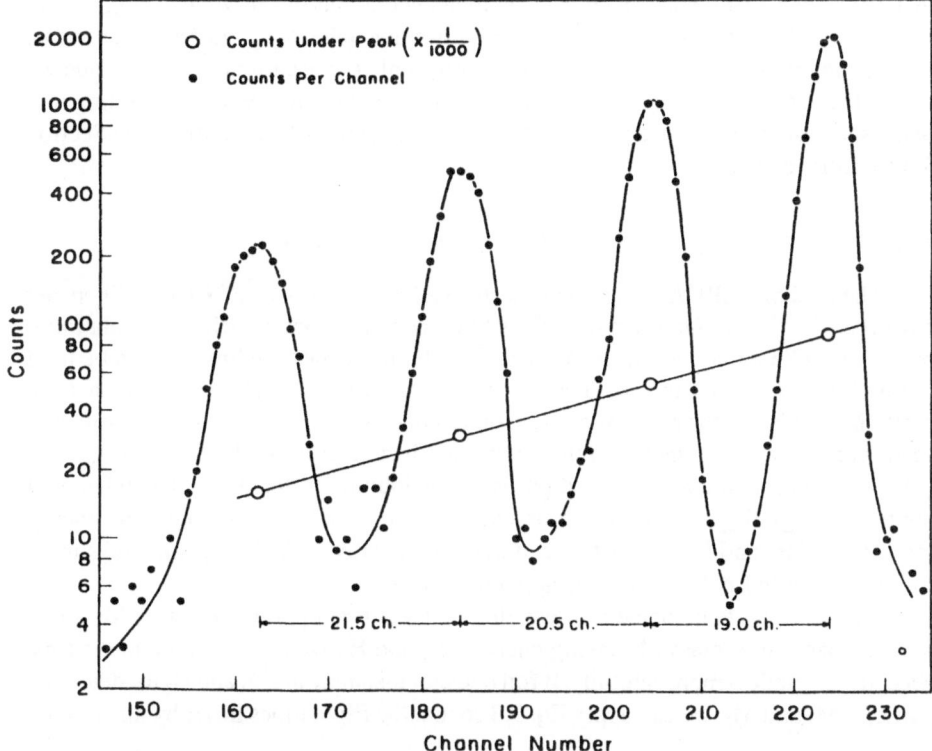

Figure 9.6. The CERDA spectrum from Fe-Mylar sandwich target (*solid circles*). The target consists of three 18-μm (0.7-mil) Fe foils interleaved with four 4-μm (0.15-mil) Mylar sheets. Peaks are from hydrogen in the Mylar. *Open circles* indicate areas of the peaks centered at their channel numbers. The fact that they lie on a straight line indicates that losses, mostly due to MS, vary exponentially with channel number. (Data from Ref. 4)

to a deviation from the 90° condition, so that one of the particles has a chance of missing the detector, and the scattering event is lost, thus reducing the coincidence yield. As shown in Section 9.6.3, a detailed calculation of coincidence yield losses due to MS is essential for the analysis of CERDA measurements to derive quantitative concentration-versus-depth profiles up to large depths.

As discussed in Section 3.3, several methods can be used to evaluate MS, and we must often pay special attention to accurately calculating MS effects. Nevertheless approaches differing from those shown in Section 3.3 can be used to estimate this effect. For example using proton-proton coincident detection, Cohen et al.,[4] analyzed an Fe-Mylar-sandwich-thin film consisting of Fe foils interleaved with Mylar sheets (Figure 9.6). Researchers observed that the area under the peaks in the total energy spectrum, arising from hydrogen in the Mylar sheets, decreases exponentially with increasing energy loss ΔE (Eq. 9.6), which is approximately proportional to the depth d. The peak area decreases by a factor of 5 over a distance of 65 μm (2.55 mil). Willemsen et al.[9] tried to avoid loosing coincidence counts due to MS by using a small area detector that provided good energy resolution and therefore good depth resolution, and a large area detector (450 mm²) to determine only the occurrence of a coincidence event. These examples show both MS effects and a method for avoiding counting loss in the spectrum rather than a procedure for evaluating the MS distribution (see Section 3.3 for more details).

9.4. MULTIPLE-ELEMENT ANALYSIS WITH CERDA

The goal of CERDA experiments performed by Moore et al.,[6] Klein, Rijken and coworkers[10,11,19] and Gebauer et al.[20] is to demonstrate that coincidence techniques provide excellent mass discrimination, and furthermore these techniques can be used for multiple-element analysis. These experiments use Eq. 2.6b, which relates the scattering angle θ to the recoil angle ϕ, to obtain mass selectivity by carefully adjusting scattering and recoil angles. Depth information is derived from the total energy E_{total} = $E_1 + E_2$ of scattered and recoiled particles. In addition to the depth dependence of the total energy E_{total}, derived in Section 9.3.3, we must take into account that energy loss, straggling, and MS differ for scattered and recoiled particles, so that the simple formula given in Eq. 9.5 cannot be applied in this case.

For an accurate coincidence detection of a scattered particle of mass M_1 and a recoiled particle of mass M_2 having energies E_1 and E_2, we must determine the time acceptance window more carefully. If foil detector distances are D_1 and D_2 for detecting particles M_1 and M_2, we can apply Eq. 9.8 so that the flight times differ by an amount:

$$\Delta t = D_1 \left(\frac{M_1}{2E_1} \right)^{1/2} - D_2 \left(\frac{M_2}{2E_2} \right)^{1/2} \tag{9.11}$$

Signals recorded within Δt produce a peak in the time spectrum, and they constitute real coincidences. Random coincidences result in a uniform background in the time spectrum. Real coincidences can be used to gate the ADCs, which record the pulse height spectra of each detector. Gated pulse height spectra therefore provide a measure of the energies of scattered and recoiled particles resulting from elastic collisions with atoms of mass M_2. Such a timing arrangement was used by Moore et $al.$[6] In those experiments an incident ion beam of energy 1 MeV/amu with a beam current of a few nA was applied to obtain ion velocities at which stopping powers are near maximum, thus giving the best depth information. Thin Ni and Au foils of 150–220 nm thickness, implanted with 40-keV Mg, K, or Cu ions at various doses up to 10^{16} cm^{-2} were analyzed. By adjusting scattering and recoil angles for a detector acceptance angle of about 5.4°, it is possible to discriminate between different masses. For example in the case of an incident ^{16}O ion beam, $\theta = \phi = 45°$ was used to select events from scattering at ^{16}O target atoms, which were present in the form of a surface oxide on the Ni foil. For $\theta = 55°$ and $\phi = 45.8°$, scattering events from implanted ^{24}Mg atoms were selected. The electronic system used is shown in Figure 9.5. In the time spectrum, the real coincidences appear as a single sharp peak about 2 nsec wide. For scattering of 15 MeV, ^{16}O ions scattered from ^{24}Mg and ^{16}O in a Ni foil, the pulse-height spectra gated with this time signal, are shown in Figure 9.7. Mass selection is achieved by positioning detectors at angles according to Eq. 2.6b. The double-peak structure in the spectra in Figure 9.7c, and 9.7d for detecting ^{16}O recoils arises from surface oxygen on both sides of the Ni foil. Different peak areas can be explained by MS effects. Compared to values determined from mass separator implant data, implant concentrations determined from these CERDA measurements are smaller, because no account was taken of real coincidence losses due to MS in the target foil or possible deviations from the Rutherford cross section. Deviations may exist because incident heavy-ions energies were close to the Coulomb barrier. Compared to conventional 2-MeV ^{4}He backscattering, CERDA using time spectra and gated pulse height spectra provides a detection limit improved by two orders of magnitude, about 10^{14} atoms · cm^{-2}.

The CERDA technique with 30-MeV alpha particles as incident ions[19] and sandwich foils consisting of three aluminum-evaporated carbon foils was investigated by Klein.[19] Due to the sample preparation technique, some hydrogen and oxygen contamination of the films were also present. The scattering angle was chosen to be fixed at $\theta = 82°$, because the elastic alpha-scattering cross-section for scattering in oxygen is extremely small at $\theta = 90°$. The recoil angle was a variable to obtain mass selectivity according to Eq. 2.6b. For recoil angles of about $\phi = 35°$, only ^{12}C recoils can be detected; for $\phi = 36°$ the set-up is sensitive to ^{13}C, and for $\theta_2 = 38°$ ^{16}O atoms can be detected. The total energy spectrum obtained for the different recoil detector positions is shown in Figure 9.8. From these spectra different carbon layers and also Al layers can be resolved. It is also possible to distinguish between isotopes ^{13}C and ^{12}C. Signals from ^{16}O contamination correlates with carbon signals, which means that carbon foils contain oxygen (or water or OH). For increasing depth a strongly decreasing count rate, i.e., a reduced coincidence efficiency, and decreasing depth resolution are evident in all spectra; this is caused by MS effects. Mass selectivity was tested by comparing ^{13}C and ^{12}C signals for a given recoil angle, and a suppression of

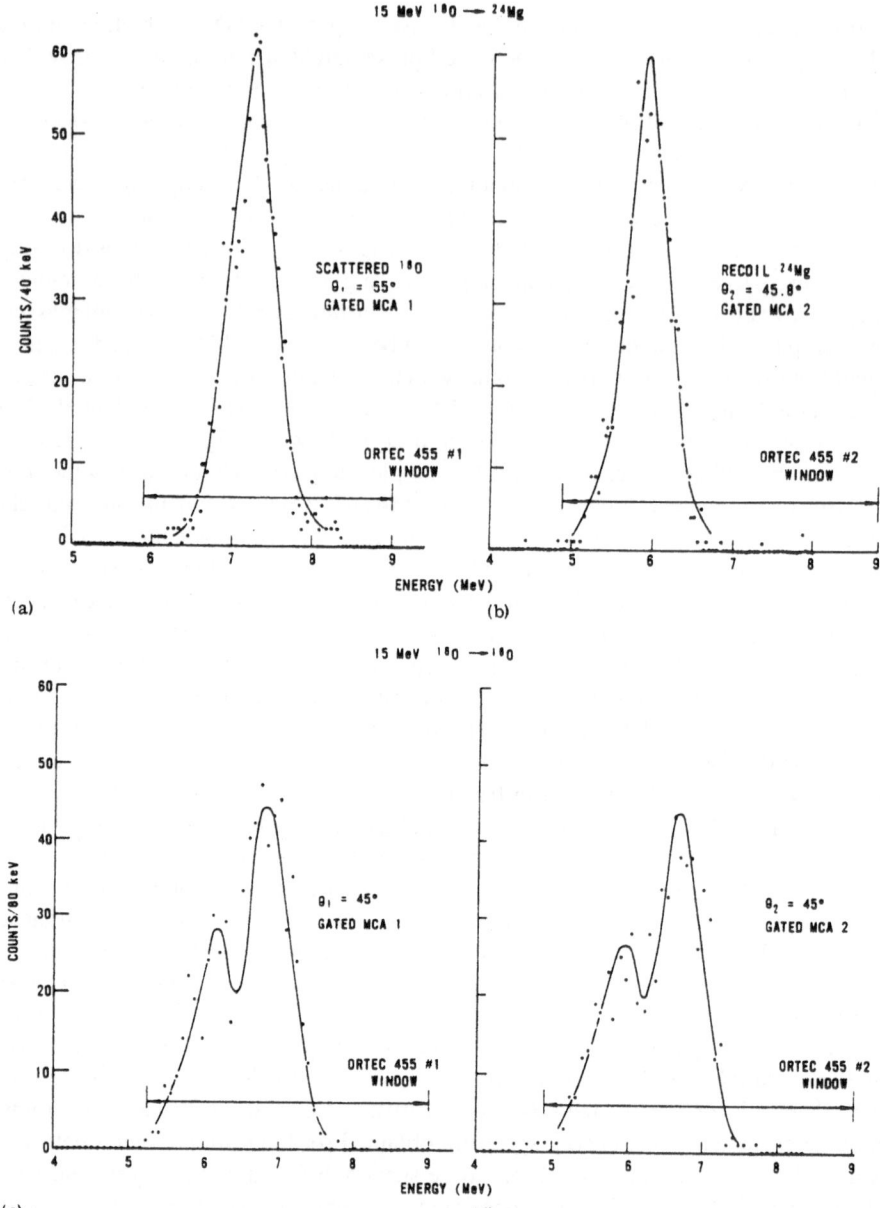

Figure 9.7. Coincidence-gated pulse height spectra recorded when 15-MeV ^{16}O is scattered from a nickel foil containing an implant of ^{24}Mg (10^{16} atoms/cm^2) and surface oxide. The doublet in the ^{16}O-^{16}O spectra, (c) and (d), corresponds to scattering from oxygen located on both surfaces of the nickel foil. (Data from Ref. 6)

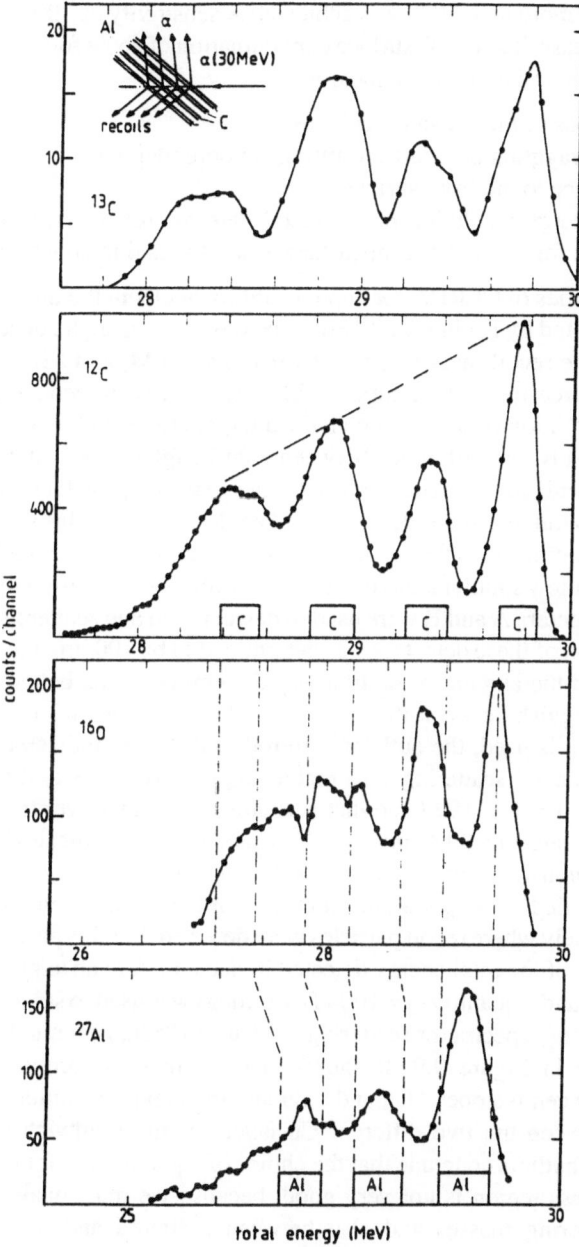

Figure 9.8. Total energy spectra obtained from a CERDA measurement at a sandwich structure consisting of several aluminum and carbon foils. The 30-MeV ^{4}He ions were used as incident projectiles. Energy scales are different to make equal intervals correspond to equal amounts of matter at the surface. Blocks indicate target composition. (Data from Ref. 19)

^{12}C by a factor of more than 3000 was achieved. A sensitivity of 300 ppm for ^{12}C was determined by measuring the ^{12}C surface contamination of an Fe foil. *Depth resolution* of the technique is limited by three factors:

- Energy loss of the heavier recoil atoms,
- Energy straggling of recoils, resulting in poorer depth resolution with increasing distance to the front surface.
- Poorer energy resolution of the recoil detector for heavier elements, depth resolution for ^{12}C at the front surface was estimated to about 50 nm.

The possibilities of CERDA for depth profiling nuclei in the middle mass region was also investigated by Klein et al.[11] For a fixed scattering angle, dependence of the recoil angle on the recoil mass is greatest when masses M_1 and M_2 are about equal. However with increasing mass numbers, differences between recoil angles decrease. For an incident ^{58}Ni ion beam and recoiled Cu target atoms with masses 63 amu and 65 amu, respectively, the difference between recoil angles is only 0.75°. To achieve mass selectivity with conventional surface barrier detectors, small entrance slits must be used, giving extremely small detector solid angles. To avoid this situation Klein et al.[11] applied one-dimensional position-sensitive detectors to increase the solid angle and additionally allow simultaneous detection of several not too different recoil masses M_2. Thus two detectors A and B were used to detect recoil and scattered particles. The position resolution of these detectors was determined to be 100 μm, measured with an alpha source, and the angular resolution was determined to be better than 0.2°. For example to distinguish between the two different Cu isotopes with mass 63 and 65 when a ^{58}Ni beam is used, the authors[11] introduced a mass indicator μ, which is a linear combination of scattering and recoil angles. We have a slightly different expression for μ, $\mu = \phi + 0.92\,\theta$, so that μ is almost constant over the full-detection angular range.* A value $\mu \approx 89.4°$ indicates ^{65}Cu and $\mu \approx 88.6°$ indicates ^{63}Cu. To assign θ and φ unambiguously we must decide whether detector A has registered a scattered ^{58}Ni ion or a recoiled Cu target atom; in other words before calculating μ, we must determine separately where recoil particles were detected. For this purpose the fraction $f_A = E_A/(E_A + E_B)$ of the total energy detected in detector A at an angle α (either θ or φ) was analyzed, and a quantity $Q = 100 \cdot f_A + \alpha$ (deg) was used to solve the ambiguity.

The total energy spectra for scattering at ^{63}Cu and ^{65}Cu, generated by a gating on μ, are displayed in Figure 9.9. It shows that the mass separation allows us to discriminate between isotopes ^{63}Cu and ^{65}Cu and the depth resolution is sufficient to discriminate between the two different Cu layers of the sandwich target used. In reference 11, the authors conclude that the choice of equal masses for detecting atoms in the middle mass region is not very good, because we must distinguish between different neighboring masses and also between scattered and recoiled atoms of comparable mass. These authors[11] suggest that it is better to choose the projectile-recoil mass ratio nearer to 2. For the case of Cu target atoms, this requires an incident Xe ion beam.

---

*Note that the definition of μ given in Fig. 1 of Ref. 11 is wrong.

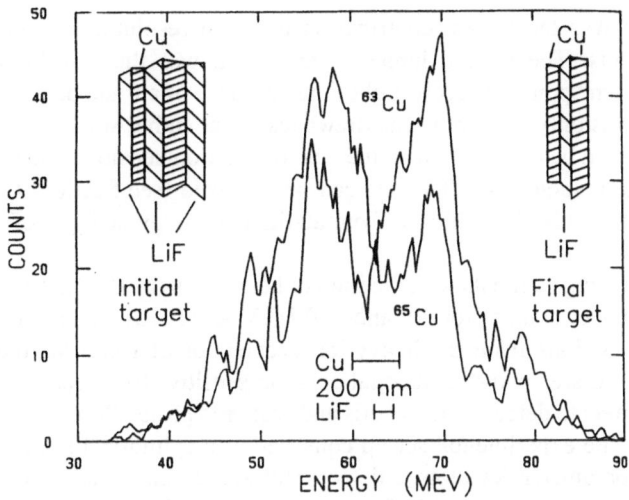

Figure 9.9.   The CERDA measurement with ^{58}Ni projectiles scattered on two thin Cu films separated by LiF. Separate total energy spectra for scattering at ^{63}Cu and ^{65}Cu isotopes obtained by gating on the mass indicator μ are shown. (Data from Ref. 11)

In Section 5.6 the mass resolution $\delta K'/\delta M_2$ (or $\delta E_2/\delta M_2$) was introduced. Figure 5.8 shows that mass resolution is increased by up to an order of magnitude if the projectile and the target atom have significantly different masses. Mass resolution is also maximized for recoil angles close to 0°. Therefore heavy-ion CERDA seems to be the best choice for detecting target atoms in the medium mass region. Such an experiment requires high-projectile energies on the order of 100 MeV. In this situation however, we must carefully judge whether the CERDA technique or a set-up using a ΔE-E telescope offers the better solution.

Time-of-flight techniques are commonly used in ERDA to improve depth resolution, which is directly related to energy resolution in the detection system (see Chapter 6). TOF detection of recoiled particles in a CERDA experiment was investigated by Rijken *et al.*[10] An incident 12.1-MeV alpha particle beam was used to recoil ^{12}C target atoms. Scattered alpha particles are detected by a surface barrier detector positioned close to the target at a scattering angle of $\theta = 101°$. This detector provides the time zero start signal. Recoils are detected by a second surface barrier detector positioned 3.5 m away from the target at an angle of $\phi = 30°$. This detector was used to record the recoil energy spectrum, and it also provided stop signals to measure the recoil TOF. A typical flight time for the 6.8-MeV ^{12}C recoil is 335 nsec. Improvement in depth resolution for the TOF measurement in comparison with the energy measurement was checked with a sandwich target consisting of three 100-nm carbon layers separated by 250-nm Al layers. By measuring the TOF spectra, the depth resolution at the front surface improved from 47 to 33 nm. In this experiment divergence of the incident beam was the largest contribution to depth resolution by the consequent kinematic effects. The second largest contribution was the distribution of recoil energies and corresponding recoil flight times because of a finite detector acceptance angle, which was only 4.2 mrad. The drawback of the rather small solid angle of the recoil detector can be overcome with the help of position-sensitive detectors, presuming that a time resolution of 0.5 nsec can be achieved (see Section 9.7). A further improvement may also be achieved from additional energy analysis of the scattered alpha particles.

Depth profiling elements with medium or heavy mass by CERDA techniques was studied by Gebauer et al.[20] for 90- and 180-MeV Ar ions as well as 365-MeV ^{129}Xe (see Chapter 13) ions from the heavy-ion accelerator of Hahn-Meitner Institut in Berlin. Two large-area two-dimensional position-sensitive (see Section 9.7) and TOF resolving counters to detect target recoils and scattered projectiles in time coincidence were applied. The experimental set-up consisted of a cylindrical scattering chamber and two multiparameter detectors, a detector telescope, and a multiwire proportional chamber. The telescope consisted of three components. The first part was a parallel-plate avalanche detector to determine the TOF start signal, the second part was a proportional counter delivering the in-plane scattering angle by charge division. The last part was a multistage ionization chamber to determine energy loss ΔE_1 and the total energy E of the detected particle. In a single-detector experiment that is a ΔE-E ERDA configuration, we determine mass, charge, angular direction, and energy of the recoiled particle, which is sufficient to derive the desired mass-resolved depth profiles. Here we discuss only measurements performed in time coincidence between scattered and recoiled particles. The correlation between the directions of flight of projectile and recoil for Si and O can be calculated from Eq. 2.6. For scattering 90-MeV Ar ions at a SiO_2 target several micrometers thick, the θ-ϕ correlation allows us to discriminate low-energy recoils like oxygen, and it can be used as a gate signal to eliminate unwanted signals from heavier elements in the target. In the experiment of Gebauer et al.[20] the θ-ϕ correlation was used to eliminate signals from Si, to resolve the layered structure of the oxygen content of the sandwich target.

Another measurement was performed in Ref. 20 with a 180-MeV Ar ion beam incident on a sandwich target consisting of two Fe layers separated by an Al foil. Here the TOF difference between projectile and recoiled target atoms was plotted versus the recoil angle θ. In this spectrum two parallel bands appear, corresponding to coincidence events from scattering at the two different Fe foils. Summing up all events along the direction of the bands in incremental steps perpendicular to the bands gives a depth profile with a depth scale in units of a normalized relative TOF difference. Conversion to a real depth scale was not carried out, because such a conversion is not straightforward, and it requires careful data evaluation (see Section 9.6).

In the case of ERDA with high-energy heavy projectiles, there are ΔE-E telescopes (see Chapter 8) with sufficient energy resolution and also position resolution. Therefore it is not desirable to use CERDA to obtain mass-selective depth profiles of elements in the middle mass region when thin-film samples only a few micrometers in thickness are necessary. The only reason to use CERDA in this case would be to filter unwanted events, especially if we were also interested in detecting light particles.

9.5. SCATTERING RECOIL COINCIDENCE SPECTROSCOPY

Chu and Wu developed the concept of SRCS, where mass $\dot{M}_2$ of recoils and depth x where scattering takes place are determined by measuring the energies of both the scattered and recoiled particles.[15,16] The principle of SRCS was illustrated by Chu and Wu for the case of depth profiling trace amounts of hydrogen in a carbon foil using an incident He ion beam. To illustrate SRCS as originally developed, we recall the concepts of binary collision using an He-H scattering. Thus for an He projectile with energy E_0 and a proton at rest, scattered energy and proton-recoiled energy just after collision are related to the scattering angle θ and recoil angle ϕ by Eqs. 2.6, 2.10, and 2.12. The relation between θ and ϕ is also plotted in Figure 2.3. For detecting amounts of hydrogen in a thin foil, the scattering problem is schematically recalled in Figure 9.10a. The energy of He ions just before scattering at depth x becomes E_0-$\Delta E_{He}(x)$, where $\Delta E_{He}(x)$ is the energy loss of He ions along path length x (see Section 4.2). At

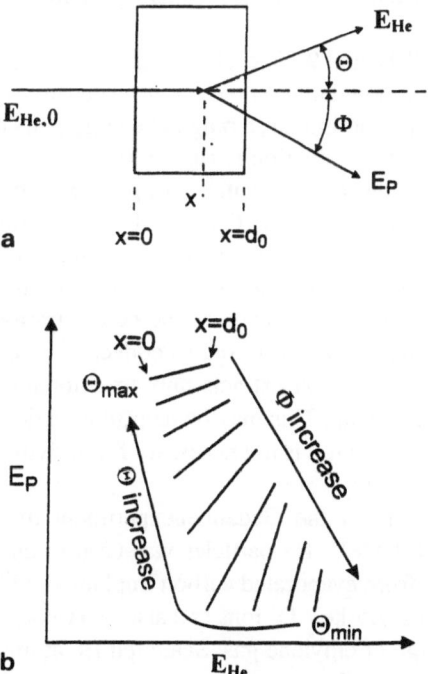

Figure 9.10.   (a) A binary collision between ^{4}He and ^{1}H in a thin foil containing trace amounts of hydrogen. (b) Biparametric (E_p, E_{He}) plot showing the region of interest to the scattering problem. (Data from Ref. 15)

depth x He ions are scattered at an angle θ, while hydrogen ions are recoiled at an angle φ. Using Eq. 4.9 we obtain the energy E_{He} of the outgoing He ions by:

$$E_{He}(d_0) \approx K[E_0 - \Delta E_{He}(x)] - \Delta E_{He}(\bar{x}) \qquad \bar{x} = \frac{d_0 - x}{\cos \theta} \qquad (9.12)$$

where $\Delta E_{He}(\bar{x})$ is the energy loss of the scattered He ion. In a similar way using Eq. 4.10, we evaluate the energy of the recoiled outgoing hydrogen recoiled atoms E_p:

$$E_p(d_0) \approx K'[E_0 - \Delta E_{He}(x)] - \Delta E_p(\hat{x}) \qquad \hat{x} = \frac{d_0 - x}{\cos \phi} \qquad (9.13)$$

Angles θ and φ can be eliminated from Eqs. 9.12 and 9.13 by using the relation between these angles (Eq. 2.6). Therefore a coincident measurement of E_{He} and E_p is sufficient to determine depth d and also mass M_2 (in our case M_H). *Measuring scattering and recoil angles in SRCS is not required*, so large detector solid angles are possible. This makes the technique very sensitive for analyzing light elements in thin films. In principle solid angles can be increased by three orders of magnitude compared to traditional RBS or ERDA experiments, where a small solid angle is required to define θ and φ.

According to Eqs. 9.12 and 9.13, each pair of energies (E_{He}, E_p) corresponds to a certain depth x. However depending on the actual scattering and recoil angles, which are not measured in these experiments, a range of energy pairs (E_{He}, E_p) corresponding to the same depth exists. This situation is shown in Figure 9.10b. The angular range of θ and φ, defined by the experimental set-up, and also film thickness define boundaries of the shaded area in the (E_{He}, E_p) plot where scattering/recoil events appear. For a given scattering and recoil angle, a contour line representing a depth scale can be drawn. Integrating over the possible range of scattering/recoil angles generates a counts-versus-depth profile, which becomes the concentration profile of hydrogen. The Jacobian transformation was proposed to convert the coordinates (E_{He}, E_p) into (x, θ). However this transformation extracts only traditional depth profiles for well-defined θ from the (E_{He}, E_p) data. This may be helpful in understanding the shape and contours of the (E_{He}, E_p) plot, but it is not necessary for deriving concentration-versus-depth profiles.

Forster *et al.* carried out the Gedankenexperiment of Chu and Wu just described.[13,14] A beam of 2-MeV ^{4}He particles was used to analyze trace amounts of hydrogen in foils made from evaporated carbon implanted with $10^{17}/cm^2$ 7.5-keV H^+ ions as well as $5 \cdot 10^{16}/cm^2$ 7.5-keV D^+ ions and also an H-containing foil produced by radio frequency discharge of ethylene gas. Scattered He atoms and recoiled H atoms were detected by two annular detectors. The maximum scattering angle for scattering at H atoms is only 14.37° (see Eq. 2.7). To avoid flooding the H-detector with He ions scattered from carbon target atoms, the angular range was reduced to 6° < θ < 15°. The beam current was also reduced to about 5–10 pA, limiting the counting rate in the detector to about 7 kHz. The corresponding recoiled H atoms were detected at recoil angles from 33°–69°.

With proper coincidence timing the FWHM of the prompt peak, corresponding to real elastic scattering events, was 12 nsec. Gating the total energy signal with this timing signal, coincidence events from scattering at hydrogen appear in the (E_{He},E_p)-diagram, as predicted by Chu and Wu[15] (Figure 9.11). A similar result was obtained for scattering He at a carbon foil implanted with deuterium.[14] In this experiment the maximum scattering angle is 30°, and the minimum detection angle for scattered He can be increased to about 14°, thus reducing events from the forward scattering of He at carbon, as a result the beam current could be increased to about 25 pA. Concentration versus depth profiles were not extracted from these data. Instead Forster *et al.*[13,14] tried to analyze total energy spectra by comparing them with Monte Carlo simulations. Because detector solid angles are large, total energy spectra are rather broad and cannot be interpreted as a depth profile. Nevertheless total energy spectra can be used to estimate the depth resolution for H and D detection in thin carbon foils, which was only about 100–120 nm. For the detection geometry used, a sensitivity limit of 10^{15} atoms/cm^2 for detecting deuterium was achieved. Forster and coworkers[13] concluded that SRCS is better suited to determining the integral amount of H or D in a thin film sample because of the limited depth resolution. This technique can also be applied if only small beam currents are required in order to avoid radiation damage of the film.

An important input parameter for quantitative analysis of SRCS measurements is total target thickness, which was determined from energy loss measurements in alpha particles from an ^{241}Am source and the stopping for ^{4}He in carbon from Ziegler *et al.*[22] With this calibration method thickness can be measured with an estimated uncertainty of ± 10%.

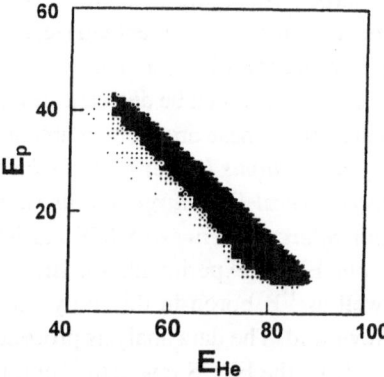

Figure 9.11. The SRCS spectrum (intensity as a function of E_p and E_{He}) for scattering 2-MeV He ions at a hydrogen containing a 150-nm thin carbon foil (25 μg/cm^2) produced by cracking ethylene gas in a radio frequency discharge. (Data from Ref. 13)

A concept similar to SRCS was applied to measure ^{6}Li or ^{10}B depth profiles in thin foils by neutron depth profiling using the reactions ^{6}Li(n,α)^{3}H and ^{10}B(n,^{4}He)^{7}Li.[23,24] In the first experiment of this type,[23] ^{6}Li-implanted Al foil was irradiated with thermal neutrons, and ^{4}He and ^{3}H reaction products were detected by two silicon surface barrier detectors placed at an equal distance of 20 mm from each side of the foil. Initial energies of reaction products are monoenergetic, with $E_{He,0}$ = 2.055 MeV and $E^3_{H,0}$ = 2.727 MeV. From the energy loss of these particles in the thin foil of depth d, the position of the nuclear reaction can be determined. Assuming constant stopping powers S_{He} and S_T for He and ^{3}H particles in a thin foil, we can find an analytic formula for the depth x as a function of E_{He} and E^3_H:

$$x \approx \Delta E^3_{H,0} \frac{E_{He,0} - E_{He}(d_0)}{S_{He}(E^3_{H,0} - E^3_H(d_0)) - S^3_H(E_{He,0} - E_{He}(d_0))} \tag{9.14}$$

Here $E_{He}(d_0)$ and $E^3_H(d_0)$ are energies of the He and ^{3}H particles emitted from the foil, and $\Delta E^3_{H,0}$ is the energy loss of ^{3}H traversing total foil thickness. Equation 9.14 contains the key concept of the SRCS coincidence technique introduced by Chu and Wu,[15] namely, a simultaneous measurement of E^3_H and E_{He} yields the depth x of the reaction event without requiring knowledge of emission angles of both particles.

9.6. ELASTIC RECOIL COINCIDENCE SPECTROSCOPY

The difficulty in measuring H or D with the SRCS coincidence technique described in Section 9.5 arises because the detector for scattered He projectiles must be placed in almost the forward direction to detect scattered He from H target atoms at angles of $\theta \approx 6-15°$. This detector can therefore be flooded by scattering events from heavier target elements like carbon. As Figure 2.3 shows it is more convenient to detect scattering events from elements somewhat heavier than He, i.e., elements ranging from ^{7}Li to ^{16}O, where scattering as well as recoil angles can be chosen in the 45°–70° range. It is also clear that several light elements can be detected simultaneously if sufficiently large detector solid angles are used. These are the fundamental ideas of ERCS. *SRCS is a technique for detecting target atoms lighter than the incident ion mass, whereas ERCS simultaneously detects several light target atoms somewhat heavier than the incident ion, this is the main difference between SRCS and ERCS.*

In the following Sections ERCS experiments for simultaneously profiling light elements ^{12}C and ^{16}O as well as ^{11}B boron in thin polymer films using an incident 2-MeV ^{4}He ion beam are reviewed. The data analysis procedure for deriving concentration versus depth profiles from the ERCS raw data is outlined.[17,18] This procedure can also be applied to analyze quantitatively measurements obtained by most other coincidence ERDA techniques, especially SRCS and multiple-element CERDA techniques discussed in Section 9.4.

9.6.1. Basic Considerations

The scattering recoil geometry is determined by the relation given by Eq. 2.6 between scattering angle θ for a projectile of mass M_1 and recoil angle ϕ for a recoil atom with mass M_2, because ERCS target atoms of interest are heavier than projectiles and a maximum scattering angle θ_{max} therefore does not exist. This gives greater flexibility in choosing the scattering/recoil geometry. Criteria for an optimized scattering geometry are given by kinematic factors and scattering cross-section.

The mass resolution of an ERCS measurement stems from the difference in the kinematic factors K or K' (Eq. 2.10 and 2.12) for different recoil elements and decreases for decreasing scattering angles. For small scattering angles the energy $E_2 = K' E_0$ transferred to the recoil atoms is small, which then limits the feasible total thickness of the foil to be analyzed. Regarding only scattering and recoil angles, the best mass resolution is obtained for scattering angles around 90° (Figure 2.3). This condition was used in the CERDA experiments of Klein *et al.*[19] (Section 9.4). Detection at scattering angles around 90° has the drawback of a nonaxially symmetric target detector configuration, which limits the maximum possible detector solid angle.

In Section 3.2 behavior of differential cross-section is discussed in detail. Regarding ERCS experiments we recall some characteristics of the Rutherford cross section. For large scattering angles, i.e., a backscattering configuration, $(d\sigma/d\Omega)_\theta$ is small but almost constant, which simplifies data evaluation, for example in a transmission ERDA experiment. Large scattering angles are less convenient for ERCS because recoil angles become small. For decreasing scattering angles, i.e., scattering angles below 90°, $(d\sigma/d\Omega)_\theta$ increases rapidly, which then requires a careful scattering yield calculation if the detector solid angle is large. Since this can be done easily with the help of computers, a configuration with scattering as well as recoil angles below 90° is favorable.

Among three possible scattering/recoil configurations, an axially symmetric forward scattering and a forward recoil appears to be the best-suited configuration. Both scattered ions and recoiled atoms are detected in a forward direction (Figure 9.12). A scattering geometry can be chosen so that both scattering and recoil angles are large enough to avoid flooding detectors due to small-angle scattering. Recoil angles of about 45°–55° and corresponding scattering angles of 60°–80° (Figure 2.3) seem to be a good compromise with regard to scattering cross-section and the kinematic factor. For recoil angles ranging from 45° to 55°, the maximum detector solid angle is $\Omega_{max} = 2\pi(\cos(45°) - \cos(55°)) \approx 0.84$ sr. Therefore the detection efficiency can be increased by about 3 orders of magnitude compared to conventional ERDA with detector solid angles of typically 10^{-3} sr. In addition the forward scattering cross-section for this configuration is about 10 times larger compared to backscattering ($\theta > 150°$) cross-section, providing an additional increase in detection efficiency by one order of magnitude compared to RBS. As previously mentioned mass resolution decreases for decreasing scattering angle θ (Figure 2.4). For the proposed detector

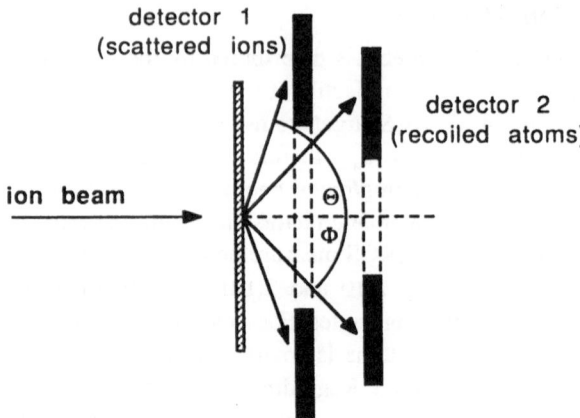

Figure 9.12.   Target detector configuration for ERCS measurements. (Data from Ref. 18)

arrangement however, this difference is still large enough to discriminate between neighboring elements.

To avoid severe energy straggling and angular straggling, the incident particle must have a sufficiently high energy, and the sample must be sufficiently thin. For ^{4}He projectiles a minimum incident beam energy of about 2 MeV is required, which can be estimated from required minimum energies for scattered and recoiled atoms and average stopping power values. A higher ^{4}He ion energy improves depth resolution and the detectable depth range and reduces angular straggling effects. However resonances in scattering cross-section and activation of target material must be considered.

9.6.2. Application: Profiling Carbon and Oxygen in Polycarbonate

Measurements described here after, were performed using a 2-MeV ^{4}He$^+$ beam collimated to about 1 mm^2, a beam current below 1 nA and a forward-scattering/forward-recoil set-up based on the configuration shown in Figure 9.12. Detectors were centered at $\phi = 48°$ to detect recoils and at $\theta = 70°$ to detect the scattered ^{4}He. Coincidence events were collected in a matrix of 64 × 64 channels or 128 × 64 energy channels. A dual-parameter multichannel analyzer was operated in an internal coincidence mode, where any detected particle can trigger the coincidence time window. The large time window used (2 μs) produces false coincidences, which may appear along the energy axes in the coincidence spectra (Figure 9.13) and reduce the true coincidence yield. This effect may be one reason for differences between measured and expected concentrations in the polycarbonate sample discussed in the following section.

Simultaneously profiling carbon and oxygen with ERCS was demonstrated by Hofsäss *et al.*[17] in 2-μm-thick self-supporting polycarbonate foils ($C_{16}O_3H_{14}$). The

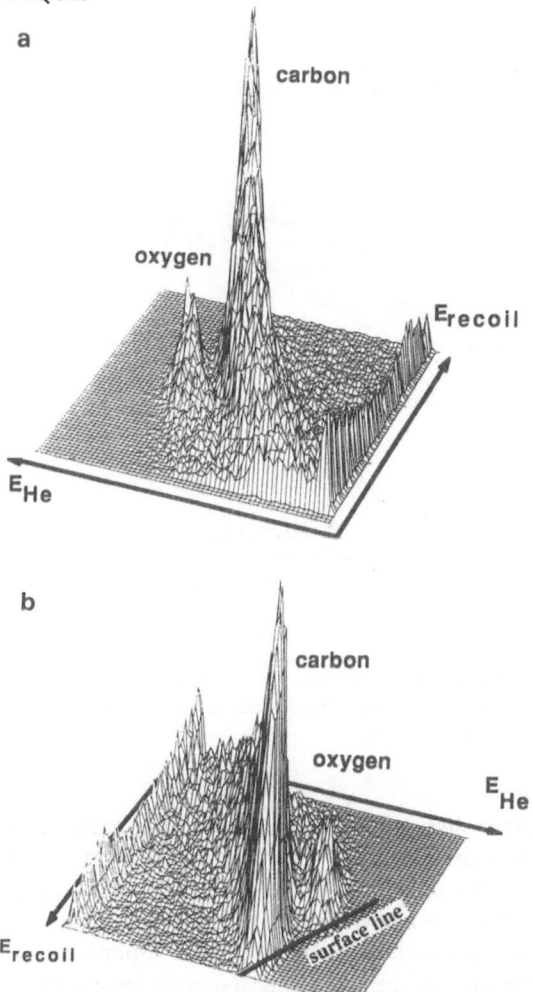

Figure 9.13. (a) The ERCS spectrum for a free-standing thin polycarbonate sample. Coincidence counts as a function of the energy of the scattered ^{4}He$^+$ and recoiled light elements are plotted. The dominant peak corresponds to recoiled carbon atoms. (b) The same spectrum as in (a) but rotated by 180° to indicate the surface line corresponding to scattering/recoil from the rear surface of the sample. The locus of the line for each element that corresponds to increasing depth of scattering extends from the surface line along the crest for each element to lower energies. Coincidence counts along the recoil energy axis are false coincidences (see Section 9.6.1). (Data from Ref. 17)

measured coincidence spectrum, i.e., coincidence counts as a function of energies of both scattered ^{4}He and recoiled light atoms, is shown in Figure 9.13, and the corresponding contour plot in Figure 9.14. Signals from recoiled C and O are well-separated in the coincidence spectrum over the whole detectable depth range of about 1 μm for C and 0.7 μm for O. Coincidence counts from O recoil atoms appear at higher ^{4}He

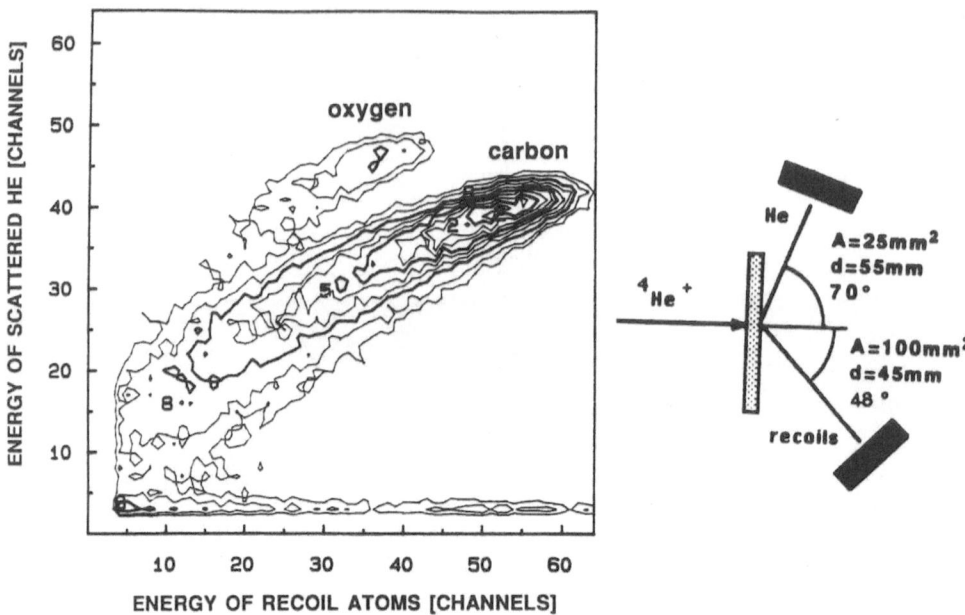

Figure 9.14. Contour plot of the ERCS spectrum shown in Figure 9.13. Contour lines are drawn in steps of 10% of the maximum coincidence yield. The diagram at right indicates Area A, distance d, and center angle of the detectors. (Data from Ref. 18)

energies because of the slightly higher kinematic factor. Scattering/recoil from the rear surface correspond to coincidence counts near the drop-off at highest scattering and recoil energies. By connecting drop-off energy pairs for each recoil element, in this case C and O, we can draw a surface line to indicate the rear surface of the sample. The locus of the line for each element, which corresponds to increasing scattering depth, extends from the surface line along the crest for each element toward lower energies.

Hofsäss *et al.*[17,18] carried out a complete study of an ERCS spectrum. Indeed profile width is determined by the range of θ and φ angles covered by corresponding detectors, whereas the cross-sectional shape is essentially determined by the shape of detector areas (for more details see Ref. 18). For an increasing range of θ and φ angles, for instance by increasing the detector area or decreasing the distance between detector and target, profiles become broader and overlap for neighboring elements. A narrow range of θ and φ results in narrow coincidence profiles well-separated in energy space. Therefore by choosing the appropriate range of θ and φ angles, it is possible to optimize either detection efficiency or mass resolution of the measurement.

Because of energy straggling in the incident beam through the sample, limited detector energy resolution, and surface roughness of the sample, the coincidence counting rate does not increase sharply at the surface line. For lower energies, i.e., greater depth, the profile is broadened because of increased energy straggling in recoiled atoms. The decrease in coincidence count rate at lower energies is a result of MS or angular straggling, which is discussed in Section 9.3.3, MS was also a severe problem in the CERDA experiments described in Section 9.3.3.

9.6.3. Data Analysis

In Chapter 11 a complete analysis is given of data-processing methods applied in ERDA spectra. Nevertheless, in this section we summarize specific aspects of numerical data analysis for ERCS spectra, as developed by Hofsäss et al.[18] The key to analyzing an ERCS coincidence spectrum is relating an energy pair $\{E_{scatter}, E_{recoil}\}$ to a depth x measured from the rear surface (the side toward the detectors) of the thin-film sample, where a scattering/recoil event took place. Because the energy loss of recoiled light atoms depends strongly on energy and the incident projectile must pass through a few micrometers of target material, it is not valid to use surface- or mean-energy approximations to calculate total energy loss. In addition the target may consist of several different layers with different stopping power dE/dx.

The stopping power curves dE/dx as a function of energy for different ions and different target materials of interest were calculated using the subroutine projected range algorithm (PRAL) of TRIM.[22] To compute the remaining energies of scattered and recoiled particles after passing a given thickness t of target material, it is convenient to cut the target into N thin slices of thickness Δt, then use the iteration:

$$E_n = \left(E_{n-1} + \frac{S_0}{m}\right)\exp\left(-m\Delta t\right) - \frac{S_0}{m} \tag{9.15}$$

where intercept S_0 and slope m of the stopping power are determined for energy E_{n-1}, assuming a linear dependence of dE/dx from E within the slice. The initial ion energy is E_0. A slice thickness of $\Delta t \approx 100$ nm allows a reasonably fast and accurate computation. The energy pair $\{E_{11}, E_{21}\}$ for scattering at depth x below the rear surface and for given scattering and recoil angles can then be calculated using Eq. 9.14 with the notations given in Figure 9.15. In a similar way energy straggling can be calculated for each slice of the target using Bohr energy straggling and applying the additivity rule.[21] To calculate the position of coincidence counts in energy space for scattering from depth x for a given detector geometry, it is in most cases sufficient to calculate only energy pairs corresponding to the maximum and minimum values of θ and ϕ, then connect these points by a straight line. The coincidence yield for a given depth x is then the summation of all coincidence counts along this line of constant depth (see Appendix B from Hofsäss et al.[18]).

The scattering cross-section for forward scattering varies strongly with the scattering angle. To calculate the coincidence yield $Y(x,x+\Delta x)$, it is therefore important to determine the detector solid angles $\Delta\Omega(\theta)$ and $\Delta\Omega(\phi)$ for small intervals of scattering angles $\theta + d\theta$ and corresponding recoil angles $\phi + d\phi$. The coincidence yield is then calculated by integrating over all scattering angles θ covered by detector 1 that have

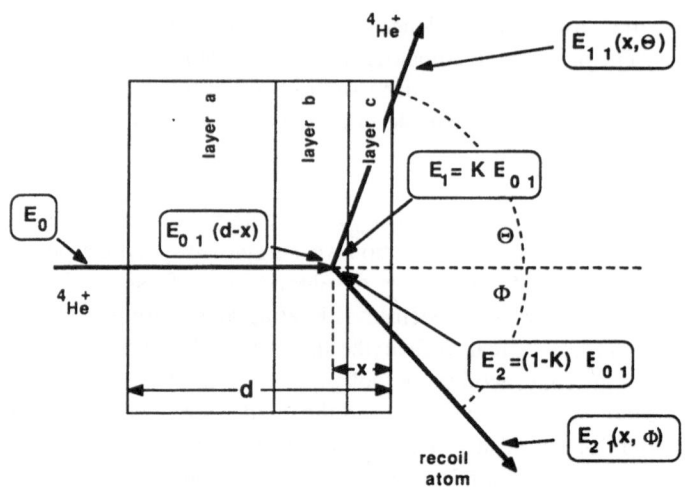

Figure 9.15. Scattering and recoil trajectories for scattering at depth x below the rear surface of a target with total thickness d consisting of several layers with different elemental composition. Energies E_{11} of the projectile and E_{21} of the recoiled atom indicated in the diagram can be calculated, and these correspond to energies measured in an ERCS experiment.

corresponding recoil angles ϕ within the area of detector 2. This integration requires an accurate calculation of the respective detector solid angles defined by the size, distance, and angular position of the scattering and recoil detector, which is described in detail in Ref. 18. The result of integrating is a term $P(x,x+\Delta x)$, which is a function of scattering cross-section, the beam current, and detector solid angles, and the coincidence yield is given by:

$$Y(x,x+\Delta x) = P(x,x+\Delta x)\, N(x,x+\Delta x) \qquad (9.16)$$

where $N(x,x+\Delta x)$ is the concentration of scattering centers in a depth interval $\{x,x+\Delta x\}$. Using this equation the concentration of recoiled atoms at depth $(x,x+\Delta x)$ can be determined from the measured coincidence yield. To derive concentration versus depth profiles, energy pairs $\{E_{scatter}, E_{recoil}\}$ must be calculated for all recoil elements of interest and for the whole depth range of interest and plotted in a two-dimensional $\{E_{scatter}, E_{recoil}\}$ diagram. For known detector geometry, energy calibration, beam energy, and beam current and by varying total thickness and the thickness of each layer of the target, we must find a good fit to an experimentally obtained contour plot like the one shown in Figure 9.14. After this is done the total coincidence yield $Y_{total}(x,x+\Delta x)$ for a given depth interval $\{x,x+\Delta x\}$ is obtained by summing coincidence events along lines of constant depth. The yield-versus-depth profile $Y_{total}(x)$ can now be corrected for scattering cross-section and solid angle using

P(x,x+Δx), which then gives a concentration-versus-depth profile for each element of interest.

Angular spread due to MS of both scattered He ions and recoiled light atoms is responsible for the decreasing coincidence count rate at lower energies corresponding to scattering at a greater depth (Section 9.3.3). Multiple scattering also leads to energy spread, and MS causes a deterioration of the depth resolution[25-27] (see Section 4.2.4.1). A decrease in coincidence efficiency due to MS with $(x)^{1/2}$ was assumed by Klein for high-energy heavy ions.[19] Correcting coincidence spectra for angular spread can be accomplished either by using calibration samples[28] or calculating the angular distribution of recoiled and scattered atoms. To calculate the decrease in coincidence count rate for increasing depth, we must know the angular spread distribution function, the dependence of angular spread on x, and the detector geometry.

Although we remarked in Section 3.3 that the MS cannot be correctly interpreted by using a Gaussian function, especially for short path lengths, we can estimate in first approximation the decrease of coincidence count rate when a Gaussian shape is used for MS.[30] Hofsäss et al.[18] calculated angular distributions based on TRIM calculations[22] as well as angular spread data tabulated by Meyer[29] and Sigmund et al.[30,31] The relation between the width of the angular spread distribution and depth or foil thickness can be determined by running TRIM for different target thicknesses and fitting a Gaussian distribution function to TRIM data. Fitted values for the width FWHM are shown in Figure 9.16 for ^{4}He, ^{12}C, and ^{16}O ions transmitted through polycarbonate foils of thickness varying from 50 to 1000 nm. For ^{12}C and ^{16}O ions FWHM increases almost linearly with film thickness, whereas for ^{4}He an $(x)^{1/2}$ dependence is observed. Values for the FWHM determined by fitting an exponential distribution are about 20% smaller than those for a Gaussian distribution. An exponential angular distribution better describes the large-angle tail of the distribution. For light-scattered ions like ^{4}He, i.e., neglecting energy loss in the target, the shape of angular distributions tabulated as a function of film thickness by Sigmund et al.[30,31] is closer to a Gaussian than to an exponential distribution. Figure 9.16c shows also an $x^{2/3}$ dependence for the width of the MS distribution. The behavior of the MS distribution as a function of the path length is discussed in Section 3.3.

The decrease in coincidence count rate for increasing depth is determined by the probability that ions emitted from the target with a given angular spread distribution will hit the detector. In the case of a circular-shaped detector with radius R at distance d to the target, this probability function can be determined by using a procedure described in Ref. 18.

Following the preceding data analysis procedure, coincidence spectra obtained for polycarbonate foils shown in Figures 9.13 and 9.14 are first converted into a spectrum giving coincidence counts as a function of depth. This is shown for C in Figure 9.17a. The coincidence counts-versus-depth profiles are then corrected for scattering cross-section and transformed into a concentration-versus-depth profile (Figure 9.17b). After correcting for annular spread, almost flat concentration profiles for carbon and oxygen, as expected for the uniform concentration of these elements in

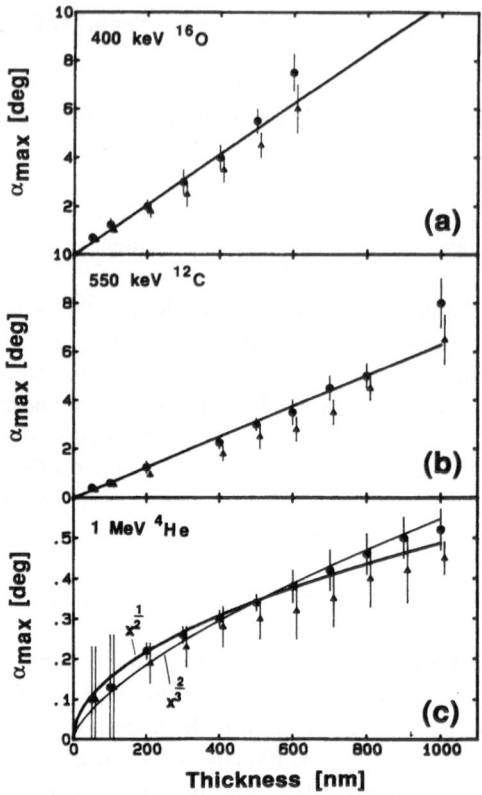

Figure 9.16. Calculated width FWHM ≈ α_{max} of angular distributions. *Full circles*: Width for (a) ^{16}O, (b) ^{12}C, and (c) ^{4}He ions transmitted through polycarbonate foils as a function of foil thickness x determined from a fit of Gaussian distributions to distributions calculated by TRIM.[25] *Triangles*: Fitted values α_{max} assuming an exponential distribution. *Thick solid lines*: $\alpha_{max} = 0.0104°$ and x (nm) for ^{16}O; $\alpha_{max} = 0.0063°$ and x (nm) for ^{12}C; $\alpha_{max} = 0.0154° \cdot (x(nm))^{1/2}$ for ^{4}He. *Thin solid line*: $\alpha_{max} = (5.342 \cdot 10^3)° \cdot (x(nm))^{2/3}$ for ^{4}He. (Data from Hofsäss and coworkers[18])

polycarbonate film, are obtained (Figure 9.17c). The measured concentration of 2.6 × 10^{22} at./cm³ is somewhat lower than the expected values of 4.43 × 10^{22} at./cm³. Moreover the measured concentration ratio $N_{carbon}/N_{oxygen} \approx 10.8$ is about a factor of 2 larger than the expected ratio of about 5.3. This factor cannot be explained by uncertainties in the MS correction, since both correction functions are close to 1 in the near-surface region. It is therefore assumed that instrumental losses due to the simple internal coincidence set-up used are responsible for this discrepancy.

Depth resolution of these ERCS measurements was about 40–50 nm. The dominating factor limiting depth resolution is the energy resolution of the surface barrier detector detecting the recoils, which varies from 15 keV for ^{4}He ions to 45 keV for ^{16}O ions.[27] Energy straggling also contributes to depth resolution. In the near-

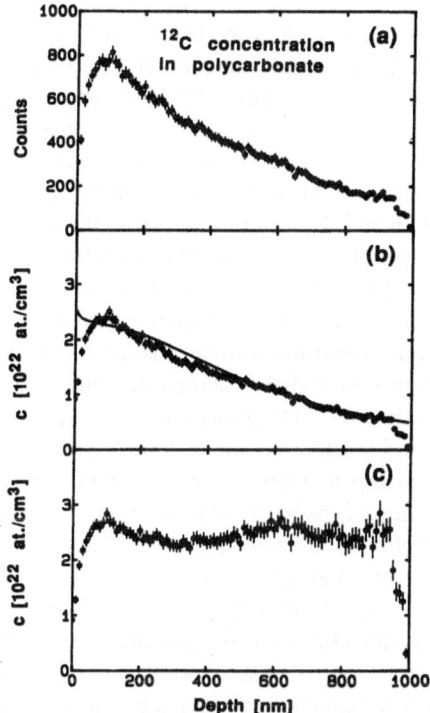

Figure 9.17. (a) Coincidence counts as a function of depth for C in polycarbonate, determined from the ERCS spectra shown in Figure 13. (b) Concentration-versus-depth profile corrected for scattering cross-section, total ^{4}He-dose, and detector solid angle. Also shown (*solid line*) is the angular spread correction function multiplied by a constant concentration $c_0 = 2.6 \times 10^{22}$ at./cm^3. (c) Final concentration-versus-depth profile including the correction for angular spread. (Data from Ref. 17)

surface region straggling of the incident ^{4}He is important, whereas at greater depth the energy straggling of recoiled atoms becomes more and more important. Energy straggling can be minimized by using thin samples; however depth resolution does not improve very much for samples below 2-μm thickness. Depth resolution is almost independent of the detected recoil mass, because improved detector energy resolution for lighter elements is compensated by a smaller stopping power. If the incident ^{4}He ion energy is increased to about 10 MeV, so that recoils like C and O atoms have energies of about 3 MeV, i.e., near the maximum of the stopping power curve, a depth resolution as low as about 20 nm seems possible.

9.7. POSITION-SENSITIVE DETECTORS FOR COINCIDENCE ERDA TECHNIQUES

Two-dimensional position-sensitive and also energy-resolving detectors appear very attractive for many ion-scattering techniques. However only a few attempts were made to use position-sensitive detectors for ERDA or in general IBA techniques to improve counting efficiency and angular resolution. The CERDA experiments using position-sensitive detectors were performed by Klein et al.[11] and Gebauer et al.[20] (see also Section 9.4). Depending on ion energy and ion mass, there are different ways of achieving position resolution with particle detectors. In the case of high ion energies, i.e., in the range of several tens of MeV, gas ionization counters using charge division read-out from a resistive anode[32] or multiwire proportional counters[20] are the best choice. For ion-scattering techniques using light projectiles like H or He with energies in the range of only a few MeV, which is typical for most RBS, ion-channeling or ERDA techniques, position-sensitive semiconductor detectors can be used.

Commercially available two-dimensional position-sensitive semiconductor detectors work on the principle of resistive charge division. A good overview of this principle is given by Laegsgaard.[33] Detectors consist of a p-i-n Si-diode of quadratic shape up to several cm^2 and a sensitive intrinsic layer several hundred micrometers thick. The p- and n-doped top and bottom layers of the pin-diode detector act as resistive layers. On each side of the detector are two electrode stripes at a distance ℓ collecting charges of electron hole pairs produced for example by ion impact at position (x,y). Charges collected at the four different electrodes in time coincidence are proportional to the quantities Ex/ℓ, $E(\ell - x)/\ell$, Ey/ℓ, and $E(\ell - y)/\ell$, where ℓ is the linear dimension of the detector. Thus the position (x,y) of the ion impact as well as the total ion energy can be derived. The relative position resolution $\Delta x/x$ and $\Delta y/y$ of a detector working with resistive charge division is given to a first approximation by its relative energy resolution $\Delta E/E$. The absolute position resolution is then simply proportional to the size of the detector, it is therefore very important to have a detector with an energy resolution as good as possible. An energy resolution of 40 keV for MeV α particles and a corresponding position resolution of 110–130 μm for a 2 cm × 2 cm commercially available detector has been obtained.[34] Schematics of a data-taking system required for the read-out of such a detector is shown in Figure 9.18. For fast and efficient data processing a multiparameter data acquisition system must be used. In addition to the set-up shown in Figure 9.18, a time signal representing TOF differences of scattered and recoiled particles can also be generated. Such a time signal was used by Moore et al.[6] for additional mass discrimination (see Figures 9.5 and 9.7 and Sections 9.3.2 and 9.4).

Detectors of this type were applied to measure channeling effects of α particles with energies of about 1.6 MeV emitted from radioactive ^8Li impurities implanted in various semiconductors. Lattice sites of ^8Li impurities can be determined from the emission distribution of α particles emitted around different crystallographic axes. With this technique the behavior of Li in different semiconductor materials can be studied.[34,35] Using position-sensitive detectors allowed the simultaneous recording of α emission distribution within an angular range of ± 3° with a resolution of about 0.1°. In this way data-taking rates up to 10 kHz were possible.

The sensitivity of proton-proton or α-α CERDA experiments similar to those performed by Cohen et al.,[4] Willemsen et al.,[9] or Klein[19] may be significantly

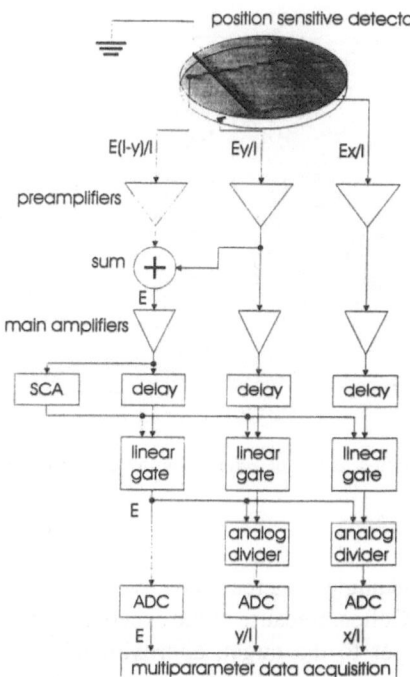

Figure 9.18. Data-taking equipment required for the read-out of position-sensitive semiconductor detectors based on the principle of resistive charge division. The three signal pulses $Ex//$, $E((-y)//$, and $Ey//$ from the position-sensitive detector are converted into energy-, x- and y-pulses by suited preamplifiers and two analog dividers. The single-channel analyzer and the linear gates discriminate a desired energy range and protect the following electronic components from unwanted coincidence events. The energy signal itself is used to monitor the proper setting of the discriminator levels and the energy resolution of the detection system. The (x,y,E) data triplets are stored in a multiparameter data acquisition system. (Data from Wahl *et al.*[34]).

improved by using position-sensitive detectors. For SRCS or ERCS experiments these detectors seem to be less favorable, because an angular resolution is not needed and energy resolution must be rather good. In some cases where good mass resolution is required, the mass resolution of an SRCS set-up may be improved with the help of position-sensitive detectors because the angular correlation between scattering angle and recoil angle given by Eq. 2.6 can also be used for mass identification.

For a proton-proton CERDA experiment a set-up using four position-sensitive detectors is proposed according to Figure 9.19. Two pairs of detectors on opposite sides detect scattering and recoil events enclosing angles of 90°. Because of the good position resolution, scattering events deviating from the 90° angle due to MS can also be resolved. Using detectors of 2 cm × 2 cm, the total active area for detecting scattered or recoiled particles is 1600 mm². For a target detector distance of 82.5 mm, as used

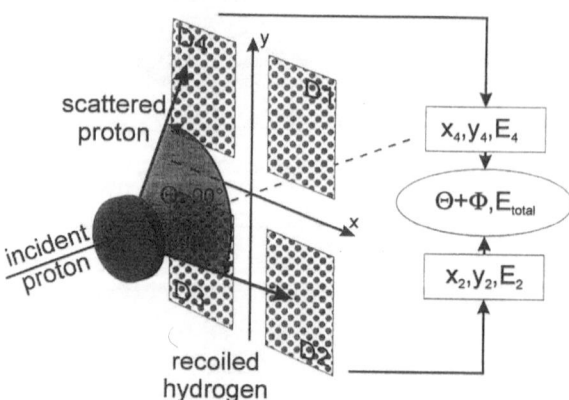

Figure 9.19. Proposed position-sensitive detector set-up for proton-proton scattering CERDA experiments.

by Cohen *et al.*[4] the solid angle is increased by two orders of magnitude compared to the 3.3×10^{-3} sr used by Cohen. Because of the excellent position resolution, detectors can be moved even closer to the target, approximately to 60 mm, which further increases the solid angle by a factor of 2. The limiting factor of such an experimental set-up is poorer energy resolution of large-area detectors and the maximum acceptable count rate for each detector, which should not exceed about 5 kHz. The incident beam current must therefore be limited to only a few pA; the proposed set-up appears to be very attractive for applications where only a small beam current is possible (for example microbeams) or necessary (radiation-sensitive samples like polymers).

9.8. CONCLUSION

Coincidence ERDA techniques can be applied to overcome the problem of mass–depth and projectile–recoil ambiguities. Detecting both scattered and recoiled particles in time coincidence allows the unambiguous identification of the mass as well as the kinetic energy of recoiled particles. A general feature of coincidence techniques is a significant reduction of background events and thus an increase of sensitivity. A further significant increase in sensitivity is achieved if large detector solid angles are used. A detection limit of a few ppm can be obtained; this is the case of single-element analysis with CERDA, discussed in Section 9.3, and in principle also for SRCS and ERCS, described in Sections 9.5 and 9.6, respectively.

The application of coincidence techniques requires at least one particle to be transmitted through the sample, and therefore it can be applied only to free-standing thin films a few micrometers thick.

Coincidence techniques that we denoted as CERDA (Sections 9.3 and 9.4) use coincidence detection essentially to discriminate a specific recoiled particle. A

CERDA spectrum is therefore similar to a conventional ERDA spectrum; i.e., the number of recoiled particles is measured as a function of recoil energy for a given well-defined recoil angle. The SRCS and ERCS techniques however are based on a different measurement principle. The key idea is to replace the measurement of scattering and recoil angles by measuring energies of scattered and recoiled particles in time coincidence. Major advantages of this principle are the ability to use large detector solid angles without sacrificing depth resolution and the possibility of simultaneously depth profiling different elements in a thin-film sample. Ideal applications of SRCS or ERCS are given when low doses of the analyzing beam are required or only a low beam current of a few picoamperes is available. Therefore SRCS or ERCS can be applied to profile light elements in polymers or depth profile with microprobes. Another application may be the analysis of multielement coatings deposited on free-standing thin substrates, such as carbon foils or ultrathin Si wafers.

The high sensitivity of CERDA measurements, where only scattered and recoiled particles of identical mass are detected, such as proton-proton or helium-helium scattering, was demonstrated in a small number of experiments discussed in Section 9.3. Especially when large-area position-sensitive detectors are used, this CERDA technique may become an attractive method of profiling light elements in a thin film at concentration levels of a few parts per million.

REFERENCES

1. Pretorius, R., Peisach, M., and Mayer, J. W., Hydrogen and deuterium depth profiling by elastic recoil detection analysis, *Nucl. Instrum. Methods Phys. Res. Sect. B* **35**, 478 (1988).
2. Pászti, F., Szilágyi, E., and Kótai, E., Optimization of the depth resolution in elastic recoil detection, *Nucl. Instrum. Methods Phys. Res. Sect. B* **54**, 507 (1991).
3. Kitamura, A., Matsui, S., Furuyama, Y., and Nakajima, T., A double-detector method for precise identification of the depth location of light atoms in ERD analysis, *Nucl. Instrum. Methods Phys. Res. Sect. B* **51**, 446 (1990).
4. Cohen, B. L., Fink, C. L., and Degnan, J. H., Nondestructive analysis for trace amounts of hydrogen, *J. Appl. Phys.* **43**, 19 (1972).
5. Smidt, F. A., Jr., and Pieper, A. G., Studies of the mobility of helium in vanadium, *J. Nucl. Mater.* **51**, 361 (1974).
6. Moore, J. A., Mitchell, I. V., Hollis, M. J., Davies, J. A., and Howe, L. M., Detection of low-mass impurities in thin films using MeV heavy-ion elastic scattering and coincidence detection techniques, *J. Appl. Phys.* **46**, 52 (1975).
7. L'Ecuyer, J., Brassard, C., Cardinal, C., Chabbal, J., Deschênes, L., Labrie, J. P., Terreault, B., Martel, J. G., and St.-Jaques, R., An accurate and sensitive method for the determination of the depth distribution of light elements in heavy materials, *J. Appl. Phys.* **47**, 381 (1976).
8. Wielunski, M., and Möller, W., A simple coincidence method of deuterium profiling using the $D(^3He,\alpha)H$ reaction, *Nucl. Instrum. Methods Phys. Res. Sect. B* **50**, 23 (1990).
9. Willemsen, M. F. C., Theunissen, A. M. L., and Kuiper, A. E. T., Hydrogen profiling by proton-proton scattering, *Nucl. Instrum. Methods Phys. Res. Sect. B* **15**, 492 (1986).
10. Rijken, H. A., Klein, S. S., and Devoigt, M. J. A., Improved depth resolution in CERDA by recoil time of flight measurement, *Nucl. Instrum. Methods Phys. Res. Sect. B* **64**, 395 (1992).

11. Klein, S. S., Mutsaers, P. H. A., and Fischer, B. E., Mass selection and depth profiling by coincident recoil detection for nuclei in the middle-mass region, *Nucl. Instrum. Methods Phys. Res. Sect. B* **50**, 150 (1990).

12. Habraken, F. H. P. M., Light-element depth profiling using elastic recoil detection, *Nucl. Instrum. Methods Phys. Res. Sect. B* **68**, 181 (1992).

13. Forster, J. S., Leslie, J. R., and Laursen, T., Scattering recoil coincidence spectrometry: A new experimental technique for profiling hydrogen isotopes in low-Z thin films, *Nucl. Instrum. Methods Phys. Res. Sect. B* **45**, 176 (1990).

14. Forster, J. S., Leslie, J. R., and Laursen, T., Depth profiling of hydrogen isotopes in thin, low-Z films by scattering recoil coincidence spectrometry, *Nucl. Instrum. Methods Phys. Res. Sect. B* **66**, 215 (1992).

15. Chu, W. K., and Wu, D. T., Scattering recoil coincidence spectrometry, *Nucl. Instrum. Methods Phys. Res. Sect. B* **35**, 518 (1988).

16. Chu, W. K., Large-angle coincidence spectrometry for neutron depth profiling, *Rad. Eff. and Defects in Solids* **108**, 125 (1989).

17. Hofsäss, H. C., Parikh, N. R., Swanson, M. L., and Chu, W. K., Depth profiling of light elements using elastic recoil coincidence spectrometry (ERCS), *Nucl. Instrum. Methods Phys. Res. Sect. B* **45**, 151 (1990).

18. Hofsäss, H. C., Parikh, N. R., Swanson, M. L., and Chu, W. K., Elastic recoil coincidence spectroscopy (ERCS), *Nucl. Instrum. Methods Phys. Res. Sect. B* **58**, 49 (1991).

19. Klein, S. S., Separate determination of concentration profiles for atoms with different masses by simultaneous measurement of scattered projectile and recoil energies, *Nucl. Instrum. Methods Phys. Res. Sect. B* **15**, 464 (1986).

20. Gebauer, B., Fink, D., Goppelt, P., Wilpert, M., and Wilpert, T., Multidimensional ERDA measurements and depth profiling of medium-heavy elements, *Nucl. Instrum. Methods Phys. Res. Sect. B* **50**, 159 (1990).

21. Chu, W. K., Meyer, J. W., and Nicolet, M. A., *Backscattering Spectrometry* (Academic, New York, 1978).

22. Ziegler, J. F., Biersack, J. P., and Littmark, U., *The Stopping and Ranges of Ions in Solids* (Pergamon, New York, 1985).

23. Parikh, N. R., Frey, E. C., Hofsäss, H. C., Swanson, M. L., Downing, R. G., Hossain, T. Z., and Chu, W. K., Neutron depth profiling by coincidence spectrometry, *Nucl. Instrum. Methods Phys. Res. Sect. B* **45**, 70 (1990).

24. Havranek, V., Hnatowicz, V., Kvitek, J., Vacik, J., Hoffmann, J., and Fink, D., Neutron depth profiling by range-angle coincidence spectroscopy, *Nucl. Instrum. Methods Phys. Res. Sect. B* **73**, 523 (1993).

25. Möller, W., Hufschmidt, M., and Kamke, D., Large depth profile measurements of D, ^{3}He, and ^{6}Li by deuteron-induced nuclear reaction, *Nucl. Instrum. Methods Phys. Res.* **140**, 157 (1977).

26. Turos, A., and Meyer, O., Depth profiling of hydrogen by detection of recoiled protons, *Nucl. Instrum. Methods Phys. Res. Sect. B* **4**, 92 (1984).

27. O'Connor, D. J., Tan, Chunyu, Application of heavy ions to high depth resolution RBS, *Nucl. Instrum. Methods Phys. Res. Sect. B* **36**, 178 (1989).

28. Cheng, H.-S., Zhou, Z.-Y., Yang, F.-C., Xu, Z.-W., and Ren, Y.-H., Depth profiling of hydrogen in thin films with the elastic recoil detection technique, *Nucl. Instrum. Methods Phys. Res.* **218**, 601 (1983).

29. Meyer, L., Plural and multiple scattering of heavy particles, *Phys. Status Solidi B* **44**, 253 (1971).

30. Sigmund, P., and Winterbon, K. B., Small-angle multiple scattering of ions in the screened Coulomb region. I. Angular distribution, *Nucl. Instrum. Methods Phys. Res.* **119**, 541 (1974).

31. Marwick, A. D., and Sigmund, P., Small-angle multiple scattering of ions in the screened Coulomb region. I. Lateral spread, *Nucl. Instrum. Methods Phys. Res.* **126**, 317 (1975).

32. Assmann, W., Hartung, P., Huber, H., Staat, P., Steffens, H., and Steinhausen, Ch. Set-up for materials analysis with heavy-ion beams at the Munich MP tandem, *Nucl. Instrum. Methods Phys. Res. Sect. B* **85**, 726 (1994).

33. Laegsgaard, E., Position-sensitive semiconductor detectors, *Nucl. Instrum. Methods Phys. Res.* **162**, 93 (1979).
34. Wahl, U., Hofsäss, H., Jahn, S. G., Winter, S., and Recknagel, E., Lattice site changes of ion-implanted ^{8}Li in InP studied by alpha emission channeling, *Nucl. Instrum. Methods Phys. Res. Sect. B* **64**, 221 (1992).
35. Hofsäss, H., Wahl, U., Restle, M., Ronning, C., Recknagel, E., and Jahn, S. G., Lattice sites of ion-implanted Li in indium antimonide, *Nucl. Instrum. Methods Phys. Res. Sect. B* **85**, 468 (1994).
36. Wahl, U., Restle, M., Ronning, C., Hofsäss, H., and Jahn, S. G., Li on bond-center sites in Si, *Phys. Rev. B: Condens. Matter* **50**, 2176 (1994).

10

Instrumental Equipment

10.1. INTRODUCTION

Chapter 10* does not present a detailed review of IBA instrumentation. The general equipment for IBA, in particular accelerators, analyzing magnets, focusing lenses, goniometers, detection devices, and related equipment are reviewed in detail in classic IBA handbooks[1,2] with abundant literature references and excellent recommendations for proper operation. Here we focus on equipment especially adapted to ERDA and topics that require more attention due to the particular requirements of ERDA. Our purpose is to give a sufficient extent of experimental details for any IBA user interested in performing elastic recoil measurements.

As a matter of fact, in Chapters 5–9, we describe and illustrate different versions of elastic recoil spectrometry. Some information concerning specific detection devices were yet given. In the following sections, we describe a typical experimental set-up required to carry out ERDA investigations. From the ion accelerator to usual detection devices, each element is briefly presented. Furthermore we emphasize the analytic constraints and the maximum performance allowed by recent technical developments.

10.2. ACCELERATOR AND RELATED EQUIPMENT

10.2.1. Accelerator

The most currently used accelerators for IBA purpose are of the electrostatic type. The literature has produced several excellent reviews about the physics of these devices, for example Refs. 3 or 4. Recent instrumental improvements are illustrated in specific documents.[5]

Three types of machines have essentially been used for ERDA applications a long time to produce the required incident ion beam: single-ended Van de Graaff accelerators,

*The authors wish to thank Ph. Trouslard for his kind contribution to this chapter.

tandem accelerators, and cyclotrons. These accelerators define respectively the three domains of ERDA measurements that can be considered. Single-ended Van de Graaff from 2–6 MV are used to perform low-energy ^{4}He$^+$-induced elastic recoil analysis.[6,7] Tandem accelerators are used to carry out heavy-ion-induced ERDA with incident beams from ^{12}C to ^{197}Au in the energy range of 0.5–2.5 MeV/amu.[8,9] Cyclotrons are generally used to perform recoil spectrometry induced by high-energy helium ions or HI-ERDA.[10]

The two main requirements for an accelerator used for IBA are its energy stabilization and the calibration of the beam energy. Energy stabilization is a determinant factor in the energetic dispersion of recoils and thus influences both mass and depth resolutions. Long-term fluctuations or drifts are generally well-controlled by modern stabilizing systems. Yet an insidious cause for energy drift may be a thermal drift of Hall-effect magnetic field probes when high power has to be applied to the analyzer magnet. Short-term energy fluctuations are more difficult to control, and a residual ripple always exists. Residual ripple depends on the stability of the magnet and its power supply, which should be at least 10^{-3} in relative value, and on the time of response of sensing and reaction electronics. Generally ripple damping is obtained by reaction to low- and high-energy deviations detected by slits following the analyzer magnet and to short-term ripple detected by a capacitive sensor. The gains of both circuits should be adjusted carefully for a minimum ripple.

Energy ripple has a secondary effect because it is converted in the analyzer magnet into an angular ripple that results in rapid displacements of the beam on the target surface and an enhancement of incertitude on the incidence angle. Since deflections outside the collision plane have less influence on incidence and recoil angles than a deflection in the collision plane, it is preferable to arrange the analysis chamber so that the collision plane is perpendicular to the median plane of the analyzer magnet. Curiously enough only a few laboratories have adopted this disposition.

Energy calibration is also determinant in calculating energy loss and cross sections that depend nonlinearly on incident particle energy. Energy calibration accuracy is thus essential. It is generally performed by using a few resonant reactions or alpha particles from a thin radioactive source to establish a few specific energies, then relies on the relation between energy and the field strength of the magnet. The reproducibility of this relation determines the long-term validity of the calibration.

10.2.2. Related Equipment

Several types of ion sources are used to produce ion beams suitable for ERDA experiments. A detailed review on ion sources was recently given by Alton.[11] Single-ended Van de Graaff are equipped with a conventional rf plasma source, able to produce intense beams of ^{4}He$^+$.[6,7] Tandems are generally equipped with charge exchange sources[8,9] duoplasmatron or sputtering sources, as presented by Neelmeijer.[12] These sources are able to produce intense highly multicharged beams like ^{35}Cl^{6+}, ^{127}I^{10+}, or ^{197}Au^{8+}.[8,9,11]

An analyzing magnet separates ions with different mass to charge ratios to select the desired species. The magnet must have minimum optical aberrations to avoid beam divergence. Its design[13] is too much a science in itself to be detailed here. The stability of the magnetic field is obtained by reaction to the signal of a sensor that is in general a Hall-effect probe.

10.3. BEAM LINE

The role of devices included in the beam line downstream the analyzing magnet is to shape the beam so that it satisfies the requirements of analysis; these are mainly beam collimation, the absence of beam divergence, adjustable beam intensity, and eventually scanning the target surface and beam intensity measurement, but they vary considerably depending on the chosen ERDA arrangement, the target material, and eventually other uses of the same beam line. In particular microbeams require more sophisticated focusing optics and collimation with associated scanning optics.

The design of focusing lenses is extensively described in specialized handbooks, e.g., Ref. 13. Two or three collimators are used both to delimit the desired beam width and insure the parallelism of the beam, which is essential for precisely defining the incidence angle. In reflexion geometry the beam diameter must not exceed 0.5 mm to keep scattering angle dispersion within reasonable limits (at 75° incidence, the largest dimension of a 0.5 mm beam is already about 2 mm). For transmission ERDA beam diameters on the order of 1–2 mm are acceptable. Because it is difficult to adjust the ion source intensity collimation in association with focusing is frequently used for fine beam intensity adjustments.

Scanning optics are used either to explore the target surface with a microbeam or to disperse beam damage, principally thermal damage, over a large enough target area. Scanning can also be accomplished by using target displacement, but this less expensive solution, although eliminating any risk of optical aberration, may not be rapid enough for certain applications.

Beam current measurement to determine the number of impinging ions generally takes place in the target chamber, but it is useful to have a beam-shape-monitoring system in the final part of the beam line. Many types of Faraday cups, Faraday cylinders, and other beam-monitoring systems are described in classical IBA handbooks.[1,2]

10.4. ANALYSIS CHAMBER

A typical IBA multipurpose vacuum chamber is shown in Figure 10.1. Its main specificity lies essentially in the presence of a large number of detectors, which allow it to perform simultaneously such different spectroscopic measurements as:

- Si-Li detector for PIXE
- Ge HP detector for PIGE
- Annular silicon surface barrier detector (SBD) for NRA

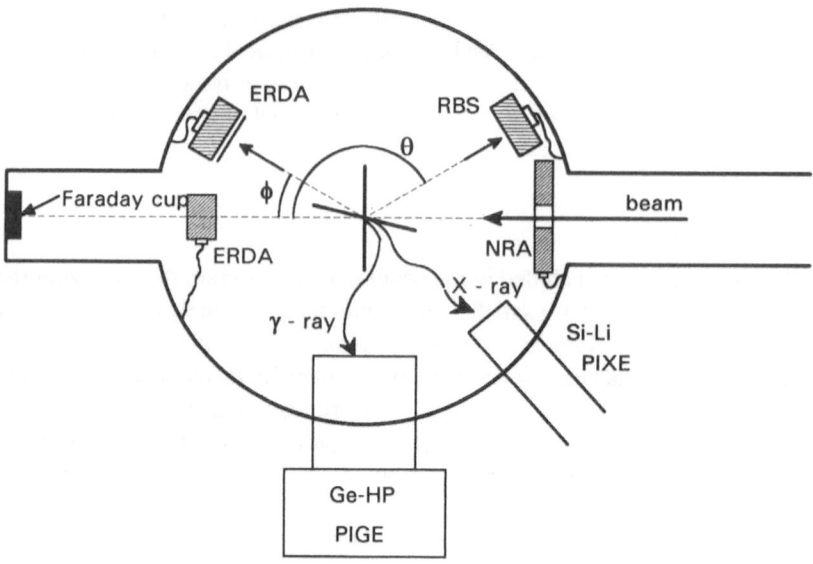

Figure 10.1. An IBA multipurpose vacuum chamber.

- Backward-geometry silicon detector for RBS and also useful for determining ion fluence in ERDA experiments
- Forward-geometry silicon detectors for reflexion and transmission ERDA

Since ERDA is very sensitive to surface contamination, a clean vacuum of at least 10^{-8} Torr is required, and trapping contaminating species is recommended. The target holder must have the quality of a precision goniometer to afford the required accuracy in determining scattering and incidence angles. It can be combined with a target exchange system.

10.4.1. Target Holder

The target holder is usually equipped with a goniometer with four free motions : three translations (x, y, z) relative to the geometrical center of the chamber and one rotation relative to an axis perpendicular to the standard ($\alpha = 0$ in Eqs. 4.4 and 4.5) collision plane. These motions are generally controlled by stepping motors, and sometimes they are computer-driven.[14] The precision required for the rotation motion as well as for the location of the forward detectors must be better than 0.1°. Two equivalent arrangements are possible for varying the angle between the sample surface and the collision plane to control the traversed length/depth ratio. This variation can be obtained either by sample tilt, adding a second rotation in the sample holder mechanism, or by rotating the detector around the beam axis. This choice is essentially determined by the conception of the chamber, and it leads to different systems of reference and geometric factors, discussed in Section 4.2.1.

10.4.2.   Beam Current Measurement

The exact measurement of beam intensity and its integration for determining total ion fluence are important requirements. For an initial adjustment of suitable intensity, taking into account conflicting requirements of sufficient counting rate and minimal beam damage to the target, mobile Faraday cups are quite appropriate. But integrating the total fluence Q that appears in the expression of yields (Eqs. 4.23 and 4.25) requires continuous measurement or high-frequency sampling of the beam current, which cannot be provided by a device occulting the beam. A Faraday cup enclosing the sample, which may be the chamber itself, is in principle an elegant solution, but it is hardly compatible with the constraints of a multipurpose arrangement. For low currents the dimensions of a Faraday cylinder would be prohibitive.

Attractive solutions are choppers or rotating tungsten wire probes[15]; the latter also allows beam-shape analysis. Alternatively we can detect ions scattered by a chopper or a rotating wire by using an SBD placed outside the beam path. However all these devices require accurate calibration against a Faraday cup.

More generally the target current returning to ground is taken as a measure of beam current. This is justified only if the target is properly polarized, typically at 250–300 V, or if polarized screens are used to repel secondary electrons, if there is no transmitted current through the target, and in the case of insulating targets, if no charge accumulation is released through dielectric break-down processes bypassing the normal current circuit.

A very convenient way of determining ion fluence is to detect simultaneously RBS and ERDA spectra in the target. The RBS yield can then be used to rescale the factor Q in ERDA yields. In transmission configurations when the range of primary ions exceeds the sample thickness, transmitted current must be measured, e.g., with a Faraday cup, in addition to the target-to-ground current. This involves neglecting the generally low current of scattered ions. It may be more suitable to resort to a chopper or a rotating tungsten wire probe.

10.4.3.   Vacuum and Trapping of Contaminants

Surface contamination is a permanent concern for ERDA. Practically it controls the possibility to analyze the near-surface region of samples. The target chamber should thus be kept under the best possible vacuum, which does not mean only the lowest possible residual pressure, at least 10^{-8} Torr for hydrogen profiling, but also a vacuum as exempt as possible of chemical species liable to be easily adsorbed on the target. Hydrogen contamination is particularly disturbing for usual ERDA applications, but we should be aware that this expression refers much more frequently to contamination by hydrogen-containing species than to molecular hydrogen; water and hydrocarbons are the most frequently encountered. Hydrocarbons can be avoided by using clean pumping systems: turbomolecular, cryogenic, and ionic pumps. The main source of water contamination is air inlet; sample exchange through a vacuum lock system is thus highly recommended, as is time-sparing. When necessary dry inert gas inlet is always preferable to air inlet.

Cryogenic trapping is an efficient way of removing contaminating species, particularly water vapor. It can be used in a differential vacuum channel between the beam line and the target chamber. Trapping surfaces that surround the target as completely as possible is an excellent solution, but they are not always compatible with a free variation of scattering and incidence angles. Conversely target cooling, while effective for reducing thermal beam damage, is limited by the risk of surface contamination below room temperature.

10.5. DETECTION DEVICES

Detection devices are an important part of the set-up, since they determine the sensitivity and resolution of measurements, once experimental conditions have been optimized for a given analysis. We give a short review of the various devices currently used in ERDA and then some data concerning typical performances.

10.5.1. Description

Several types of detection systems are available for ERDA. In addition to their description, we give some comments about energy calibration and such associated devices as collimator and absorber foils, which have a strong influence on the accuracy of IBA.

10.5.1.1. Silicon Barrier Detectors

Classical detectors, such as silicon surface barrier detectors, are currently used for ERDA as well as $^4\text{He}^+$-induced recoil spectrometry,[16] as in the case of HI-ERDA.[17,18] The theory and technology of silicon detectors and associated electronics are extensively described in classic handbooks.[19,20] Their main advantages are their small size and an intrinsic efficiency practically equal to unity for any ion species, which simplifies interpreting yields (cf. Eq. 4.26, where the ratio ξ_j/ξ_i disappears). On the other hand, they are rather sensitive to irradiation damage, and their energy response is only nearly linear.

Silicon surface barrier (SB) and passivated implanted planar silicon (PIPS) detectors have a metallic window or a dead layer that are traversed by ions before entering the active detector material. Dead layers of PIPS detectors are typically equivalent to 50-nm silicon, and they induce a significant energy loss before detection. Moreover electron-hole pair formation energy in the detector depends on ionization density, which varies with the ion-stopping power. Thus the detected energy E_{det} is lower than the energy E_{part} of the incident particle, and the former depends on the detected species[21,22]:

$$E_{det} = \varepsilon_0 \int_0^{E_{part}-\Delta E_w-\Delta E_n} \frac{d E}{\varepsilon_0 - kS(E)} \qquad (10.1)$$

where ΔE_w is energy lost in the window or dead layer, ΔE_n is energy dissipated without electron-hole formation in nuclear collisions (i.e., the part of the ballistic energy transfer that does not contribute to electron-hole pair creation by electronic stopping of recoils), ε_0 is the formation energy for one electron-hole pair (3.67 eV in silicon), and k is a factor scaling the decrease in electron-hole pair creation energy with increasing stopping power in the detected ion. When the term kS(E) is neglected, Eq. 10.1 reduces to:

$$E_{part} = E_{det} + \Delta E_w + \Delta E_n \qquad (10.2)$$

Both energy loss in the dead layer and mass dependence of the electron-hole pair formation energy have to be taken into account in the energy calibration of the system (see Section 11.4.3). A method for finding effective window thickness and the relative energy per electron-hole pair for various incident particles is given in Ref. 22. In the k = 0 approximation, effective window thickness can also be determined with a fair precision by tilting the detector and measuring the resulting energy shift.

10.5.1.2. Energy Calibration of the Detection System

Energy calibration of the system must be determined very carefully for accurate profile reconstitution, as emphasized in Section 11.4.3. Since the effective electron-hole pair creation energy depends on the ion-stopping power, calibration must be performed *separately* for each ion species. Nevertheless provided E_{part} is not exceedingly small, energy loss is initially almost entirely due to electronic stopping, and nuclear energy loss is concentrated near the end of the primary and secondary ion paths. Then ΔE_n is almost constant for a given ion species. If in addition k = 0, it represents a shift of the whole energy scale that is canceled by energy calibration and energy calculation using this calibration.

Now ΔE_w is strongly energy-dependent; as a consequence it cannot be overlooked, so a straightforward calibration using E_{det} values associated to corresponding channel numbers is a rather rough approximation, as shown in Figure 10.2. Such a calibration is indeed almost linear, but discrepancies appear at intermediate energies, and they may become rather large when energies are extrapolated outside the range of calibration energies. This occurs because the stopping power is not linear as a function of energy. It is much more nearly exact to associate channel numbers with approximated E_{part} values evaluated using

$$E_{part} = E_{det} + \Delta E_w = E_{det} + t_w S(E_{det}) \qquad (10.3)$$

where ΔE_w is deduced in first approximation from the window thickness t_w. This procedure consists of assuming k = 0, then use Eq. 10.2 to take advantage of the previously mentioned possibility of treating ΔE_n as a constant shift in the energy scale.

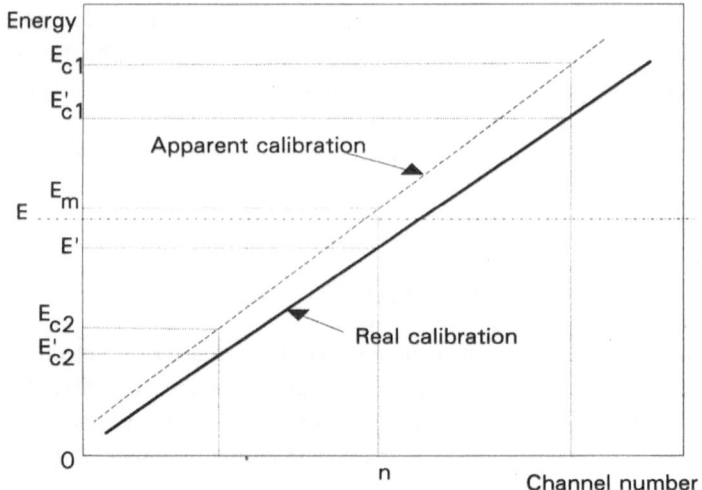

Figure 10.2. Influence of energy loss in the detector dead layer on energy calibration of the detection system. Only two calibration points corresponding to energies E_{c1} and E_{c2} are shown. The real calibration is linear and relates channels to detected energies that are lower than E_{c1} and E_{c2} by the energy lost in the dead layer, which itself depends on the stopping power at these energies. When energies are interpreted according to *apparent* calibration, which relates channels to energies of calibration particles impinging on the detector, a particle impinging with energy E is detected in channel n with energy E' according to the *real* calibration, but it is interpreted as though its energy were E_m, somewhat different from E, according to *apparent* calibration.

10.5.1.3. Detection Collimators and Absorber Foil

Detection collimators are generally placed in front of surface barrier detectors to well define detection angle and prevent ions from reaching the nonhomogeneous edge of the detector. Circular or rectangular collimators are frequently used. Rectangular slits are preferable for minimizing the energy spread because the deviation perpendicular to the nominal collision plane corresponds to a smaller deviation from the nominal recoil angle than an equal deviation in the collision plane. Their optimal aspect ratio is given by:

$$L = 2(WD \tan \phi_0)^{1/2} \qquad (10.4)$$

where L and W are the slit length and width, respectively, D the detector to beam spot distance, and ϕ_0 the nominal recoil angle. More accurately the slit should approach as nearly as possible a section of a cone defined by a constant recoil angle. On the basis of precise calculations including energy loss effects on energy shifts over the slit area, Brice and Doyle[23] recommend a curved shape resembling a bent rectangle with an aspect ratio defined by:

$$L = 4(WD \tan \phi_0)^{1/2} \tag{10.5}$$

and with a radius of curvature at the slit center:

$$R_c = D \tan \phi_0 \tag{10.6}$$

These curved slits provide a 50% improvement in energy resolution over rectangular slits at least over a 500-nm depth range. Determination of detection solid angles for usual collimator shapes is discussed in Appendix B.

In reflexion geometry, an absorber foil is also placed in front of the detector for stopping scattered projectiles and recoils heavier than the species to be analyzed. The thickness of this absorber foil must be as perfectly constant as possible in order to avoid any path length variation that would degrade the energy resolution (cf. section 4.3.2). Mylar or polyethylene are particularly suitable from this point of view.

10.5.1.4. Other Detection Devices

In addition to simple barrier detectors, many specific detection devices were also used or specially developed since the beginning of the eighties, for example, the *ExB* filter,[24] the telescope,[25,26] the electrostatic spectrometer,[27] the magnetic spectrometer,[28-30] the gas ionization chamber,[31] and the TOF spectrometer.[32,33] Specific detection configurations were also introduced to improve performances allowed, such as coincidence spectrometry, as presented by Hofsäss[34] and Klein.[35] Note that gas ionization chambers and electrostatic time detectors for TOF-ERDA (cf. Section 6.5.1) require us to take into account their variable efficiency for different ion species.

The basic principles, intrinsic performances, and some application examples of TOF spectrometers, *ExB* filters, $\Delta E - E$ telescopes, and coincidence spectrometry were discussed in details in Chapters 6–9. Therefore we present here only the characteristics of the other detection devices.

An electrostatic spectrometer was described in 1994 by Kruse and Carstanjen.[27] It is composed (see Figure 10.3) of four quadrupole lenses that focus particles emitted parallel to the optical axis onto the entrance slit of an analyzer. Then an electrostatic analyzer with a 100° cylindric sector field of 70-cm radius and 20-mm gap width images particles of different energies onto different positions of a standard linear position-sensitive detector.

Using a magnetic spectrometer for ERDA measurements was proposed by several authors, for example Gossett,[28] Dollinger[29] and Boerma.[30] A magnetic spectrometer separates particles as a function of their mass/charge ratio, but it must also focus the particles of different energies onto a detector. This requires multiple optical corrections. As an example Figure 10.4 shows the Q3D spectrometer (quadrupole followed by three dipoles) used in Garching. Dipoles 2 and 3 are devoted to producing correction fields (quadrupolar, hexapolar, octupolar, and decapolar) perpendicular to the beam direction to compensate for kinematic effects. The energy resolution of the system is as low as 4×10^{-4}, although a large solid angle of 5 msr is used.[25]

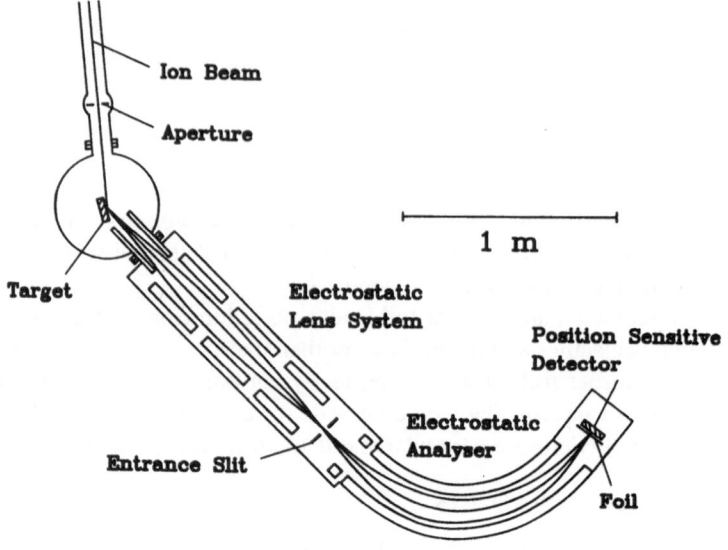

Figure 10.3.   High-resolution ERDA set-up using an electrostatic analyzer. (From Ref. 27)

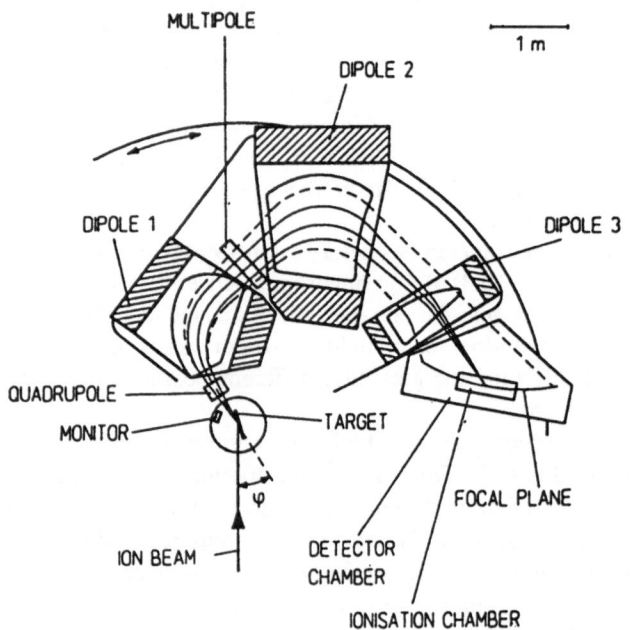

Figure 10.4.   Experimental arrangement for high-resolution depth profiling in ERDA by using a Q3D magnetic spectrograph. (From Ref. 29)

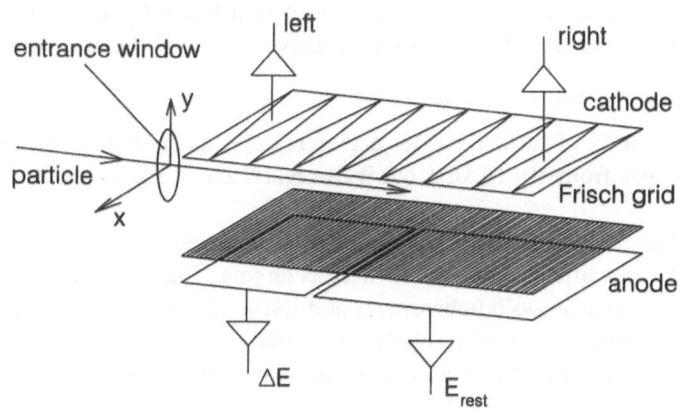

Figure 10.5. Ionization detector layout and signal generation. (From Ref. 31)

Using a gas ionization chamber is discussed in Chapter 8, which is devoted to the telescope ERDA technique. A typical device described recently by Assmann and coworkers[31] is shown in Figure 10.5. The cathode electrode and entrance window of this detector are at ground potential. The distance between the Frisch grid and the cathode is 10 cm, and the distance between the grid and the anode is 2 cm. The length of the active gas cell is 28 cm. The entrance window consists of a stretched 50-$\mu g/cm^2$ polypropylene foil supporting a 75% transmission grid. The cathode-to-grid voltage is around 500 V, and the grid-to-anode voltage is around 1000 V for a gas pressure of about 50 mbar. The effective solid angle of this system is 7.5 msr.[31]

Gas ionization chambers have the considerable advantage of being practically insensitive to irradiation damage except for their windows. Another interesting feature is the relatively low energy loss and straggling they induce, as compared to thin-silicon detectors. A gas chamber is typically equivalent to 10-μm silicon or even less, depending on the ion energy. However the rather large volume of a gas chamber may require a special target chamber.

Klein and coworkers recently proposed a new method of separating recoil nuclei and scattered alpha particles by using pulse shape discrimination.[37] This method consists of suppressing pulses caused by scattered incident particles. In the case described in Ref. 37, the pulse heights from 13.4-MeV alpha particles and 4-MeV recoils nearly coincides. Their typical PSD set-up is composed of a low-biased 500 Ω cm silicon detector at an angle of about 30° with respect to the incident beam direction, with a depleted layer of about 10 μm. Pulses are discriminated by using a time-to-amplitude converter started by the timing output of a fast amplifier and stopped by a timing single-channel analyzer in the zero-crossing mode.

Position-sensitive detectors were also tested to determine simultaneously charged particle position and energy as reviewed by Laegsgaard in 1979.[38] Resistive charge division is used to discriminate pulses produced at different positions in the detector; a special section is dedicated to position-sensitive detectors* (Section 9.7).

10.5.2 Usual Performances

The resolution of surface-barrier detectors depends on their size and quality. It generally ranges from 10–16 keV, but it can reach about 7 keV for the best quality detectors.

Note that performances of a solid-state detector are not guaranteed for ever. Radiation damage affects detectors as well as targets. The dose rate is considerably lower, but the exposure to bombardment and ion implantation is accumulative over many experiments. A loss of resolution is already observable for an accumulated fluence of 10^8–10^9 He/cm^2 and complete failure around 10^{11} He/cm^2.[39] Heavier ions are liable to cause even more damage. On the whole the life span of a solid-state detector is not expected to exceed a few years.

By using conventional MeV ^{4}He$^+$-induced recoil spectrometry, two geometry options are available. Analyzable depth varies from a few hundred nanometers in reflexion geometry to several micrometers in transmission geometry, strongly depending on the incident energy value.[6,40] Thus in a polymer film, analyzable depth ranges from 2.2 to 6.2 µm for ^{4}He$^+$ energies from 1.8 to 3 MeV.[40] A current depth resolution of about 25–35 nm can be reached at the sample surface in reflexion geometry.[6] In transmission geometry using 3 MeV ^{4}He$^+$ ions, total depth resolution varies from 38 nm at the sample surface to 29 nm at a 6-µm depth in a polymer sample for example.[40]

An improvement in depth resolution to 10 nm was shown by Pászti and coworkers through a suitable choice of both the recoil angle and the nature and thickness of the absorber foil.[41] The detection limit for hydrogen or deuterium analysis is around 50 weight ppm in usual cases,[6] but a value as low as 1 at. ppm was reported for transmission ERDA of hydrogen in thin films.[42]

High-energy alpha particles allow us to obtain an ERDA probing depth of several micrometers in reflexion geometry. A strong improvement in the depth resolution to 4–5 nm was obtained by using a TOF spectrometer and the coincidence method.[43]

HI-ERDA offers a wide range of performances, depending on the mass and energy of the incident ion beam and the type of detection device used. Typical depth resolution values are in the range of 20–80 nm.[8,10,29,31–33,35,44–56] Some authors (for example Thomas[57] and Dollinger[58]) obtained 1-nm depth resolution under exceptionally favorable conditions. The detection limit is typically around 10^{15} at./cm^2.[57]

10.6. CONCLUSION

Classical multipurpose IBA vacuum chambers are very well-suited for performing conventionnal ERDA measurements in glancing or transmission geometry. The

*Note that PSD is indifferently used for both pulse shape discrimination and position sensitive detector.

choice of a suitable experimental arrangement is essentially related to the type of target under consideration (thin film or bulk sample, smooth or rough surface, etc.) and to the particular purpose of the analysis. The latter determines what factors require more attention, for example, mass–depth or recoil–projectile ambiguities, depth resolution, or detection limit. Coincidence spectroscopy and ΔE – E telescope need only a careful adaptation of the electronics of the data acquisition system. Specific modifications are required only for using particular detection devices, for example, TOF spectrometers, *ExB* filters, or magnetic or electrostatic spectrometers

Technical improvements are continuously progressing in the instrumentation field both in terms of materials constituting the detector itself and associated electronic devices. Nevertheless a depth resolution of about 1 nm and a detection limit of nearly a few tenths of a monolayer appear to be optimum limits. We cannot stress too much the importance of a high and clean vacuum, since surface contamination rules the limits of the method for the near-surface region.

The last aspect in ERDA evolution seems strongly linked with progress in data analysis procedures, as discussed in Chapter 11. However limitations of these procedures themselves depend on the status of equipment and instrumentation in addition to such unsuperable physical limits as sensitivity, depth resolution, mass selectivity, mass-depth and recoil-projectile ambiguities.

REFERENCES

1. Chu, W.-K., Mayer, J. W., and Nicolet, M.-A., *Backscattering Spectrometry* (Academic, New York, 1978).
2. Bird, J. R., and Williams, J. S., *Ion Beams for Materials Analysis* (Academic, New York, 1989).
3. Scharf, W., *Particle Accelerators and Their Uses* (Harwood, Chur, Switzerland, 1991).
4. Bromley, D. A., ed., *Large Electrostatic Accelerators*, special issue of *Nucl. Instrum. Methods. Phys. Res.* **122**, 1 (1974).
5. *Pelletron: Accelerators/Ion Beam Systems* (product catalog) (National Electrostatics Corp., Middleton, Wisconsin, 1994).
6. Tirira, J., Contribution à l'étude de la collision hélion-4 proton et à la spectrométrie de recul élastique, thesis, CEA report CEA-R-5529, 1990.
7. Pászti, F., Kótai, E., Mezey, G., Manuaba, A., Pocs, L., Hildebrandt, D., and Strusny, H., Hydrogen and deuterium measurements by elastic recoil detection using alpha particles, *Nucl. Instrum. Methods. Phys. Res. Sect. B* **15**, 486 (1986).
8. Arnoldbik, W. M., de Laat, C. T. A. M., and Habraken, F. H. M. P., On the use of a *dE-E* telescope in elastic recoil detection, *Nucl. Instrum. Methods. Phys. Res. Sect. B* **64**, 832 (1992).
9. Whitlow, H. J., Johansson, E., Ingemarsson, P. A., and Hogmark, S., Recoil spectrometry of oil-additive-associated compositional changes in sliding metal surfaces, *Nucl. Instrum. Methods. Phys. Res. Sect. B* **63**, 445 (1992).
10. Rijken, H. A., Klein, S. S., and de Voigt, M. J. A., Improved depth resolution in CERDA by recoil time of flight measurement, *Nucl. Instrum. Methods. Phys. Res. Sect. B* **64**, 395 (1992).
11. Alton, G. D., Ion sources for accelerators in materials research, *Nucl. Instrum. Methods. Phys. Res. Sect. B* **73**, 221 (1993).
12. Neelmeijer, C., Grötzschel, R., Hentschel, E., Klabes, R., Kolitsch, A., and Richter, R., Ion beam analysis of steel surfaces modified by nitrogen ion implantation, *Nucl. Instrum. Methods. Phys. Res. Sect. B* **66**, 242 (1992).

13. Carey, D. C., *Optics of Charged Particle Beams* (Harwood, Chur, Switzerland, 1987).
14. Kótai, E., Computer methods for analysis and simulation of RBS and ERDA spectra, *Nucl. Instrum. Methods. Phys. Res. Sect. B* **85**, 588 (1994).
15. Section des accélérateurs, Centre d'Etudes Nucléaires de Grenoble, Sondes tournantes pour analyse des distributions spatiales de courant dans les faisceaux de particules, *Le Vide* **157**, 34 (1972).
16. Tirira, J., Trocellier, P., Mosbah, M., and Metrich, N., Study of hydrogen content in solids by ERDA and radiation-induced damage, *Nucl. Instrum. Methods. Phys. Res. Sect. B* **56/57**, 839 (1991).
17. Doyle, B. L., and Wing, N. D., The Sandia nuclear microprobe, Sandia report SAND 82-2393, 1982.
18. Nölscher, C., Brenner, K., Knauf, R., and Schmidt, W., Elastic recoil detection analysis of light particles (^{1}H - ^{16}O) using 30-MeV sulfur ions, *Nucl. Instrum. Methods. Phys. Res.* **218**, 116 (1983).
19. Tait, W. H., *Radiation Detection* (Butterworth, London, 1980).
20. Knoll, G. F., *Radiation Detection and Measurement* (Wiley, New York, 1989).
21. Hösler, W., and Darji, R., On the nonlinearity of silicon detectors and the energy calibration in RBS, *Nucl. Instrum. Methods. Phys. Res. Sect. B* **85**, 602 (1994).
22. Langley, R. A., Study of the response of silicon barrier detectors to protons and α-particles, *Nucl. Instrum. Methods. Phys. Res.* **113**, 109 (1973).
23. Brice, D. K., and Doyle, B. L., A curved detection slit to improve ERD energy and depth resolution, *Nucl. Instrum. Methods. Phys. Res. Sect. B* **45**, 265 (1990).
24. Ross, G. G, Terreault, B., Gobeil, G., Abel, G., Boucher, C., and Veilleux, G., Inexpensive quantitative hydrogen depth profiling for surface probes, *J. Nucl. Mater.* **128/129**, 730 (1984).
25. Prozesky, V. M., Churms, C. L., Pilcher, J. V., Springhorn, K. A., and Behrisch, R., ERDA measurement of hydrogen isotopes with a DE-E telescope, *Nucl. Instrum. Methods. Phys. Res. Sect. B* **84**, 373 (1994).
26. Hult, M., Whitlow, H. J., Ostling, M., Lundberg, N., Zaring, C., Cohen, D. D., Dytlewski, N., Johnston, P. N., and Walker S. R., RBS and recoil spectrometry analysis of $CoSi_2$ formation on GaAs, *Nucl. Instrum. Methods. Phys. Res. Sect. B* **85**, 916 (1994).
27. Kruse, O., and Carstanjen, H. D., High-depth resolution ERDA of H and D by means of an electrostatic spectrometer, *Nucl. Instrum. Methods. Phys. Res. Sect. B* **89**, 191 (1994).
28. Gossett, C. R., Use of a magnetic spectrometer to profile light elements by elastic detection, *Nucl. Instrum. Methods. Phys. Res. Sect. B* **15**, 481 (1986).
29. Dollinger, G., Elastic recoil detection analysis with atomic depth resolution, *Nucl. Instrum. Methods. Phys. Res. Sect. B* **79**, 513 (1993).
30. Boerma, D. O., Labohm, F., and Reinders, J. A., Design of a magnetic spectrograph for surface, interface, and thin-layer analysis, *Nucl. Instrum. Methods. Phys. Res. Sect. B* **50**, 291 (1990).
31. Assmann, W., Hartung, P., Huber, H., Staat, P., Steffens, H., and Steinhausen, Ch., Set-up for materials analysis with heavy ion beams at the Munich MP tandem, *Nucl. Instrum. Methods. Phys. Res. Sect. B* **85**, 726 (1994).
32. Goppelt, P., Gebauer, B., Fink, D., Wilpert, M., Wilpert, Th., and Bohne, W., High-energy ERDA with very heavy ions using mass and energy-dispersive spectrometry, *Nucl. Instrum. Methods. Phys. Res. Sect. B* **68**, 235 (1992).
33. Arai, E., Zounek, A., Sekino, M., Takemoto, K., and Nittono, O., Depth profiling of porous silicon surface by means of heavy-ion TOF-ERDA, *Nucl. Instrum. Methods. Phys. Res. Sect. B* **85**, 226 (1994).
34. Hofsäss, H. C., Parikh, N. R., Swanson, M. L., and Chu, W. K., Elastic recoil coincidence spectroscopy (ERCS), *Nucl. Instrum. Methods. Phys. Res. Sect. B* **58**, 49 (1991).
35. Klein, S. S., Mutsaers, P. H. A., and Fischer, B. E., Mass selection and depth profiling by coincident recoil detection for nuclei in the middle-mass region, *Nucl. Instrum. Methods. Phys. Res. Sect. B* **50**, 150 (1990).
36. Löffler, M., Scheerer, H. J., and Vonach, H., The ion optical properties of the Munich Q3D-spectrograph investigated by means of a special experimental ray-tracing method. *Nucl. Instrum. Methods. Phys. Res.* **111**, 1 (1973).
37. Klein, S. S., Rijken, H. A., Tolsma, H. P. T., and de Voigt, M. J. A., Elastic recoil selection by pulse shape analysis, *Nucl. Instrum. Methods. Phys. Res. Sect. B* **85**, 660 (1994).

38. Laegsgaard, I., Position-sensitive semiconductor detectors, *Nucl. Instrum. Methods. Phys. Res.* **162**, 93 (1979).
39. Chu, W.-K., Mayer, J. W., and Nicolet, M.-A., *Backscattering Spectrometry* (Academic, New York, 1978), p. 171.
40. Tirira, J., Trocellier, P., and Frontier, J. P., Analytical capabilities of ERDA in transmission geometry, *Nucl. Instrum. Methods. Phys. Res. Sect. B* **45**, 147 (1990).
41. Pászti, F., Szilágyi, E., and Kótai, E., Optimization of the depth resolution in elastic recoil detection, *Nucl. Instrum. Methods. Phys. Res. Sect. B* **54**, 507 (1991).
42. Wielunski, L., Benenson, R., Horn, K., and Lanford, W. A., High-sensitivity hydrogen analysis using elastic recoil, *Nucl. Instrum. Methods. Phys. Res. Sect. B* **15**, 469 (1986).
43. Rijken, H. A., Klein, S. S., van IJzendoorn, L. J., and de Voigt, M. J. A., Elastic recoil detection analysis with high-energy alpha beams, *Nucl. Instrum. Methods. Phys. Res. Sect. B* **79**, 532 (1993).
44. Gebauer, B., Fink, D., Goppelt, P., Wilpert, M., and Wilpert, Th., Multidimensional ERDA measurements and depth profiling of medium-heavy elements, *Nucl. Instrum. Methods. Phys. Res. Sect. B* **50**, 159 (1990).
45. Goppelt, P., Biersack, J. P., Gebauer, B., Fink, D., Bohne, W., Wilpert, M., and Wilpert, Th., Investigation of thin films by high-energy ERDA, *Nucl. Instrum. Methods. Phys. Res. Sect. B* **80/81**, 142 (1993).
46. Whitlow, H. J., Possnert, G., and Petersson, C. S., Quantitative mass and energy-dispersive elastic recoil spectrometry: Resolution and efficiency considerations, *Nucl. Instrum. Methods. Phys. Res. Sect. B* **27**, 448 (1987).
47. Siegele, R., Davies, J. A., Forster, J. S., and Andrews, H. R., Forward elastic recoil measurements using heavy ions, *Nucl. Instrum. Methods. Phys. Res. Sect. B* **90**, 606 (1994).
48. Siegele, R., Haugen, H. K., Davies, J. A., Forster, J. S., and Andrews, H. R., Forward elastic recoil measurements using heavy ions, *J. Appl. Phys.* **76**, 4524 (1994).
49. Habraken, F. H. M. P., Light-element depth profiling using elastic recoil detection, *Nucl. Instrum. Methods. Phys. Res. Sect. B* **68**, 181 (1992).
50. Klein, S. S., Separate determination of concentration profiles for atoms with different masses by simultaneous measurement of scattered projectile and recoil energies, *Nucl. Instrum. Methods. Phys. Res. Sect. B* **15**, 464 (1986).
51. Martin, J. W., Cohen, D. D., Dytlewski, N., Garton, D. B., Whitlow, H. J., and Russell, G. J., Materials characterisation using heavy-ion elastic recoil time of flight spectrometry, *Nucl. Instrum. Methods. Phys. Res. Sect. B* **94**, 277 (1994).
52. Ross, G. G., and Terreault, B., High-precision depth profiling of light isotopes in low-atomic-mass solids, *J. Appl. Phys.* **51**, 1259 (1980).
53. Oura, K., Naitoh, M., Morioka, H., Watamori, M., and Shoji, F., Elastic recoil detection analysis of coadsorption of hydrogen and deuterium on clean Si surfaces, *Nucl. Instrum. Methods. Phys. Res. Sect. B* **85**, 344 (1994).
54. Arai, E., Funaki, H., Katayama, M., Oguri, Y., and Shimizu, K., TOF-ERD experiments using a 10-MeV ^{35}Cl beam, *Nucl. Instrum. Methods. Phys. Res. Sect. B* **64**, 296 (1992).
55. Nagai, H., Hayashi, S., Aratani, M., Nozaki, T., Yanokura, M., Kohno, I., Kuboi, O., and Yatsurugi, Y., Reliability, detection limit, and depth resolution of the elastic recoil measurement of hydrogen, *Nucl. Instrum. Methods. Phys. Res. Sect. B* **28**, 59 (1987).
56. Arnoldbik, W. M., and Habraken, F. H. M. P., Elastic recoil detection, *Rep. Prog. Phys.* **56**, 859 (1993).
57. Thomas, J. P., Fallavier, M., Ramdane, D., Chevarier, N., and Chevarier, A., High-resolution depth profiling of light elements in high-atomic-mass materials, *Nucl. Instrum. Methods. Phys. Res.* **218**, 125 (1983).
58. Dollinger, G., Faestermann, T., and Maier-Komor, P., High-resolution depth profiling of light elements, *Nucl. Instrum. Methods. Phys. Res. Sect. B* **64**, 422 (1992).

Numerical Methods for Recoil Spectra Simulation and Data Processing

11.1. INTRODUCTION

An ERDA experiment results in a spectrum, but the final purpose is in general to determine one or several concentration profiles. The purpose may also be to determine such physical parameters as cross sections or stopping powers, using a target of well-known composition. This is a very similar problem, although the unknown variables are not the same. Considering the problem of profile determination, it is true that detected energy is related to the depth at which collisions occurred, and yield is related to the concentration at this depth. But there is many a slip 'twixt the cup and the lip. The energy is the result of collision kinematics and energy loss in the sample and eventually in an absorber foil. This latter depends on both target and foil compositions that determine their stopping power. Similarly the yield depends not only on composition but on scattering cross-section, which itself depends on energy loss before the collision. Moreover the yield from a target constituent located at a given depth does not appear in the spectrum at a well-defined energy, but it is spread around this energy by straggling, MS effects, and experimental energy dispersion. Finally if several isotopes or elements with neighboring masses are detected simultaneously, their contributions may overlap in some energy range, and then it is impossible at first sight to identify them in the total yield.

Note that the depth never appears alone in expressions of energy loss and yield (see for example Eqs. 4.8 and 4.21) but always as terms $N\delta x$, where N is the target atomic density, which appears explicitly in Eq. 4.21 and implicitly in Eq. 4.8 through the stopping power. As a consequence ERDA, like RBS, cannot really give access to concentration-versus-depth profiles but only to concentration versus a traversed amount of matter, expressed in atoms·cm^{-2} or equivalently in μg·cm^{-2}. The conversion into concentration-versus-depth profiles is possible only if the target density is precisely known. Yet for brevity in the following discussion we use the expressions depth profile and energy–depth relation, but note that except in Section 11.4.2, the word depth represents a traversed amount of matter (at·cm^{-2} or µg·cm^{-2}).

From another point of view, there is in principle no great skill in predicting the recoil energy spectrum when complete knowledge of the target description and experimental set-up is assumed to be available. Such a computation is called spectrum *simulation*. It can be realized using the corpus of physical laws presented in the first chapters of this book and abundant data compilations from the literature on stopping powers and eventually non-Rutherford cross sections. It is clear that the choice of the best available data determines the quality of the treatment of spectra. Nevertheless this choice is independent of processing methods, and it can be discussed separately. Following the physical processes experienced by the projectile and recoils from the ion impact on the target surface until detection of the recoil, it is possible to compute the entire recoil energy spectrum,[1–19] at least with the precision of the present physical knowledge. An historical review of simulation codes was recently given by Kótai.[17]

Spectrum simulation is most certainly useful in many cases for predicting the shape of the spectrum that would result from a given experiment and thus to check before performing the experiment that desired information will be accessible and such experimental parameters as the scattering angle, beam energy, and so on, are appropriately chosen. But what is more generally desired is to solve the inverse problem, i.e., find as nearly as possible the target description corresponding to a given experimental spectrum. We call this operation *profile extraction* or *profile determination*. This is a considerably more difficult task, not only because of the large number of unknown variables involved but also because of the ambiguities underlined in Section 5.2—mass–depth ambiguity, eventually projectile–recoil ambiguity—and because of the nonlinear form of different physical processes, typically energy straggling and MS, that forbids an analytic deconvolution.

There is some disagreement among authors on the appellation of different operations. Some authors (e.g. Ref. 16) call *synthesis* what we previously defined as *simulation*, and reserves the word simulation for what we call *profile extaction* or *profile determination*. In the following pages, we systematically use the expressions *simulation* and *profile extraction* or *profile determination* as previously defined. We reserve the appellation *profile reconstitution* for the particular method of iterative profile extraction[20] described in Section 11.4.1.4.

Different methods of extracting concentration profiles from spectra were developed.[1,3,6–8,10,11,16–21] These rely most generally on successive attempts to approach the experimental spectrum by simulating the spectrum resulting from a hypothetical description of the target. Two methodologies are possible: The first consists in seeking a fit of the spectrum by interactive changes to the initial description; the second attempts to optimize profiles by a semiautomatic procedure. Whatever the strategy adopted, we resort to simulation in the procedure and in checking results. Simulation algorithms are thus at the basis of any data processing of recoil spectra.

In Chapter 11 the process of spectrum simulation is first explained using the basic method that is common, with some practical variants, to almost all simulation algorithms. An alternative, physically equivalent, method, the so-called retrograde method,[14] is then presented with a short discussion of its possible advantages. We describe some possible approaches to the question of concentration profile determination, with particular attention to practical difficulties, possible sources of incertitude or misinterpretation, and experimental requirements for a reliable and accurate profile extraction. These sections are developed in the case of classic (transmission or reflection) ERDA. Adaptations required by other ERDA variants (*ExB*, coincidence, TOF-ERDA, telescope) are discussed in Section 11.6. Indeed concerning the physical processes inside the target, the treatment is essentially the same for any experimental set-up, and the main differences appear in the treatment of the detection process. Furthermore we give the main characteristics of different available algorithms and softwares and some examples of their use (Section 11.5).

11.2. SIMULATION PROCESS: BASIC METHOD

The most natural way of simulating a spectrum is to follow step-by-step the physical processes in their chronological order, from impact to detection of the scattered atom. In fact this procedure is applicable to the *ideal* path (cf. Section 4.2) of ions and recoils, overlooking all phenomena that involve energetic or angular dispersion: straggling, MS, energetic and angular beam ripple, finite-detection solid angle, and detector resolution. We can thus build an *ideal* simulated spectrum. To build a *realistic* spectrum, dispersion effects can be evaluated as functions of depth along the mean path, and superimposed on the *ideal* spectrum at a final stage of the computation. Almost all simulation softwares[1-19] proceed this way.

Initial data for a simulation are composed of a description of the target, different experimental parameters, and a set of stopping power and eventually non-Rutherford cross-sections data or formulae fitting these data. Data and formulae for the stopping power as a function of composition and when relevant non-Rutherford cross-sections are taken from the literature (cf. Chapters 3 and 4, respectively) for each component of the target and absorber foil. As for stopping powers, they may also be computed using a dedicated software, such as PRAL.[22] It is clear that the accuracy of stopping power and cross-section data are essential for the precision of calculations. This topic is abundantly discussed in Sections 2.4, 3.2, and 4.3.3 and does not require more comments here.

The target description represents its assumed composition as a function of depth. In most cases, it is presented as a succession of homogeneous slabs, commonly called layers or blocks. Their thickness is chosen as small as necessary to account for all details of composition variation with depth. Sometimes the description can be presented as a mathematical formula, e.g., in the case of a Gaussian diffusion or implantation profile.

Experimental data include mass and atomic number of ions and target elements, projectile energy, geometrical parameters (scattering angle, incidence angle, detector acceptance solid angle), data pertaining to energy calibration of the multichannel analyzer and to detection resolution, the absorber foil nature and thickness, and the number of incident particles.

Let us now examine the successive steps of a classical spectrum simulation. Note that ERDA treatment softwares are most generally derived from algorithms for Rutherford backscattering, which rely on almost the same physical concepts and have begun to be developed formerly. Nevertheless ERDA involves non-Rutherford collisions, a variety of different set-ups, and due in particular to the use of grazing incidences, more attention is required for such processes as straggling and MS than in the case of RBS. For clarity and according to the general architecture of current algorithms, we first describe (Sections 11.2.1–11.2.4) the simulation of an *ideal* spectrum as previously defined, then we consider dispersion phenomena and the simulation of a *realistic* spectrum (Sections 11.2.5–11.2.7).

11.2.1. Energy Loss of the Incident Ion

The stopping power of the target depends on both local target composition and ion energy. At any energy the resulting stopping power is a combination of the stopping powers of target components, which is generally assumed to conform to Bragg's rule. In some materials (cf. Section 4.3.3), this may be too rough an approximation, and other combination of formulae or corrections may be applied. In each slab of the target description, composition is constant, but variation in the stopping power with energy cannot generally be neglected over the slab thickness. The energy decrease must then be computed step-by-step, taking a new value of the stopping power at each step.

This numerical integration of the energy loss can be performed equivalently by successive energy or depth steps. In the case of depth steps, slabs are subdivided into so-called sublayers, the thickness of which must be chosen with care for the variation of the stopping power over the path of the ion through the sublayer to be really negligible. This requirement is easier to meet with a constant energy step, but then a distinct treatment is necessary at each slab boundary where the stopping power may change abruptly. In any case, quite small steps are required for good precision, particularly when at grazing incidence, the ion path to sublayer thickness ratio is large. Yet since we are rarely interested in very localized details of the target at large depths and since energy straggling and MS unavoidably smear out these details, larger steps can be taken as depth increases to avoid a too time- and memory-consuming computation.

Ion energy at the (n+1)th step is thus deduced from the preceding energy value (or beam energy at the first step) according to:

$$E_0(x_{n+1}) = E_0(x_n) - \frac{\Delta x N(x) \varepsilon [E_0(x_n)]}{\cos \theta_1} \tag{11.1}$$

where $N(x)$ is the number of atoms per volume unit at depth x, ε is the stopping cross-section evaluated as a function of local composition and ion energy $E_0(x_n)$, Δx is the depth step equal to $x_{n+1} - x_n$, θ_1 is the incidence angle or conversely:

$$x(E_{n+1}) = x(E_n) + \frac{\Delta E \cos \theta_1}{N(x)\varepsilon[E_n]} \tag{11.2}$$

when using an energy step ΔE. Equations 11.1 and 11.2 are simply discretized forms of Eq. 4.8 using Eq. 4.1 for the expression of the path length increment. Equations 11.1 and 11.2 can be replaced by more or less sophisticated expressions to improve the accuracy by taking into account the variation of the stopping power over the space or energy step. Some of these expressions are discussed in Section 11.5.

In principle, it is only a matter of preference to proceed to the following operations immediately after each of these steps until completing the calculation of the sublayer's contribution to the total *ideal* spectrum or to save results for all sublayers up to the analyzable depth before proceeding further. However the first solution is less memory-consuming, and what is more important, it allows us to stop calculating as soon as the initially unknown analyzable depth is reached (i.e., when the energy of recoils becomes less than some detection limit before reaching the target surface).

11.2.2. Collision Events and Recoil Yields

Once the incident ion energy at a given depth has been determined, it is trivial to multiply it by the kinematic factor K' (Eq. 2.12) to obtain the energy of any type of recoil atom just after the scattering event. When scattered projectiles are also detected (e.g., coincidence-ERDA, *ExB* filter, simultaneous RBS detection, etc.), their energy after collision is obtained similarly by using factor K (Eq. 2.10).

To determine the number of recoils emitted in the direction of the detector from a given depth, the scattering cross-section must be found for each isotope liable to recoil, while taking into account the proper incident ion energy. Depending on the scattering couple and energy, the simple Rutherford expression can be used or corrected under nearly Rutherford conditions by using for example, the formula proposed by L'Ecuyer *et al.*[23] or replaced by an interpolation of literature data in the case of a fully non-Rutherford scattering (cf. Section 3.1). Furthermore, the cross section is multiplied by the number of scattering centers present in the sublayer. This provides the *ideal* yield produced by each target isotope j in the sublayer and the corresponding interval of recoil energies obtained by applying the kinematic factor to the energy interval of ions impinging on the sublayer:

$$Y_{n,j} = \frac{N(x)F_j(x)Q\delta\sigma_j[\phi,E_0(x_n)]\Delta\Omega\Delta x}{\cos \theta_1} \tag{11.3}$$

In this discretized form of Eq. 4.20, modified using Eq. 4.1, the atomic fraction $F_j(x)$ of isotope j was introduced to express the partial yield of this isotope. After computing subsequent energy losses, this yield represents the *contribution* of the amount of the considered isotope present in the sublayer to the total *ideal* spectrum. For heavy elements isotopes may not have to be distinguished, so the fractions $F_j(x)$ of isotopes can be reduced to atomic fractions of elements; but for brevity in the following pages we use the word isotope without mentioning the alternative word element.

If required scattered projectile yield can be computed by replacing the recoil differential cross-section by the scattered ion differential cross-section, and the detector solid angle by the relevant value. Equation 11.3 must be applied separately for all types j of scattering centers, because each type generates a yield of scattered ions having a different energy. However the energy interval in which this yield will appear is still to be determined, and for this purpose, we must still compute the energy loss of recoils.

11.2.3. Energy Loss of Recoils

The principle for computing energy loss along the outward path of recoils (or scattered projectiles) does not differ from calculating energy loss on the inward path except that it must be repeated for each recoil type, since it encounters a different stopping power, and for each initial recoil energy value, i.e., the recoil energy produced in each sublayer. This part of the calculation is consequently rather time-consuming. Naturally the incidence angle must be replaced by the emergence angle in Eq. 11.1 or 11.2, using Eqs. 4.4–4.7, depending on the chosen reference angles. Whether a transmission or reflection geometry is used, i.e., whether the description of the traversed material is the same as on the inward path or not, does not make any real difference. In the latter case, there is some simplification of secondary importance.

Energy loss in the absorber foil can be treated as though it were an additional slab. In the low-energy range, it may be necessary also to take into account energy loss in the dead layer of the detector, which is not negligible, since it is equivalent to a few tens of nanometers of silicon (cf. Section 10.5.1.1).

11.2.4. Total Spectrum

After calculating energy losses on the outward path and in the absorber, final energy intervals associated with the counting yields computed in Section 11.2.2, are now determined. If several recoil types exist, their *contributions* have to be summed to build the total yield. Since there is no reason for energy intervals containing *contributions* from different sublayers to coincide for different types of recoils, individual yields may be redistributed to predetermined energy channels. This cumbersome operation is avoided when using the retrograde method (Section 11.3).

However we must still introduce the dispersion phenomena to convert the *ideal* spectrum into a *realistic* one. This must be done before summing the *contributions*, since dispersion effects on the outward path depend on the particle species.

11.2.5. Energy Straggling

Energy straggling is computed in most cases according to Bohr's formula (Eq. 2.18). This model allows the description of energy straggling contribution to the energy distribution using a single parameter, the Gaussian standard deviation Ω_B^2. This approximation is particularly convenient because Gaussians can be combined by summation in quadrature of standard deviations to calculate the total energy spread. Moreover an extension of Bohr's model for compounds[24] leads to a quite simple expression for straggling over a traversed thickness t in a compound of stoichiometry $A_m C_n$ of elements with atomic numbers Z_A and Z_C:

$$(\Omega_B^{A_m C_n})^2 = 4\pi(Z_1 e^2)^2 N_{A_m C_n}(mZ_A + nZ_C)t = N_{A_m C_n}\left\{\left[\frac{m(\Omega_B^A)^2}{N_A}\right] + \left[\frac{n(\Omega_B^C)^2}{N_C}\right]\right\} \quad (11.4)$$

However as developed in Section 2.5, the Bohr formalism is relevant only in the 0.01–0.2 fractional energy loss domain. Some algorithms[16-17] use more accurate expressions for the description of straggling outside this range, particularly for fractional energy losses larger than 0.2 where errors using the Bohr formalism would be considerable.

For a nonhomogeneous target, increments of Ω_B^2 corresponding to each traversed sublayer must be computed and summed over the entire traversed thickness t. This calculation can conveniently be performed in parallel with the step-by-step calculation for energy loss.

In the case of reflection geometry, straggling on inward and outward paths is generated by the same target material. This allows us to simplify the treatment using Eq. 11.5:

$$\Omega_s^2 = K'^2 \Omega_{in}^2 + \Omega_{out}^2 \quad (11.5)$$

which relates total straggling to straggling accumulated on the way in and the way out. Since in the Bohr formalism, straggling is independent of the moving ion energy, Ω_{out}^2 can be deduced from Ω_{in}^2 by substituting the atomic number of the recoil for the atomic number of the incident ion and multiplying t by the geometrical factor. Thus Eq. 11.5 reduces the calculation of total straggling to the step-by-step computation of straggling on the inward path only and a few trivial operations. Eventually the result must still be incremented by straggling in the absorber foil.

Note that energy broadening due to straggling on the inward path is reduced by the scattering process (factor K'^2 in Eq. 11.5). Indeed this reduction applies to the entire energy distribution.

11.2.6. Multiple Scattering

Calculating the contribution of MS to energy dispersion follows essentially the same procedure as straggling (e.g., Ref. 9), and it can also be performed in parallel with calculating energy loss. In most algorithms explicitly taking MS into account, its contribution is obtained using Möller's formalism. A more sophisticated treatment is adopted in RBX.[17]

11.2.7. Total Energy Dispersion

Other contributions to energy dispersion from detector resolution, the finite collection angle, beam energy fluctuations, and beam walk on the target due to its energy ripple are generally considered as a single Gaussian term, sometimes called *system resolution*, characterized by its standard deviation Ω^2_{syst}. The standard deviation is easily deduced from the FWHM of the signal from a very thin target.

The last operation before summing elemental *contributions* from various target isotopes consists in combining *system resolution*, straggling, and MS terms corresponding to each isotope and each collision depth, then convoluting the corresponding *ideal* yield with the resulting distribution. This is particularly easy when the three contributions are supposed to be Gaussian, since a quadrature summation instead of the intricate exact combination (see Section 4.2.4.1) gives the standard deviation of the total dispersion. Most softwares use this principle.[9,10,12,20] Since a Gaussian quite rapidly approaches its asymptotical zero value, the convolution algorithm involves only a very limited number of significant terms.

Bohr's formalism is not a good model for straggling outside a limited range of energy loss (cf. Section 2.5). In the low fractional energy loss domain (Section 2.5.1.1), the straggling distribution is non-Gaussian and asymmetric, and for MS, it is clearly non-Gaussian. Quadratic summation is thus an approximation; moreover recent discussion of MS[25,26] show that the resulting lateral and angular spreads cannot really be considered independent (cf. Section 3.3.4). As a consequence addition in quadrature of both terms is not justified, and it may overestimate the total MS contribution. This argument gives rise to so too much intricacy (although it would not change the general software architecture) that it is not taken into account in current algorithms.

Note that contributions from straggling and MS are zero at the target surface, and they become important only if they become significant in comparison with the system resolution. Moreover the accuracy of the simulation depends only on departures from the real distributions. Straggling with the Landau–Vavilov range makes so small a contribution that it can generally be approximated by Bohr's formula without significant consequences. Deviations from Bohr's formula at large depths are more noteworthy but less difficult to take into account, since the Gaussian model is still valid. Moreover these deviations grow only progressively with depth. The error due to assimilating MS to a Gaussian term and summing lateral and angular spread in quadrature is important at low depths. However total MS is small up to relatively large depths, as shown in the example in Section 3.3.3, so that these approximations have a limited effect. The ability of current algorithms to reproduce experimental spectra fairly well seems to indicate that these approximations are not too rough, at least in the usual range of analyzed depths. The question of accuracy of energy spread models nevertheless deserves particular investigation to improve ERDA interpretation techniques, as stressed in Sections 4.2.4.2 and 11.4.3.

11.3. ALTERNATIVE SIMULATION PROCESS: RETROGRADE METHOD

11.3.1. Principle

The *retrograde* method proposed by Serruys[14] for Rutherford backscattering and directly applicable to ERDA as a consequence of the similarity of the physical processes involved, consists of substituting for the natural sequence of events an imaginary sequence in which time is reversed (hence its self-derisory appellation) for the computation over the outward path. While the natural sequence:

1. Energy loss on the inward path
2. Collisional energy transfer
3. Energy loss along the outward path, including the absorber

leads to the classic simulation procedure shown in Figure 11.1, the sequence used for the retrograde procedure in Figure 11.2 is the following:

1. The impinging ion progressively loses energy along the inward path.
3'. Simultaneously the recoil follows an inversed chronology, in which as though emitted by the detector, it gains energy while penetrating the target with a *negative* stopping power.
2'. When the recoil energy to projectile energy ratio equals the recoil kinematic factor for the considered recoil type, the scattering event occurs at the depth reached by both the projectile and the recoil.

The following observations are the basis of the retrograde procedure:

As developed in Section 11.3.2, simulation is generally performed to compare its result with an experimental spectrum in an attempt to find the target description that matches the experimental spectrum. When following the classic simulation procedure, final energy intervals in which the yield was computed do not coincide with energy channels used to record the experimental spectrum. This situation does not facilitate the comparison, and it may require a supplementary, eventually somewhat approximative, rescaling of one or the other spectrum.

It is not easy to determine at which depth calculations should be stopped, since final energies appear only at the end of the classic procedure, and thus

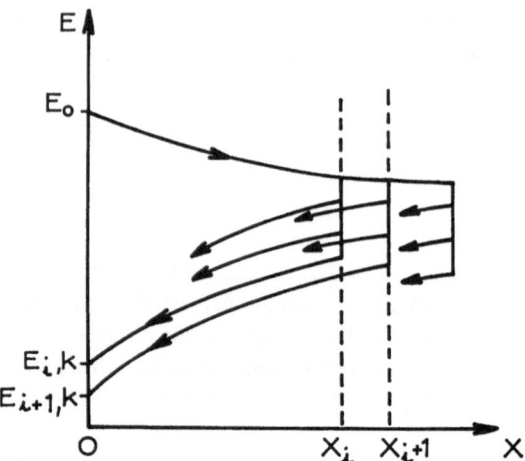

Figure 11.1. Classic simulation procedure in particle energy versus penetration coordinates (reflection geometry). Recoil energies are calculated starting from each discrete penetration value; energy just after collision is deduced for each species from preliminarily computed projectile energy.

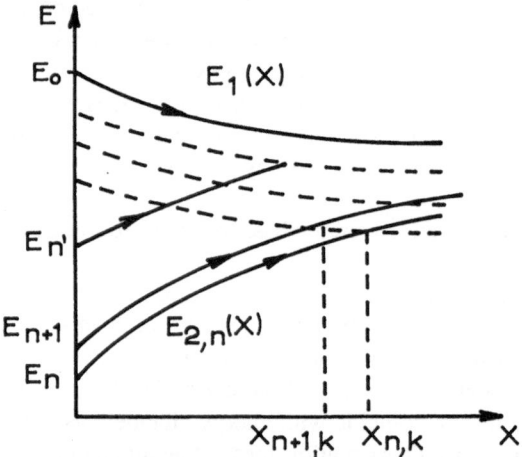

Figure 11.2. Simulation procedure by the retrograde method in the same representation as Figure 11.1. Dashed curves represent recoil energy after collision deduced from the projectile energy for different species. Recoil energies are computed from fixed final energy values along with projectile energy at each depth step (recoil curves are represented for only one recoiling species). Collisions are determined by the intersection of recoil curves with the dashed curved corresponding to the recoiling species. The yield produced by species k between depths $x_{n+1,k}$ and $x_{n,k}$ can thus be directly assigned to the final energy interval $[E_n, E_{n+1}]$.

unnecessary calculations may be performed for scattering events beyond the analyzable depth.

The identity of the stopping power for scattered ions produced at different depths but reaching the surface at the same final energy is proper to RBS, and it is transferable only to the case of scattered particles in coincidence ERDA.

11.3.2. Energy Loss and Collision Kinematics

Energy loss calculations proceed in the same way as for the classic method except for the following differences. At each depth step, starting from the target surface, the energy change is evaluated for the projectile and for each final energy of each recoil species, except if the corresponding collision depth has already been reached. For recoils energy is incremented instead of decreased in agreement with the inversion of the particle movement. It is possible to choose a scale of final energies that coincides with channels of the multichannel analyzer (MCA), thus simplifying the comparison of simulated and experimental spectra. Naturally only final energies lower or equal to $K'E_0$ are treated. When an absorber foil is present, it is treated as a first layer of the target in which the projectile energy is kept constant.

As soon as impinging energy at a new depth is calculated, resulting recoil energies are also computed by applying the kinematic factor, then compared to current recoil energies. If a collision occurred during the step, i.e., when the current energy of a recoil species becomes larger than the recoil energy corresponding to the current projectile energy, the exact depth where it occurred is interpolated with the corresponding incident and emerging energies to determine cross section and the number of scattering centers involved.

As soon as the collision is found for a recoil of given mass and final energy, this recoil can be disregarded. After this occurs for each energy and each recoil species, the calculation has precisely covered the analyzable depth for each recoil species.

11.3.3. Recoil Yields

The depth interval between collisions determined for two successive final energies precisely defines the sublayer that produces the *ideal* yield of the considered species between these two energies (cf. Figure 11.2). This yield is thus given by Eq. 11.3, where Δx is now the depth interval $x_{n,k} - x_{n+1,k}$ between these two collision points. The treatment of straggling, MS, and experimental dispersion factors does not differ from what was described for the basic method.

If the scale for final energies is initially taken as the energy scale defined by MCA channels, the retrograde method thus directly realizes a perfect coincidence of the discrete energy scales for both spectra to be compared. Since in ERDA processes after collision involve different recoil species (if not only one), instead of one species scattered on different target species in RBS, the possibility described in Ref. 14 of treating these processes once for all target elements is not applicable in ERDA. Yet it is still applicable to scattered projectiles in coincidence ERDA.

11.4. PROFILE EXTRACTION FROM EXPERIMENTAL SPECTRA

11.4.1. Methods

Two main approaches are possible for retrieving concentration profiles from energy spectra. The first consists of simulating the spectrum corresponding to an initial guess of the target composition, then changing the guess until satisfactory agreement is reached. The second approach, known as *spectrum scaling*, consists of reducing the spectrum by using scaling factors determined for each channel so that the yield per channel data appears as though the incident particle energy and the effective stopping power were constant throughout the sample.[7] The transformation of the yield versus energy form into a concentration versus depth profile is then straightforward. But this is only possible if the contributions from different recoil species do not overlap or can be separated by substraction and for contents in recoil species low enough to be neglected or roughly estimated by calculating scaling factors. For cases that do not satisfy these conditions, an iterative generalization of the spectrum scaling method was proposed for RBS,[20] and it can be transposed to ERDA.

All these methods originate in data-processing techniques developed for RBS. The similarity of physical relations allows an easy enough adaptation even if the choice of cross-section data is much less straightforward and MS description requires much more attention. But the main obstacle facing their application to ERDA is that contrary to what happens in RBS, the recoil yield is generally produced by only a few species present in the target. There are examples of unrealistic results that can result from neglecting hydrogen when interpreting RBS spectra, e.g., Ref. 27. When the absence of helium can be reasonably supposed, the uniqueness of the absent element, hydrogen, may allow us to take it into account in RBS,[27] but for conventional ERDA, the matrix may contain several elements that are not detected. Simulations or scaling thus must rely either on a matrix description obtained from other analytical techniques or on standard samples of known composition. From this point of view, the advantage of detecting all recoiled species with techniques such as TOF-ERDA, telescopes, mass spectrometers, or in some cases *ExB* filters, is striking. When such set-ups are not available, the complementariness of RBS and ERDA may provide the missing information. In principle standard samples allow us to deduce absolute values of concentrations from profiles obtained in relative value. But as discussed in Section 11.4.1.1., this may not be so simple as it first seems.

11.4.1.1. Interactive Modification Method

A very elementary but rather commonly used way of approaching depth profiles consists of interactive, user-defined modifications to the target description based on discrepancies observed between simulated and experimental spectra. Such a procedure can hardly improve more than the main parameters of a simple target model composed of a stack of homogeneous layers: layer thickness, atomic fraction in a layer. Such methods encounter two considerable limitations:

- They rely almost necessarily on the hypothesis that the target can be reasonably described by a limited number of homogeneous layers, which may be deprived of any physical meaning in many cases, particularly when a continuous concentration gradient is present.

- The tedious interactive quest of suitable parameters may well have the user's patience as an ultimate limit.

11.4.1.2. Multiparametric Fit

This is a more elaborate solution to the profile extraction problem. It is adapted whenever physical considerations allow us to postulate a determined mathematical shape of depth profiles. For instance Fick's laws of atomic diffusion predict Gaussian or error function profiles, respectively, for thin or thick initial layers of a diffusing impurity. In such cases the profile is entirely determined by two parameters: the total amount of impurity and a characteristic profile width related to the diffusion coefficient. From a limited number of simulations with different parameter values, multiparametric minimization of a norm of the difference between experimental and simulated spectra (least-squares expression, or χ^2 maximum likelihood) leads to an interpolation of best-fitting values of parameters within a user-defined accuracy.

Since the postulated mathematical expression for the depth profile allows us to consider an unlimited number of concentration values in the analyzed depth range, such a fitting method is able to treat continuous concentration gradients that would require much too large a number of layers for the interactive modification method. The fitting method can also give direct access to such physical parameters as diffusion coefficients. This efficient technique[3,21] is applicable as long as the target model depends on a small number of parameters. It is clear that the validity of results is related to the physical relevancy of the postulated mathematical profile description.

11.4.1.3. Spectrum-Scaling Method

The spectrum scaling method[7] consists in determining scaling factors for each channel of the spectrum by interpolating tabulated cross sections and effective stopping powers recorded during a classic simulation. The effective stopping power is defined as the derivative of final energy relative to scattering depth. These factors remove effects of cross section and stopping power variations inside the sample from the spectrum. The interpretation of the rescaled spectrum becomes as simple as though the cross section and effective stopping power were constant throughout the sample. It is then straightforward to transform the energy scale to depth and the yield scale into concentration.

However such a method can be applied only if contributions from different elements do not overlap or if overlapping signals can be separated by background substraction. In the latter case a preliminary knowledge of the profile shape to be subtracted is necessary, and results depend on the reliability of this shape. For instance to analyze thin layers deposited on homogeneous substrates, some authors rely on the nominal composition of the substrate, but any unexpected modification can cause errors.

11.4.1.4. Iterative Profile Reconstitution Method

The so-called *profile reconstitution* method was proposed for RBS analysis.[20] Its very general principle applies to multielemental analysis with a minimum of assumptions, and it is devised for reaching the best possible *final*-depth resolution. Although there is no example of its application to ERDA up to now, the very similar physical background of both techniques suggests that it is easily transferable to ERDA. Similar to other softwares for ERDA,[1,16,17] this method consists of an iterative process of correcting an initial guess about the profiles. Its original features are that the spectrum is treated channel per channel and profile corrections are determined directly from the difference between experimental and simulated spectra without the intervention of the user's intuition. Although the initial guess is more or less necessarily described by a stack of homogeneous layers or by simple mathematic formulae (e.g., Gaussian or error function), there is no other limit on modifying profiles than the sum of atomic fractions kept equal to unity.

Concentration corrections are determined for one target element, assuming momentarily atomic fractions to be exact, from the ratio of the contribution of the considered element to the difference between the experimental spectrum and the sum of simulated contributions from other elements, according to[20]:

$$F_i^{m+1}(x_{n,i}^m) = F_i^m(x_{n,i}^m) \frac{\left(Y_n - \sum_{j \neq i} Y_{n,j}^m\right)}{Y_{n,i}^m} \qquad (11.6)$$

where $F_i^m(x_{n,i}^m)$ is the atomic fraction of element i estimated at the mth iteration at the depth $x_{n,i}^m$ corresponding to channel n for this element, as estimated at the mth iteration; Y_n is the experimental yield in channel n, and $Y_{n,i}^m$ is the contribution of element i to channel n simulated at the mth iteration. Corrections for other elements are deduced from Eq. 11.6 according to:

$$F_j^{m+1}(x) - F_j^m(x) = a_j(F_i^m(x) - F_i^{m+1}(x)) \qquad (11.7)$$

where coefficients a_j are user-defined with the condition $\Sigma_{j \neq i} a_j = 1$ for consistency. The choice of these coefficients represents a substitution rule that may as well be preferential as congruent, and it relies on the only strong assumption of the method, namely, that there exists a simple correlation between concentration changes in different elements. This choice is revisable at any moment, and the form of Eq. 11.7 is general enough to allow for most usual situations.

Equation 11.6 implicitly uses cross sections and stopping powers derived from the target composition at the mth iteration. Although these parameters are still approximative and despite the fact that the correction is thus applied to a depth that corresponds only approximately to the considered channel, the resulting inaccuracy of profile corrections decreases with successive iterations, while atomic fractions approach the true values. In fact the error associated with these approximations tends to zero near the surface, from which convergence propagates toward larger depths. For full convergence, the iterative process must be repeated over all unknown elements, recycling eventually after the last element. It is noteworthy that unsuitable choices for a_j do not allow convergence, and thus they are easily identified.

Note that when simulations are performed including the most accurate available models for energy-spreading terms, such a method realizes as nearly as possible the process imagined in Section 4.2.4.2 (Eqs. 4.47 and 4.48). Indeed the practically impossible "deconvolution" of the experimental yield is replaced by a fit of a convoluted signal to experimental data (in fact, since energy spread is depth-dependent, the operation is not really a convolution in the mathematical meaning of the term). This allows us to approach quite nearly the optimal *final* resolution, as defined in Section 4.2.4.2, and to obtain realistic profiles instead of profiles broadened by energy spreading.

The main difficulty in transferring this technique to ERDA resides in *missing elements*, i.e., target components that are not present in the spectrum. It is necessary either to determine their profiles from other analysis techniques to compute accurately composition-dependent stopping powers or resort to hypothetical relations between their atomic fractions. The problem of missing elements is extensively discussed by Arnoldbik and Habraken.[28] According to these authors, the global fraction of missing elements, considered as a molecular compound of known stoichiometry, can be derived from comparison with a standard sample, but this is far from solving the problem in its general form.

In the present state of the algorithm, stopping powers derived from concentrations determined at iteration m are used throughout the computation of new concentrations of iteration $m+1$. An improvement might consist in using stopping powers derived from concentrations calculated at iteration $m+1$ for sublayers between the surface and the sublayer under consideration, as proposed by Li and Al-Tamimi.[5] This modification accelerates the convergence of the algorithm, and it could possibly reduce the noise level in the resulting concentration profiles. It can be applied to both inward and outward paths when using the retrograde method, although it was applied only to the outward path in the classic algorithm of Li and Al-Tamimi.

11.4.2. Converting into Depth Profiles

As stated in Section 11.1, ERDA allows us to determine only concentrations as a function of the traversed amount of matter (at·cm^{-2} or µg·cm^{-2}). It is sometimes more convenient to obtain concentrations versus depth. Converting profiles requires knowledge of the target density, which may be a function of depth when concentration profiles are not flat. As a consequence this operation may include some incertitude about density, and the final depth resolution of concentration versus depth profiles is generally worse than for corresponding profiles of concentration versus traversed amount of matter.

When analyzing small amounts of impurities in materials of well-known density, typically implants in high-purity silicon, matrix density can be considered unchanged, so the conversion is straightforward. Converting into depth of traversed amounts of matter is also possible when a known relation exists between density and composition. If atomic fractions $F_j(T)$ are determined as functions of the traversed amount of matter T and if there is a reliable expression $\rho(F_1, \ldots, F_n)$ for the density as a function of atomic fractions, it is possible to obtain atomic fractions $F_j(x)$ as functions of depth simply by writing

$$x = \int_0^T \frac{\overline{M}}{\rho[F_1(\tau), \ldots, F_n(\tau)]} \, d\tau \qquad \overline{M} = \sum_{j=1}^n M_j F_j(\tau) \tag{11.8}$$

or in discrete form:

$$x_k = \sum_{i=1}^k \frac{\overline{M}_i}{\rho[F_1(T_i), \ldots, F_n(T_i)]} \, \Delta T_i \qquad \overline{M}_i = \sum_{j=1}^n M_j F_j(T_i) \tag{11.9}$$

For instance this principle was used by Maugis and Serruys[29,30] in studying intermetallic compounds formed by annealing an Al thin layer deposited on a Ti substrate. In this system the density varies linearly with Al content from pure Ti to 72% Al. Authors could thus determine the position in depth of the $Al_3Ti/Al_{23}Ti_9$ interface found by RBS and confirm the presence of the $Al_{23}Ti_9$ phase from a comparison with sputtering thermoionization mass spectrometry (STIMS) results, from which concentration is obtained as a function of depth.

11.4.3. Experimental Parameters

The validity of profile determinations clearly depends on the accuracy of the experimental parameters used in calculations. The accuracy of beam energy, scattering, incidence and emergence angles is only a question of equipment quality (energy stabilizers and goniometers), and it is generally fairly good. On the other hand, incertitudes of energy calibration of the MCA, system resolution, total ion dose or integrated beam current, and detector solid angle appear as critical.

Energy calibration consists in relating MCA channels to the *energy of detected particles*. The latter is not exactly the *detected energy of particles*, and calibration is not independent of the detected target species. Both difficulties are discussed in Section 10.5.1. As a result energy calibration is less straightforward than it appears at first sight. We must take into account energy loss in the detector dead layer, as explained in

Section 10.5.1.2, and this should be done for each detected species. Naturally for coherence this energy loss must also be considered in simulating recoil energies (cf. Section 11.2.3).

Inaccurate energy calibration has the following consequences:

• Departures from the true relation between channels and recoil energy, particularly in energy ranges outside the interval between reference energies used to determine calibration, result in departures from the true depth scale, hence in errors in determining depths.

• If simulated and experimental edges corresponding to energy of recoils from the target surface do not coincide, either some experimental yield above the simulated edge cannot be interpreted, which truncates the profile near the surface, or no experimental yield appears in the near-surface region of the simulation, i.e., a spurious empty layer is introduced into the profile. Moreover in both cases the discordance between simulation and experiment is liable to produce instabilities in semiautomatic techniques, such as profile reconstitution algorithm (see Ref. 20 for more details).

System resolution, composed of detector resolution, beam energy fluctuations, finite beam size, and acceptance angle of the detector, is dominant in energy spreading in spectra for low depths; deeper in the target straggling and MS become important. For these latter, we must rely on calculation, the precision of which is discussed in Sections 2.5, 3.3, and 4.2.3.3, but system resolution is generally determined experimentally as the FWHM of the recoil peak from a very thin target. If this resolution is underestimated, sharp profile features smear out because profile extraction algorithms attempt to reproduce the experimental broad features by progressive concentration changes. If resolution is overestimated, the algorithm is unable to reproduce the sharpest features in the experimental spectrum. This may lead to violent instabilities in the algorithm (e.g., Ref. 20) similar to Gibbs instability in numerical Fourier analysis. In both cases the *final* depth resolution is degraded.

Measurements of both the integrated beam current Q and the detection solid angle $\Delta\Omega$ are subject to considerable incertitude. In addition to the limited confidence that one may have in the calibration of current integrators and its stability, secondary electron emission and the presence of multicharged or neutral ions in the beam contribute to incertitude in the integrated current; moreover Q may be extremely difficult to measure properly for insulating targets. Determining the true detection solid angle is far from straightforward, since the elliptical section of the beam at the target surface and the circular aperture of the detector are not coplanar. A fairly precise method based on a Monte Carlo calculation was developed by Tirira and coworkers[15,31] and it is included in PYROLE, a code by Trouslard inspired by Ref. 31. This method still requires precise knowledge of the beam impact area and shape.

Some authors[12,20] prefer to eliminate these incertitudes by omitting the term $Q \cdot \Delta\Omega$ in their calculations, then normalizing the simulated spectrum to the experimental one. In Ref. 20 this normalization is corrected at each iteration of the profile reconstitution procedure. However such a normalization is possible only if the spectrum represents the total amount of matter in the analyzed volume. This is generally the case for RBS if the target does not contain hydrogen or helium, but not for ERDA if only a few constituents are analyzed. In the latter case there remains an unknown constant multiplier to determine to obtain absolute concentration profiles. A convenient and frequently used solution is to normalize spectra on the basis of a simultaneously recorded RBS yield of a pure element. We can also resort to standard samples of known stoichiometry or to well-determined concentrations of some components of the target or to stoichiometry relations between them. However as developed in Chapter 14, it is particularly difficult to find reliable and stable standards for hydrogen isotopes.

11.4.4. Counting Statistics

11.4.4.1. Counting Noise

In previously described simulations, statistical departures in yields relative to the mean value predicted by scattering cross-section are overlooked. In addition to uncertainties in parameters and physical laws used in the simulation, these departures introduce another difference between experimental and simulated spectra (cf. Section 4.2.3.4). This is of no importance when visually comparing simulations with experiments, but it becomes a genuine difficulty if directly reconstituting profiles.

Scattering events do not occur as a deterministic process according to an unvariable number of occurrences fixed by cross sections, but as a stochastic process ruled by a Poisson distribution law. As a consequence an experimental spectrum presents fluctuations in the order of the square root of its intensity due to the statistical dispersion of detected yields. These fluctuations are sometimes referred to as counting noise.[32] These are of particular importance in ERDA because of the relatively low number of counts generally recorded in comparison, e.g., with RBS. Counting noise should not be confounded with background detection events, another sort of noise. This latter can be treated only by substraction. The low-energy range where it is important (electronic noise from the detector) is generally excluded for spectra interpretation and even truncated to avoid flooding MCA's.

Counting noise intensity is nearly proportional to the square root of the yield; and, as concerns its frequential character (the real part of its Fourier transform relative to energy) it is nearly "blank noise" (deprived of any preferential frequency). Although it is fairly well characterized from a statistical point of view by these two properties, it cannot be adequately implemented in simulations because its phase characteristics (the imaginary part of its Fourier transform relative to energy) are purely random and thus defy prediction. In other words it is theoretically possible to simulate a realistic counting noise, but this noise can never coincide with the particular noise of the experimental spectrum under consideration, because its peaks and valleys have no reason to coincide.

The impossibility of fitting counting noise has direct implications on profile determination. As stressed in Section 4.2.4.2, noise is one of the main components of the *final* depth resolution. In the reconstitution method, another difficulty arises: The difference between experimental spectrum and the sum of simulated contributions in Eq. 11.6 includes experimental counting noise but no corresponding simulated noise. Since this difference is taken as an estimate of the yield of only one element, the full amplitude of noise is transferred to the profile of this element, which amplifies the noise/signal ratio proportional to the inverse of its atomic fraction.[20,33] This process of transfer and amplification occurs at each iteration, and it is limited only by phase coincidence in already well-fitted regions of the spectrum. When the noise level becomes unacceptable after some iterations, the only possibility is to smooth the resulting profiles and take them as a new but more accurate guess. This suggests that preliminary noise reduction would improve the result. The fact that counting noise always degrades the final depth resolution whatsoever is an additional argument in favor of noise filtering.

11.4.4.2. Noise Filtering

Smoothing data generally result in aesthetic improvement but a loss of information, since averaging techniques typically level-off peaks and valleys. Note that grouping channels with the purpose of simplifying data analysis consists in arithmetically averaging groups of channels; choosing an exaggeratedly large channel width is strictly equivalent. Smoothing procedures based on more sophisticated methods than the various forms of averaging avoid such inconveniences as peak truncature. Among these, the deconvolution method of Ref. 34 very efficiently filters noisy data without truncature effects even in the case of sharp peaks or strong discontinuities. Nevertheless these methods rely on mathematic techniques, and it is not clear that they are built on physical properties of scattering spectra. It is a possible explanation for some spurious shifts of broad peaks that appear in Ref. 34.

Taking these properties as starting point, Serruys[32] developed a frequential-filtering method based precisely on the frequential-filtering properties of energy-dispersing phenomena (system resolution, straggling, and MS). As a matter of fact, energy spread broadens the sharp features of ideal spectra, thus attenuating below any detectable level the high-frequency range of their Fourier transform relative to energy (the word frequency is used here for convenience, by analogy with a Fourier transform relative to time). Consequently only counting noise can contribute to this high-frequency range. Its extent can be easily and safely underestimated by observing that it is determined by the FWHM of energy spreading at the surface, which is easily measurable. It is thus possible to cancel out high-frequency components in counting noise by frequential filtering and thereby reduce considerably the global noise level without perturbing the useful signal. Since it is determined only by the width of energy spreading at the surface, the filter can be built once and for all for a given experimental set-up, independently of target composition. Applications of this technique[29,30,32] do not show truncature or distortion of sharp features. They allow identification and interpretation of small features that cannot be distinguished from counting noise without such a treatment.[29,30]

11.4.5. Particular Target Behaviors

As mentioned in Section 4.3, real targets frequently do not have the ideal properties assumed in the first chapters of this book or in the models underlying usual analysis algorithms. Among the nonideal features presented in Section 4.3, only lateral nonuniformity and charge accumulation retain our attention here. Deviations from Bragg's rule are easily treated by substituting data or formulae in the computation of stopping cross-sections, provided suitable knowledge of the target matrix is available. Since these deviations strongly depend on the material under consideration, suitable data and treatment must be found for each particular case. When target changes occur from beam damage, the spectrum represents an integral over time or dose of a variable signal. It is generally untractable if data are not recorded as a function of time or ion dose. For a time- or dose-resolved record, a classical treatment applies. Channeling and blocking ERDA are such particular and intricate problems and so generally avoided by experimentalists that we do not discuss them here.

11.4.5.1. Lateral Nonuniformity

Lateral nonuniformity is accessible to simulation, each impinging ion behaves as though the target were homogeneous and identical to what it encounters along its path. Detected yield is thus a linear combination of spectra from the target description along the path of each ion. Since the number of ions is always very large (at least 10^{12}), we can identify the distribution of possible paths with the distribution of effectively explored paths. Many paths can also be considered as equivalent within acceptable incertitude, so the linear combination is reduced to a limited number of terms. For instance, given a rough surface with 100-nm peak-to-peak amplitude, if we assume a path length incertitude of 5 nm to be acceptable, the combination can be reduced to 20 terms.

In general changes in the target description between different impact points can be described by a single parameter, e.g., the local thickness of a rough coating layer. The simulation is built using a model of nonuniformity consisting of a finite number of target descriptions corresponding to discrete values of the parameter, and of surface fractions representing the probability of an ion encountering the value of the parameter approximated by one of these discrete values. Figure 11.3 shows this subdivision of the beam area for the case of a rough overlayer. The spectrum is simulated as a linear combination of spectra simulated for each value of the parameter weighted by the corresponding surface fraction.

A similar procedure is used in RUMP (see Ref. 3 and Section 11.5.1.1) to simulate rough interfaces with a Gaussian model of roughness and in PERM,[35] a software for RBS analysis with a more general model. Note that the simplified Gaussian model is quite convenient for checking before the experiment that a given degree of roughness

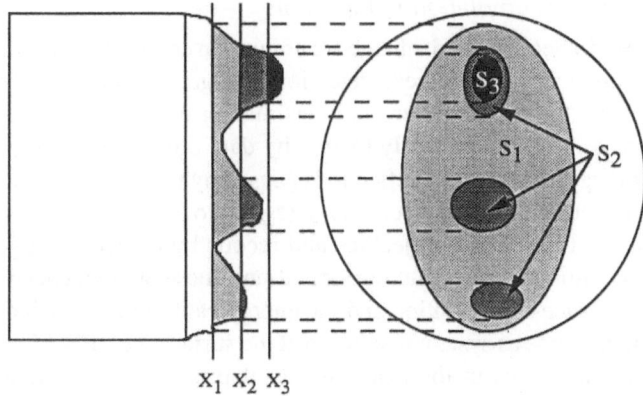

$$x_1 \; x_2 \; x_3$$

Figure 11.3. Level-line representation of surface fractions S_n in the case of surface roughness. In this occurrence the distance x between the surface and a reference plane is taken as the parameter describing lateral nonuniformity. For unit total area S_n is the probability of finding x in the interval $[x_n, x_{n+1}]$. (Reproduced from Ref. 35)

is acceptable for the considered target material. The model can also be used to check experimental conditions by comparing simulations with and without roughness. However the procedure by linear combination is simple enough only if a single parameter controls variations in the target description from one impact point to another. The procedure was developed for RBS, where the impact and emergence points of an ion are not too distant, and as a consequence it is reasonable to assume that inward and outward paths are not significantly different. In the case of reflection ERDA, this assumption clearly breaks down, so descriptions of inward and outward paths are more or less uncorrelated. The validity of the model with a single parameter is questionable in this case.

The inverse problem of analyzing RBS spectra from a nonuniform target is examined by Calmon[36] and more recently by Marin *et al.*,[35] which determine surface fractions corresponding to a small number of discrete values of a single parameter characterizing a nonuniformity in the frame of morphological model is given by other techniques (typically scanning or transmission electron microscopy). The method used consists of finding analytically surface fractions that minimize the difference between experimental and simulated spectra. The transferability of this method to ERDA, particularly in reflection geometry, is questionable, because it relies on the same assumption of equivalent inward and outward paths. Moreover it is clear that in any case this sort of characterization of lateral nonuniformity cannot be performed when unknown concentration gradients are present in the same depth range.

11.4.5.2. Charge Accumulation in Insulators

As described in Section 4.3.5, charge accumulation causes different perturbations in the analysis process. Among these detection background due to hot electrons or photons emitted by the target and errors in estimating the beam current from target current measurement cannot be really treated by data processing. In this section we consider only energy changes and deflections caused by the electric field, assuming a low enough beam current for the electrostatic regime to be stationary.

Repulsion or attraction of projectiles and recoils by a charged target results in energy shifts that must be taken into consideration. These are evidenced by discrepancies between surface edge position and the energy calibration, which can be of the same order of magnitude as system resolution at the surface, i.e., 12–15 keV.[36] Depth scale errors and anomalies in the near surface of profile are then similar to what happens with an erroneous energy calibration (see Section 11.4.5.1). It is much more difficult to describe deflections due to the oblique incidence of particles onto equipotentials in the absence of precise knowledge of the curvature of equipotentials. Particles detected at an angle ϕ relative to the incident beam have been scattered under a somewhat different but unspecified angle.

A treatment of acceleration and deceleration of particles by the electric field was proposed for RBS in Ref. 36 and recently improved in Ref. 37. Impinging energy is taken as $E_0' = E_0 - eV_S$, and final energy is modified according to $E_2'' = E_2' + qV_S$, where E_0 is the beam energy, V_S is the target surface potential, e is the electronic charge (assuming He$^+$ ions), and E_2' is final energy calculated with impinging energy E_0' from the scattered particle (or recoil) reaching the surface with a charge q (q = e in Ref. 36). For $V_S = 0$ the final energy is E_2, which can be calculated from energy calibration using a conducting target. Parameters q and V_S are obtained from the difference $E_2'' - E_2$ measured for ions scattered by two different elements respectively at the surface of the insulating target (E_2'') and of a conducting reference sample (E_2). Values for q are fractional, and these should be interpreted as the mean charge of particles emerging with different charge states. In the case of RBS, very small values of q, typically less than 0.1, allow us to neglect the He^{++} state and identify q with the fraction of He$^+$ state. It was observed that for a very pure bulk TiO_2 target, a linear combination of simulations with q = 1 and q = 0, respectively, weighted by q and 1 − q gave a better fit of experimental spectra than a single simulation with the mean charge q.

This treatment is in principle transferable to ERDA, but several difficulties arise. First recoils may have more than two charge states. Secondly, this treatment does not take into account deflections, which are possibly negligible in RBS at normal incidence and a large backscattering angle, since equipotentials are probably approximately flat over the area of a millimetric beam. But in reflection ERDA ion paths may be strongly inclined relative to the equipotentials. If deflections are significant, possibly more important than repulsion and attraction, the previously mentioned error on the scattering angle results in some error on all terms depending on scattering geometry: kinematic factor, path length/depth ratios, traversed amount of matter in a depth interval Δx. There is currently no method for taking these complex effects into account. In transmission geometry charge accumulation in the back face due to secondary-electron emission introduces a surface potential probably different from the front-face potential, so the preceding treatment also breaks down.

11.4.6. Mass–Depth Ambiguity

We can wonder whether the solution to profile extraction is unique or in other words whether we do not risk obtaining profiles that fit the experimental spectrum quite exactly, but differ from real profiles. This question is clearly related to mass–depth ambiguity and eventually to projectile–recoil ambiguity, the two possible causes for misinterpretation. It should in principle concern only conventional ERDA or other arrangements when all target elements are not detected.

Variations of concentration profiles at a tight depth scale and with low enough amplitude have no detectable effect on the spectrum because they are smeared out by energy-spreading effects. Differences at this scale are not due to a multiplicity of solutions, but only to doubt about the solution in the limits of the *final* resolution. On the other hand it is not excluded a priori that some part of the spectrum could be attributed to an element to which it does not belong. However several conditions are necessary for such a confusion to remain undetected. First there is generally at least one region of the spectrum where only one element can be present at energies higher than the edge of other elements. Once the profile is described in the corresponding depth range, there is generally a range where only a single other element remains undetermined, and so on. Misinterpretation is thus possible only if these ranges become too narrow. This results from a lack of mass resolution, which may be due either to an unfavorable target composition or to unsuitable experimental parameters. In principle any confusion between two elements, concerning eventually only a fraction of their yield, should result in discrepancies between the final simulation and experiment because the simulated variation of stopping powers and cross sections do not reproduce the real variation. However errors may be too small or partially cancel each other so that these discrepancies remain unobserved. Finally if confusion between elements occurred or equivalently if one element is omitted in the simulation, the fit of the experimental spectrum is obtained with erroneous concentrations and the atomic fractions probably no longer sum up to unity. But this inconsistency can be hidden when some target components are not detected.

It is thus difficult to give a clear-cut answer to the preceding question. Yet we remark that erroneous profiles due to confusion or element omission cannot fit an experimental spectrum detected under different enough experimental conditions (beam energy, incidence angle) where stopping powers are significantly different (e.g., Ref. 27).

11.5. ALGORITHMS AND PROGRAMS

Software for simulating ERDA spectra and extracting depth profiles have been developed by many laboratories. Some of the software is quite popular, and it can be obtained from its authors at negligible cost. However most software remains confidential either because it was written for specific features of a given set-up or because it is still in an experimental stage. This situation presents several inconveniences: First the validity of experimental papers is sometimes difficult to appreciate because hypotheses and approximations included in the software used for data processing are insufficiently known. Secondly few comparative studies between existing algorithms have been performed. As a consequence no software can be considered a standard, which makes comparing experimental results difficult. Thirdly much work is certainly spent developing algorithms that already exist but are not easily available.

11.5.1. Typical Software

11.5.1.1. RUMP Simulation Algorithm

RUMP is certainly the most widespread among ERDA algorithms. It was initially written by Doolittle[2] for RBS and later adapted for ERDA.[6] Further development was performed by Doyle and Brice.[7]

Integrating energy loss over depth is the most time-consuming part of simulating *ideal* spectra because the number of calculations increases as the square of the number of sublayers. For this reason the number of sublayers in RUMP is as small as possible. After passing through a sublayer energy is approximated from energy at the beginning of the sublayer using a Taylor development of energy loss to the second-order derivative with respect to energy of the stopping power; the latter and its derivatives are evaluated at the initial conditions, i.e., at the beginning of the sublayer. This development is shown to be accurate enough to allow the use of relatively large sublayers. The yield at both interfaces of the sublayer is computed using a generalization of Eqs. 4.34 and 4.35 of Wei-Kan-Chu *et al.*[38] for the case of several sublayers, wherein the stopping power of the backscattered ion is replaced by a value for the recoil where appropriate. The total yield generated by the sublayer is found by integrating over energies a development to third-order in energy of the scattering cross-section. The shape of the spectrum between sublayer edges is then approximated by a parabolic fit.

Straggling is simulated by convoluting a linearized approximation of the contribution of each sublayer—and eventually each recoil species—with a Gaussian, according to Bohr's approximation. This is not too rough an approximation, but a more criticable point is the possibility of increasing arbitrarily the standard deviation to manage the additional spread due to deviations from Bohr's model and MS. It is thus possible, and some authors did not waver in doing so, to fit spectra exhibiting broad fronts with a target description composed of a stack of homogeneous layers, ignoring rough interfaces or interdiffusion layers and "adding more straggling" instead.

RUMP computes the cross-section scattering according to a single expression of its energy dependance:

$$\frac{d\sigma}{d\Omega} = C_0 + \frac{C_2}{E^2} \tag{11.10}$$

starting from a value corresponding to the beam energy, calculated using Eqs. 3.16 and 3.17 for proton–helium collisions. Since coefficients of Eq. 11.10 are given only for a few determined angles, values for intermediate angles are obtained by interpolating between data of Tirira *et al.*[39] This procedure was criticized by Lucas and Tirira[40] on two grounds: First the general angle dependence results from a fit of values by the authors, but it may not accurately represent details of the angular variation in the intervals between these values. Secondly an E^{-2} dependence does not exactly fit experimental data for proton–helium cross sections. If the reference energy (beam energy) is taken as 2.0 MeV, RUMP values become smaller than experimental ones below 1.4–1.5 MeV, which is still acceptable because these values correspond approximately to the limit of analyzable depth with a 160° scattering angle.

RUMP is basically a simulation program. Profile determination using RUMP generally proceeds by interactive modifications of the initial guess of the target description. This latter is generally composed of homogeneous layers, but the possibility also exists of choosing among some current profile shapes, such as Gaussian or error function profiles. Moreover a fitting procedure based on a multiparametric χ^2 minimization[3] allows finding the best possible parameters of these latter (see, for instance, Figure 12.16 and with a slightly different algorithm, Figure 13.9). The possible drawbacks of such profile determinations are developed in Ref. 20 and Section 11.8. In the RUMP-like software by Doyle and Brice,[7] spectrum scaling is used for extracting profiles (e.g., Figures 1.5, 12.8, and 13.17).

SENRAS,[8] the simulation and evaluation of nuclear analysis spectra, software was developed by Vizkelethy essentially for nuclear reaction analysis but also for ERDA. It contains an abundant library of cross sections. The simulation algorithm is very similar to RUMP. Application examples[8] show increasing discrepancies at lower energies due to the omission of MS in the simulation.

11.5.1.2. RBX Simulation and Analysis Program

RBX is a general software developed by Kótai[17] that relies on the same basic principles as RUMP. Some differences exist, such as the linear interpolation of stopping power in the depth width of a brick, which must be small enough for this reason. A very positive feature is that RBX now includes the DEPTH routine by Szilágyi and Pászti[41] for complete and accurate determination of total energy dispersion, including all relevant effects and particularly MS according to the formalism of Szilágyi.[42] A very complete library of cross sections is also included, with an edition tool to maintain it. Profile extraction can be performed by an iterative procedure modifying the concentration in predefined layers in which the effects of the concentration gradient is neglected. The authors mention a precision of 5–10%.

11.5.1.3. LORI Analysis Program

LORI[1] is another analysis program developed at the University of Aarhus that is quite similar in principle to the preceding ones. It includes a careful treatment of energy dispersion using the formalism of Möller for the MS. As in RBX profile extraction is performed by iterative correction of concentrations in predefined sublayers.

11.5.1.4. GABY Simulation and Profile Optimization Algorithm

GABY[9–11,15] is another simulation program based on the classic method. It includes a versatile profile optimization procedure for determining profiles. Energy loss is treated by using the surface approximation (Eq. 4.15) over very thin sublayers with the new projectile energy as incident energy at each depth step. For thin enough sublayers, this first-order approximation is accurate enough. The ideal yield is computed by using an integral form instead of the generalized product used in RUMP. This approach is more exact from the physical point of view, but it necessarily involves some incertitude due to the numerical integration. As in RUMP scattering cross-sections are calculated by using data from Tirira and Bodart.[39] But the really original feature of GABY is the very accurate determination of spreading effects with which the ideal yield is convoluted. This treatment[9] includes not only detector resolution, geometric dispersion, and straggling but also MS effects according to Möller's formalism. Both straggling and MS are computed step-by-step on inward and outward paths and the path in the absorber foil, then summed quadratically for each path. For MS in the absorber, only the angular effect is taken into account, then other contributions to total energy resolution are added quadratically. The simulation algorithm is thus as complete and accurate as possible from the physical point of view, and it shows good agreement with experiments on standards of known composition. Only the summation of MS contributions in quadrature may be somewhat approximative in light of recent discussions (cf. Section 4.2.4.1), but this feature is common to all current software. A very similar simulation method was proposed by Wang et al.[12] (Figures 12.5 and 12.6).

The optimization procedure[10] for profile determination consists of a nonlinear least-square algorithm for finding concentrations in different sublayers that correspond to the minimum of the Poisson maximum-likelihood χ^2 function describing the difference between simulated and experimental spectra.[21] An initial guess of the profile is required, and it may have to be replaced by a more appropriate one when the unknown profile is not simple. Such a semiautomatic procedure is able to determine profiles that could hardly be composed by someone using homogeneous slabs or simple mathematical functions (see, e.g., Figures 1.9, 12.4, 12.9, 14.4, 14.5, and 14.14).

11.5.1.5. GISA Simulation and Analysis Program

This very versatile and user-friendly software was developed by Saarilahti and Rauhala.[16] It allows interactive profile extraction (*analysis*, according to the authors' vocabulary) from RBS spectra, but the abundant data base for non-Rutherford cross sections makes it quite attractive for ERDA interpretation, although no example of this possible application is given by the authors.

The software includes a choice of several possible formulae for ion stopping and an extended library for cross sections, with the interesting possibility of the user completing it later. Straggling is treated by Bohr's theory corrected by the Lindhard–Scharff formalism. However the simple Bragg's rule is applied for compounds and mixtures; in addition channeling can be treated by using the classic MS formalism.

Starting from an initial target definition, for which a large choice of descriptions is available, changes in stoichiometry, layer thicknesses, or profile shape are selected by the user with the help of screen images where the contribution of various elements is represented with the total spectrum. Multiple linear iterations are used to adjust profiles until the simulation deviates from the experimental spectrum less than a user-defined value. A cross-reference table linking simulated and experimental spectra, rebuilt at each iteration, allows to speed up these interactive modifications. Another original feature is that the sublayer thickness used for calculations is not constant nor optimized by the software, which allows to take larger steps on the outward path than on the inward path. The RBS examples[16] show quite satisfactory fits with a fine depth resolution even for intricate profiles in multielemental targets that reflect the accuracy of the simulation procedure.

11.5.2. Hardware

Most of the software just discussed are written in FORTRAN or PASCAL, and they run on personal computers (PCs); some are developed for workstations. The increasing memory and speed of PCs presently make them perfectly suitable. A simulation with RUMP or similar software requires about 1 second, and sophisticated iterative profile extraction algorithms do not require more than a few seconds per iteration. There is thus decreasing advantage in resorting to large machines, although workstations offer powerful graphic tools, which are particularly suitable for advanced image processing, e.g., for three-dimensional TOF-ERDA representations. However graphic tools are also the main cause of difficulties in compatibility and transferability.

11.6. ADAPTATION TO OTHER ERDA VARIANTS

The previously described softwares are essentially devised for classic reflection or transmission ERDA. The ERDA variants described in Chapters 6–9 are designed to resolve mass–depth and recoil–projectile ambiguities, two of the main difficulties encountered in profile determination. These variants also have particular features that must be incorporated into analysis software. In addition TOF ERDA and telescopes provide yields as functions of both energy and velocity (or energy and mass after transforming data) that may require three-dimensional graphics.

11.6.1. Time of Flight ERDA

Data processing for TOF-ERDA is quite specific to this technique,[18,19] although it is similar to data processing for Telescope-ERDA. For this reason it is discussed in Section 6.6 with the typical software PAW,[18] and we limit ourselves to some general remarks in Chapter 11.

The ability of TOF-ERDA to separate masses is in principle a great simplification since at first sight the profile of each element may be treated separately. Practically the analysis is still rather intricate: we must first convert yield versus energy and velocity data into yield versus energy for each mass, which is simple enough with Eq. 6.3; see the example in Figure 6.9. Three-dimensional graphs help find criteria for gating counts between neighboring masses; this gating becomes difficult when large enough neighboring masses are present (e.g., Figure 6.7). For example interference occurs between scattered iodine with barium recoils from high T_c superconductors.[19] Similarly a minimum energy of 71 MeV for iodine ions is necessary to resolve As and Ga.[19]

In first approximation the profile of each element can be deduced in relative value from corresponding yield versus energy data and then in absolute value by normalizing the sum of concentrations to unity. But stopping powers as a function of depth, and as a consequence collision energies determining cross sections, depend on the whole target composition, so an independent treatment of various elements is only a rough approximation. Indeed the profile of all other elements must be taken into account to determine one element just as for classic ERDA. In other words mass–depth ambiguity and resulting possibilities of misinterpretation are excluded, but profile extraction is still a multiparametric problem.

In addition to physical processes that are part of classic ERDA analysis, processes proper to TOF apparatus must be taken into account, and these may be difficult to determine accurately,[19] in particular straggling and MS in carbon foils and the effects of thickness fluctuations in carbon foils. Another difficulty involves taking into account properly effects of the detector dead layer—the fact that silicon detector calibration is not independent of ion mass and the variable efficiency of carbon foil time detectors for various masses. Two possibilities may be considered[18]: First we can use the time signal associated with an assumed mass (deduced from data gating) to determine the true energy; secondly we can calibrate the detector for each recoil species by using pure element targets. The second solution appears to be safer because imperfections in the timing detector (carbon foil wrinkles, etc.) are not involved.

Figure 11.4 shows typical profile extraction for a multielemental target. Other examples are given in Figures 1.6, 6.7, and 6.9. A comparison of profiles extracted from a rather weak signal using PAW and RUMP is shown in Figure 11.5. Profiles exhibit only unsignificant differences.

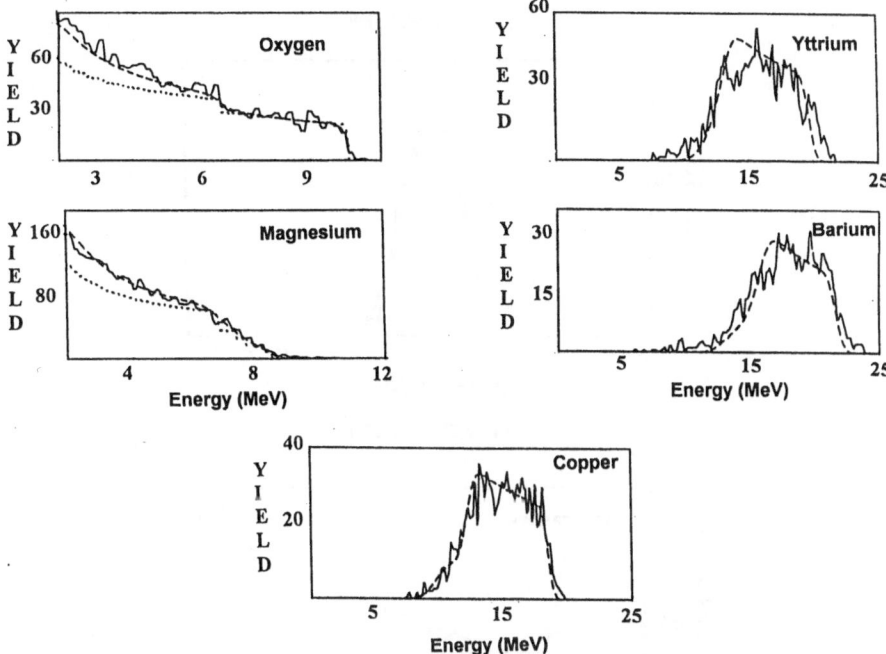

Figure. 11.4. Individual elemental profiles determined using PAW from the TOF-ERDA spectrum of an YBaCuO high-T_c superconductor thin film deposited on MgO. Both O and Mg profiles contain an additional profile (*dotted line*) constructed with no effects due to straggling, MS, roughness, or detector resolution. (Reproduced from Ref. 19)

11.6.2. ERDA with ExB Filters

This technique offers the same advantage as TOF-ERDA for resolving the mass–depth and recoil–projectile ambiguities. Yield versus energy is normally obtained separately for each mass without requiring data transformation. In some cases, e.g., for helium-4 and deuterium, mass separation is not obtained in the complete energy range, but it is still helpful to exclude ambiguity on part of the analyzable depth. Nevertheless limitations of this advantage are the same as for TOF-ERDA, which means that profile determination still requires a multiparametric treatment.

An important obstacle to accurately process data may be detecting particles scattered by the slit edges of the filter or by electric field electrodes. This signal is mixed with the signal from recoils directly reaching the detector, and it becomes important when a large fraction of projectiles is scattered in the forward direction, as occurs with heavy target elements. No simple discrimination is possible, and the recoil yield is obtained through a background substraction that is necessarily a source of incertitude.[43]

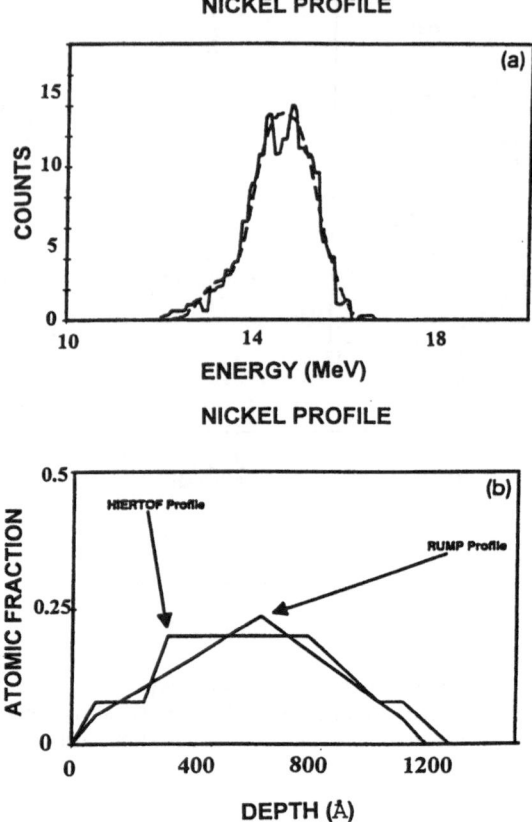

Figure 11.5. (a) Nickel signal extracted from a TOF-ERDA experiment with a simulated fit according to the profile determined by using a TOF-ERDA-dedicated software (PAW). (b) Nickel profile used for the simulation and the profile obtained using RUMP. Comparison of both profiles exhibits only unsignificant differences. (Reproduced from Ref. 19)

As discussed in detail in Section 7.3.1, another difficulty is that detected yields must be corrected for the fraction of neutrals and eventually of undetected charge states by using charge-state fractions that are difficult to measure accurately and may depend on target composition. Several examples of profile extraction from *ExB* spectra are given in Figures 1.7, 7.6, 7.9–7.12, 7.15, 7.16, and 14.9.

11.6.3. ERDA with ΔE-E Telescopes

The treatment of spectra recorded with ΔE-E telescopes is very nearly similar to the treatment for TOF-ERDA. Indeed data are obtained as a biparametric or three-dimensional representation of yield versus energy loss ΔE in the thin detector and residual energy E_R instead of TOF and residual energy for TOF-ERDA. Hence the reader is referred to Sections 6.6 and 11.7.1 for details.

The definition of domains in the $\Delta E - E_R$ plane corresponding to each recoil mass can be performed in the same way as for domains in the $t_f - E_r$ plane for TOF-ERDA. By adding ΔE to E_R to obtain the total energy, yield versus total energy spectra, similar to classic ERDA spectra, can be plotted for each element. Another possibility is to treat the yield directly as a function of E_R, considering ΔE equivalent to energy loss in the absorber foil used in classical ERDA. In both cases the problem is reduced to interpreting of several elemental spectra, as shown in Figures 8.6 and 8.8. Treatments for classic ERDA are applicable with the considerable simplification offered by eliminating the mass–depth ambiguity (e.g., Ref. 28). However as for TOF-ERDA, the entire set of spectra must be considered as a whole to determine profiles thoroughly.

A noticeable difference in the analysis of classic spectra is that straggling and MS in the thin detector is larger than in an usual absorber, and these can introduce a larger absolute incertitude. However using the total energy calculated as $\Delta E + E_R$ eliminates straggling in the thin detector for this variable.

11.6.4. Coincidence ERDA

Outward paths of both the scattered ion and the recoil must be included in calculations for coincidence techniques, but in both cases the principles are the same as for classic ERDA. Mass discrimination resulting from the complete determination of E_1, E_2, θ, and ϕ offers the same advantage as in TOF-ERDA, *ExB* and telescope ERDA. However the biparametric relation of scattering/recoil events to both scattered ion and recoil energies, the eventually large detection angles with the possible use of position- sensitive detectors, and the problem of false coincidence and coincidence losses require quite specific data treatment, which is developed in detail in Section 9.6.3 (see also Figures 9.14 and 9.17).

11.7. CONCLUSION

Concluding this chapter we regret that only a few software products are easily accessible among those that exist. This reason and the difficulty of having access to the details of various formalisms make it difficult to compare different methods significantly. While each software has its own salient features that provide more accuracy or easier use on some particular topic, none can probably be proposed as a standard. It is remarkable that there scarcely exist publications comparing different analysis methods except for occasional comparisons (e.g., Figure 11.5) of a particular sample, not necessarily chosen specifically for a comparative study. Moreover in such cases, there is no guarantee that both softwares were used optimally.

Note that the choice of reference data, physical models, and approximations is not necessarily the main question when comparing data-processing algorithms. Indeed stopping power or cross-section libraries, formulae for the stopping power of compounds (Bragg's rule or models taking into account deviations from this rule), straggling or the MS may be changed at leisure according to further improvement without disturbing the software architecture. Important differences reside in methods underlying data interpretation.

Whether the target description should be discretized into many sublayers to allow low-order approximations or into a few sublayers, to take advantage of higher order developments, may also be of secondary importance, because the same accuracy can be reached in both cases. This question arises only if the limited number of sublayers becomes an obstacle in realistically describing the target. It also assumes some importance if we consider that this choice determines the architecture of the program and as a consequence the accessibility of its contents to any user and the possibility of performing modifications easily to improve it or extend its possibilities to particular cases. From this point of view simplifications resulting from the retrograde method of simulation[14] are an advantage.

There is also less and less advantage in sacrificing accuracy for the sake of saving time. As already stated the accuracy of models adopted for energy-dispersion processes is with counting statistics one of the main requirements for an optimal *final* resolution of profile determinations. With the increasing speed and memory capacity of computers, additional time required for using the most accurate expressions of physical laws is no longer a considerable argument. It is a conclusion common to chemical kinetics, stock management, logistics, and even politics that the rate of a process composed of *successive* steps is controlled by the slowest one. Nowadays the label time consuming cannot be applied to computerized calculations (milliseconds, seconds) but rather to human operations: programming (weeks to years) and interactive interventions in executing softwares (seconds, minutes, exceptionally hours).

The main question is of a more epistemological nature: Is it preferable to take data as they appear, accepting the intricacies of energy spreading and statistical-counting fluctuations as unsuperable limits to data interpretation or attempt to reach a better *final* resolution using all the available knowledge about physical processes and signal treatment theories (e.g., Ref. 44)?

The first option may appear safer, but it leads to profiles including a broadening due to energy spread. Although this broadening can be evaluated accurately enough (see Section 4.2.4.1), the depth extent of profile features remains unspecified and eventually subject to discussion. Another reason for accepting the width of energy spreading as a limit of depth resolution is that details would be hardly distinguishable from counting noise at a tight depth scale, but this argument ignores the possibility of resorting to a noise-filtering procedure justified by the physical properties of scattering spectra (see Section 11.4.4.2 and Ref. 32). An assumption that is not fundamentally related to this option but generally accepted simultaneously is that profiles must be approximated by simple mathematical shapes (steplike, Gaussian, etc.). Indeed we cannot expect much improvement from more general forms in the frame of this option. But as discussed in more detail in Ref. 20, these simple shapes may be qualitatively irrelevant while providing a quantitatively satisfactory fit. For example approximating a layer with a continuous gradient of composition by a large enough number of homogeneous layers may very nearly fit a spectrum, but it is nonsense from the physicochemical point of view. This latter argument is probably the most relevant one in favor of the other option.

Techniques similar to the profile reconstitution method, which exclude as far as possible a priori assumptions concerning profile shapes and target description, are in principle able to evidence profile features on the basis of experiment alone.[29,30] They can in principle reconstitute any feature of concentration profiles without the help of mathematical models and what is more important, the user's intuition. Most certainly their convergence is not warranted in any case.[33] Moreover results generally require a careful check of the relevancy of small features, since incertitudes are still present in the calculation for stopping powers and cross sections as well as energy-spreading terms and residual counting or background noise. Nevertheless these techniques can be expected to offer a way to a better *final* resolution, hence toward more realistic and precise profiles. It is clear that since two of the authors of this book have ventured in this direction,[20,21,32] we should leave this question open to discussion.

REFERENCES

1. Børgesen, P., Bøttiger, J., and Möller, W., Ranges of 10–30 keV deuterons implanted into solids, *J. Appl. Phys.* **49**, 4401 (1978).
2. Doolittle, L. R., Algorithms for the rapid simulation of Rutherford backscattering, *Nucl. Instrum. Methods Phys. Res. Sect. B* **9**, 344 (1985).
3. Doolittle, L. R., A semiautomatic algorithm for Rutherford backscattering analysis, *Nucl. Instrum. Methods Phys. Res. Sect. B* **15**, 227 (1986).
4. Benenson, R. E., Wielunski, L. S., and Lanford, W. A., Computer simulation of helium-induced forward-recoil proton spectra for hydrogen concentration determinations, *Nucl. Instrum. Methods Phys. Res. Sect. B* **15**, 453 (1986).
5. Li, W.-K., and Al-Tamimi, Z., A computer program for the analysis of RBS spectra, *Nucl. Instrum. Methods Phys. Res. Sect. B* **15**, 241 (1986).
6. Doolittle, L. R., Ph.D. diss., Cornell University, 1987.
7. Doyle, B. L., and Brice, D. K., The analysis of elastic recoil detection data, *Nucl. Instrum. Methods Phys. Res. Sect. B* **35**, 301 (1988).
8. Vizkelethy, G., Simulation and evaluation of nuclear reaction spectra, *Nucl. Instrum. Methods Phys. Res. Sect. B* **45**, 1 (1990).
9. Tirira, J., Trocellier, P., Frontier, J. P., and Trouslard, P., Theoretical and experimental study of low-energy ^{4}He-induced ^{1}H elastic recoil with application to hydrogen behavior in solids, *Nucl. Instrum. Methods Phys. Res. Sect. B* **45**, 203 (1990).
10. Tirira, J., Trocellier, P., Frontier, J. P., Massiot, Ph., Costantini, J. M., and Mori, V., 3D hydrogen profiling by elastic recoil detection analysis in transmission geometry, *Nucl. Instrum. Methods Phys. Res. Sect. B* **50**, 135 (1990).
11. Tirira, J., Trocellier, P., and Frontier, J. P., Analytical capabilities of ERDA in transmission geometry, *Nucl. Instrum. Methods Phys. Res. Sect. B* **45**, 147 (1990).
12. Wang, Y., Liao, C., Yang, S., and Zheng, Zh., A convolution analysis method for hydrogen concentration profiles by elastic recoil detection, *Nucl. Instrum. Methods Phys. Res. Sect. B* **47**, 427 (1990).

13. Guo, X. S., Lanford, W. A., and Rodbell, K. P., A 2-dimensional RBS simulation program for studying the edges of multilayer integrated circuit components, *Nucl. Instrum. Methods Phys. Res. Sect. B* **45**, 157 (1990).
14. Serruys, Y., Simulation of Rutherford backscattering spectra: "retrograde method," *Nucl. Instrum. Methods Phys. Res. Sect. B* **61**, 221 (1991).
15. Tirira, J., Frontier, J. P., Trocellier, P., and Trouslard, P., Development of a simulation algorithm for energy spectra of elastic recoil spectrometry, *Nucl. Instrum. Methods Phys. Res. Sect. B* **54**, 328 (1991).
16. Saarilahti, J., and Rauhala, E., Interactive personal computer data analysis of ion backscattering spectra, *Nucl. Instrum. Methods Phys. Res. Sect. B* **64**, 734 (1992).
17. Kótai, E., Computer methods for analysis and simulation of RBS and ERDA spectra, *Nucl. Instrum. Methods Phys. Res. Sect. B* **85**, 588 (1994).
18. Whitlow, H. J., Possnert, G., and Petersson, C. S., Quantitative mass and energy-dispersive elastic recoil spectrometry: Resolution and efficiency considerations, *Nucl. Instrum. Methods Phys. Res. Sect. B* **27**, 448 (1987).
19. Martin, J. W., Cohen, D. D., Dytlewski, N., Garton, D. B., Whitlow, H. J., and Russell, G. J., Materials characterisation using heavy-ion elastic recoil time of flight spectrometry, *Nucl. Instrum. Methods Phys. Res. Sect. B* **94**, 277 (1994).
20. Serruys, Y., Concentration profile reconstitution from Rutherford backscattering spectra, *Nucl. Instrum. Methods Phys. Res. Sect. B* **74**, 565 (1993).
21. Tirira, J., Bodart, F., Serruys, Y., and Morciaux, Y., Optimization algorithm for elastic recoil spectra simulation, *Nucl. Instrum. Methods Phys. Res. Sect. B* **79**, 527 (1993).
22. Ziegler, J. F., Biersack, J. P., and Littmark, U., *The Stopping and Ranges of Ions in Solids* (Pergamon, New York, 1985).
23. L'Ecuyer, J., Davies, J. A., and Matsunami, N., How accurate are absolute Rutherford backscattering yields, *Nucl. Instrum. Methods Phys. Res.* **160**, 337 (1979).
24. Chu, W.-K., Mayer, J. W., and Nicolet, M.-A., *Backscattering Spectrometry* (Academic, New York, 1978), pp. 51–52.
25. Amsel, G., L'Hoir, A., and Battistig, G., Projected small-angle multiple-scattering angular and lateral spread distributions and their combination. I. Basic formulae and numerical results, submitted to *Nucl. Instrum. Methods Phys Res. Sect. B*.
26. Szilágyi, E., Pászti, F., and Amsel, G., Theoretical approach of depth resolution in IBA geometry, *Nucl. Instrum. Methods Phys. Res. Sect. B* **100**, 103 (1995).
27. Serruys, Y., Sakout, T., and Gorse, D., Anodic oxidation of titanium in 1-M H_2SO_4 studied by Rutherford backscattering, *Surf. Sci.* **282**, 279 (1993).
28. Arnoldbik, W. M., and Habraken, F. H. P. M., Elastic recoil detection. *Rep. Prog. Phys.* **56**, 859 (1993).
29. Maugis, P., Formation de composés intermétalliques Ti-Al en couches minces par interdiffusion réactive, thesis, Université Paris-Sud-Orsay, (1994).
30. Maugis, P., and Serruys, Y., Al-Ti reactive interdiffusion studied by STIMS and RBS, *Vacuum* **45**, 413 (1994).
31. Tirira, J., Contribution à l'étude de la collision helion-4 proton et à la spectrométrie de recul élastique. CEA-Report R-5529 (1990).
32. Serruys, Y., Rational smoothing applied to Rutherford-backscattering spectrometry, *Nucl. Instrum. Methods Phys. Res. Sect. B* **44**, 473 (1990).
33. Serruys, Y., Convergence de la méthode de reconstitution de profils de concentration à partir des spectres de rétrodiffusion Rutherford, Note technique 92/216, DTA/CEREM/DECM/SRMP, CEA, Saclay, France (1992).
34. Schiettekatte, F., Marchand, R., and Ross, G. G., Deconvolution of noisy data with strong discontinuity and uncertainty evaluation, *Nucl. Instrum. Methods Phys. Res. Sect. B* **93**, 334 (1994).
35. Marin, N., Serruys, Y., and Calmon, P., Extraction of lateral nonuniformity statistics from Rutherford-backscattering spectra, *Nucl. Instrum. Methods Phys. Res. Sect. B* **108**, 179 (1996).

36. Calmon, P., Contribution de l'analyse RBS à l'étude des effets d'irradiation sur la diffusion dans les verres d'oxydes, Rapport CEA R-5560 (1991).
37. Marin, N., Contribution à l'étude de la diffusion d'impuretés métalliques dans un film polymère (Kapton) sous et hors irradiation, Rapport CEA R-5713 (1996).
38. Chu, W.-K., Mayer, J. W., and Nicolet, M.-A, *Backscattering Spectrometry* (Academic, New York, 1978), p. 104.
39. Tirira, J., and Bodart, F., Alpha-proton elastic-scattering analyses up to 4 MeV, *Nucl. Instrum. Methods Phys. Res. Sect. B* **74**, 496 (1993).
40. Lucas, S., and Tirira, J., private communication.
41. Szilágyi, E., and Pászti, F., Theoretical calculation of the depth resolution of IBA methods, *Nucl. Instrum. Methods Phys. Res. Sect. B* **85**, 616 (1994).
42. Szilágyi, E., Some problem of ion beam analysis, thesis, Eötvös University, Budapest (1993).
43. Roux, B., Chevarier, A., Chevarier, N., Wybourn, B., Antoine, C., Bonin, B., Bosland, P., and Cantacuzene, S., High-resolution hydrogen profiling in superconducting materials by ion beam analysis (ERD *ExB*), *Vacuum* **47**, 629 (1995).
44. Max, J., *Methodes et techniques de traitement du signal et applications aux mesures physiques* (Masson, Paris, 1985).

12

Applications of Elastic Recoil Spectrometry to Hydrogen Determination in Solids

12.1. INTRODUCTION

A large number of application examples of ERDA spectrometry were given in previous chapters concerning the determination of hydrogen isotopes as depth profiling other light elements (Chapters 5–9). These examples are classified on the basis of the ERDA spectrometry involved. In Chapter 12 we focus on hydrogen isotope determination using conventional ERDA. We illustrate the wide range of application fields covered by this type of investigation.

Four main application fields are distinguished for ERDA; each has been strongly allied with technological, industrial, and scientific development during the last few years: polymers, materials for microelectronics and optoelectronics, thin films, and elemental transport phenomena near interfaces.

We begin by presenting application examples of ERDA in polymer sciences because polymers are among the most hydrogen-rich media that can be found. We successively examine surface properties of polymers, polymer blends, and the evolution of polymer composition induced by irradiation. The following section deals with the application of ERDA in the field of new materials processed for microelectronics and optoelectronics purposes. The third section concerns thin films applications. In the fourth section we show that ERDA is also a performing method to follow elemental transport mechanisms and particularly hydrogen transfer near interfaces induced by corrosion, wear, etc. Then we illustrate some other research topics in which ERDA has been successfully used to perform composition diagnosis in various media. Finally we discuss some studies dealing with hydrogen behavior under beam irradiation in various media except for polymers.

Last but not the least, a detailed bibliography allows anyone interested in ERDA to carry on a well-defined experiment.

12.2. APPLICATIONS IN POLYMER SCIENCES

Characterizing how polymer molecules behave at free polymer surfaces at inter-faces between incompatible polymers and at interfaces with inorganic solid substances is very critical for understanding and improving the performances of polymers in numerous application fields. For example the adhesion of two polymers strongly depends on exchange mechanisms occurring at their interface between polymer segments.[1] Forward-recoil spectrometry is one of the most attractive methods for investigating these aspects of polymer science quantitatively.[1]

12.2.1. Surface Properties

Modifications of polymeric surfaces by RF or microwave was studied by Chou and coworkers.[2] A resonant forward-recoil technique was used to monitor deuterium incorporation in the surface after treatment. Figure 12.1 shows resonance forward-re-coil spectra with 2.11-MeV $^4He^+$ for untreated H_2O and D_2O plasma-treated polyimide films, respectively. The degree of hydroxylation, measured in terms of D uptake, was found to increase monotonically, then reach a plateau with prolonged exposure time when the absorbed microwave power and operating pressure were kept constant.

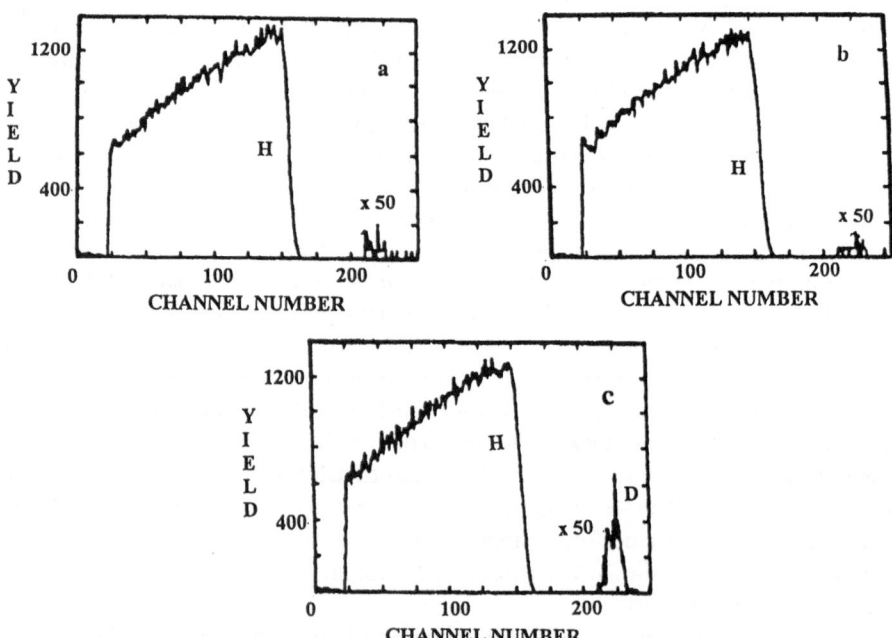

Figure 12.1. Resonance forward recoil spectra of (a) a control or untreated, (b) a H_2O plasma treated, and (c) a D_2O plasma-treated PI films. (Data from Ref. 2)

Investigation of polymer molecules diffusion through polymer interfaces has been abundantly carried out using ERDA.[3,6] For example Composto and coworkers[6] identified tracer diffusion coefficients of deuterated poly(xylenyl)ether chains in protonated polystyrene and poly(xylenyl)ether. These coefficients were 3.9×10^{-5} and 1.0×10^{-6} cm^2·s^{-1} at 295 °C, respectively.

12.2.2. Polymer Blends

Hydrogen and deuterium profiles have been measured in various polymer blends.[7,12] Green and Russell[13] studied the segregation of deuterated polystyrene/polymethylmetacrylate copolymer at the interface of polystyrene and polymethylmetacrylate homopolymer by using ERDA with 2.8-MeV ^{4}He$^+$ ions (Figure 12.2). The same group also studied interface properties of copolymers/Al or Si structures.[14]

12.2.3. Polymer Irradiation

Ion-implantation is one of the methods currently used to modify physical properties of polymers and to improve their electrical, optical, and mechanical performances. Many authors measured the evolution of electrical conductivity, optical transparency, corrosion resistance, and wear resistance of different polymers after irradiation by electron or low-energy light ions or high-energy heavy ions. Then they interpreted these results in terms of structural and compositional modifications.[15,20]

Trocellier and coworkers proposed for the first time using ERDA in transmission geometry to study the composition degradation of thin polyimide films bombarded

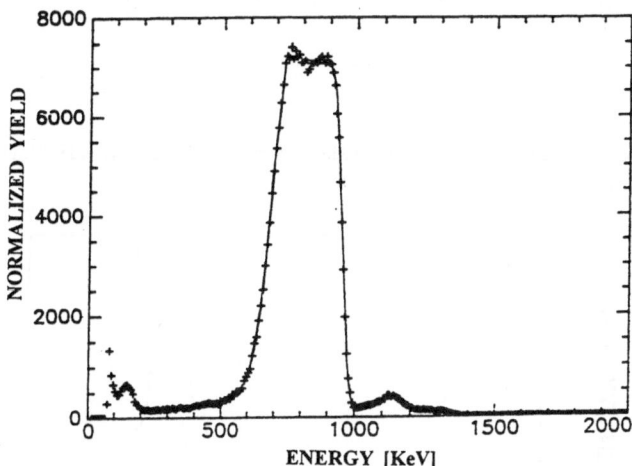

Figure 12.2. Typical ERDA spectrum of yield versus energy of P (d-S-b-d-MMA) block copolymer chains segregated at the interface of the PS and PMMA homopolymers. The profile that lies in the energy range from 600–1000 keV is hydrogen from homopolymers; the other profile, which lies between 1000–1400 keV, is deuterium from copolymer chains. (Data from Ref. 13)

with high-energy heavy ions.[18] Figures 12.3 and 12.4 present recoil spectra and derived hydrogen distributions from a 12.5-µm polyimide film irradiated by 50-MeV ^{32}S ions. The formation of local hydrogen-rich inclusions was evidenced by scanning large areas of the polymer surface.

12.3. APPLICATIONS TO SEMICONDUCTOR MATERIALS

Electronic devices generally consist of successive thin layers (oxides, nitrides, silicides, metals, polymers, or doped semiconductor-based media) coated on a single crystalline substrate (Si, Ge, or AsGa). These structures are ideal candidates for analysis by ion beam techniques, as already shown by several authors.[21-23]

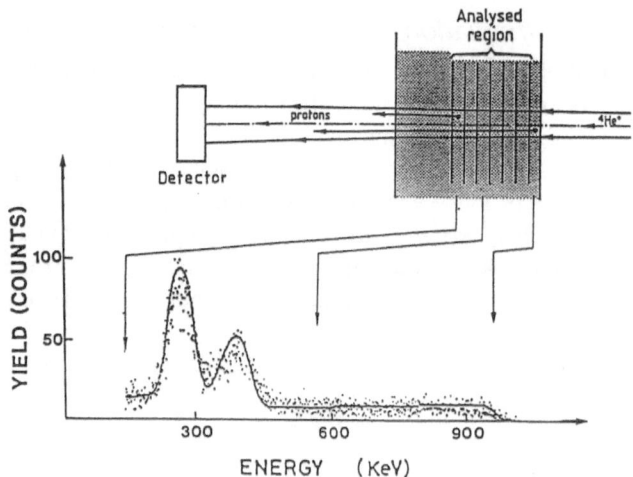

Figure 12.3. Experimental recoil spectra for 50 MeV ^{32}S pre-irradiated polyimide (film thickness = 12.5 µm, ^{4}He^{+} microbeam energy = 2.05 MeV). (Data from Ref. 18)

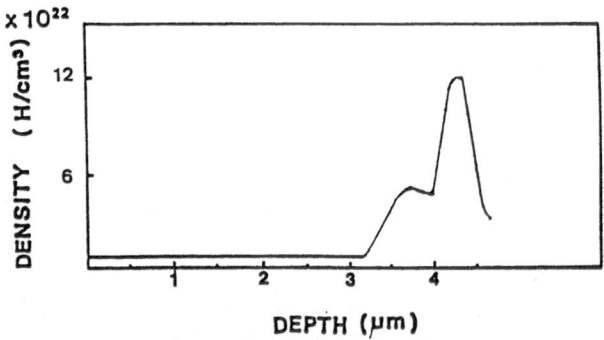

Figure 12.4. Hydrogen atomic density depth profile derived from the energy spectrum in Figure 12.3 using the ERDA computer code GABY. (Data from Ref. 18)

12.3.1. Silicon-Based Media

Silicon, hydrogenated silicon, silicon dioxide, silicon nitride, and silicon carbide systems have been abundantly developed and studied for microelectronics and optoelectronics purposes.[24-48]

Wang and coworkers used ERDA to profile hydrogen in semi-insulating polysilicon films. They succeeded correlating H distributions with reactant gas flow ratio and annealing temperature[49] (Figures 12.5 and 12.6).

Barbour and coworkers[50] prepared silicon nitride and oxynitride films using SiH_4 and N_2 gas mixtures from an electron cyclotron resonance (ECR) plasma source. Rutherford backscattering and elastic recoil spectra of these samples are shown in Figures 12.7 and 12.8. The ECR $Si_{2.8}N_4$ sample has a low H concentration nearly equal to the concentration obtained for a high-temperature chemical-vapor-deposited sample (5.8 at. %).

Compagnini and coworkers[51] determined the optical and electrical properties of hydrogenated amorphous silicon carbide films (a-Si_xC_{1-x}:H), synthesized by implantating carbon and hydrogen ions in silicon. RBS and forward recoil spectrometry were used to measure the stoichiometry of synthesized films (Figure 12.9).

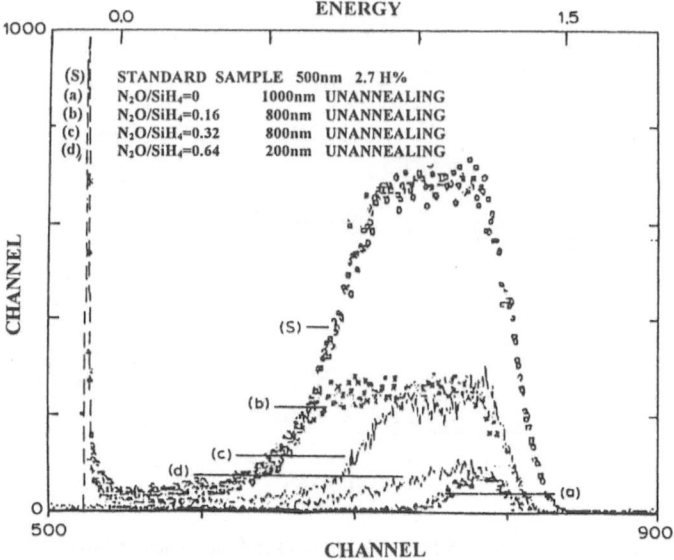

Figure 12.5. Hydrogen recoil spectrum for various reactant gas flow ratio. (Data from Ref. 49)

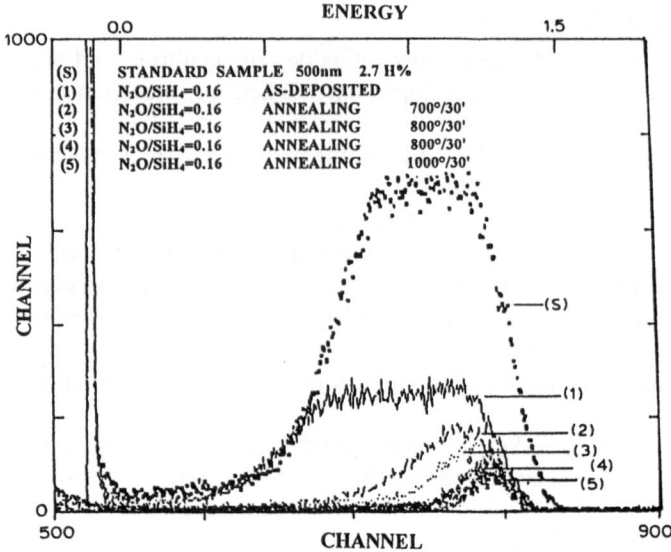

Figure 12.6. Hydrogen recoil spectrum for various annealing temperatures. (Data from Ref. 49)

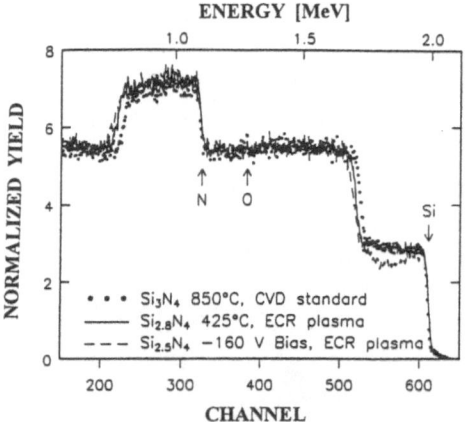

Figure 12.7. The RBS spectra using 3.5-MeV ^{4}He$^+$ ions taken from samples tilted at 45° relative to the incident ion beam. The scattering angle was 164°. At this energy the nitrogen scattering cross-section is twice as large as the nitrogen Rutherford cross section. (Data from Ref. 50)

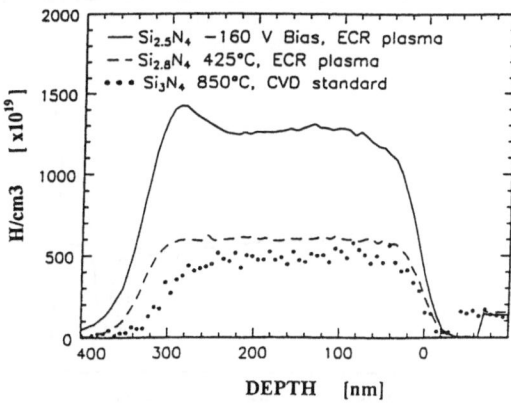

Figure 12.8. Hydrogen concentration profiles determined from the ERDA spectra in Figure 12.7. Based on a density of 9×10^{22} atoms/cm^3, the H contents are 5.8% for Si_3N_4, 6.7% for $Si_{2.8}N_4$, and 14% (16% at 300 nm) for $Si_{2.5}N_4$. (Data from Ref. 50)

12.3.2. Germanium-Based Media

Graeff and coworkers studied the H and D incorporation and release processes in RFC-sputtered hydrogenated amorphous germanium.[52]

Feuser and coworkers have showed that complex defects formed in germanium by indium implantation can be identified as vacancies trapped by the indium probe by combining results from RBS, ERDA, and $\gamma-\gamma$ perturbated angular correlation techniques.[53] More generally hydrogen diffusion in amorphous hydrogenated germanium can be investigated using ERDA.[54,55]

12.3.3. Other Media

The stoichiometry of thin, insulating Ta_2O_5 films deposited on silicon substrates for ultralarge-scale integration was determined by combining RBS and ERDA by Shimizu and coworkers.[56] Adopting an analogous experimental approach, Gordon and Riaz measured the composition of aluminum nitride films coated on silicon[57]; results are shown in Figures 12.10 and 12.11. Both the nitrogen/aluminum ratio and the hydrogen content decrease with increasing temperature. Moreover the refractive indexes of the films increase regularly, and their initial amorphous structure becomes progressively polycrystalline.

12.4. APPLICATIONS TO THIN FILMS

The complete determination of thin-film composition requires combining different analytic techniques, including ion beam methods; the RBS–ERDA combination appears to be very attractive.[58,59]

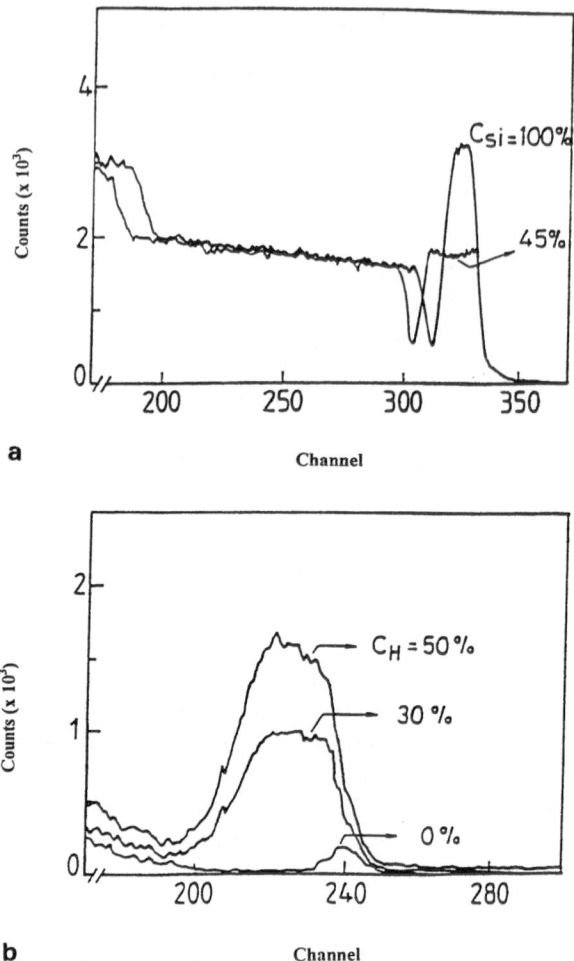

Figure 12.9. (a) Rutherford-backscattering spectra of unimplanted silicon on sapphire and of a sample implanted with carbon ions (x = 0.45). (b) Forward recoil spectra of amorphous silicon-carbon alloys containing different H concentrations. (Data from Ref. 51)

12.4.1. Metal and Inorganic Compound Layers

Titanium, zirconium, and hafnium nitrides exhibit metallic behavior, extreme hardness, high melting points, remarkable chemical resistance, good optical transparency, and high-temperature superconductivity; their thin films have many potential applications. The ERDA investigations performed by Fix and coworkers[60,61] show that their hydrogen content is strongly temperature-dependent.

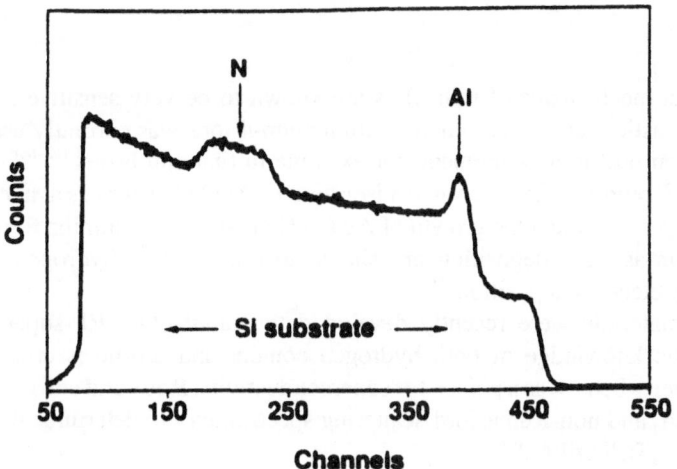

Figures 12.10. The RBS spectrum of an aluminum nitride film deposited at 200° C on silicon (2-MeV ^{4}He$^+$ beam). (Data from Ref. 57)

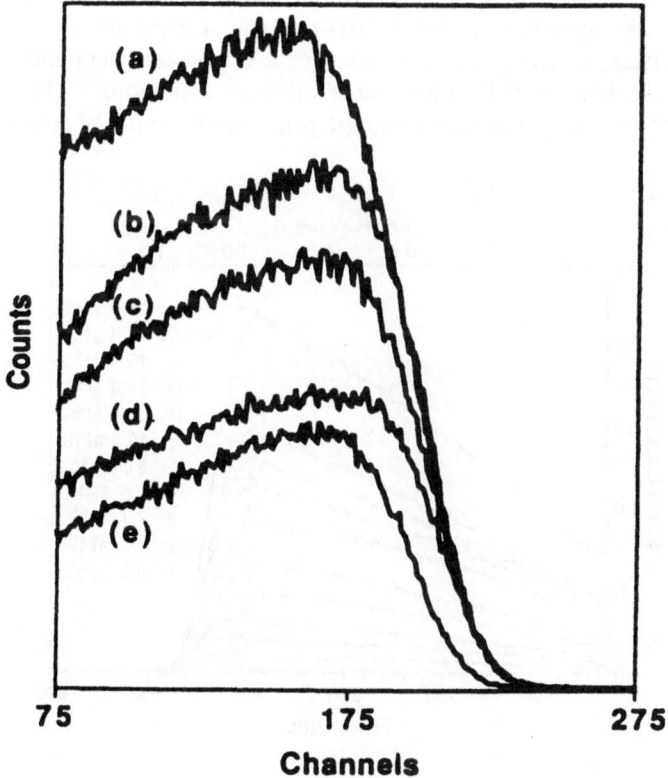

Figure 12.11. Forward recoil spectra for aluminum nitride films deposited at (a) 100 °C, (b) 200 °C, (c) 300 °C, (d) 400 °C, and (e) 500 °C on silicon. (Data from Ref. 57)

Growth mechanisms of thin films are known to be very sensitive to hydrogen presence. Elastic recoil spectrometry with helium-4 ions was currently used to carry out hydrogen profile measurement, for example in titanium layers.[62–64] Naitoh and coworkers[65] reported that hydrogen adsorption on Si (111) surfaces at a level around 1.5 monolayer promoted the growth of Ag (111) crystallites. Alumina films prepared by ion-beam-assisted deposition are shown to contain less hydrogen than those prepared by electron deposition.[66]

New materials were recently developed to manufacture RF-superconducting cavities. The knowledge of both hydrogen content and profile is crucial because hydride species behave as a poison for superconductivity. Roux and coworkers applied RBS, ERDA, and non-Rutherford-scattering spectrometry to determine the composition of $[Nb_{1-x}Ti_x]N$ films.[67]

12.4.2. Amorphous Carbon Films and Diamondlike Carbon Layers

Such films were well-studied, including their preparation modes and physical and chemical properties. Hydrogenated carbon containing such metals as B, Ti, Ta, and W,[68] as well as amorphous hydrogenated carbon films[69] possess excellent tribological characteristics. Many authors reported ERDA results on these two compound families.[70,77] For example, Wang and coworkers prepared tungsten containing amorphous hydrogenated carbon (W-C:H) films under different conditions[78] (Figure 12.12). Researchers found that physical properties depend mainly on the W content.

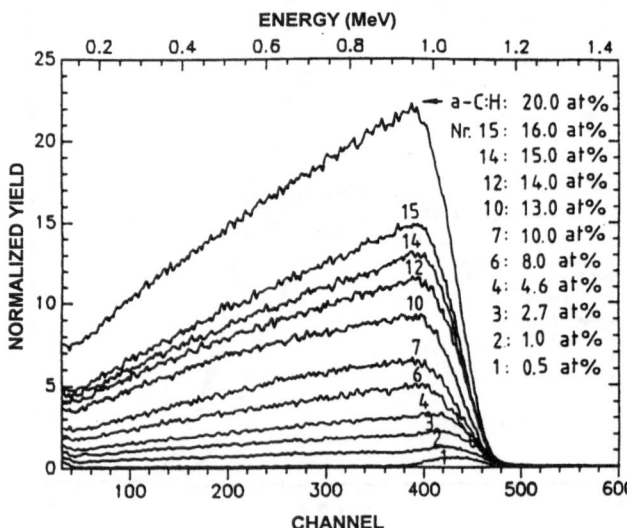

Figure 12.12.   Elastic recoil detection spectra of W-C:H films prepared under different conditions. (Data from Ref. 78)

Boutard and coworkers studied the hydrogen isotope exchange in a-C:H/D layers submitted to 8–11 keV H^+ and D^+ irradiations,[79,80] as shown in Figure 12.13. They showed a strong initial hydrogen release and a tendency to resaturation with increasing ion fluence. Perrière and coworkers indicate that the atomic exchange rate appears to depend on the energy of incident ions.[81]

Diamondlike carbon (DLC) films are metastable materials, composed essentially of trigonal and tetragonal carbon atoms in an amorphous structure. The DLC films are generally prepared by direct current or RF plasma-assisted chemical vapor deposition, sputtering, or ion beam deposition. Physical properties of DLC films are very specific: extreme hardness, low friction coefficient, high internal compressive stresses, high optical transparency, high electrical resistivity, and chemical inertness.[82] These properties are known to depend strongly on deposition conditions and film composition.[83]

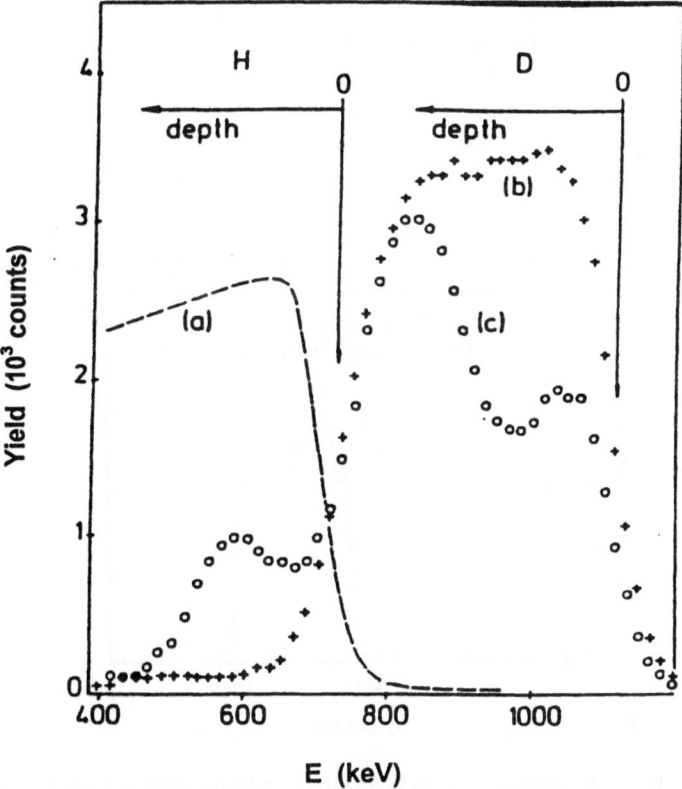

Figure 12.13. Energy spectrum of recoiled protons and deuterons obtained with 2.6-MeV $^4He^+$ (incident angle = 75°, detection angle = 30°): (a) proton reference from unimplanted a:C-H layer, (b) deuteron reference from unimplanted a:C-D layer, (c) implanted spot of the same layer in (b) with 11-keV H^+ at a nominal fluence = 1.8×10^{17} ions/cm². (Data from Ref. 79)

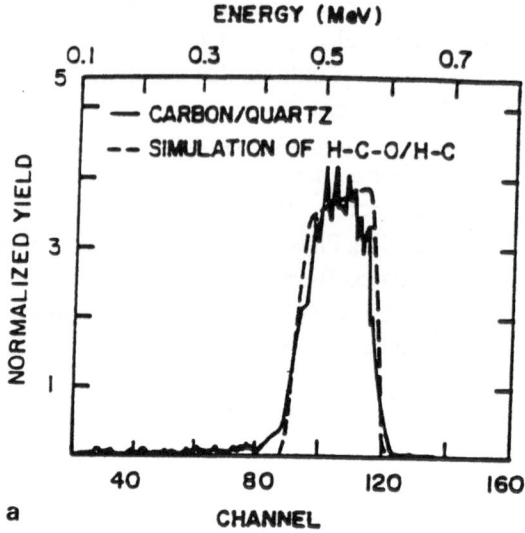

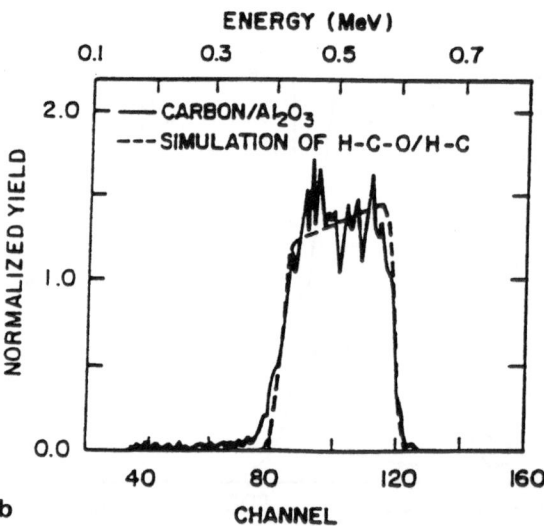

Figure 12.14. Forward recoil scattering spectra for a thick ion-beam-sputtered film deposited at 77 K: (a) layer thickness = 70 nm, H content = 6.5 at. %, system base pressure = 2.6×10^{-7} Torr; (b) layer thickness = 100 nm, H content = 2.5 at. %, system base pressure = 8×10^{-8} Torr. (Data from Ref. 84)

The hydrogen content of DLC films varies from 0–50 at. %. It influences film structure and consequently its properties.

Cuomo and coworkers[84] determined the hydrogen content of ion-beam-sputtered DLC layers using $^4He^+$ ions with an incident energy of 2.3 MeV. Researchers show that greater hydrogen incorporation at a higher base pressure (Figure 12.14b) is due to residual H_2 and water vapor.

Some authors performed hydrogen measurements on DLC films to improve their deposition process and particularly to lower both the substrate temperature and the annealing temperature[85] or the bias voltage applied to the substrate.[86]

Mechanical properties of DLC films were investigated in terms of compressive stress, hardness and Young's modulus determination by Dekempeneer and coworkers[87] or in terms of friction coefficients by Wu.[88] When compressive stress, hardness, Young's modulus, and wear resistance increase, the local sp^2/sp^3 ratio is found to reach a minimum.[87] This fact does not imply that the corresponding ratio of volume fractions of sp^3 and sp^2 phases is a minimum but more probably that hydrogen-bonded sites are heterogeneously distributed. By measuring the hydrogen depth distribution, Nabot and coworkers[89] reports the absence of a sharp interface for DLC films coated on glass, steel, and silicon prepared by dual-sputtering and ion beam-assisted deposition.

Peebles and coworkers combined several analytic techniques—scanning electron microscopy, atomic force microscopy, transmission electron microscopy, Raman spectroscopy, and elastic recoil spectrometry—to characterize diamond layers coated

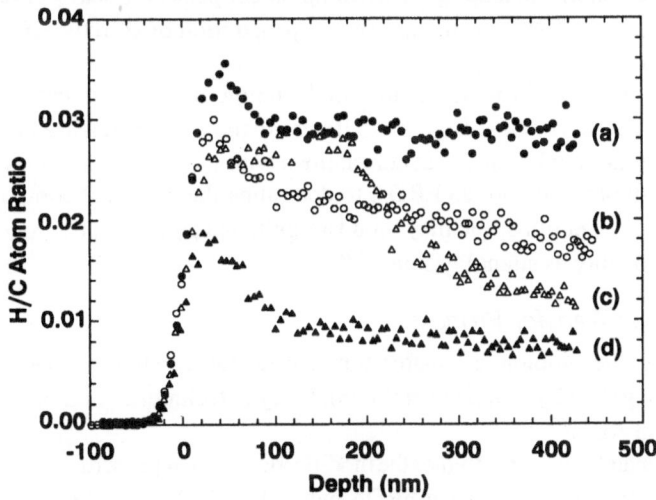

Figure 12.15. Hydrogen/carbon atom ratio in the diamond film as a function of film depth, as measured by elastic recoil: (a) Filled-in circles represent the intermediate zone; (b) open circles represent the outer zone; (c) open triangles represent the outer zone of another sample; and (d) filled-in triangles represent an edge sample. (Data from Ref. 90)

on silicon.[90] Evolution of the hydrogen/carbon ratio shows that the hydrogen content is higher near the surface than in the bulk of the layer. It also appears that hydrogen increases continuously from the edge to the middle of the silicon wafer (Figures 12.15d and 12.15a).

12.5. STUDY OF INTERFACE REACTIONS

Elastic recoil spectrometry is well-adapted for studying hydrogen transport near interfaces.[91] Much work was performed on surface corrosion[92–98] and the tribological properties of materials.[99–103]

12.5.1. Surface Corrosion Mechanisms

The dissolution rate for a mineral, glass, or ceramic strongly depends on the number of different surface cationic sites and the attachment of hydrogen to these centers.

Arnold and coworkers, studying the alteration mechanism of labradorite feldspar by using ERDA, shows the influence of pH on the surface hydrogen accumulation (Figure 12.16a–c).[92]

Matzke and coworkers developed combined RBS and ERDA to investigate modifications in the chemical durability of nuclear waste glasses and simulated nuclear fuels induced by ion implantation.[93,94] Figure 12.17 shows that damage induced by irradiation (displacements of atoms, incorporation of new atoms, mobility of alkali ions, fracturation, etc.) enhances glass leaching: larger peaks for heavy glass constituents (Ti, Fe, U) in the RBS spectra and deeper penetration of H in the ERDA spectra displayed in the insert.[94]

Freire reported the influence of ion implantation on the chemical durability of $Na\beta''$-alumina single crystals: Both hydration rate and sodium release are enhanced after ion implantation without direct correlations with ion fluence and mass.[95,96]

Several authors also applied ERDA to determine the chemical composition and density of SiO_2 sol-gel films, and they found a significant amount of hydrogen remains in the films, probably as silanol groups.[97,98]

12.5.2. Wear Transfer Process

Measuring the amount of matter transferred during sliding contact between different media generally requires combining analytic techniques: ion beam methods, XPS, SIMS and AES.[99,100]

Neelmeijer and coworkers studies the modification of wear properties of steel surfaces after nitrogen incorporation by using ERDA induced by 30-MeV ^{35}Cl ions, as shown in Figure 12.18.[101] Researchers found that wear resistance increases systematically with increasing nitrogen ion fluence.

ERDA was also used to investigate the mechanical qualities of oil additives, such as zinc dialkyldithiophosphate and di-tert nonyl pentasulphide in the case of sliding

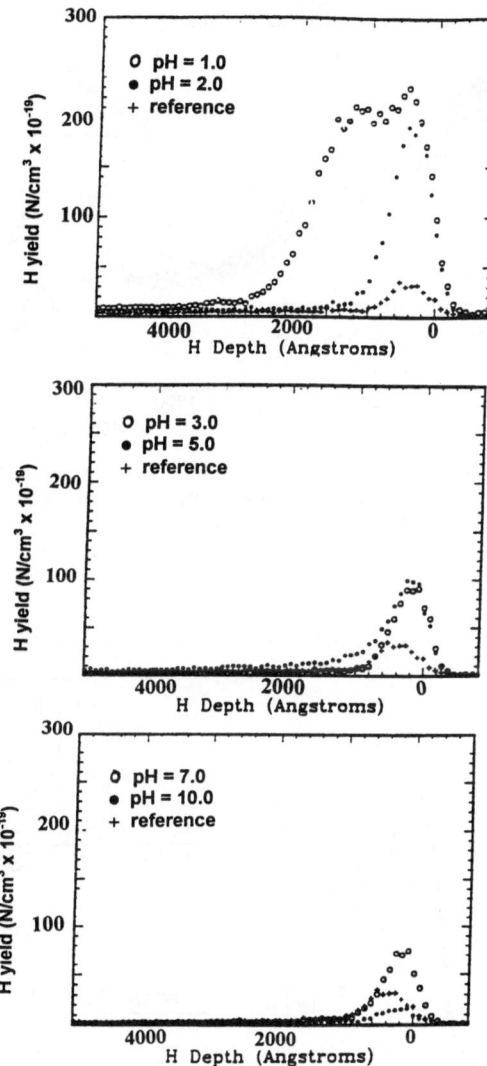

Figure 12.16. The H recoil yield versus depth for feldspar crystals reacted at the indicated pH values for 264 h at 45 °C. (Data from Ref. 92)

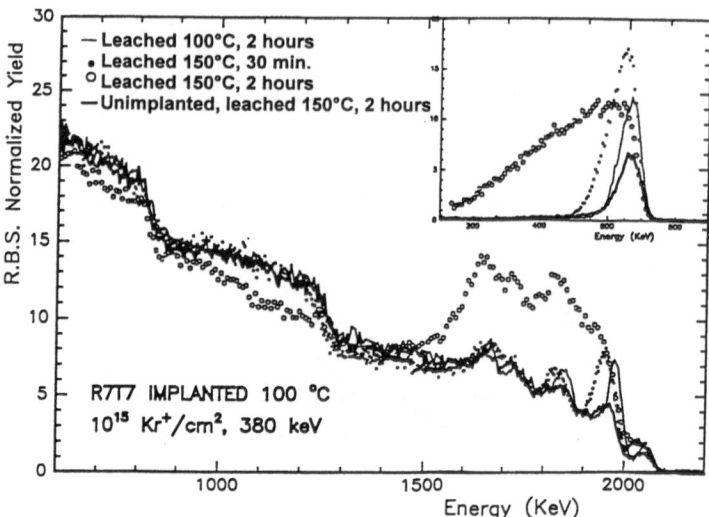

Figure 12.17. The RBS and ERDA experimental spectra of samples not implanted and implanted at 100 °C after leaching at different temperatures and times. (Data from Ref. 94)

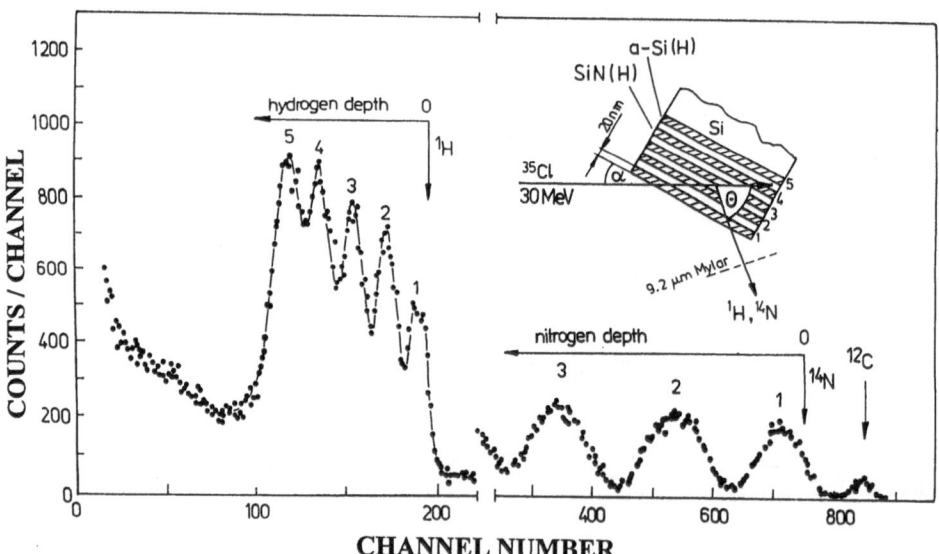

Figure 12.18. The ERDA conventional energy spectrum of a sandwich structure produced by depositing SiN(H) and a-Si:H layers of 20-nm thickness; $\alpha = 20°$, $\theta = 30°$. (Data from Ref. 101)

steel surfaces,[102] and to evaluate incorporating hydrogen in nitride steels exposed to glow discharge.[103]

12.6. OTHER APPLICATION FIELDS

Measuring hydrogen content in the near-surface region is often required to control different steps in a preparation process or a surface treatment and to evaluate any contamination effect.[104,105] Freire and coworkers describe near-surface modifications of glass composition induced by a *dc* potential, combining resonant NRA for sodium and ERDA for hydrogen profiling.[106]

Several authors reported ERDA results for the surface stoichiometry of slightly hydrated WO_3 layers,[107] the implantation of deuterium in B_4C and Be,[108,109] the proton exchange in $LiNbO_3$ crystals,[110] and the retention of hydrogen in TiC[111] or in nickel deuteride.[112] Hydrogen distribution within melt inclusions trapped in volcanic quartz was determined by Mosbah and coworkers[113] to understand better the mechanisms of volcanic dynamism and to establish the balance of material exchange processes. For the first time researchers saw hydrogen-rich fluid inclusions in a natural glass from Pantelleria Island (Figure 12.19). Much work has also been published about the surface composition of fusion reactor limiters.[114–118] For example Antoni and coworkers

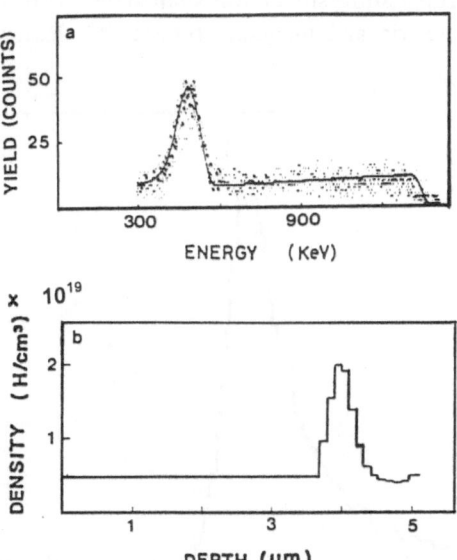

Figure 12.19. (a) Evidence of a hydrogen-rich fluid inclusion in glass (E_0 = 3 MeV, i = 0.7 nA, beam diameter = 30 μm, t = 1800 s). (b) Hydrogen atomic density profile derived from the spectrum. (Data from Ref. 113)

determines the hydrogen content of a glassy carbon limiter exposed to five discharges on the ion drift of the toroidal device ETA BETA II (Figure 12.20).[119]

ERDA can also be used to study the reordering of aluminum-adsorbed silicon surfaces, such as $\sqrt{3} \times \sqrt{3}$ Al/Si (111) induced by hydrogen sorption.[120] At room temperature this structure evolves toward 1×1 Al(H)/Si (111) for hydrogen coverage corresponding to 3.9×10^{14} H/cm^2. Saturation occurs with 1.3×10^{15} H/ cm^2, which means a clean 7×7 surface. By heating at 400 °C, almost all hydrogen is desorbed, so the surface corresponds to a 1×1 Al(H)/Si(111), at 700 °C the initial $\sqrt{3} \times \sqrt{3}$ Al surface is recovered.

12.7. STUDY OF HYDROGEN BEHAVIOR UNDER IRRADIATION

This subject is discussed in more detail in Chapter 14, which is devoted to ion beam damaging. Nevertheless in a large variety of materials, ion beam irradiation may induce hydrogen motion and consequently affect hydrogen depth distribution. Elastic recoil spectrometry, which demonstrates good sensitivity and excellent depth resolution, was successfully applied to this research topic.[121]

Ascheron published a review paper in 1991 on various irradiation effects induced by protons in A^{III}B^V compounds.[122] Tirira and coworkers reported hydrogen loss in polyimide films increasing with increasing microbeam density.[123] Laser-induced hydrogen desorption in crystalline silicon was studied by Boivin and coworkers, using ERDA with crossed electric and magnetic filtering.[124] Researchers showed that

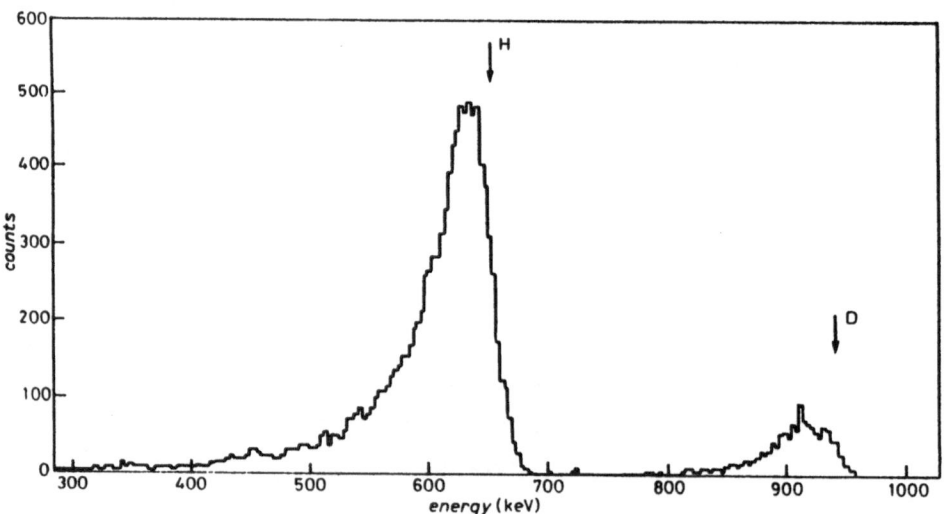

Figure 12.20. The ERDA spectrum of a vitreous graphite sample exposed to five discharges on the ion drift side. (Data from Ref. 119)

hydrogen desorption can be complete (residual hydrogen content < detection limit of the technique ≈ 0.5 at. %).

12.8. CONCLUSION

Considering all the examples discussed here, including data presented in Chapters 5–9, it is clear that ERDA has become a very mature analytic method. Its flexibility and performance level give it a relevant place among the-ion beam-based techniques. Recent and future developments in fast electronics and detection systems still contribute to improve its use in ion beam laboratories all over the world.

REFERENCES

1. Russell, T. P., The characterization of polymer interfaces, *Ann. Rev. Mater Sci.* **21**, 249 (1991).
2. Chou, N. J., Marwick, A. D., Goldblatt, R. D., Li, L., Coleman, G., Heidenreich, J. E., and Paraszczak, J. R., An isotope approach to characterization of microwave-water-plasma-modified polyimide surfaces, *J. Vac. Sci. Technol. A* **10**, 248 (1992).
3. Tead, S. F., Kramer, E. J., Russell, T. P., and Volksen, W., Interdiffusion at polyimide interfaces, *Polymer* **33**, 3382 (1992).
4. Shull, K. R., Dai, K. H., Kramer, E. J., Fetters, L. J., Antonietti, M., and Silelescu, H., Diffusion by constraint release in branched macromolecular matrices, *Macromolecules* **24**, 505 (1991).
5. Shull, K. R., Kramer, E. J., Bates, F. S., and Rosedale, J. H., Self-diffusion of symmetric diblock copolymer melts near the ordering transition, *Macromolecules* **24**, 1383 (1991).
6. Composto, R. J., Kramer, E. J., and White, D. M., Reptation in polymer blends. *Polymer* **31**, 2320 (1990).
7. Zhao, X., Zhao, W., Sokolov, J., Rafailovich, M. H., Schwarz, S. A., Wilkens, B. J., Jones, R. A. L., and Kramer, E. J., Determination of the concentration profile at the surface of d-PS/h-PS blends using high-resolution ion-scattering techniques, *Macromolecules* **24**, 5991 (1991).
8. Dai, K. H., Kramer, E. J., and Shull, K. R., Interfacial segregation in two-phase polymer blends with diblock copolymer additives: The effect of homopolymer molecular weight, *Macromolecules* **25**, 220 (1992).
9. Krausch, G., Dai, C. A., Kramer, E. J., Marko, J. F., and Bates, F. S., Interference of spinodal waves in thin polymer films, *Macromolecules* **26**, 5566 (1993).
10. Green, P. F., Adolf, D. B., and Gilliom, L. R., Dynamics of polystyrene/poly(vinyl methyl ether) blends, *Macromolecules* **24**, 3377 (1991).
11. Mills, P. J., Green, P. F., Palmstrom, C. J., Mayer, J. W., and Kramer, E. J., Analysis of diffusion in polymers by forward recoil spectrometry, *Appl. Phys. Lett.* **45**, 957 (1984).
12. Bruder, F., and Brenn, R., Measuring the binodal by interdiffusion in blends of deuterated polystyrene and poly(styrene-co-4-bromostyrene), *Macromolecules* **24**, 5552 (1991) and Spinodal decomposition in thin films of a polymer blend, *Phys. Rev. Lett.* **69**, 624 (1992).
13. Green, P. F., and Russell, T. P., Segregation of low molecular weight symmetric diblock copolymers at the interface of high molecular weight homopolymers, *Macromolecules* **24**, 2931 (1991).
14. Green, P. F., and Russell, T. P., Adsorption of copolymer chains from a melt onto a flat surface, *Macromolecules* **25**, 783 (1992).
15. Lewis, M. B., and Lee, E. H., Residual gas and ion beam analysis of ion-irradiated polymers, *Nucl. Instrum. Methods Phys. Res. Sect. B* **61**, 457 (1991).
16. Calcagno, L., and Foti, G., Interdiffusion in polystyrene crosslinked by keV ion irradiation, *J. Appl. Phys.* **71**, 3216 (1992).

17. Wang, Y., Mohite, S. S., Bridwell, L. B., Gield, R. E., and Sofield, C. J., Modification of high-temperature and high-performance polymers by ion implantation, *J. Mater. Res.* **8**, 388 (1993).

18. Tirira, J., Frontier, J. P., Trocellier, P., and Trouslard, P., Development of a simulation algorithm for energy spectra of elastic recoil spectrometry, *Nucl. Instrum. Methods Phys. Res. Sect. B* **54**, 328 (1991).

19. Wang, W., Lin, S., Bao, J., Rong, T., Wan, H., and Sun, J., The n-type doping of polyaniline films by ion implantation, *Nucl. Instrum. Methods Phys. Res. Sect. B* **74**, 514 (1993).

20. Kaiser, R. I., Lauterwein, J., Müller, P., and Roessler, K., Energy density effects in the formation of organic residues in frozen methane by MeV ions, *Nucl. Instrum. Methods Phys. Res. Sect. B* **65**, 463 (1992).

21. Earwaker, L. G., Briggs, M. C., Nasir, M. I., Farr, J. P. G., and Keen, J. M., Characterisation of thin layers in processed silicon, *Nucl. Instrum. Methods Phys. Res. Sect. B* **66**, 23 (1992).

22. Gusinskii, G. M., Kudryavtsev, I. V., Kudoyarova, V. Kh., Naïdenov, V. O., and Rassadin, L. A., A method for investigation of light-element distribution in the surface layers of semiconductors and dielectrics, *Semicond. Sci. Technol.* **7**, 881 (1992).

23. Ascheron, C., Lehmann, D., Neeljmeijer, C., Schindler, A., and Bigl, F., Hydrogen depth profiling in bevelled proton-implanted semiconductors by ^{4}He ERDA and ^{19}F NRA, *Nucl. Instrum. Methods Phys. Res. Sect. B* **63**, 412 (1992).

24. Meda, L., Cerofolini, G. F., Ottaviani, G., Tonini, R., Corni, F., Balboni, R., Anderle, M., Canteri, R., and Dierckx, R., Evidence for molecular hydrogen in single-crystal silicon, *Physica B* **170**, 259 (1991).

25. Ramm, J., Beck, E., Zueger, A., Dommann, A., and Pixley, R. E., Hydrogen cleaning of silicon wafers: Investigation of the wafer surface after plasma treatment, *Thin Solid Films* **228**, 23 (1993).

26. Naitoh, M., Morioka, H., Shoji, F., and Oura, K., Coadsorption of hydrogen and deuterium on Si(100) surfaces studied by elastic recoil detection analysis, *Surf. Sci.* **297**, 135 (1993).

27. Umezawa, K., Yamane, J., Kuroi, T., Oura, K., and Hanawa, T., Nuclear reaction analysis and elastic recoil detection analysis of the retention of deuterium and hydrogen implanted into Si and GaAs crystals, *Nucl. Instrum. Methods Phys. Res. Sect. B* **33**, 638 (1988).

28. Bae, Y. W., Du, H., Gallois, B., Gonsalves, K. E., and Wilkens, B. J., Structure and chemistry of silicon nitride and silicon carbonitride thin films deposited from ethylsilazane in ammonia or hydrogen, *Chem. Mater.* **4**, 478 (1992).

29. Markwitz, A., Bachmann, M., Baumann, H., Bethge, K., and Krimmel, E., Hydrogen profiles of thin PVD silicon nitride films using elastic recoil detection analysis, *Nucl. Instrum. Methods Phys. Res. Sect. B* **68**, 218 (1992).

30. Niu, H., Wu, S., Huang, S., Lin, J., and Deng, R., Hydrogen depth profiling of SiN_x films by the detection of recoiled protons, *Nucl. Instrum. Methods Phys. Res. Sect. B* **79**, 536 (1993).

31. Ermolieff, A., Sindzingre, T., Marthon, S., Martin, P., Pierre, F., and Pecoud, L., PECVD silicon oxides as studied by XPS, RBS, ERDA, IRS, and ESR, *Appl. Surf. Sci.* **64**, 175 (1993).

32. Whitlow, H. J., Andersson, A. B. C., and Petersson, C. S., Thermally grown SiO_2 film standards for elastic recoil detection analysis, *Nucl. Instrum. Methods Phys. Res. Sect. B* **36**, 53 (1989).

33. Finger, F., Kroll, U., Viret, V., Shah, A., Beyer, W., Tang, X. M., Weber, J., Howling, A., and Hollenstein, Ch., Influences of a high-excitation frequency (70 Mhz) in the glow discharge technique on the process plasma and the properties of hydrogenated amorphous silicon, *J. Appl. Phys.* **71**, 5665 (1992).

34. Paquin, L., Leclerc, G., and Wertheimer, M. R., Electrochemical effects in $metal_1$-a-Si:H-$metal_2$ structures, *Can. J. Phys.* **68**, 1396 (1990).

35. Diawara, Y., Currie, J. F., Najafi, S. I., Brebner, J. L., Cochrane, R. W., and Gujrathi, S. C., Caractérisation des dispositifs photovoltaïques au a-Si:H, *Can. J. Phys.* **69**, 530 (1991).

36. Stolze, F., Zacharias, M., Schippel, S., and Garke, B., Compositional investigation of sputtered amorphous SiO_x:H, *Solid State Commun.* **87**, 805 (1993).

37. Zellama, K., Labidi, H., Germain, P., Von Bardeleben, H. J., Chahed, L., Theye, M. L., Roca i Cabarrocas, P., Godet C., and Stoquert, J., Systematic study of light-induced effects in hydrogenated amorphous silicon, *Phys. Rev. B: Condens. Matter* **45**, 13314 (1992).

38. Amato, G., Della Mea, G., Fizzotti, F., Manfredotti, C., Marchisio, R., and Paccagnella, A., Hydrogen bonding in amorphous silicon with use of the low-pressure chemical-vapor-deposition technique, *Phys. Rev. B: Condens. Matter* **43**, 6627 (1991).

39. Cerofolini, G. F., Meda, L., Balboni, R., Corni, F., Frabboni, S., Ottaviani, G., Tonini, R., Anderle, M., and Canteri, R., Hydrogen-related complexes as the stressing species in high-fluence, hydrogen-implanted, single-crystal silicon, *Phys. Rev. B: Condens. Matter* **46**, 2061 (1992).

40. Van den Boogaard, M. J., Arnoldbik, W. M., Habraken, F. H. P. M., and Van der Weg, W. F., Deuterium diffusion in a-Si:H studied with elastic recoil detection, *J. Non Cryst. Solids* **137/138**, 29 (1991).

41. Tang, X. M., Weber, J., Baer, A., and Finger, F., Annealing-temperature influence on the dispersive diffusion of hydrogen in undoped a-Si:H, *Phys. Rev. B: Condens. Matter* **42**, 7277 (1990).

42. Mathe, E. L., Naudon, A., Elliq, M., Fogarassy, E., and de Unamuno, S., Influence of hydrogen on the structure and surface morphology of pulsed ArF excimer laser-crystallized amorphous silicon thin films, *Appl. Surf. Sci.* **54**, 392 (1992).

43. Asano, A., Slow structural transitions of hydrogen in hydrogenated amorphous silicon during low-temperature annealing, *Physica B* **170**, 277 (1991).

44. Demichelis, F., Crovini, G., Perri, C. F., Tresso, E., Giamello, E., and Della Mea, G., Hydrogen evolution in amorphous silicon carbide, *Physica B* **170**, 149 (1991).

45. Laidani, N., Capelletti, R., Elena, M., Guzman, L., Mariotto G., Miotello, G. A., and Ossi, P. M., Spectroscopic characterization of thermally treated carbon-rich $Si_{1-x}C_x$ films, *Thin Solid Films* **223**, 114 (1993).

46. Herremans, H., Grevendonk, W., van Swaaij, R. A. C. M. M., van Sark, W. G. J. H., Berntsen, A. J. M., Arnoldbik, W. M., and Bezemer, J., Structural, compositional, and optical properties of hydrogenated amorphous silicon-carbon alloys, *Philos. Mag. B* **66**, 787 (1992).

47. Kudoyarova, V. Kh., Gusinsky, G. M., Rassadin, L. A., and Kudryavtsev, I. V., Hydrogen depth profile measurement in a a-$Si_{1-x}C_x$:H films by elastic recoil detection, *Appl. Surf. Sci.* **50**, 173 (1991).

48. Jean, A., Chaker, M., Diawara, Y., Leung, P. K., Gat, E., Mercier, P. P., Pepin, H., Gujrathi, S., Ross, G. G., and Kieffer, J. C., Characterization of a-SiC:H films produced in a standard plasma-enhanced chemical vapor deposition system for X-ray mask application, *J. Appl. Phys.* **72**, 3110 (1992).

49. Wang, Y., Huang, B., Cao, D., Cao, J., Zhu, D., and Shen, K., Analysis of hydrogen in oxygen-doped polysilicon by $^4He^+$ -H elastic recoil detection, *Nucl. Instrum. Methods Phys. Res. Sect. B* **84**, 111 (1994).

50. Barbour, J. C., Stein, H. J., Popov, O. A., Yoder, M., and Outten, C. A., Silicon nitride formation from a silane-nitrogen electron cyclotron resonance plasma, *J. Vac. Sci. Technol. A* **9**, 480 (1991).

51. Compagnini, G., Calacagno, L., and Foti, G., Properties of fully implanted Si_xC_{1-x}:H alloys, *Nucl. Instrum. Methods Phys. Res. Sect. B* **80/81**, 978 (1993).

52. Graeff, C. F. O., Freire, F. L., Jr., and Chambouleyron, I., Hydrogen and deuterium incorporation and release processes in RF-sputtered hydrogenated or deuterated amorphous germanium films, *Philos. Mag. B* **67**, 691 (1993).

53. Feuser, U., Vianden, R., Alves, E., da Silva, M. F., Szilágyi, E., Pászti, F., and Soares, J. C., Vacancy-acceptor complexes in germanium produced by ion implantation, *Nucl. Instrum. Methods Phys. Res. Sect. B* **59/60**, 1049 (1991).

54. Scott, W. P., Jones, J., Turner, W. A., and Wickboldt, P., Structural properties of amorphous hydrogenated germanium., *J. Non-Cryst. Solids* **141**, 271 (1992).

55. Graeff, C. F. O., Freire, F. L., Jr., and Chambouleyron, I., Hydrogen diffusion in RF-sputtering a-Ge:H thin films, *J. Non-Cryst. Solids* **137/138**, 41 (1991).

56. Shimizu, K., Katayama, M., Funaki, H., Arai, E., Nakata, M., Ohji, Y., and Imura, R., Stoichiometry measurement and electrics characteristics of thin-film Ta_2O_5 insulator for ultralarge-scale integration, *J. Appl. Phys.* **74**, 375 (1993).

57. Gordon, R. G., Riaz, U., and Hoffman, D. M., Chemical vapor deposition of aluminum nitride thin films, *J. Mater. Res.* **7**, 1679 (1992).

58. Wu, H. Z., Chou, T. C., Mishra, A., Anderson, D. R., and Lampert, J. K., Characterization of titanium nitride thin films, *Thin Solid Films* **191**, 55 (1990).
59. Zdaniewski, W. A., Wu, J., Gujrathi, S. C., and Oxorn, K., Preparation and characterization of sputtered TiB$_2$ films, *J. Mater. Res.* **6**, 1066 (1991).
60. Fix, R., Gordon, R. G., and Hoffman, D. M., Chemical vapor deposition of titanium, zirconium, and hafnium nitride thin films, *Chem. Mater.* **3**, 1138 (1991).
61. Gordon, R. G., Hoffman, D. M., and Riaz, U., Low-temperature atmospheric pressure chemical vapor deposition of polycrystalline tin nitride thin films, *Chem. Mater.* **4**, 68 (1992).
62. Yamada, Y., Kasukabe, Y., Nagata, S., and Yamaguchi, S., Spontaneous hydrogenation of Ti films evaporated on NaCl substrates I, *Jpn. J. Appl. Phys.* **29**, L1888 (1990).
63. Krist, Th., Briere, M., and Cser, L., H in Ti thin films, *Thin Solid Films* **228**, 141 (1993).
64. Sugizaki, Y., Furuya, T., and Satoh, H., Influence of nitrogen implantation on the hydrogen absorption by titanium, *Nucl. Instrum. Methods Phys. Res. Sect. B* **59/60**, 722 (1991).
65. Naitoh, M., Shoji, F., and Oura, K., Hydrogen-termination effects on the growth of Ag thin films on Si(111) surfaces, *Surf. Sci.* **242**, 152 (1991).
66. Bhattacharya, R. S., Rai, A. K., and McCormick, A. W., Ion-beam-assisted deposition of Al$_2$O$_3$ thin films, *Surf. Coat. Technol.* **46**, 155 (1991).
67. Roux, B., Chevarier, A., Chevarier, N., El Bouanani, M., Gerlic, E., Stern, M., Bosland, P., and Guemas, F., Ion beam analysis using alpha particles and protons for compositional determination of niobium-superconducting compound films, *Nucl. Instrum. Methods Phys. Res. Sect. B* **64**, 184 (1992).
68. Wang, M., Schmidt, K., Reichelt, K., Dimigen, H., and Hübsch, H., Characterization of metal-containing amorphous hydrogenated carbon films, *J. Mater. Res.* **7**, 667 (1992).
69. Okada, M., Kono, T., Tanaka, K., Sato, M., and Fujimoto, K., Properties and structure of carbon films prepared by ion beam deposition, *Surf. Coat. Technol.* **47**, 233 (1991).
70. Wang, M., Schmidt, K., Reicheilt, K., Jiang, X., Hübsch, H., and Dimigen, H., The properties of W-C:H films deposited by reactive RF sputtering, *J. Mater. Res.* **7**, 1465 (1992).
71. Von Seggern, J., Wienhold, P., Esser, H. G., Winter, J., Gorodetsky, A., Grashin, S., Gudowska, I., and Ross, G. G., Properties of a-C/B:H films relevant to plasma-surface interactions, *J. Nucl. Mater.* **176/177**, 357 (1990).
72. Godet, C., Schmirgeld, L., Zuppiroli, L., Sardin, G., Gujrathi, S. C., and Oxorn, K., Optical properties and chemical reactivity of hydrogenated amorphous boron thin films, *J. Mater. Sci.* **26**, 6408 (1991).
73. Paynter, R. W., Ross, G. G., Boucher, C., Pageau, J. F., and Stansfield, B., Analysis of a-C/B:H layers deposited in TdeV, *J. Nucl. Mater.* **196–198**, 553 (1992).
74. Raveh, A., Martinu, L., Gujrathi, S. C., Klemberg-Sapieha, J. E., and Wertheimer, M. R., Structure-property relationships in dual-frequency plasma deposited on hard a-C:H films, *Surf. Coat. Technol.* **53**, 275 (1992).
75. Compagnini, G., Calcagno, L., and Foti, G., Hydrogen effect on atomic configuration of keV-ion-irradiated carbon, *Phys. Rev. Lett.* **69**, 454 (1992).
76. Matsunami, N., A study of carbon thin films by ion beams, *Nucl. Instrum. Methods Phys. Res. Sect. B* **64**, 800 (1992).
77. Freire, F. L., Jr., Achete, C. A., Franceschini, D. F., Gatts, C., and Mariotto, G., Characterization of hard amorphous carbon films implanted with nitrogen ions, *Nucl. Instrum. Methods Phys. Res. Sect. B* **80/81**, 1464 (1993).
78. Wang, M., Schmidt, K., Reichelt, K., Dimigen, H., and Hübsch, H., The properties of titanium-containing amorphous hydrogenated carbon films, *Surf. Coat. Technol.* **47**, 691 (1991).
79. Boutard, D., Möller, W., and Scherzer, B. M. U., Hydrogen isotope exchange in amorphous hydrocarbon layers: Plasma-deposited a-C:H and ion-saturated carbon, *Rad. Eff. and Defects in Solids* **114**, 281 (1990).
80. Boutard, D., Möller, W., and Scherzer, B. M. U., Isotopic exchange in hard amorphous carbonized layers, *J. Appl. Phys.* **67**, 163 (1990).

81. Perrière, J., Laurent, A., and Enard, J. P., Study of the growth mechanisms of amorphous carbon films by isotopic tracing methods, *Mater. Sci. Eng. B* **11**, 347 (1992).

82. Grill, A., Patel, V., and Meyerson, B. S., Temperature and bias effects on the physical and tribological properties of diamondlike carbon, *J. Electrochem. Soc.* **138**, 2362 (1991).

83. Grill, A., and Patel, V., Characterization of diamondlike carbon by infrared spectroscopy, *Appl. Phys. Lett.* **60**, 2089 (1992).

84. Cuomo, J. J., Doyle, J. P., Bruley, J., and Liu, J. C., Ion-beam-sputtered diamondlike carbon with densities of 2.9 g/cc, *J. Vac. Sci. Technol. A* **9**, 2210 (1991).

85. Long, X., Peng, X., He, F., Liu, M., and Lin, X., The hydrogen concentration of diamondlike amorphous carbon films by elastic recoil detection analysis, *Nucl. Instrum. Methods Phys. Res. Sect. B* **68**, 266 (1992).

86. Tanaka, K., Okada, M., Kohno, T., Yanokura, M., and Aratani, M., Hydrogen content and bonding structure of diamondlike carbon films deposited by ion beam deposition, *Nucl. Instrum. Methods Phys. Res. Sect. B* **58**, 34 (1991).

87. Dekempeneer, E. H. A., Jacobs, R., Smeets, J., Meneve, J., Eersls, L., Blanpain, B., Roos, J., and Oostra, D. J., RF-plasma-assisted chemical vapour deposition of diamondlike carbon: Physical and mechanical properties, *Thin Solid Films* **217**, 56 (1992).

88. Wu, R. L. C., Miyoshi, K., Vuppuladhadium, R., and Jackson, H. E., Physical and tribological properties of rapid thermal-annealed diamondlike carbon films, *Surf. Coat. Technol.* **54/55**, 576 (1992).

89. Nabot, J. P., André, B., and Païdassi, S., Diamondlike carbon films prepared by various ion-beam-assisted techniques, *Surf. Coat. Technol.* **43/44**, 71 (1990).

90. Peebles, D. E., and Pope, L. E., Analytical and mechanical evaluation of diamond films on silicon, *J. Mater. Res.* **5**, 2589 (1990).

91. Kuiper, A. E. T., and Habraken, F. H. P. M., Ion beam analysis of interface reactions, *Nucl. Instrum. Methods Phys. Res. Sect. B* **64**, 739 (1992).

92. Arnold, G. W., Westrich, H. R., and Casey, W. H., Application of ion beam analysis (RBS and ERD) to the surface chemistry study of leached minerals, *Nucl. Instrum. Methods Phys. Res. Sect. B* **64**, 542 (1992).

93. Matzke, Hj., and Turos, A., Mechanisms and kinetics of leaching of UO_2 in water, *Solid State Ionics* **49**, 189 (1991).

94. Matzke, Hj., Della Mea, G., Dran, J. C., Rigato, V., and Bevilacqua, A., Chemical and physical modifications in waste glasses ion implanted at different temperatures, in *Modifications Induced by Irradiation in Glasses* (P. Mazzoldi, ed.) (Elsevier Science, Amsterdam, 1992), pp. 25–31.

95. Freire, F. L., Jr., Hydrogen analysis of the surface layers in sodium β''-alumina crystals by elastic recoil detection, *Jpn. J. Appl. Phys.* **30**, 1428 (1991).

96. Freire, F. L., Jr., Effects of ion implantation on the hydration of $NA\beta''$-alumina crystals, *J. Phys. D Appl. Phys.* **25**, 974 (1992).

97. Keddie, J. L., and Giannelis, E. P., Ion beam analysis of silica sol-gel films: Structural and compositional evolution, *J. Am. Ceram. Soc.* **73**, 3106 (1990).

98. Guglielmi, M., Colombo, P., Rigato, V., Battaglin, G., Boscolo-Boscoletto, A., and De Battisti, A., Compositional and microstructural characterization of RuO_2-TiO_2 catalysts synthesized by the sol-gel method, *J. Electrochem. Soc.* **139**, 1655 (1992).

99. Lappalainen, R., Wang, E. L., and Hirvonen, J. P., Application of nuclear methods to the study of transfer wear of PTFE on stainless steel and silicon, *Surf. Int. Anal.* **16**, 385 (1990).

100. Ronkainen, H., Likonen, J., and Koskinen, J., Tribological properties of hard-carbon films produced by the pulsed vacuum arc discharge method, *Surf. Coat. Technol.* **54/55**, 570 (1992).

101. Neelmeijer, C., Grötzschel, R., Hentschel, E., Klabes, R., Kolitsch, A., and Richter, E., Ion beam analysis of steel surfaces modified by nitrogen ion implantation, *Nucl. Instrum. Methods Phys. Res. Sect. B* **66**, 242 (1992).

102. Whitlow, H. J., Johansson, E., Anders Ingemarsson, P., and Hogmark, S., Recoil spectrometry of oil-additive-associated compositional changes in sliding metal surfaces, *Nucl. Instrum. Methods Phys. Res. Sect. B* **63**, 445 (1992).

103. Rickards, J., Trejo-Luna, R., Ortiz, M. E., Andrade, E., Chavez, E., Zironi, E. P., del Castillo, H., and Sanchez, M., Hydrogen pick-up in ion nitrided steels, *Mater. Sci. Technol.* **9**, 536 (1993).

104. Rigato, V., and Della Mea, G., Thin-film stoichiometry and impurity concentration analysis studied by nuclear techniques, *Nucl. Instrum. Methods Phys. Res. Sect. A* **334**, 203 (1993).

105. Ila, D., Jenkins, G. M., Holland, L. R., Thompson, J., Evelyn, L., and Hodges, A., Measurement of accumulated contaminants in glassy carbon by RBS, ERD, and NRA, *Nucl. Instrum. Methods Phys. Res. Sect. B* **64**, 439 (1992).

106. Freire, F. L., Jr., Lepienski, C. M., and Achete, C. A., Characterization by Rutherford backscattering, elastic recoil, and nuclear reaction analysis of near-surface modifications of glasses submitted to a DC potential, *Nucl. Instrum. Methods Phys. Res. Sect. B* **68**, 227 (1992).

107. Wagner, W., Rauch, F., Ottermann, C., and Bange, K., Analysis of tungsten oxide films using MeV ion beams, *Nucl. Instrum. Methods Phys. Res. Sect. B* **68**, 262 (1992).

108. Jimbou, R., Saidoh, M., Ogiwara, N., Ando, T., Morita, K., and Muto, Y., Retention of deuterium implanted into B_4C-overlaid isotropic graphites and hot-pressed B_4C, *J. Nucl. Mater.* **196–198**, 958 (1992).

109. Kawamura, H., Ishituka, E., Sagara, A., Nakata, H., Saito, M., and Hutamura, Y., Retention of deuterium implanted in hot-pressed beryllium, *J. Nucl. Mater.* **176/177**, 661 (1990).

110. Rottschalk, M., Bachmann, T., and Witzmann, A., Investigation of proton-exchanged optical wave guides in $LiNbO_3$ using elastic recoil detection, *Nucl. Instrum. Methods Phys. Res. Sect. B* **61**, 91 (1991).

111. Fournier, D., Ross, G. G., and Saint-Jacques, R. G., The influence of the implantation temperature on the hydrogen and deuterium retention in TiC, *Nucl. Instrum. Methods Phys. Res. Sect. B* **52**, 154 (1990).

112. Petitpierre, O., Möller, W., and Scherzer, B. M. U., Low-temperature dynamic depth profile measurements of implanted deuterium in nickel: Precipitation of nickel deuteride, *J. Appl. Phys.* **68**, 3169 (1990).

113. Mosbah, M., Clocchiatti, R., Tirira, J., Gosset, J., Massiot, P., and Trocellier, P., Study of hydrogen in melt inclusions trapped in quartz with a nuclear microprobe, *Nucl. Instrum. Methods Phys. Res. Sect. B* **54**, 298 (1991).

114. Ross, G. G., Paynter, R. W., and Michaud, D., Hydrogen and impurity fluxes in the scrape of layer of the Tokamak de Varennes, *J. Nucl. Mater.* **176/177**, 962 (1990).

115. Ishii, M., Nakashima, K., Tajima, I., and Yamamoto, M., Investigation of hydrogen-plasma-etched Si surfaces, *Jpn. J. Appl. Phys.* **31**, 4422 (1992).

116. Nygren, R. E., Doyle, B. L., Walsh, D. S., Ottenberger, W., Brooks, J. N., and Krauss, A. Low-energy helium implantation in evaporated nickel surfaces. *J. Nucl. Mater.* **196–198**, 558 (1992).

117. Morita, K., and Muto, Y., Isotopic effect in thermal reemission of hydrogen from graphite at elevated temperatures, *J. Nucl. Mater.* **196–198**, 963 (1992).

118. Hasebe, Y., Sonobe, M., and Morita, K., Isotopic effect in ion-induced reemission of hydrogen implanted into graphite, *J. Nucl. Sci. Technol.* **29**, 859 (1992).

119. Antoni, V., Bagatin, M., Buffa, A., Della Mea, G., Desideri, D., Freire, F. L., Jr., Mazzoldi, P., and Romanato, F., Edge plasma investigation on the reversed field pinch ETA BETA II, *J. Nucl. Mater.* **176/177**, 1076 (1990).

120. Naitoh, M., Ohnishi, H., Ozaki, Y., Shoji, F., and Oura, K., Hydrogen-induced reordering of the Si(111)-$\sqrt{3} \times \sqrt{3}$-Al surface studied by ERDA/LEED, *Appl. Surf. Sci.* **60/61**, 190 (1992).

121. Ross, G. G., and Richard, I., Influence of the ion-beam-induced desorption on the quantitative depth profiling of hydrogen in a variety of materials, *Nucl. Instrum. Methods Phys. Res. Sect. B* **64**, 603 (1992).

122. Ascheron, C., Proton beam modification of selected $A^{III}B^V$ compounds, *Phys. Status Solidi A* **124**, 11 (1991).

123. Tirira, J., Trocellier, P., Mosbah, M., and Metrich, N., Study of hydrogen content in solids by ERDA and radiation-induced damage, *Nucl. Instrum. Methods Phys. Res. Sect. B* **56/57**, 839 (1991).
124. Boivin, R., Ross, G. G., and Terreault, B., A method of pulsed-laser desorption of hydrogen, *J. Appl. Phys.* **73**, 1936 (1993).

BIBLIOGRAPHY

Arnoldbik, W. M., Marée, C. H. M., and Habraken, F. H. P. M., Deuterium diffusion into plasma-deposited silicon oxynitride films, *Appl. Surf. Sci.* **74**, 103 (1994).

Arnoldbik, W. M., Marée, C. H. M., Maas, A. J. H., van den Boogaard, M. J., Habraken, F. H. P. M., and Kuiper, A. E. T., Dynamic behavior of hydrogen in silicon nitride and oxynitride films made by low-pressure chemical vapor deposition, *Phys. Rev. B: Condens. Matter* **48**, 5444 (1993).

Denisse, C. M. M., Smulders, H. E., Habraken, F. H. P. M., and van der Weg, W. F., Oxidation of plasma-enhanced chemical-vapour-deposited silicon nitride and oxynitride films, *Appl. Surf. Sci.* **39**, 25 (1989).

Habraken, F. H. P. M., Geerlings, E. L. J., Tijhaar, R. H. G., Slomp, A., and van der Weg, W. F., Thermal nitridation of SiO_xH_y films, *J. Appl. Phys.* **62**, 2573 (1987).

Krooshof, G. J. P., Habraken, F. H. P. M., van der Weg, W. F., van den Hove, L., Maex, K., and De Keermaecker, R. F., Study of the rapid thermal nitridation and silicidation of Ti using elastic recoil detection I: Ti on Si, *J. Appl. Phys.* **63**, 5104 (1988).

Krooshof, G. J. P., Habraken, F. H. P. M., van der Weg, W. F., van den Hove, L., Maex, K., and De Keermaecker, R. F., Study of the rapid thermal nitridation and silicidation of Ti using elastic recoil detection II: Ti on SiO_2, *J. Appl. Phys.* **63**, 5110 (1988).

Kuiper, A. E. T., Willemsen, M. F. C., Mulder, J. M. L., Oude Elferink, J. B., Habraken, F. H. P. M., and van der Weg, W. F., Thermal oxidation of silicon nitride and silicon oxynitride films, *J. Vac. Sci. Technol. B* **7**, 455 (1989).

Kuiper, A. E. T., Willemsen, M. F. C., Theunissen, A. M. L., van de Wijgert, W. M., Habraken, F. H. P. M., Tijhaar, R. H. G., and van der Weg, W. F., Hydrogenation during thermal nitridation of silicon dioxide, *J. Appl. Phys.* **59**, 2765 (1986).

Oude Elferink, J. B., Habraken, F. H. P. M., van der Weg, W. F., Dooms, E., Heyns, M., and De Keersmaecker, R., Rapid thermal nitridation of SiO_2 films, *Appl. Surf. Sci.* **39**, 219 (1989).

Vanden Brande, P., Lucas, S., Winand, R., Weymeersch, A., and Renard, L., Determination of the chemical and physical properties of hydrogenated carbon deposits produced by DC magnetron reactive sputtering, *Surf. Coat. Technol.* **68/69**, 656 (1994).

13

Elastic Recoil Spectrometry Using High-Energy Ions for Hydrogen and Light Element Profiling

13.1. INTRODUCTION*

Elastic recoil detection using low-energy ion beams in the MeV range was successfully applied to depth profile measurements in target materials. As discussed in previous chapters, low-energy ERDA is particularly suitable for analyzing elements lighter than projectiles, which means hydrogen isotopes for helium-4 ions. Nevertheless using an adequate detection system, for example, an *ExB* detector, it is possible to perform a profile for elements heavier than the projectile. Otherwise several methods based on elastic recoil are still available to extend both the range of analyzable elements and the total analyzed depth.

The first consists in increasing the energy of the incident helium beam to increase energy separation between contributions from light components of the target medium to the total recoil spectrum. Several authors did much work in this field, and they showed that it is possible to profile simultaneously H, B, C, N, and O in various media[1-4] by using a high ^{4}He energy beam. The second way consists of increasing the mass of the incident ion beam. This technique is called HI-ERDA. The selectivity of the method (mass resolution) is strongly increased, and recoil spectra can be interpreted in terms of multielemental depth distributions.[5-8] Note that the first reported example of elastic recoil analysis is based on the use of 25–40 MeV ^{35}Cl ion beams to determine hydrogen and lithium distributions in LiF and LiOH thin layers.[9] The achieved depth resolution was about 30 nm and sensitivity was as small as 10^{14} at./cm^2. Note also that Stoquert and coworkers have reported extensive use of heavy ions (^{127}I up to 240 MeV) for ERDA investigation of thin films of dielectric ($SiO_xN_yH_z$), amorphous semicon-

*The authors wish to acknowledge Walter Assmann for his critical reviewing of this chapter.

ductors (a-GaAs:H) and superconductors (YBaCuO, BiSrCaCuO). Moreover these researchers were the first to mention the approximately constant depth resolution and sensitivity values for all elements from H to Bi.

The energy of ions used for HI-ERDA measurements must be chosen below the Coulomb barrier to justify the hypothesis of the classic elastic projectile/target nucleus interaction. It is clear that the use of high-energy heavy-ion beams does not imply that recoil spectrometry can be performed only in transmission geometry with thin layers. In fact the experimental geometry arrangement strongly depends on the target to be investigated. On the one hand, much work was done in transmission geometry, for example, the study of thin lithium compounds targets[9] or the analysis of BN con-tamined foils by Nölscher.[10] On the other hand, glancing geometry is currently being used, as in work reported by Dytlewski,[11] for the analysis of such optoelectronic and semiconductor devices as boron-silicium-cobalt-based multilayers. In these experi-ments a beam of 77-MeV $^{127}I^{10+}$ generated recoil atoms detected using a TOF and total energy detector device. The detection device for HI-ERDA must be suitably adapted both to the experimental configuration (nature and energy of the incident ion beam, type of target) and required performances (optimum depth resolution, unambiguous mass separation, large analyzable depth).

In Chapter 13 we first discuss some general considerations about high-energy heavy-ion recoil spectrometry. A second section is devoted to describing the experi-mental HI-ERDA arrangement. Capabilities are discussed in terms of both energy and depth resolution. A survey of literature for application examples allows us to show the typical usefulness of recoil spectrometry using high-energy recoil ions and to conclude about the advantages and limitations of this technique.

13.2. GENERAL CONSIDERATIONS

13.2.1. Kinematics

Although basic kinematic considerations of elastic collision were discussed in Chapter 2, some comments can be added about the projectile mass M_1 used, which is generally much larger than the target nucleus mass M_2 ($M_2 \ll M_1$). In this case recoil energy (Eq. 2.13) can be written in the following simplified form:

$$E_2 \approx 4E_0 \cos^2 \phi \, \frac{M_2}{M_1} \qquad (13.1)$$

As in Figure 2.5, the variation of the recoil energy is plotted versus the recoil angle ϕ and the mass ratio M_2/M_1, when $M_1 > M_2$ in Figure 13.1. Note that the recoil energy of a particular mass M_2 (Eq. 13.1) decreases when the mass of the incident particle increases. Therefore high energies are preferably used when heavy ions are chosen as incident particles to obtain a detectable energy (energy detection limit) for recoil particles. This energy is enhanced when the recoil angle ϕ is close to 0°

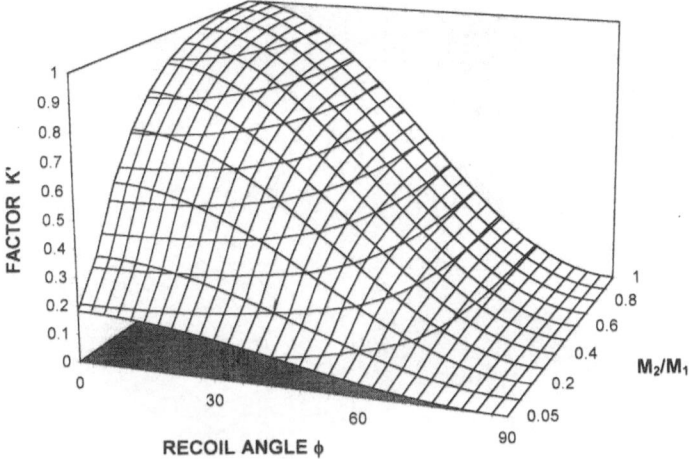

Figure 13.1. Variation of the kinematic factor K' versus recoil angle ϕ and mass ratio M_2/M_1 when $M_1 > M_2$.

(transmission geometry). Indeed recoil energy is maximum for $\phi = 0°$, and a small variation of ϕ around 0° has only a slight influence on E_2. For example Eq. 13.1 is a reasonable approximation for a ^{35}Cl ion beam up to mass 7 (Li), while it is good up to mass 24 (Mg) for a heavier projectile, such as ^{127}I.[12-14]

13.2.2. Rutherford Cross Section for Recoil Scattering with High-Energy Heavy Ions

As mentioned in the introduction, the highest useful beam energy is limited by the Coulomb barrier energy E_{co} for describing elastic scattering between heavy incident ions and target particles as a Rutherford collision. In fact incident energy must be chosen somewhat below E_{co} to avoid not only deviations from pure elastic scattering but also background scattering due to inelastic collisions and capture reactions. It is well-known that for high-energy projectiles, the simple Rutherford model is no longer applicable. A Coulomb potential added to a nuclear potential can be used to describe this type of collision. Thus for each projectile–target couple, it becomes necessary to analyze how effects of nuclear force affect elastic scattering. When we choose to stay on the pure Coulomb domain, the Rutherford cross section is used for HI-ERDA experiments. When the elastic collision deviates only slightly from Rutherford scattering (around 3%, see Section 3.3), we can still use Eq. 3.8 to describe the elastic cross section using the simple Rutherford model. Particular threshold energies of non-Rutherford nuclear cross section for IBA can be calculated by using Eq. 3.10.

If $M_1 >> M_2$ Eq. 3.8 can be simplified as:

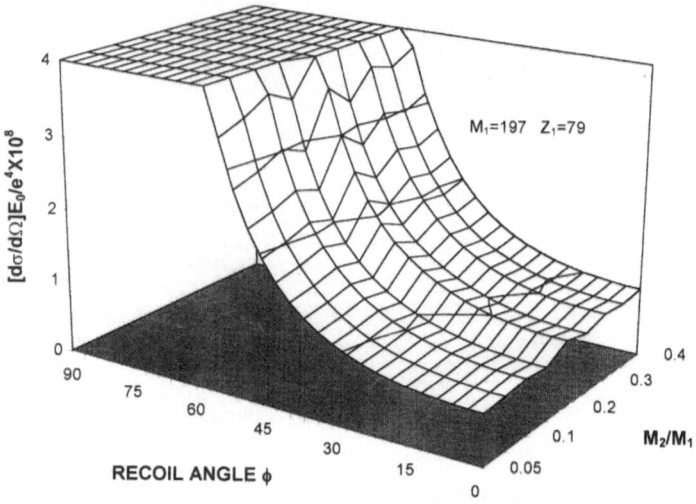

Figure 13.2. Variation of the Rutherford cross section as a function of the recoil angle ϕ and the mass ratio M_2/M_1 when $M_1 > M_2$ (sharp lines do not have any physical reason, they are only due to graphical artefact).

$$\left(\frac{d\sigma}{d\Omega}\right)_{\phi} \approx \left(\frac{Z_1 Z_2\, e^2}{2E_0}\right)^2 \frac{(M_1/M_2)^2}{\cos^3 \phi} \tag{13.2}$$

Figure 13.2 shows an expanded view of a Rutherford cross-section scattering (Eq. 13.2.) when $M_1 \gg M_2$ as a function of the recoil angle ϕ and mass ratio M_1/M_2.

In the case of HI-ERDA experiments, some useful remarks can be added:

- Since the ratio between charge and mass (Z_2/M_2) exhibits only slight oscillations from 0.4–0.5 for all elements except H, the cross section is nearly independent of the recoil mass.[13]
- For H recoils $Z_2/M_2 = 1$, and the cross section is increased by about a factor of 4, relatively to any other recoil atom.
- Due to the quadratic dependence of $(d\sigma/d\Omega)_{\phi}$ on the atomic number of the projectile, the cross-section recoil and therefore the sensitivity of all elements increases with Z_1^2. Moreover heavier beams allow currents as low as 50–100 pA or integrated doses of less than 10^{14} ions/cm^2 (depending on the detector solid angle) to be used.
- Using 136-MeV ^{127}I beam, Siegele and coworkers showed it is possible to well separate light elements such as C, N, and O in stainless steel (Figure 13.3).[13] Si and Ni can only be separated at higher energies. Furthermore selectivity is also improved when Z_1 increases as reported by Assmann[15] for target elements around Si in Al/Cu multilayers or by Goppelt and coworkers[16] for Ni isotopes in Al/Ni multilayers.

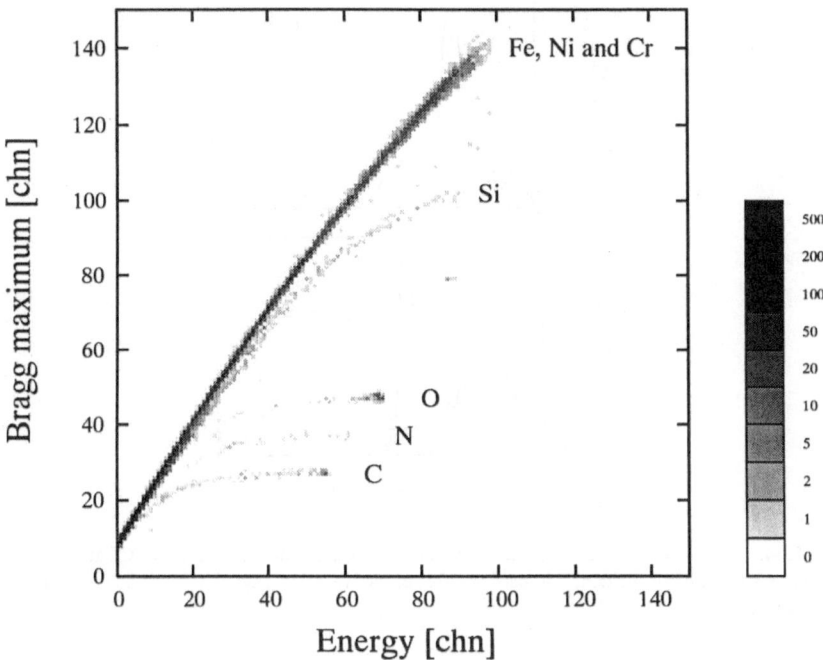

Figure 13.3. Spectrum of stainless steel showing a two-dimensional plot of maximum energy loss and energy of recoils. (Data from Ref. 13)

- Due to the $1/\cos^3 \phi$ dependence, the cross section increases with increasing recoil angle (e.g., fivefold from 30° to 60°), but the recoil energy is simultaneously reduced.

13.2.3. Cross Section for Recoil Scattering with High-Energy ^{4}He Ions

The classical Rutherford model gives a good first approach for elastic cross sections between ^{4}He particles and light elements in the vicinity of 1-MeV incident energy. Since the energy of ^{4}He particles increases, Rutherford scattering is no longer an applicable model. Scattering high-energy ^{4}He particles on light elements has non-Rutherford behavior, and cross-section resonances exist for different ^{4}He–nucleus elastic collisions at high energy.

Elastic cross-section resonances have already been measured for different ^{4}He target pairs as a function of incident energy. A typical example of this non-Rutherford behavior is shown in Figure 13.4[17,18] for cross-section values of carbon-^{4}He and oxygen-^{4}He collisions as a function of ^{4}He energy (from 10 to 18 MeV). A resonance of about 180 mb/sr is shown for carbon at 12 MeV with a flat top about 200 keV wide; this value corresponds approximately to the energy loss induced by a 1800-nm-thick

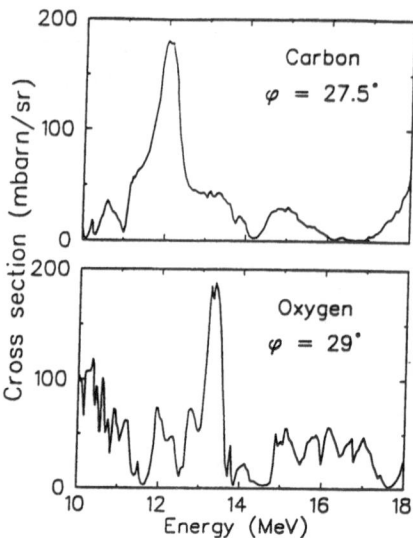

Figure 13.4. Cross-section scattering in the laboratory system for ^{4}He-carbon (^{12}C) and ^{4}He-oxygen collisions as a function of incident ^{4}He particles for respective recoil angles of 27.5° and 29°. (Data from Refs. 17 and 18)

Si-C target. In such a case a reasonable constant value can be used for the cross section around this energy for routine HI-ERDA measurements. Note that to perform ERDA using high-energy ^{4}He, the literature must be surveyed for suitable cross-section resonances for each particular collision couple.

13.3. EXPERIMENTAL ARRANGEMENT FOR HI-ERDA

In light of the discussion in Chapter 5 of the mass–depth ambiguity, using a conventional absorber foil to stop both high-energy helium-4 ions or heavy ions in front of a surface barrier detector cannot completely solve this ambiguity for high-energy ERDA. Straggling of different recoiled atoms (thick absorber foils would be required), nuclear reactions between incident ions and other target atoms and eventually between detected ions and absorber constituents would lead to an overlap between their signals. However surface layers thinner than 100 nm can be analyzed for elements from hydrogen to oxygen in some materials with a suitable combination of primary ions, incident energy, scattering geometry, and absorber thickness.[8] To use the most appropriate detection technique for HI-ERDA, several methods were developed for a particular problem. Section 10.5 reviews different detection techniques and discusses their respective limits and advantages. This section discusses some guidelines based on resolution capabilities using high-energy projectiles. Indeed since energy loss, energy straggling, MS in the sample and in the detectors can be reduced, energy resolution can be minimized to enable us to profile light elements up to a depth of 3 or 4 μm without mass ambiguity and with adequate energy resolution.

In terms of detection devices, it is still possible to use silicon SBDs to undertake RBS or ERDA experiments with high-energy helium[4] and even heavier ions.[19,20] The SBD is currently coupled with another detection system to discriminate different signals, for example, a TOF device[11,21] (Chapter 6). Other ways of discriminating scattered particles from recoiled atoms are pulse shape discrimination[4,18] or coincidence detection (Chapter 9). Note that the lifetime of a solid detector is dramatically reduced due to strong damage from high-energy particles, so the radiation damage accumulation must be measured. More generally in the case of heavy-ion beams, other detection devices may be employed, such as ΔE-E telescopes,[22] magnetic spectrometers,[23] or ionization chambers.[15]

For data processing the detection of only one parameter (recoil energy) is possible in a limited number of cases, especially for transmission ERDA on thin foils or for thin layers, where only light elements (B, C, O) are of interest. The general use of HI-ERDA makes two-parameter measurements necessary—either TOF spectrometry or ΔE-E telescope detection; two-parameter data acquisition systems are now currently available in large accelerator laboratories anyway. Multiparameter data- processing methods were discussed in Section 6.6.

13.4. DETECTION CAPABILITIES

In HI-ERDA identifying the nature of the recoil species is based on detection systems with particle identification properties using either mass or atomic number discrimination. High-energy heavy ions produce recoils with energies well above 1 MeV/amu; thus Z identification by different energy loss can be used. To determine the limits of the method, some detection device properties must be well-controlled, for example energy resolution and depth resolution. No systematic review is available on this subject, but some authors reported the main characteristics of their own detection device. Consequently we present here only some points relative to the main characteristics of HI-ERDA.

13.4.1. Energy Resolution

As shown before (Section 4.2.4.1), energy resolution depends essentially on the energy spread of the beam, kinematic broadening due to beam divergence, energy broadening due to the spread in the recoil angle (geometrical factor), detector resolution, energy-broadening effects correlated with the thermal motion of the analyzed atoms (Doppler effect, especially important in hydrogen analysis, as shown in Section 15.2.2), and energy spread by energy loss straggling and MS, which themselves depend on the chosen geometry.

From an experimental point of view, a large detection angle in the analysis system is required to obtain acceptable counting statistics, a tolerable irradiation damage, and short acquisition time. A large acceptance angle implies strong kinematic recoil energy spread with recoil angle ϕ. For a small acceptance angle, Eq. 4.32 is in good agreement with experimental data. However when a more precise evaluation of energy resolution is required, the kinematic energy shift that must be taken into account can be developed for larger order terms as:

$$\frac{\Delta E_2}{E_2} \approx -2 \tan \phi \, \Delta\phi + (\tan^2 \phi - 1) \, \Delta\phi^2 + \frac{4}{3} \tan \phi \, \Delta\phi^3 + \frac{1}{3} (1 - \tan^2 \phi)\Delta\phi^4 \qquad (13.3)$$

Therefore this kinematic energy broadening is smallest close to the 0° recoil angle (transmission geometry), when the linear term in Eq. 13.3 is vanishing and only higher order terms in $\Delta\phi$ remain. In the case of reflection geometry with a large enough acceptance angle, this energy shift must be corrected for each recoil angle. Depending on the type of detection system, several measurements were carried out to study this energy spread parameter. For example Figure 13.5 shows different correction stages of kinematic energy shifts as studied by Dollinger and coworkers for a particular magnetic spectrometer.[23] Besides, Assmann and coworkers,[15] using a position-sensitive ionization chamber, carried out a particular correction to obtain a kinematic

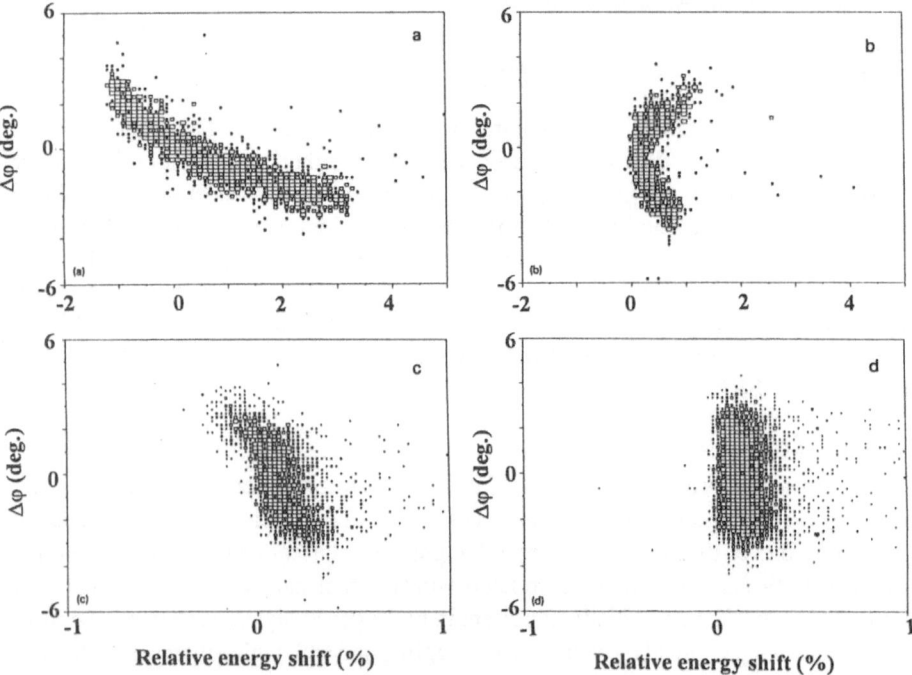

Figure 13.5. Angle variation versus relative energy shift ($\Delta E_2/E_2$) of recoil $^{12}C^{6+}$ ions from a 2.2 μg/cm^2 thick carbon foil target bombarded by 60-MeV ^{58}Ni ions measured with a magnetic spectrograph, (a) without correction, (b–d) with first- to third-order correction using multipole fields. (Data from Ref. 3)

energy shift less than 1%, which requires an angular resolution around 0.2° ($\Delta\Omega \approx 7.5$ msr, $\phi = 37.5°$, and $\Delta\phi \approx 10$ mrad). This example shows that using detectors with both a small solid angle and a small acceptance angle is sufficient for a small kinematic energy shift and therefore a good energy resolution. It is also possible to tilt the detector to compensate for kinematic shifts.[24] Stoquert and coworkers demonstrated that it is possible to cancel kinematic and path length effects at a certain depth (depth-focusing effect) and consequently to enhance the energy resolution of the technique.[5]

However as remarked before, a compromise must be found between different parameters influencing the energy spread and other considerations, for example acceptable counting statistics, tolerable irradiation damage, and short acquisition time.

In Section 4.2.4.1, which discusses total energy resolution in detail, we showed that parameters must be optimized when high-depth resolution is desired. Table 13.1 shows typical values for energy resolution contributions due to different origins. These values were obtained in the case of 120-MeV ^{197}Au ions impinging on a carbon sample with a Q3D magnetic spectrograph at the Munich tandem accelerator.[3]

13.4.2. Depth Resolution

Depth resolution strongly depends on total energy resolution (Eq. 4.33), on both the nature and energy of incident ions and recoil particles (energy loss effect), and on

Table 13.1. Parameters of Energy Resolution for a Typical System Used in HI-ERDA for 120-MeV ^{197}Au Incident Ions Impinging on a Carbon Target with $\theta_1 = 80°$ and $\phi = 15°$[a]

Energy Spread Contributions	$\Delta E/E$
Ion beam energy	$\leq 3 \times 10^{-4}$
Effect of the ion beam divergence	4.5×10^{-4}
Detector resolution	5×10^{-4}
Kinematic shift in each plane	4.5×10^{-4}
Doppler effect at 300 K	
for H	4×10^{-4}
for C	1×10^{-4}
Energy loss straggling per $(nm)^{1/2}$	7×10^{-4}
Multiple scattering per $(nm)^{1/2}$	3×10^{-4}
Path length differences due to the Q3D acceptance	
for H per nm	8×10^{-5}
for C per nm	2×10^{-4}
Total energy resolution	
at surface	9.5×10^{-4}
1 nm below	1.3×10^{-3}
10 nm below	3.3×10^{-3}
Energy broadening due to C recoil at 1 nm	1.4×10^{-3}

[a] Data from Ref. 3.

the depth where the collision occurs. Some data have been published considering several incident ion/target/recoiling atom systems; for example, Dollinger and co-workers reported a depth resolution as low as 8 nm at the surface and 20 nm at a depth of 115 nm for hydrogen measurement in a thin carbon film bombarded with 120-MeV ^{197}Au in transmission geometry. [3] In the case of carbon/boron multilayers deposited on thick silicon substrates, researchers obtained a depth resolution of about 0.7 nm at the surface in reflection geometry (incidence angle = 80°, recoil angle = 15°). Obviously depth resolution decreases with analyzed depth, so they achieved 2.5 nm at a 15-nm depth and 4 nm at a 40-nm depth.

Another example is reported by Rijken, [4] who compared some high-energy projectiles by using a pulse shape discriminator and coincidence spectrometry. Figure 13.6 shows the depth resolution obtained in a silicon target for carbon recoil under 10- and 30-MeV helium and chlorine beams. We see that 30-MeV ^{35}Cl ions give the best depth resolution up to about 600 nm in depth. For 10-MeV ^{35}Cl and 10-MeV ^{4}He projectiles, depth resolution is of the same order of magnitude up to 300 nm. Nevertheless depth resolution for 10-MeV ^{35}Cl ions is slightly better than for ^{4}He ions. However MS and lateral energy spreading effects increase drastically above 300 nm for ^{35}Cl ions. The advantage of a large stopping power value for incident ions then almost vanishes. For recoil angles close to 0° (transmission geometry), the kinetic energy shift is minimized (Eq. 2.12), leading to an improvement in depth resolution (Eq. 4.33). In reflection geometry a typical recoil angle is around 30°, leading to an

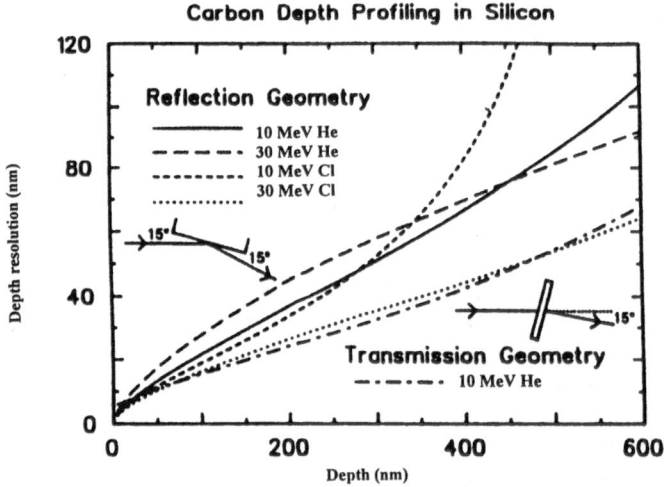

Figure 13.6. Fundamental limit of depth resolution as a function of depth for carbon in silicon. Beam direction, sample orientation, and recoil detection angle are illustrated for both geometries (reflection and transmission). For calculations in transmission geometry, a foil thickness of 1 μm is taken into account. (Data from Ref. 4)

increase in the lateral energy spread, thus depth resolution is strongly damaged. Using relatively high-energy ^{4}He ions (10–15 MeV) is a possible option for analyzing moderately thick samples in transmission geometry (around 70 μm of Si).[4] This effect is clearly illustrated on Figure 13.6. Note that it is necessary to choose as well as possible every parameter involved in energy resolution to obtain suitable depth resolution.

13.4.3. Limitations

Dollinger and coworkers recently reviewed the limits in ERDA with heavy ions; in particular they have examined the role of sputtering effects, elemental loss by effusion process, MS and secondary reactions. It is currently claimed that both radiation damage and the total amount of energy deposited within the target material by heavy incident ions are larger than for light projectiles. However these physical processes cannot be quantified as a function of depth in a simple way. Damage effects depend essentially on the nature of the target, the type of projectile, and incident energy. Thus the pertinence of recorded data largely depends on the extent of induced secondary effects, as reported by Rijken and coworkers.[4] Relatively high cross sections permit us to use small integrated doses with beam currents about 1 nA. Indeed a compromise has to be found for the value of the total dose applied to the sample, taking into account the small solid angle of detection generally acceptable. Moreover detecting recoil species with energies higher than 1 MeV/amu poses the problem of detector degradation versus operating time. It is clear that ionization chambers that do not suffer so highly as solid-state detectors should be preferred.

When high-energy ions are used, depth resolution of 1 nm can be reached. On such a scale it is necessary to understand how the lowest measurable atomic content in the target depends on irradiation damage in the sample matrix. In fact several physical processes can result in modifying the original depth profile in the target sample. To evaluate the lowest detectable atomic concentration of a typical element (Z_2), a simple approach considering only the process of ballistic damaging can be developed. We can compare the number N_D of atoms allowed to be displaced by nuclear interaction without a large variation in the original depth profile with the number N_{ERD} of observed recoil particles for a reasonable statistical number of counts. N_{ERD} can be obtained from Eq. 4.20 as $N_{ERD} \approx \Delta x_D [\delta N(x, NF_2)/\delta x]$, where N is matrix atomic density, Δx_D analyzed depth interval, and F_2 is the atomic fraction of an element Z_2. The total number N_D of atoms that can be displaced due to the scattering interaction is given in first approximation by:

$$N_D = \sigma_D (E_0, Z_m, M_m) Q N \Delta x_D \qquad (13.4)$$

In this equation Z_m, M_m are matrix atomic numbers and matrix atomic weight, respectively (eventually, we can use an average value for Z_m, M_m), and σ_D is the total average cross section for the displacement of an atom, which can be written as[3]:

$$\sigma_D \, (E_0, \, Z_m, \, M_m) = \pi \, \frac{(Z_1 Z_m e^2)^2}{4 E_0 E_D} \, \frac{M_1}{M_m} \left(1 + \ln \left(\frac{\Delta E_{max}}{2 E_D} \right) \right) \tag{13.5}$$

In Eq. 13.5 E_D is the threshold energy for the displacement of an atom from the lattice site, and ΔE_{max} is the maximum energy of displaced atoms able to induce secondary collisions in the damage cascade. Hence using Eqs. 13.4 and 13.5 and the number of recoil atoms detected (Eq. 4.20), we can write the ratio N_D / N_{ERD} [3]:

$$\frac{N_D}{N_{ERD}} \approx \frac{\sigma_D \, (E_0, \, Z_m, \, M_m)}{F_2 \, (d\sigma \, (E_0) / d\Omega) \, \Delta\Omega} \tag{13.6}$$

When $M_1 \gg M_2$ the factor F_2 can be estimated from Eq. 13.6 by:

$$F_2 \approx \frac{4\pi N_{ERD}}{N_D \, E_D} \left(\frac{Z_m}{Z_2} \right)^2 \frac{M_2^2}{M_m \, M_1} \frac{E_0}{\Delta\Omega} \left(1 + \ln \frac{\Delta E_{max}}{2 E_D} \right) \tag{13.7}$$

The lowest detectable concentration can be roughly estimated from Eq. 13.7 for a depth resolution of about 1 nm. Indeed an atomic fraction of 1% can be expected under the following assumptions[3]: $N_{ERD} \approx 10$ (number of recoils identified per atomic layer), $N_D \approx 2 \times 10^{13}$, which is about 10% of the atoms in an atomic layer covered by a beam spot of $\approx 0.2 \, cm^2$, with $Z_m \approx 10$, $(E_0 \, (x) \, / \, M_1) \approx 0.6 \, MeV/amu$, $M_2 \approx 2 \, Z_2$, $\Delta E_{max} \approx 10 \, keV$, and $E_D \approx 25 \, eV$. Note that factor F_2 as previously obtained takes only the ballistic process into account. An accurate evaluation of this factor requires us to consider such other physical processes as thermal- and radiation-induced diffusion.

We see that it is possible to obtain simultaneously an acceptable sensitivity (10^{14} at/cm^2) and a depth resolution of about 1 nm using relatively low-energy incident ions and large acceptance angle detectors. Clearly large acceptance angles introduce an additional energy spread, as previously discussed (Section 13.4.1). However a relatively large acceptance angle can be supported by using for example a rectangular slit with optimized height and width for a minimum increase of $\Delta\phi$ (Section 10.5.1.3).

Heavy ions commonly used for HI-ERDA have an energy on the order of 1 MeV/amu. Hence one of the major limitations, in terms of sample damaging, may be attached to the sputtering phenomenon. On the basis of the sputtering theory developed by Sigmund,[26,27] the erosion yield of a given target bombarded by a given ion beam can be derived from the nuclear stopping power and the surface binding energy. At 1 MeV/amu the nuclear energy loss is around 10^{-3} of the electronic energy loss. For example in Si, 4-MeV ^{4}He has a 16-eV/Å electronic and 1.3×10^{-2}-eV/Å nuclear energy loss; for 32-MeV ^{32}S losses are 4.0×10^2 and 7.0×10^{-1}-eV/Å, respectively, and for 130-MeV ^{127}I losses are 1.2×10^3 and 6.8 eV/Å, respectively. The sputter yield should be about a factor of 520 more with I than with He. But due to the Z^2 cross-section dependence, the recoil cross-section is about a factor 700 larger for I than He. In conclusion as long as only nuclear effects are dominating, HI-ERDA has no real disadvantage compared with conventional ERDA.[25]

Of course the practical situation is much more complex, and the radiation durability of the target must be taken into account to find the real limits of the method. Other secondary effects, mainly electronic energy loss processes, occur and induce chemical bond breaking, volatile element migration, and ion-induced desorption (see Chapter 14).

13.5. APPLICATION EXAMPLES

This section discusses typical examples of high-energy elastic recoil analysis using both ^{4}He and heavy projectiles. Several examples are shown when high-energy ^{4}He particles from 5 to 30 MeV were used.[1–4] Some results are presented of HI-ERDA applications for medium-mass ion beams (A < 100) with energy in the range of 20–50 MeV[7,28–32] and for heavy-mass ion beams (A > 120) at energies higher than 100 MeV.[11–13,15,16]

13.5.1. ERDA Using High-Energy ^{4}He Ions

High-energy helium-4 ions (30 MeV) were used by the Eindhoven group[1] to characterize the surface distribution of a-C-Al sandwiches, as shown in Figure 13.7, using coincidence spectrometry in transmission geometry (for complementary details also see Figure 9.8). The selectivity of the method is suitably optimized to separate carbon isotopes; the achieved depth resolution is about 50 nm. Sensitivity may reach the ppm region with suitable improvements in the detection system.

Comparing ERDA of low Z elements (C, N, O) in semiconductor materials by high-energy ^{4}He particles (10–15 MeV) and HI-ERDA, Rijken and his coworkers showed that the depth/mass ambiguity can be avoided in an Si_2O_3N layer coated on a silicon substrate. Researchers obtained a depth resolution of about 15 nm and an analyzable depth of 250 nm for oxygen. Some other interesting works were performed using high-energy ^{4}He particles (see for example Refs. 24, 33, and 34).

13.5.2. ERDA Using High-Energy Heavy Ions

Habraken and coworkers[8] investigated multilayer structures using recoil spectrometry induced by 30-MeV ^{28}Si ions. Figure 13.8 shows the ERDA spectrum of an 80-nm $SiO_xN_yH_z$ film on a 30-nm SiO_2 layer coated on a thick Si substrate using an 8.9-µm Mylar absorber foil in front of the silicon detector. Energy separation of the elements of interest is fairly good, decreasing with increasing total layer thickness. Surface layers up to 100 nm can be easily characterized for elements from H to O in Si.

Reflection HI-ERDA using a 30-MeV ^{35}Cl^{6+} ion beam was combined with RBS and deuteron-induced nuclear reaction analysis by Neelmeijer and coworkers to evaluate the wear behavior of stainless steel tools after N implantation. The C, N, and

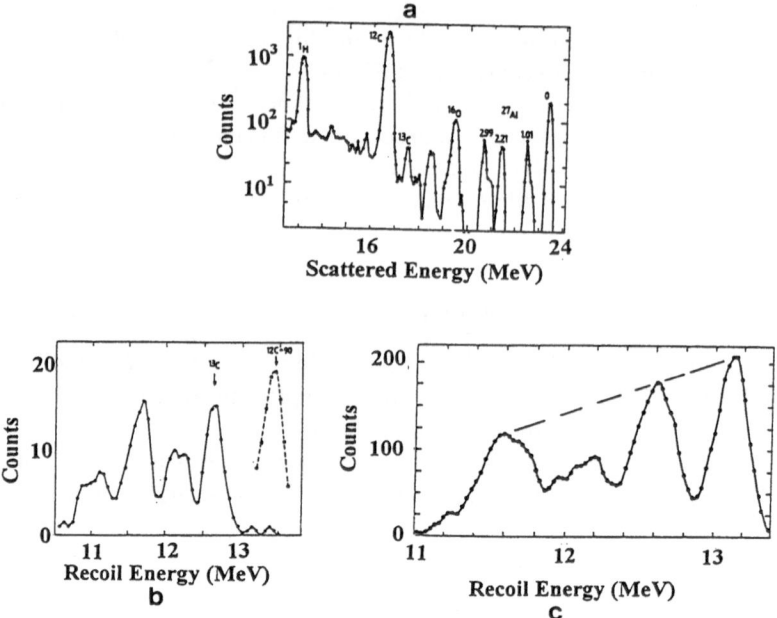

Figure 13.7. The 30-MeV ^{4}He particle scattering/recoil data on an Al/C sandwich target: (a) elastic scattering spectrum (some states of ^{27}Al are excited in inelastic scattering and their energies are given); (b) recoil energy spectrum for ^{13}C; (c) recoil energy spectrum for ^{12}C. (Data from Ref. 1)

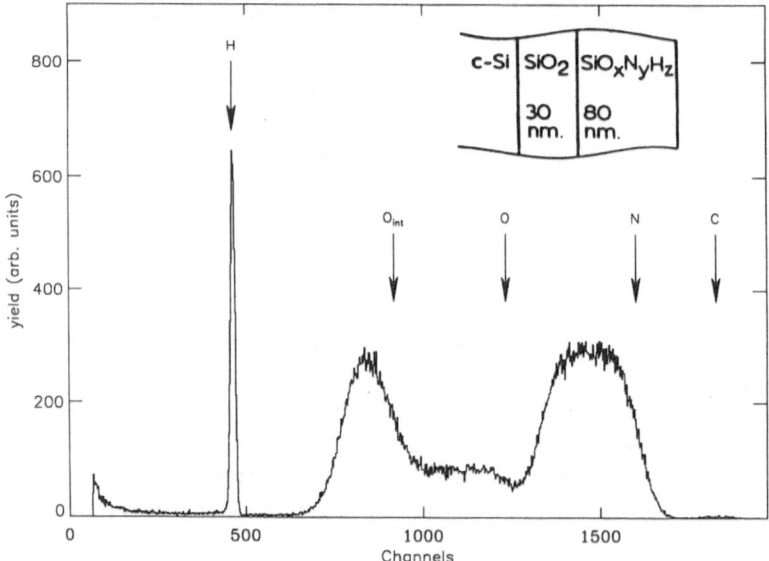

Figure 13.8. The ERDA spectrum of a 80-nm silicon oxynitride film on 30-nm SiO$_2$ on Si. Arrows denote surface positions of various elements. Spectra were taken using a 30-MeV ^{28}Si primary beam, a recoil angle of 34°, the angle of incidence was 27° with respect to the surface. (Data from Ref. 8)

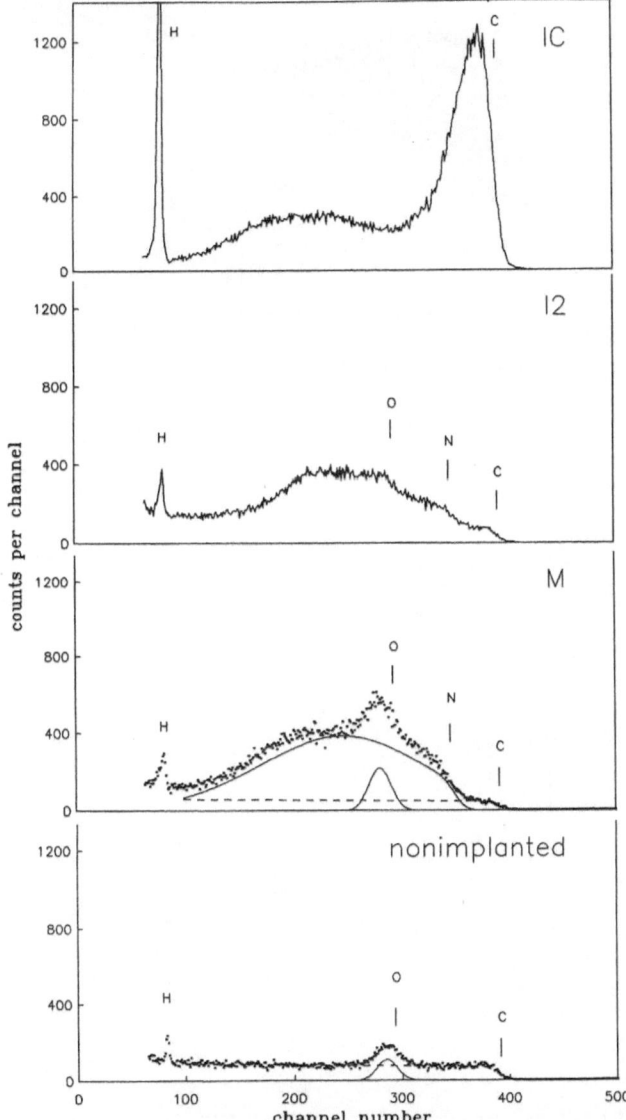

Figure 13.9. Energy spectra of C, N, and O recoiled atoms measured by conventional ERDA using a 30-MeV $^{35}Cl^{6+}$ and a 9.2-μm Mylar foil absorber. Samples: ^{14}N (50 kV, 8×10^{17} at./cm^2) implantation into 210Cr46 tool steel; M/I: implantation with and without mass separator, respectively; IC: increased carbon partial pressure. Dashed and solid lines in the nonimplanted and M spectra represent fitting results for C and N/O, respectively. (Data from Ref. 31)

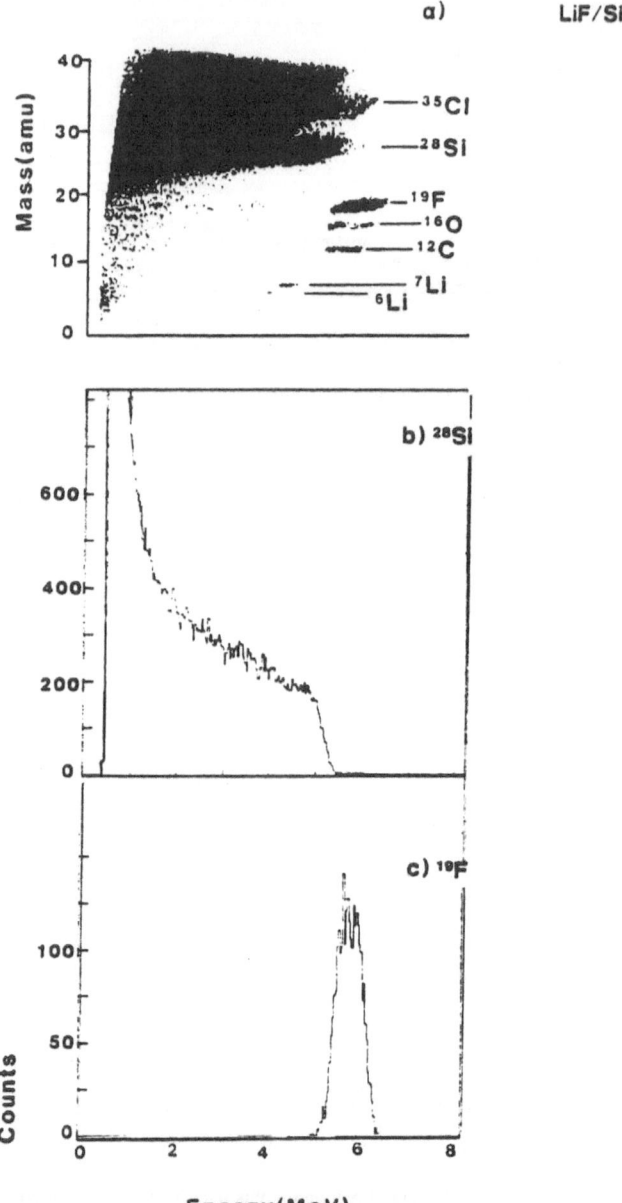

Figure 13.10. The ERDA of a LiF/Si target using a beam of 10-MeV ^{35}Cl: (a) energy mass-chart of detected recoils (the density of points corresponds to the number of registered particles), (b–e) energy spectra of Si, F, O, and ^{6,7}Li recoils. The energy scale given at the bottom of the Figure was calibrated by 5.482-MeV ^{4}He particles. The energy scale for each nuclide is not given in the figure because it differs from nuclide to nuclide due to the pulse height defect (PHD) in the surface barrier detector. (Data from Ref. 19)

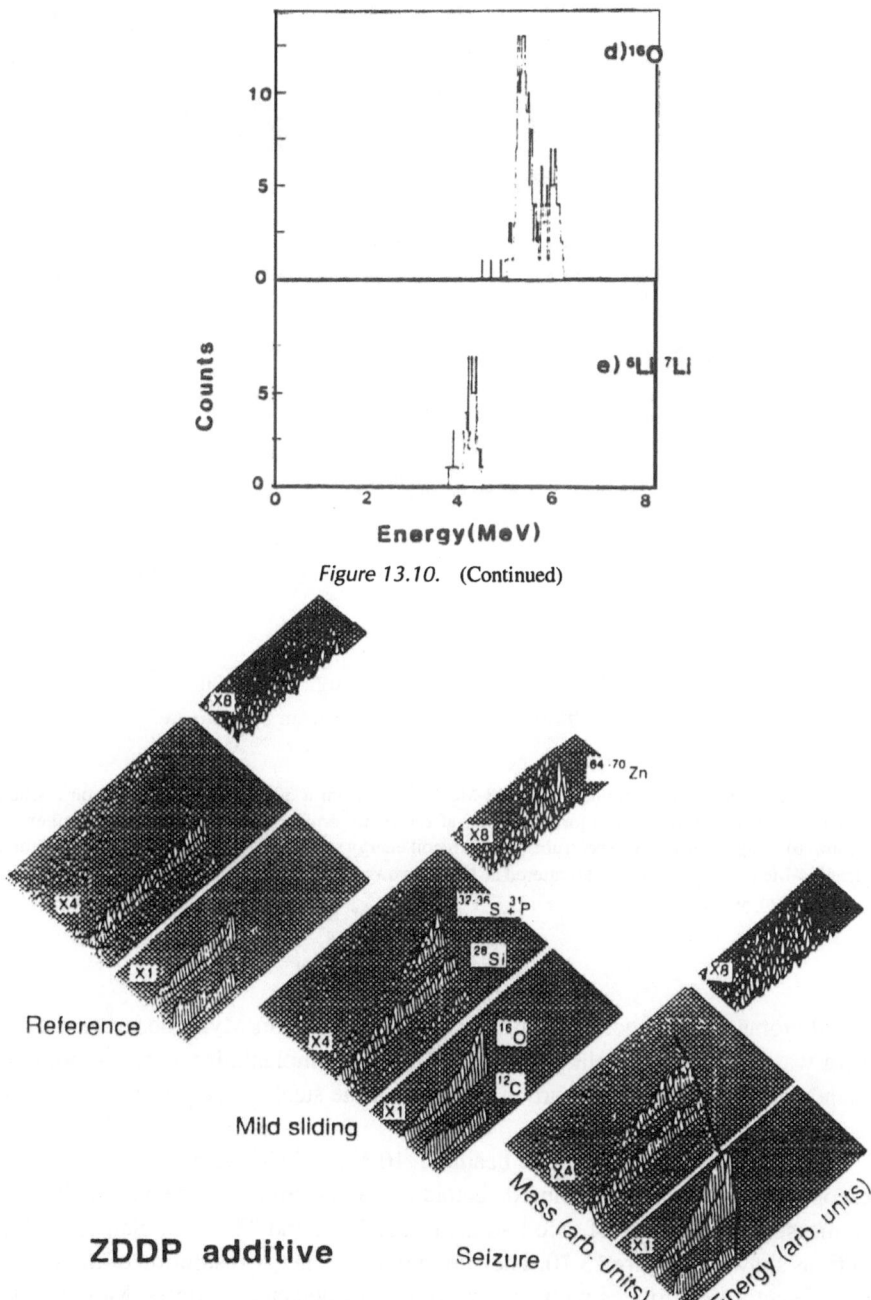

Figure 13.10. (Continued)

Figure 13.11. Recoil data obtained using 48-MeV $^{81}Br^{8+}$, showing two-dimensional mass–energy spectra for the reference sample and samples subjected to mild sliding and seizure in oil containing ZDDP additive. The solid line in the plot corresponding to seizure indicates that the position of the surface and lower energies correspond to greater depths. Spectra were normalized with respect to the Fe signal from a depth interval deep within the bulk. (Data from Ref. 32)

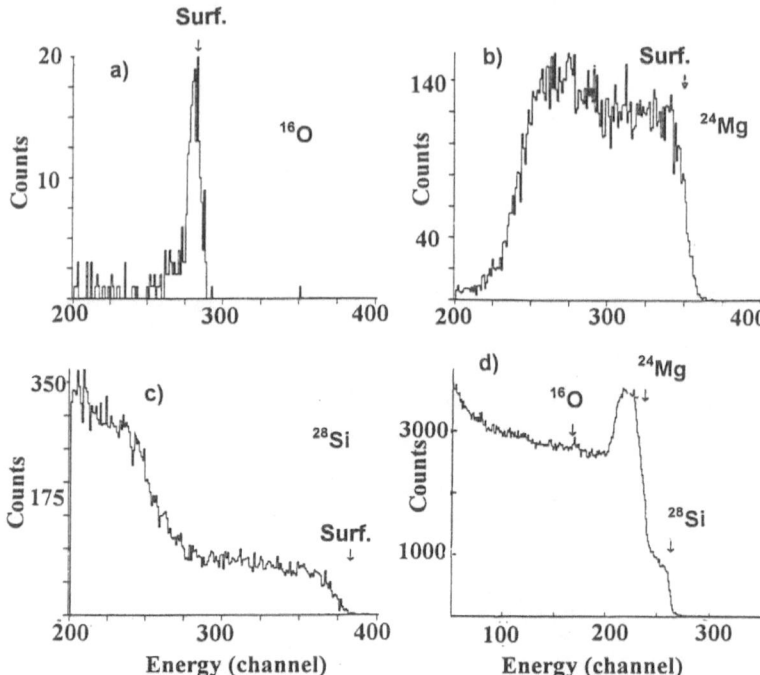

Figure 13.12. Recoil data obtained using 48-MeV $^{79}Br^{8+}$ from a 300-nm Mg overlayer on a silicon substrate after annealing in vacuum for 30 minutes at 250° C to form a Mg_2Si layer: (a) ^{16}O recoil energy spectrum, (b) ^{24}Mg recoil energy spectrum, (c) ^{28}Si recoil energy spectrum, (d) RBS spectrum for normally incident 2.4-MeV ^{4}He-particles backscattered at 168°. Arrows denote surface positions for ^{16}O, ^{24}Mg, and ^{28}Si. (Data from Ref. 29)

O recoil atoms were detected (Figure 13.9) using a 9.2-μm Mylar absorber. Carbon growth was found to take place during the nitrogen implantation process. Nitrogen implantation and nonmetallic carbon covering of the steel surface strongly increases wear resistance.

Arai and coworkers, using a beam of 10-MeV $^{35}Cl^{5+}$ and a TOF-E counter telescope consisting of two timing detectors and a 300-mm^2 surface barrier detector, have measured the energy of recoil atoms from LiF/Si and $^{10}BF_2^-$ -implanted SiO_2/Si targets, as shown in Figure 13.10. They point out that energy resolution of the surface barrier detector deteriorates with an increasing Z of detected particles. Moreover the detection efficiency of the telescope is nearly 1 for atoms with a Z larger than 9 but decreases rapidly with the atomic number for smaller values of Z. Data show a discrepancy between real stoichiometries and expected values due to water adsorption on the target surface.

Mass and energy-dispersive recoil spectrometry induced by 48-MeV $^{81}Br^{8+}$ ions was employed by Whitlow and coworkers[32] to characterize changes partially in surfaces of bearing steel submitted to lubricated sliding wear. Figure 13.11 gives the two-dimensional mass–energy spectra for the reference sample and samples submitted to mild sliding and seizure in an oil-containing zinc dialkyl dithiophosphate (ZDDP) additive. The surface layer formed in the case of mild sliding appears to be strongly enriched in P, S, and Zn and slightly enriched in O. After seizure the surface film has lost C, Si, and Fe and won O, P, and Zn.

The formation of Mg_2Si layers after electron beam deposition of 300-nm Mg on a silicon substrate was followed by HI-ERDA TOF spectrometry, using 48-MeV $^{79}Br^{8+}$.[29] Figure 13.12 shows the mass–energy plots for ^{16}O, ^{24}Mg, and ^{28}Si and the RBS spectrum from a piece of the sample previously defined after annealing in a vacuum at 250° C for 30 minutes. A homogeneous layer of composition $Mg_{2.28\pm0.11}Si$ was formed due to silicon diffusion toward the surface.

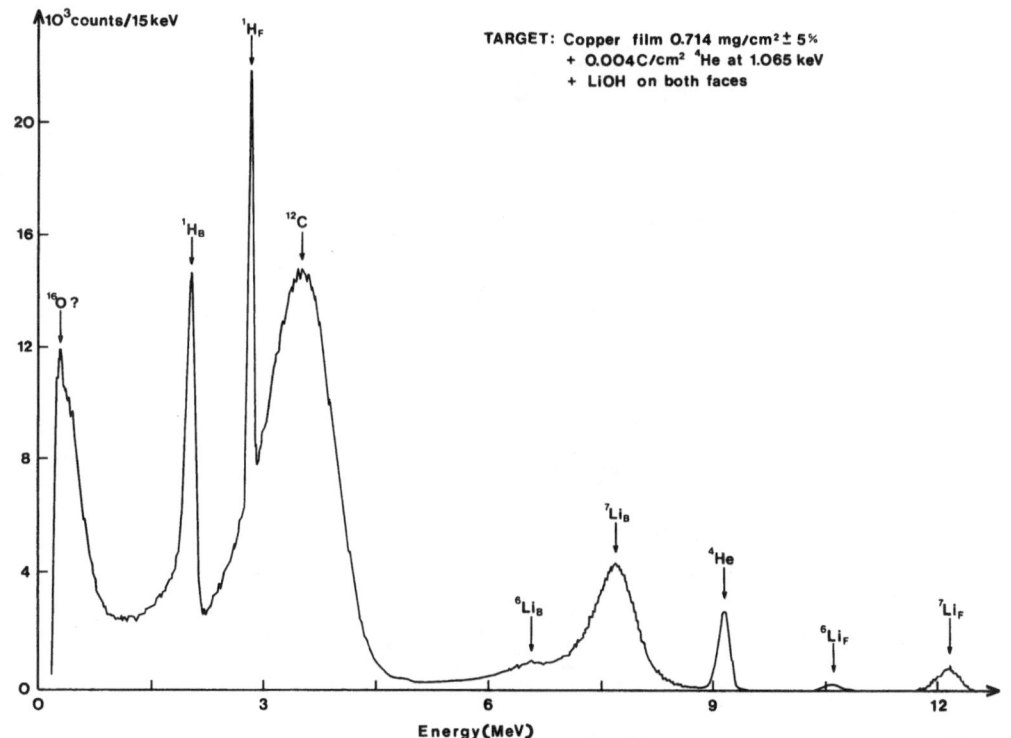

Figure 13.13. The ERDA energy spectrum recorded for a 30-MeV ^{35}Cl beam incident on a target of LiOH on Cu implanted with He. The F and B refer to front and back of the target. (Data from Ref. 6)

A typical example of HI-ERDA in transmission geometry is shown in Figure 13.13. Terreault and coworkers[6] investigated Cu thin films covered on both faces with LiOH and implanted with He. Separating ^{6}Li from ^{7}Li is clearly evident by using 35-MeV ^{35}Cl ions. The mass resolution is 1 amu for a 0.3-µm profile, with sensitivity in the 10^{-9}–10^{-8} g/cm^2 range; depth resolution is improvable up to 10 nm.

Using a 48-MeV ^{81}Br^{8+} beam in conjunction with a telescope (incidence angle = 67.5°), Whitlow studied the evolution of both thickness and stoichiometry in SiO$_2$/Si

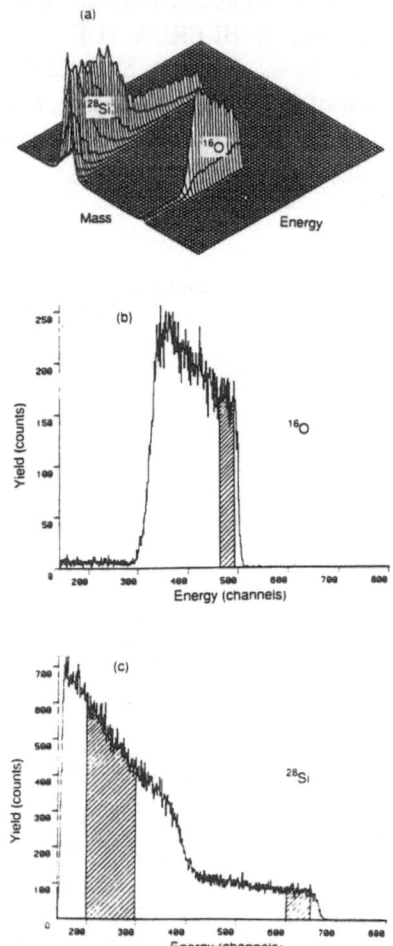

Figure 13.14. Recoil data using 48-MeV ^{81}Br^{8+} from the data set corresponding to the largest ion fluence (total ion fluence 2×10^{14} ions/cm^2): (a) isometric plot of the sorted mass–energy matrix from the SiO$_2$/Si standards, (b) recoil energy spectrum for ^{16}O recoils, (c) recoil energy spectrum for ^{28}Si recoils. Regions used to determine surface heights and silicon substrate height are indicated by shading. (Data from Ref. 30)

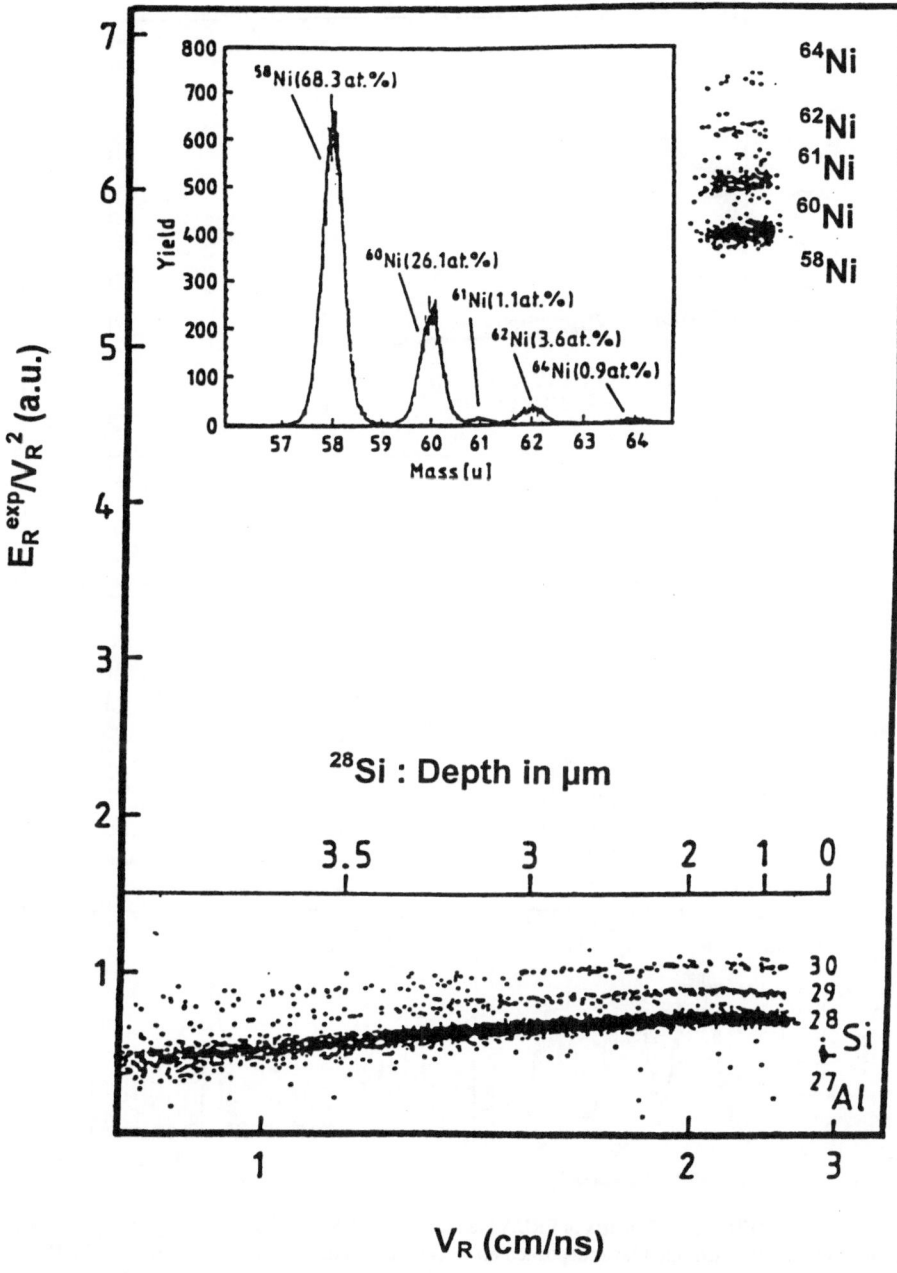

Figure 13.15. Time-of-flight separation of nickel isotopes in an Al/Ni/Si sandwich, obtained using 340-MeV ^{129}Xe ions (incidence = 22.5°, detection = 41°). The E_R^{exp} and v_R, respectively, correspond to measured energies and velocities of the recoil ions. (Data from Ref. 16)

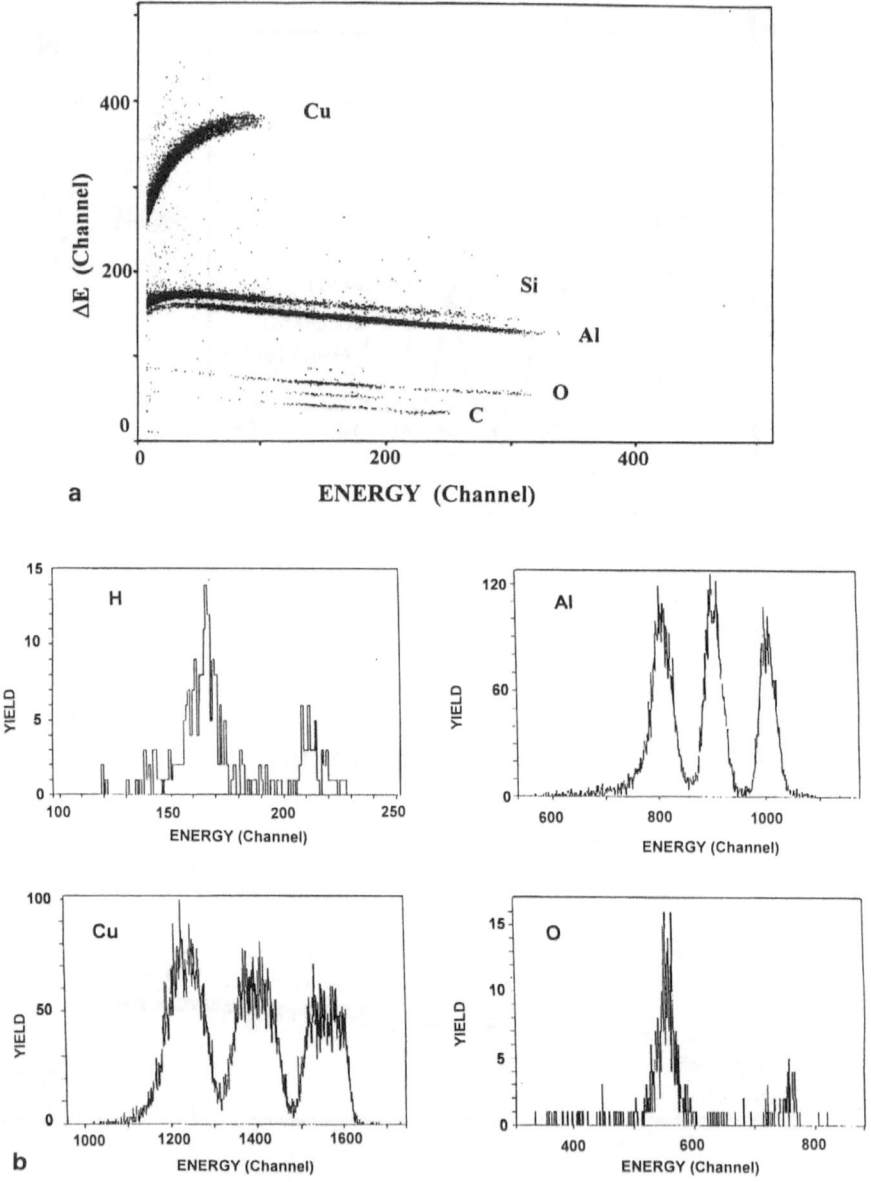

Figure 13.16. (a) The (ΔE, E) matrix of ERDA induced by a 200-MeV [197]Au beam in glancing incidence for an Al/Cu (15 nm each) multilayer deposited on silicon. (b) Recoil energy spectrum for Al, Cu, H, and O. (Data from Ref. 15)

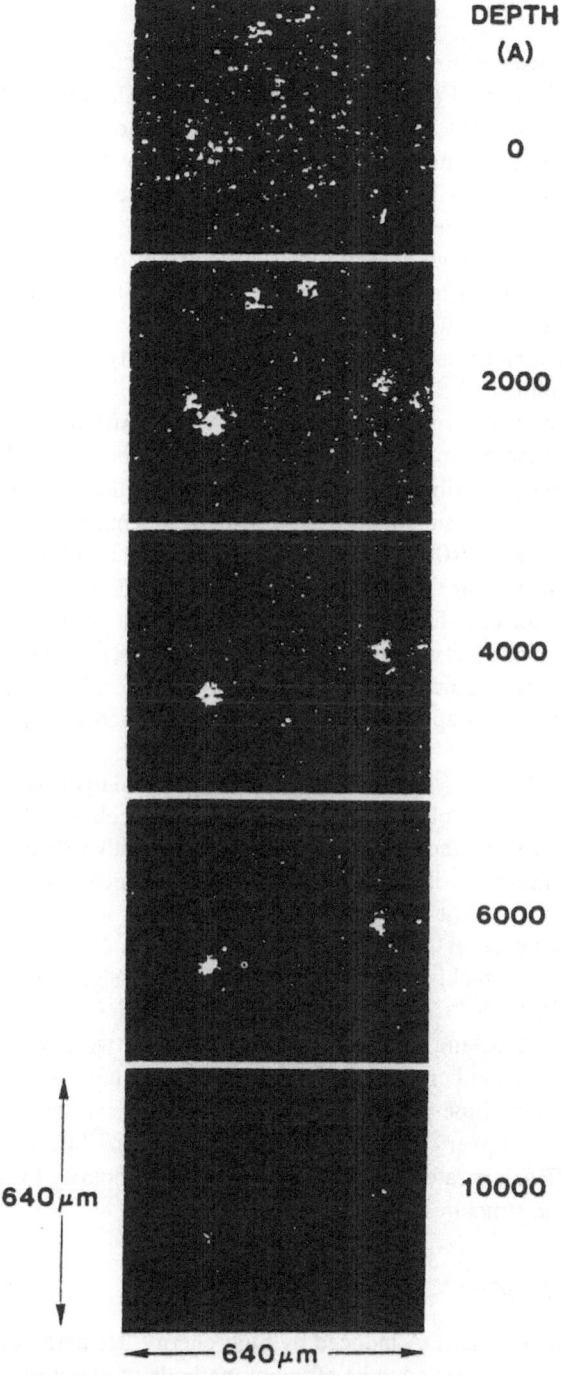

DEPTH
(A)

0

2000

4000

6000

640 μm

10000

640 μm

Figure 13.17. Mapping and depth profile of H in an Si sample exposed to H plasma recorded using 20-MeV ^{28}Si-induced ERDA. Bright regions indicate the presence of H. (Data from Ref. 7)

structures.[30] Fluences up to 2×10^{14} ions/cm^2 lead to a statistically nonsignificant oxygen loss, less than 5% (Figure 13.14). They show that detection efficiencies of the recoil telescope are constant within ± 5 % over the energy range of 9–13 MeV for ^{16}O recoils and ± 3 % over the energy range of 13–18 MeV for ^{28}Si recoils. Moreover this efficiency variation was attributed to dependency of the secondary electron yield on electronic stopping in the carbon foil.

Using an extremely high-energy (340-MeV) ^{129}Xe ion beam, Goppelt and co-workers have studied the distribution of silicon and nickel in 128-nm Al and a 360-nm Ni/Si(100) sandwich as shown in Figure 13.15.[16] They showed that silicium isotopes (28, 29, and 30) and nickel isotopes (58, 60, 61, 62, and 64) are completely separable up to depths extending to 3.5 μm.

Assmann and coworkers proposed to use a 200-MeV ^{197}Au ion beam at 3.5° incidence with an ionization chamber to characterize Al/Cu multilayers deposited on Si.[15] Figure 13.16 shows the energy distribution of Al recoil ions obtained for a 15-nm Al-15-nm Cu sandwich. The Cu profile is simultaneously measured. A depth resolution better than 10 nm and a sensitivity below 10^{14} at./cm^2 can be achieved in this configuration. Hydrogen and oxygen are located near the interfaces Al/Cu (Figure 13.16b). It is clear that depth resolution is quite constant from H to Cu. Some other interesting works devoted to HI-ERDA and also conducted with high-energy alpha particles, medium-mass energy ions, or high-energy and heavy-mass ions are listed in the References.[35–43]

HI-ERDA was rarely applied in conjunction with microbeams (less than 5×5 μm^2) in opposition to low-energy $^4He^+$ (see Chapter 12). The first example is reported by Doyle,[7] where 20-MeV $^{28}Si^{4+}$ ions were used to perform hydrogen depth-profiling data in silicon exposed to H plasma (Figure 13.17). Using a classic reflexion geometry (incidence angle = 75°, surface barrier detector covered with a 10-μm Al foil), Doyle observed two columnar distributions of H ≈ 50 μm in diameter and ≈ 1 μm deep and additional H features present only near the surface.

Two other elastic recoil techniques, channeling ERDA[44] and blocking ERDA,[45] must be mentioned. Janicki and coworkers have successfully applied channeling ERDA using 15–30 MeV ^{35}Cl beams to determine dopant localization (B and BF$_2$) in a silicon crystal.[44] Blocking ERDA is based on using a two-dimensional position-sensitive ionization chamber. Assmann and coworkers have shown recoil atoms starting in single-crystalline materials in the direction of crystal axes or planes are blocked and give rise to very distinct angular distributions.[45] Both methods may be applied in many IBA laboratories in the future to study impurity location and distribution in crystalline structures.

13.6. CONCLUSION

In the 10 past years ERDA induced by high-energy ^4He particles and heavy-ion ERDA (HI-ERDA) have proved to be efficient methods of measuring depth distributions of light elements quantitatively in the near-surface region of solids. In contrast with ERDA induced by low-energy particles, HI-ERDA is not a method adapted to characterizing trace elements due to its medium sensitivity. A depth resolution as low as 1 nm can be achieved, and analyzed depth is generally greater than 1 µm.

A current practice of HI-ERDA can be reasonably considered only in a nuclear physics environment. A solid background in the field of detection devices and components is also required. Moreover software improvements are required to improve the accuracy of calculations, and both cross section and stopping power databases must be completed.

REFERENCES

1. Klein, S. S., Separate determination of concentration profiles for atoms with different masses by simultaneous measurement of scattered projectile and recoil energies, *Nucl. Instrum. Methods Phys. Res. Sect. B* **15**, 464 (1986).
2. Rijken, H. A., Klein, S. S., and de Voigt, M. J. A., Improved depth resolution in CERDA by recoil time of flight measurement, *Nucl. Instrum. Methods Phys. Res. Sect. B* **64**, 395 (1992).
3. Dollinger, G., Faestermann, T., and Maier-Komor, P., High-resolution depth profiling of light elements, *Nucl. Instrum. Methods Phys. Res. Sect. B* **64**, 422 (1992).
4. Rijken, H. A., Klein, S. S., Van IJzendoorn, L. J., and de Voigt, M. J. A., Elastic recoil detection analysis with high-energy alpha beams, *Nucl. Instrum. Methods Phys. Res. Sect. B* **79**, 532 (1993).
5. Stoquert, J. P., Guillaume, G., Hage-Ali, M., Grob, J. J., Ganter, C., and Siffert, P., Determination of concentration depth profiles by elastic recoil detection with a ΔE-E gas telescope and high-energy incident heavy ions, *Nucl. Instrum. Methods Phys. Res. Sect. B* **44**, 184 (1989).
6. Terreault, B., Martel, J. G., Saint Jacques, R. G., and L'Ecuyer, J., Depth profiling of light elements in materials with high-energy ion beams, *J. Vac. Sci. Technol.* **14**, 492 (1977).
7. Doyle, B. L., and Wing, N. D., The Sandia nuclear microprobe, Sandia report SAND82-2393 (1982).
8. Habraken, F. H. P. M., Light-element depth profiling using elastic recoil detection, *Nucl. Instrum. Methods Phys. Res. Sect. B* **68**, 181 (1992).
9. L'Ecuyer, J., Brassard, C., Cardinal, C., Chabbal, J., Deschênes, L., Labrie, J. P., Terreault, B., Martel, J. G., and Saint Jacques, R., An accurate and sensitive method for the determination of the depth distribution of light elements in heavy materials, *J. Appl. Phys.* **47**, 381 (1976).
10. Nölscher, C., Brenner, K., Knauf, R., and Schmidt, W., Elastic recoil detection analysis of light particles (^{1}H - ^{16}O) using 30-MeV sulfur ions, *Nucl. Instrum. Methods Phys. Res.* **218**, 116 (1983).
11. Dytlewski, N., Cohen, D. D., Whitlow, H. J., Ostling, M., Zaring, C., Johnston, P., and Walker, S., in *Proc. of the First French-Australian Workshop on the Applications of IBA*, Saclay (France), 17–21 May 1993 (J. P. Frontier, J. Trochon, and P. Trocellier, eds.), (Saclay), pp. 173–86.
12. Siegele, R., Haugen, H. K., Davies, J. A., Forster, J. S., and Andrews, H. R., Forward elastic recoil measurements using heavy ions, *J. Appl. Phys.* **76**, 4524 (1994).
13. Siegele, R., Davies, J. A., Forster, J. S., and Andrews, H. R., Forward elastic recoil measurements using heavy ions, *Nucl. Instrum. Methods Phys. Res. Sect. B* **90**, 606 (1994).
14. Davies, J. A., in *Proc. of the IBMM'95 Workshop on the Applications of IBA*, Lucas Heights, Sydney 1–3 Feb. 1995, to be published.
15. Assmann, W., Huber, H., Steinhausen, Ch., Dobler, M., Glückler, H., and Weidinger, A., Elastic recoil detection analysis with heavy ions, *Nucl. Instr. Meth. Phys. Res. Sect. B* **89**, 131 (1994).
16. Goppelt, P., Gebauer, B., Fink, D., Wilpert, M., Wilpert, Th., and Bohne, W., High-energy ERDA with very heavy ions using mass and energy-dispersive spectrometry, *Nucl. Instrum. Methods Phys. Res. Sect. B* **68**, 235 (1992).
17. Carter, E. B, Mitchell, G. E, and Davis, R. H., Elastic scattering of alpha particles by ^{12}C in the bombarding energy range 10–19 MeV, *Phys. Rev.* **133**, B1421 (1964); Mehta, M. K., Hunt, W. E., and

Davis, R. H., Scattering of alpha particles by oxygen, II: Bombarding energy range 10–19 MeV, *Phys. Rev.* **160**, 791 (1967).

18. Rijken, H. A., Detection methods for depth profiling of light elements using high-energy alpha particles, thesis, University of Eindhoven (1993).

19. Arai, E., Funaki, H., Katayama, M., Oguri, Y., and Shimizu, K., TOF-ERD experiments using a 10-MeV ^{35}Cl beam, *Nucl. Instrum. Methods Phys. Res. Sect. B* **64**, 296 (1992).

20. Comedi, D., and Davies, J. A., Pulse height response of Si surface barrier detectors to 5–70 MeV heavy ions, *Nucl. Instrum. Methods Phys. Res. Sect. B* **67**, 93 (1992).

21. Knapp, J. A., Barbour, J. C., and Doyle, B. L., Ion beam analysis for depth profiling, *J. Vac. Sci. Technol. A* **10**, 2685 (1992).

22. Arnoldbik, W. M., de Laat, C. T. A. M., and Habraken, F. H. P. M., On the use of a dE-E telescope in elastic recoil detection, *Nucl. Instrum. Methods Phys. Res. Sect. B* **64**, 832 (1992).

23. Dollinger, G., Elastic recoil detection analysis with atomic depth resolution, *Nucl. Instr. Meth. Phys. Res. Sect. B* **79**, 513 (1993).

24. Klein, S. S., Rijken, H. A., van Dijk, P. W. L., and de Voigt, M. J. A., High-resolution elastic recoil depth profiling using alpha-particle beams: CERDA-TOF, *Nucl. Instrum. Methods Phys. Res. Sect. B* **85**, 655 (1994); Klein, S. S., Rijken, H. A., Tolsma, H. P. T., and de Voigt, M. J. A., Elastic recoil selection by pulse shape analysis, *Nucl. Instrum. Methods Phys. Res. Sect. B* **85**, 660 (1994).

25. Dollinger, G., Bergmaier, A., Faestermann, T., and Frey, C. M., in *Proc. of the Twelfth International Conference on Ion Beam Analysis*, Tempe (Arizona), 22–26 May, to be published in a special issue of *Nucl. Instrum. Methods Phys. Res.*; High-resolution depth profile analysis by elastic recoil detection with heavy ions, to be published in *Fresenius J. Anal. Chem.*

26. Sigmund, P., Theory of sputtering I. Sputtering yield of amorphous and polycrystalline targets, *Phys. Rev.* **184**, 383 (1969).

27. Sigmund, P., Sputtering by ion bombardment: theoretical concepts, in *Sputtering by Particle Bombardment I*, Topics in Applied Physics, vol. 47 (R. Behrisch, ed.) (Springer, Berlin, 1981). pp. 9–71.

28. Thomas, J. P., Fallavier, M., Ramdane, D., Chevarier, N., and Chevarier, A., High-resolution depth profiling of light elements in high-atomic-mass materials, *Nucl. Instrum. Methods Phys. Res.* **218**, 125 (1983).

29. Whitlow, H. J., Possnert, G., and Petersson, C. S., Quantitative mass and energy-dispersive elastic recoil spectrometry: Resolution and efficiency considerations, *Nucl. Instrum. Methods Phys. Res. Sect. B* **27**, 448 (1987).

30. Whitlow, H. J., Andersson, A. B. C., and Petersson, C. S., Thermally grown SiO_2 film standards for elastic recoil detection analysis, *Nucl. Instrum. Methods Phys. Res. Sect. B* **36**, 53 (1989).

31. Neelmeijer, C., Grötzschel, R., Hentschel, E., Klabes, R., Kolitsch, A., and Richter, E., Ion beam analysis of steel surfaces modified by nitrogen ion implantation, *Nucl. Instrum. Methods Phys. Res. Sect. B* **66**, 242 (1992); Neelmeijer, C., Grötzschel, R., Klabes, R., Kreissig, U., and Sobe, G., Study of carbon and oxygen incorporation in reactively sputtered Cr-Si-Al films using ERDA, *Nucl. Instrum. Methods Phys. Res. Sect. B* **64**, 461 (1992).

32. Whitlow, H. J., Johansson, E., Ingemarsson, P. A., and Hogmark, S., Recoil spectrometry of oil-additive-associated compositional changes in sliding metal surfaces, *Nucl. Instrum. Methods Phys. Res. Sect. B* **63**, 445 (1992).

33. Van IJzendoorn, L. J., Rijken, H. A., Klein, S. S., and de Voigt, M. J. A., Elastic recoil detection of light elements (C, N, O) with high-energy (10–15 MeV) He beams, *Appl. Surf. Sci.* **70/71**, 58 (1993).

34. Maas, A. J. H., Klein, S. S., Simons, D. P. L., and de Voigt, M. J. A., Recoil selection by pulse shape discrimination in a multi-parameter data acquisition system, in *Proc. of the Twelfth International Conference on Ion Beam Analysis*, Tempe (Arizona), 22–26 May 1995, to be published in a special issue of *Nucl. Instrum. Methods Phys. Res. Sect. B*.

35. Nagai, H., Hayashi, S., Aratani, M., Nozaki, T., Yanokura, M., Kohno, I., Kuboi, O., and Yatsurugi, Y., Reliability, detection limit, and depth resolution of the elastic recoil measurement of hydrogen, *Nucl. Instrum. Methods Phys. Res. Sect. B* **28**, 59 (1987).

36. Martin, J. W., Cohen, D. D., Dytlewski, N., Garton, D. B., Whitlow, H. J., and Russell, G. J., Materials characterisation using heavy-ion elastic recoil time of flight spectrometry, *Nucl. Instrum. Methods Phys. Res. Sect. B* **94**, 277 (1994).

37. Arai, E., Zounek, A., Sekino, M., Takemoto, K., and Nittono, O., Depth profiling of porous silicon surface by means of heavy-ion TOF-ERDA, *Nucl. Instrum. Methods Phys. Res. Sect. B* **85**, 226 (1994).

38. Oura, K., Naitoh, M., Morioka, H., Watamori, M., and Shoji, F., Elastic recoil detection analysis of coadsorption of hydrogen and deuterium on clean Si surfaces, *Nucl. Instrum. Methods Phys. Res. Sect. B* **85**, 344 (1994).

39. Goppelt, P., Biersack, J. P., Gebauer, B., Fink, D., Bohne, W., Wilpert, M., and Wilpert, Th., Investigation of thin films by high-energy ERDA, *Nucl. Instrum. Methods Phys. Res. Sect. B* **80/81**, 142 (1993).

40. Green, P. F., and Doyle, B. L., Silicon elastic recoil detection studies of polymer diffusion: Advantages and disadvantages. *Nucl. Instr. Meth. Phys. Res. Sect. B* **18**, 64 (1986).

41. Gujrathi, S. C., in *Metallization of polymers* (American Chemical Society, New York, 1990), pp. 88–109.

42. Hult, M., Whitlow, H.J., Östling, M., Lundberg, N., Zaring, C., Cohen, D. D., Dytlewski, N., Johnston, P. N., and Walker, S. R., RBS and recoil spectrometry analysis of $CoSi_2$ formation on GaAs, *Nucl. Instrum. Methods Phys. Res. Sect. B* **85**, 916 (1994).

43. Maas, A. J. H., Klein, S. S., Rademakers, F. P., Minnaert, A. W. E., and de Voigt, M. J. A., Focusing in high-resolution time of flight elastic recoil detection analysis, in *Proc. of the Twelfth International Conference on Ion Beam Analysis*, Tempe (Arizona), 22–26 May 1995, to be published in a special issue of *Nucl. Instrum. Methods Phys. Res. Sect. B*.

44. Janicki, C., Hinrichsen, P. F., Gujrathi, S. C., Brebner, J., and Martin, J. P., An ERD/RBS/PIXE apparatus for surface analysis and channeling, *Nucl. Instrum. Methods Phys. Res. Sect. B* **34**, 483 (1988).

45. Assmann, W., Ionization chambers for materials analysis with heavy-ion beams, *Nucl. Instrum. Methods Phys. Res. Sect. B* **64**, 267 (1992).

BIBLIOGRAPHY

The following papers included in the *Proc. of the Twelfth International Conference on Ion Beam Analysis*, Tempe (Arizona), 22–26 May 1995 will be published in a special issue of *Nucl. Instrum. Methods Phys. Res. Sect. B*.

Aoki, Y., Goppelt–Langer, P., Yamamoto, S., Takeshita, H., Naramoto, H., Light- and heavy-element profiling using heavy-ion beams.

Assmann, W., Davies, J. A., Forster, J. S., Huber, H., and Siegele, R., ERD with extremely heavy-ion beams.

Assmann, W., Huber, H., Reichelt, Th., Schellinger, H., and Wohlgemuth, R., ERDA of TiN_xO_y-Cu/Al solar absorbers with heavy ions above 150 MeV.

Avasthi, D. K., Kabiraj, D., Subramaniyam, E. T., Mehta, G. K., and Jain, A., Depth profile of boron-implanted stainless steel by ERDA using a detector telescope.

Bair, A. E., Atzmon, Z., Russell, S. W., Alford, T. L., and Mayer, J. W., Comparison of elastic resonance and elastic recoil detection in the quantification of carbon in SiGeC.

Barbour, J. C., Walsh, D. S., and Doyle, B. L., Low-velocity high-Z projectiles for elastic recoil detection analysis.

Baumann, S. M., Hitzman, C. J., Kirchhoff, J. F., and Strossman, G. S., ERD/RBS, SIMS, and FTIR analysis of silicon nitride and diamondlike carbon films.

Behrisch, R., Prozesky, V. M., Huber, H., and Assmann, W., Hydrogen desorption induced by heavy-ions during surface layer analysis with HIERDA.

Dytlewski, N., and Evans, P. J., Heavy-ion time-of-flight ERDA of high-dose metal-implanted germanium.

Endisch, D., Giessler, K. H., Laube, M., Rauch, F., and Stamm, M., ERD analysis of thin polymer films with improved depth resolution using 8.5-MeV ^{12}C ions and glancing angles.

Konishi, Y., Konishi, I., Sakauchi, N., Hirakimoto, A., Hayashi, S., Asari, M., and Suzuki, J., Measurement of hydrogen content in DLC thin films by ERDA.

Kruse, O., Plieninger, R., and Carstanjen, H. D., High-depth resolution ERDA of light elements with an electrostatic spectrometer for MeV ions.

Marée, C. H. M., Kleinpenning, A., Vredenberg, A. M., and Habraken, F. H. P. M., Ion beam analysis of electropolymerized porphyrin layers.

Parikh, N. R., Patnaik, B. K., Neuman, Ch., Swanson, M. L., and Ishibashi, K., Improvement of depth resolution in ERDA using deflecting electric and magnetic fields.

Prozesky, V. M., Huber, H., Assmann, W., and Behrisch, R., Evaluation of spectra of energetic particles not fully stopped in ΔE-E telescopes.

Schiettekatte, F., Chevarier, A., Chevarier, N., Plantier, A., and Ross, G. G., Quantitative depth profiling of light elements by means of the ERD ExB technique.

Sellschop, J. F., and Smallman, C. G., 3D-ERDA microscopy of implanted H-distributions in diamond.

Siegele, R., Assmann, W., Davies, J. A., and Forster, J. S., Simultaneous multielement analysis with a simple ERD detector and extremely heavy-ion beams.

Turgeon, S., and Paynter, R. W., Empirical modeling of ion-beam-induced hydrogen depletion.

Walsh, D. S., and Doyle, B. L., Hydrogen profiling using enhanced cross-section elastic recoil detection with the ^{1}H(^{12}C, p)^{12}C reaction.

Wielunski, L. S., Multiple-scattering contribution to low-energy background in hydrogen elastic recoil detection.

Yagi, H., Tanida, K., Hatta, A., Ito, T., and Hiraki, A., Hydrogen profile in homoepitaxially grown CVD diamonds by ERDA method.

14

Ion-Beam Damaging Effects

14.1. INTRODUCTION

Characterization of the near-surface region of any solid by IBA in the MeV range cannot really be considered a nondestructive investigation because the basic physical processes involved are themselves damaging. These processes of atomic collision and energy loss of ions in matter unavoidably include excitation of electronic levels, scattering of incident ions, recoil of target atoms, and eventually the transmutation of target atoms by nuclear reaction. Energy deposited by incident particles also appears as heat released in the target. In insulators projectile implantation and expulsion of particles can result in electrostatic charge accumulation.

In this chapter we use the expressions radiation damage, electronic damage, ballistic damage, and thermal damage when appropriate in relation to the elementary processes under consideration, but in fact these expressions represent only different aspects of phenomena covered by the general concept of *radiation damage*. However three important parameters must be distinguished: impulsion, temperature and electrical charge.

It is of primary importance to discuss beam damaging in connection with elemental analysis because damage can simultaneously affect the target material (for example, visible tracks left in the material as a consequence of ion beam investigation), the analytic process itself (for example, irradiation-induced surface roughness), the detection process (for example, hot electrons and photons emitted by the target), data processing (for example, unaccounted shift in calibration due to charge accumulation near the target surface), and even the relevance of the whole analysis (for example, loss of volatile species, redistribution of mobile elements or ion mixing).

The macroscopic and microscopic consequences of ion beam irradiation of a solid can be easily observed. For example, in the case of polymer irradiation, these effects are chain scissions, cross-links, molecular emission, double-bond formation and amorphisation of the structure.[1-4] In insulators radiation effects can be summarized by the creation and diffusion of local defects, amorphisation, ion-mixing processes, and such physical modifications as swelling.[5-10] In alkali halides color center formation constitutes the most visible effect.[11] For metal thin films, sputtering effects, irradiation-induced swelling or shrinking, and modification of the coating/substrate interface have been reported.[12–14]

With regard to damage by ion beam bombardment of common materials, two main classes of interactions are involved: interactions between incident ions and electrons in the target medium (electronic processes) and collisions of the incident ions with target atoms nuclei (ballistic processes). As a result two sorts of defects are produced[15]:

- Point defects that may affect either the crystalline structure (e.g., vacancies and interstitials), the electronic structure (e.g., color centers), or the chemical structure (e.g., bond breaking)
- Extended defects that generally result from an accumulation of point defects (e.g., vacancy clusters, bubbles, dislocation loops,etc.).

A third class of interactions, transmutation of target atoms by nuclear reactions, has much less important consequences as well as the advantage to be completely predictable.

Electronic effects constitute the relevant process for MeV light ions (protons, deuterons, ^{3}He$^+$, ^{4}He$^+$) along their path within the solid except near the end of their range, where ballistic effects begin to predominate. For high-energy heavy ions, ballistic effects are considerable all along the trajectory of the incident ions within the target,[15] though electronic excitations may be as considerable as for light ions.

The effects of ion beam irradiation on the physical, chemical, and mechanical properties of solids have proved to be of a great interest in scientific and technological fields ranging from astrophysics and astronautics to metallurgy and semiconductor devices applications. This increasing interest results from the fact that our universe is permanently crossed by radiation (electromagnetic waves and particles) and practical applications have been developed for radiation effects in science, medicine, and technology. Some of these effects were studied by scientists before the theoretical basis of charged particle interactions with condensed matter was established. Other radiation effects developed from recent technical developments, particularly those involving nuclear power and materials processing by electron, photon, and ion beam irradiation.[16] *However, ion beam modifications during analysis are always a nuisance.*

We cannot review here the immense domain of ion beam damaging effects. Our aim is to point out only features important for ERDA and possible ways of limiting their consequences. Section 14.2 is devoted to general considerations of ion beam damaging effects and the primary consequences of material irradiation by ion beams with energy less than 1 MeV/amu, stressing charge accumulation effects (see also Section 4.3.5) and durable structural damaging (which may be particularly dramatic for archeological samples). Elemental losses are treated in Section 14.3. Section 14.4 discusses the primary solutions to limiting beam damaging to carry out relevant IBA investigations in solids. Section 14.5 examines the problem of stability in standard samples, with particular reference to hydrogen standards.

14.2. BASIC CONSIDERATIONS ON ION BEAM DAMAGING

When a given material is exposed to radiation, a possible primary radiation damage event corresponds to the displacement of one or more atoms and the subsequent formation of vacancy/interstitial pairs or broken bonds. Displacement can occur in two different ways—ballistically by transferring kinetic energy from an incident particle (electron, neutron, proton, other ion) to a collided atom or radiolytically by converting radiation-induced electronic excitations and ionizations into a bond-breaking, atomic rearrangement and finally motion of atoms or radicals.[17,18]

Note the typical orders of magnitude of the different processes involved. For example consider a thin film of Ta_2O_5 containing 3.25×10^{17} atoms of ^{18}O per cm^2, bombarded by a 630-keV proton beam (0.1 μC or 6.25×10^{11} protons) in a vacuum chamber equipped with a silicon surface barrier detector located at 175° sustending a solid angle of 25 msr. About 1.5×10^6 scattered protons and 300 alpha particles created through the $^{18}O(p,\alpha)^{15}N$ nuclear reaction will be detected in this angle, simultaneously 3.25×10^6 ^{18}O atoms will be transmuted. Correlatively, nearly 2×10^8 atoms of both tantalum and oxygen will be sputtered.[12,19]

As a second example, in the case of the irradiation of a 25-μm thin polyester sample ($C_{10}H_8O_4$) coated with a 100-nm gold layer by a 3-MeV $^4He^+$ microbeam (100 μm^2, 1 nA, 1000 s), about 2.4×10^4 recoil protons will be detected in transmission geometry.[20] But for such an investigation, respective elemental losses in various polymers range from 10 to 30% for carbon and nitrogen and 20 to 60% for hydrogen and oxygen,[21–23] depending on the experimental configuration adopted. All these results indicate that more than 10% of the total number of chemical bonds can be destroyed within the irradiated area (see at end of the Section 14.2.1).

Electronic excitations and ionizations can result in permanent or long-lived charged defects (e.g., color centers in glasses and ionic crystals). More frequently they disappear through de-excitation processes, which finally result in heat release. Lattice vibrations due to collisions below the threshold for atomic displacements also result in heat release.

14.2.1. Thermal Effects

One consequence of irradiation of a solid by an ion beam is heat deposition in the irradiated volume and a resulting temperature increase within and around this zone. Two important factors must be considered: the dimension of the irradiated zone (from a few μm^2 to a few mm^2) and target thickness in comparison with the ion range. For example with a conventional 2-MeV $^4He^+$ millibeam (1 nA/mm^2), total thermal energy deposition in silicon (projected range = 7.5 μm) is around 270 mW/mm^3. In comparison for a microbeam (10 pA/μm^2), total energy deposition reaches 2.7 kW/mm^3. The main controlling parameter is the ability of the target material to remove the heat deposit. If any mechanism for heat removal is absent and the incident ion range is greater than the target thickness, the rate of temperature increase dT/dt can be approximated by:

$$\frac{dT}{dt} = \left(\frac{1}{mC_p}\right)\frac{dQ}{dt} = \frac{(dE/dx)\,i}{\pi\,r_0^2\,\rho\,C_p} \tag{14.1}$$

where m is the target mass, r_0 is beam radius, ρ is target density, C_p is specific heat of the target, dE/dx is energy loss rate averaged over the entire thickness of the target, and i is beam current. For a 4.5-MeV proton beam (diameter = 3 mm, intensity = 10 nA) bombarding a paper sample ($C_p = 0.42$ J/mol·K), the increase rate of temperature is 370 °C/mn, which is enough to ignite paper in less than 1 minute.[24] For a stainless steel sample ($C_p = 25$ J/mol·K), temperature increase is only 6 °C/mn. These orders of magnitude are only theoretical calculations. Measuring a local temperature or evaluating a temperature increase throughout an irradiated area appears to be a very difficult experimental task, even by infrared pyrometry, because of the minimum investigated area required.

There are three possible mechanisms for heat removal: radiation, conduction, and convection. The radiation mechanism assumes grey body behavior and obeys the Stefan–Boltzmann equation with a T^4 temperature dependence.[25,26] Energy dissipated by radiation is expressed by:

$$E = C\,\varepsilon\,(T^4 - T_0^4) \tag{14.2}$$

where C is a constant (5.7×10^{-8} J·m^{-2}·K^{-4}·s^{-1}), T and T_0 are, respectively, the temperatures (K) of the specimen and its surrounding, ε is emissivity, which ranges from 1 for an ideal black body to 0.6 for common substances to 0.05 for polished metals. For a wood sample, the amount of heat removed by radiation is negligible, less than 0.1 mW/cm^2. It is clear that radiation removal is relevant only for very high temperatures, intolerable for most solids.

In the case of conductive cooling through specimen support, simple equations based on electron microprobe studies were proposed by Talmon and Thomas[27] to describe uniform heat deposit through a thin specimen. The model was further adapted to the case of ion microbeams.[26,27,29] This model assumes that the only heat sink is the sample edge and energy loss within the sample is constant because the beam range is much larger than target thickness. In this radial conduction situation, local temperature increase can be expressed as:

$$\Delta T = Q(1 + 2\text{Ln}(R/r))\frac{r^2}{4\kappa} \tag{14.3}$$

where Q is uniform heat generated per unit volume, r is beam radius, R is distance between the center of the beam spot and the specimen support, and κ is heat conductivity of the material.

Applying this formalism to stainless steel and wood samples, we find a local temperature increase of about 0.5 and 10 K, respectively; these values appear too low and quite unrealistic. McColm and Cahill propose considering the nonuniform character of the energy deposit along the incident beam path within the sample.[29,30] The mathematical treatment is much more complex in this model, and it gives only upper and lower limits to the temperature profile of a target bombarded by a 4.5-MeV proton microbeam. For a stainless steel sample, the surface temperature is around 40 °C, although it can reach 65 °C near the end of the incident ion range. For a wood specimen, the surface temperature is about 330–430 °C, and the end-of-range temperature is around 500–1300 °C.

For thin samples supported at the edge, heat production is limited to energy loss in the sample, and it may be much less than total beam energy. This advantage disappears for thick targets, where cooling the back face is generally adopted. If we consider a heat sink with fixed temperature and the same area as the irradiated zone, the heat flux Φ is constant over the distance D between irradiated zone and sink and proportional to the temperature gradient. Hence the temperature difference between the irradiated zone and sink is simply:

$$\Delta T = \frac{D\, d T}{dx} = \frac{D\, \Phi}{\kappa} \qquad (14.4)$$

It appears immediately that we must minimize D and maximize heat conductivity κ for reducing the sample heating. If the heat flux is allowed to diverge toward a heat sink larger than the irradiated area, Eq. 14.4 provides an upper limit for the temperature difference.

But in practice Eqs. 14.1–14.4 are not applicable if thermal conduction is affected by imperfect thermal contacts. In this case, which is comparable to the case of a very large electrical resistance in a series chain, the temperature gradient is essentially concentrated in the weakest thermal contact, where a large temperature drop may take place. Then thermal conductivity of the rest of the circuit is of secondary importance.

A third method of heat removal involves a convective process and implies the presence of a gaseous fluid near the solid surface. This may be the case in external beam analysis when samples are immersed in ambient air or submitted to flowing helium. These conditions are never encountered in light-element analysis by ERDA due to the required vacuum.

In conclusion note that the main thermal damage resulting from ion beam irradiation of a solid corresponds to elemental redistributions within and around the irradiated zone. Several mechanisms are thus involved, such as diffusion processes, surface desorption, or elemental release from the volume. Moreover these redistributions are generally associated with such physical and mechanical effects as local fusion, volume change, density variation, or fracture.[15]

14.2.2. Electronic Defects and Bond Breaking

Electronic interactions often create electronic defects and as a consequence changes in electrical and optical properties and chemical modifications through bond-breaking.

Changes in electrical or optical properties, typically the formation of color centers, are not a major concern for analysis, but these may leave traces in precious samples. This is exemplified in the complete opaqueness reached by radiological protections made of lead-rich glasses when not compensated by appropriate additives or in the grayish color of ancient Rhenan crystal beakers due to exposure to sunlight. On the other hand bond-breaking may form mobile and eventually volatile species and thus alter the chemical composition during analysis.

In the case of hydrogenated materials, even in conditions where pyrolysis is excluded, thermal effects may explain hydrogen loss and molecular emission under irradiation subsequent to bond breaking. Table 14.1 gives some energy values for the most common chemical bonds occurring in current organic media.[31] Note that a 2-MeV $^4He^+$ beam with a current density about 100 nA/cm^2 brings 200 mW/cm^2 to the target material. Taking into account an average penetration depth of about 10 μm, total energy transferred reaches 200 W/cm^3. This value appears large even though only a small fraction of the thermal energy dissipated in the solid is transformed into chemical

Table 14.1. Energy of the Most Common Chemical Bonds in Organic Materials[a]

Bond	Energy (kcal/mole, 298K)	Energy (eV)
C-H[b]	100–111	4.3–4.8
O-H	87–106	3.8–4.6
C-C[c]	67–102	2.9–4.4
C=C	172	7.5
C≡C	230	10
N-H	88–107	3.8–4.6
C-O	57–63	2.5–2.7
C-N	70	3
C=N	147	6.4
C≡N	213	9.2
O-O	30–38	1.3–1.6
N=O	145	6.3
C=O	177	7.7
C-F	125	5.4

[a]From Ref. 31.
[b]For example, 105 kcal/mole for CH_3-H and 111 kcal/mole for C_6H_5-H.
[c]For example, 67 kcal/mole for CH_3CO-$COCH_3$, 76 kcal/mole for C_6H_5-CH_3, and 90 kcal/mole for CH_3-CH_3.

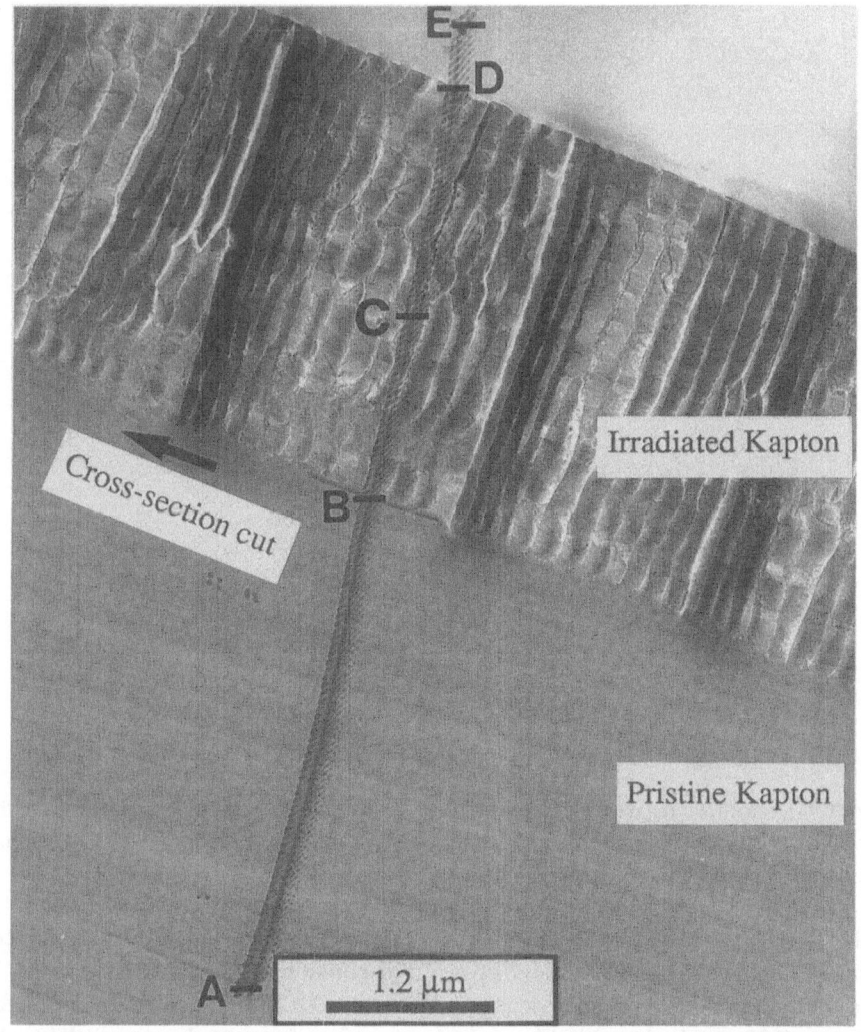

a

Figure 14.1. Consequences of irradiating a 75-μm-thick polyimide film by 1-MeV $^4He^+$ ions (2.5×10^{15} ions/cm^2 s, t = 60 s). (a) Image of a 200-nm-ultrathin transverse section obtained by scanning transmission electron microscopy, (b) C, N, and O distributions measured by energy-dispersive X-ray spectrometry. (Data from Ref. 33)

energy. Considering the initial energy loss value of 2-MeV $^4He^+$ in a common polymer (200 eV/nm),[32] it is easy to understand how ions penetrating a polymer can induce broken hydrogen bonds (average energy = 4.2 eV). For example in polyester, assuming that all energy transferred within the first nanometer is used to break chemical bonds, a total irradiation dose of 10^{13} protons (1.6 μC) would produce

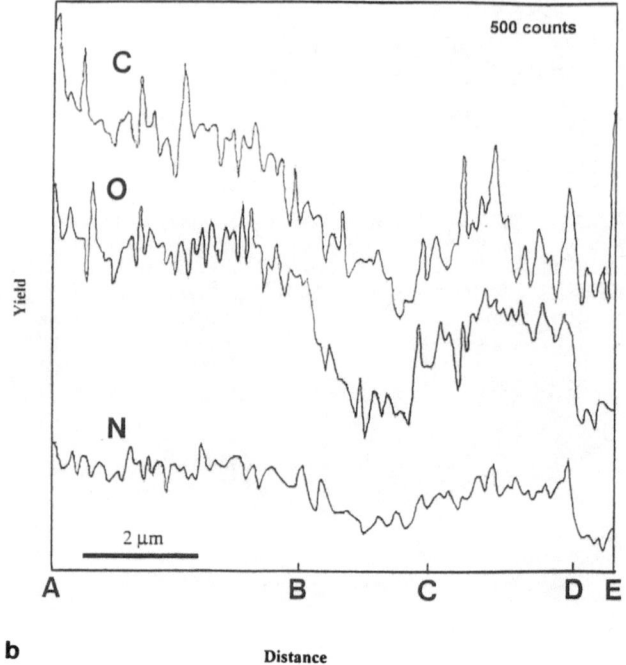

Figure 14.1. Continued

5×10^{14} broken C-H, among 3.5×10^{15} C-H bonds present in this portion of the irradiated volume. This simple calculation shows that up to 15% of the C-H bonds would be destroyed under these conditions.

Recently Marin and coworkers showed that irradiation of a 75-μm-thick poly-imide film by a 1-MeV ^{4}He$^+$ beam at 25 °C (2.5×10^{15} ions/cm^2·s, t = 60 mn), induces film blackening at a depth around 4.2 μm, corresponding to the incident ion range.[33] The transmission electron microscopy image of an ultrathin section of the irradiated polymer (200 nm) is shown in Figure 14.1a. A series of fracture lines, parallel to the sectioning direction, clearly appears. In this region the polymer film has lost its elastic properties and become brittle. Carbon, oxygen, and to a lesser extent nitrogen distributions, measured by energy-dispersive X-ray spectrometry (EDXS), exhibit strong depletions (especially in the area between B and C in Figure 14.1b, where electronic stopping dominates).

Structural modification by electron excitation processes under irradiation is demonstrated by the work of Biron and coworkers, who shows that in a PbO-B$_2$O$_3$ glass, electron irradiation induces the precipitation and coagulation of a new phase, composed of lead droplets.[34] The rate of coagulation depends on both electron flux and temperature. A strong decrease in glass viscosity by up to 10 orders of magnitude was correlatively observed,[34] which implies a considerable enhancement of diffusion rates. While the appearance of a new phase is irreversible, viscosity decrease and diffusion enhancement appear only *during* irradiation. Since they occur even at very low electron energies (60 keV), they can be unambiguously attributed to electronic excitation effects.

Chemical modification due to broken bonds may also result in permanent changes in other properties. As a matter of fact, color alteration may be the plague of IBA applied to inks, dyes, or pigments.

14.2.3. Ballistic Damage

The occurrence of ballistic damage during charged particle irradiation of common materials is clearly illustrated in the following example. Studying the effect of ion implantation on electrochemical properties of pure metals, Serruys and coworkers found that the implantation profile of Xe ions (10 keV) in Ti does not correspond to the theoretical prediction by TRIM. This discrepancy is attributed to the progressive erosion of the Ti surface (sputtering) by incident Xe ions during the implantation process.[35]

The primary consequences of ballistic damage are: diffusion enhancement by formation of point defects eventually correlated with a thermal gradient; crystallographic transformations (preferential elemental loss or creation of a new phase, destabilization of metastable phases); segregation induced by modifying the vacancies/interstitials equilibrium, ion mixing (apparition of solid solutions and alloys); sputtering.[15]

Point defect formation is related to the nuclear stopping power of the ion. But note that ballistic damaging depends not only on the number of defects created by an ion but also on the number of ions necessary to obtain sufficient counting statistics. The simple argument that heavy ions produce more defects is not directly applicable without considering their cross section.

Other factors also influence the effective damaging. First, volumic density of defect creation depends on the detailed structure of the collision cascade and contributes to determine the short-term recombination of vacancies and interstitials. Secondly, recent simulations by molecular dynamics[36] show that in crystalline solids recombination of point defects is much more efficient in the first picosecond after the formation of a cascade than was previously believed. As a consequence observed effects are due much less to defects liable to contribute to atomic mobility than predicted by calculating the initial defect formation rate using, e.g., TRIM. Atomic replacements formed during the initial and recombination phases produce chemical disorder, which is liable to induce phase transformations.

The sputtering phenomenon was mathematically described by Sigmund[37] after several modeling attempts to use semiempirical approaches by Goldman and Simon,[38] Pease,[39] Pleshitsev[40] and Kaminsky.[41] According to Sigmund, the elemental sputtering yield Y_i can be written as:

$$Y_i = \frac{\lambda \, \alpha \, S_n(E)}{U_{si}} \qquad\qquad (14.5)$$

with $\lambda = 0.042$, a dimensionless factor; $S_n(E)$, the nuclear stopping power; U_{si}, the sublimation energy of the target atom species i (from 0.9 eV for K to 8.7 eV for W), and α, a function of M_i/M_1, the ratio of target atom i and incident ion masses. The main parameters involved in the sputtering process are the nature and energy of the incident ion, composition of the target, and nature and chemical environment of the sputtered element.[42] Sputtering can generally be neglected for MeV light-ion beam bombardment of solid targets because the typical order of magnitude for Y_i is about 10^{-4} atoms/incident ion for protons.[12] Nevertheless when characterizing thin films with heavy ions, sputtering effects must be considered.

14.2.4. Charge Accumulation Effects

Perturbations of the analytical process by charge accumulation are the focus of Section 4.3.5. But charge accumulation in the near-surface region of insulating materials irradiated by charged particles may also cause target alterations during analysis. This charge accumulation is due to both ion implantation and secondary electron and ion emission, resulting in huge electric fields near the surface of insulating targets. Several consequences may result, ranging from target modification (elemental migration under electric field) to severe target damaging (dielectric breakdown).[15]

Dielectric breakdown is the most spectacular among these effects. Even if it occurs along the surface, which is the most frequent case, it may drag ionic species from relatively deep regions of the solid and more generally induce the desorption of volatile species. At the surface it leaves craters or grooves where sublimation and melting have changed the composition, and eventually hillocks of redeposited matter. In the bulk of the target, breakdowns leave arborescent tracks that behave like macroscopic optical defects, diffusion short circuits, and potential paths for breakdowns at lower potentials. In such brittle materials as glasses, rupture may occur in conjunction with thermal stresses.

Under less violent dose rates, some ionic or covalent compounds may expel ions or neutrals,[43] thus changing their stoichiometry near the surface. More frequently we observe electric-field-driven diffusion (electromigration) and subsequent loss by beam-induced desorption or sputtering[10,44–55] under proton, helium, or heavy-ion irradiation. This is particularly the case of alkali ions in glasses, which have noticeable mobility even at room temperature, but also of calcium and heavier components, such as metal ions, at least with valences I or II.[56] Radiation-induced enhancement of diffusion coefficients (cf. Section 14.3.2) makes these elemental redistributions even more important. This process attracted much attention and was modeled by several authors.[10,48,50,52,53] We stress that this process is highly influenced by electrostatic conditions imposed at the surface.[54] In fact electromigration should be suspected in any insulator submitted to an ion beam in vacuum.

14.3. ELEMENTAL LOSSES

Elemental losses during IBA of solids do not only affect volatile elements having a very low vaporization heat, such as hydrogen, nitrogen, oxygen, but mobile species are also concerned, for example, weakly bonded cations and anions (alkali and halogen elements). Nevertheless the behavior of any element depends strongly on its chemical environment and partly on the structure in which it is included.

14.3.1. Hydrogen Loss from Polymers

Much work was done on hydrogen behavior in polymers under MeV ion irradiation. Cholewa and coworkers[28] studied the effects of 3-MeV protons and 2-MeV ^{4}He

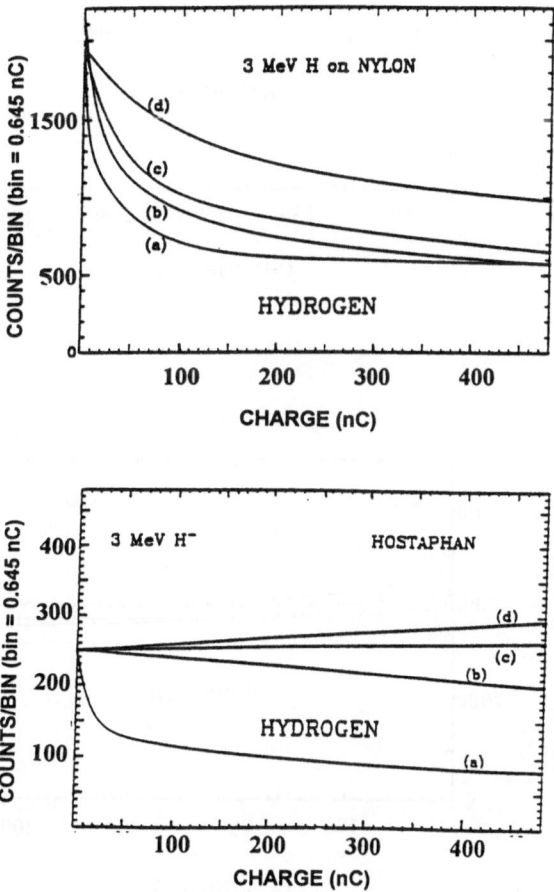

Figure 14.2. Concentration of hydrogen in nylon and Hostaphan (Mylar) foils after irradiation with a 3-MeV H$^+$ beam focused to 6 μm diameter for (a) unscanned beam, (b) 20 × 20 μm^2, (c) 50 × 50 μm^2, and (d) 100 × 100 μm^2 scan area, respectively. (Data from Ref. 28)

beams on nylon and Mylar foils (with a thickness range of 1–2.5 μm). Figure 14.2 shows the hydrogen loss detected in nylon and Mylar versus total charge for a focused 6-μm proton beam. Figure 14.3 gives similar results for a focused 6-μm ^{4}He beam. These data suggest that using high-frequency beam scanning can play an important role in limiting elemental losses in the investigated area.

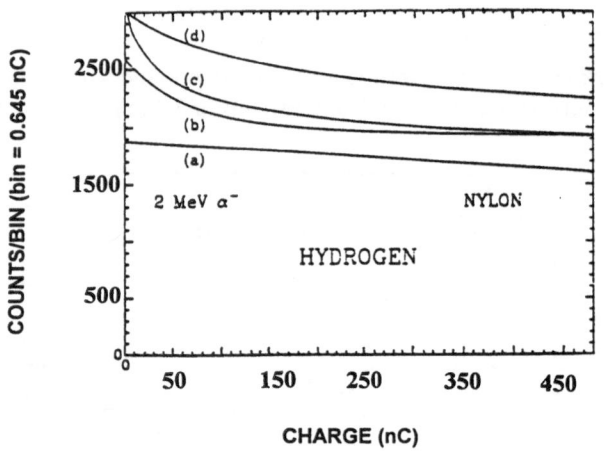

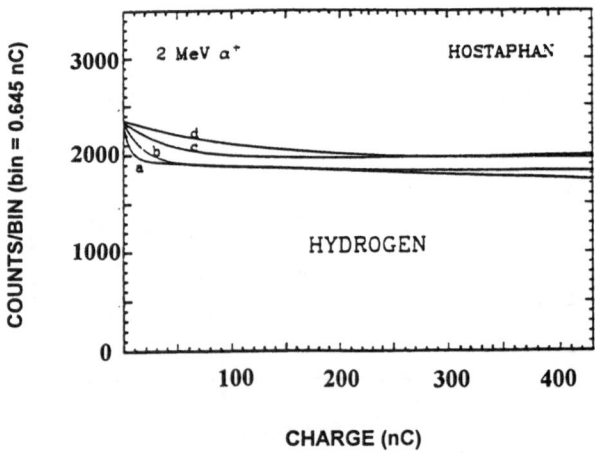

Figure 14.3. Concentration of hydrogen in nylon and Hostaphan (Mylar) foils after irradiation with a 2-MeV ^{4}He$^+$ beam focused to 6 μm diameter for (a) unscanned beam, (b) 20 × 20 μm^2, (c) 50 × 50 μm^2, and (d) 100 × 100 μm^2 scan area, respectively. (Data from Ref. 28)

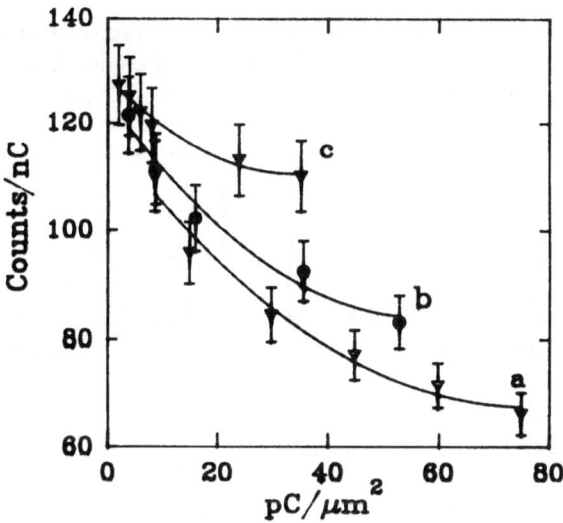

Figure 14.4. Hydrogen content in polyimide film as a function of $^4He^+$ microbeam current density: (a) 14.8 μA/cm², (b) 7.8 μA/cm², and (c) 2.6 μA/cm². (Data from Ref. 22)

Themner[57] reported hydrogen loss ranging from 10–15% for polyester films of 3.5-μm thickness irradiated by a 2.55-MeV proton microbeam with a size of 10×20 μm² (1 nA). Tirira and coworkers discusses hydrogen loss in polyimide films irradiated by 1.8–3 MeV ^4He microbeams. Hydrogen loss increases with increasing microbeam density,[22] as shown in Figure 14.4. In an extension of previous work on polyimide films, Tirira also demonstrates that hydrogen is able to move toward the sample surface under irradiation, as shown in Figure 14.5. High hydrogen concentration zones created under krypton bombardment are both able to evolve in terms of geometric structure and to migrate under microbeam irradiation. Nevertheless thermal effects are involved in these processes as well as mechanical effects due to high local stresses.

Lewis and coworkers[3] have reported the formation of several gaseous compounds by irradiating Teflon or Kapton with helium, nitrogen, and silicon ion beams in the energy range of 0.2–2.0 MeV (0.05–2 MeV/amu). Teflon released residual gases as CF and CF_3, and Kapton yielded H_2, CO, and CO_2. Figure 14.6 shows a mass spectrum characteristic of the residual gaseous environment after 200 keV silicon irradiation of Teflon and Kapton films.

14.3.2. Irradiation-Enhanced Diffusion of Hydrogen in Solids

Diffusion of hydrogen under irradiation is a common effect in insulators; moreover hydrogen diffusion coefficients are fairly high in any type of material, especially in polymers, as reported by Venkatesan and coworkers in 1984.[4] They found that diffusion coefficients of several molecular species, including H_2, in polymethylmetacrylate are in the range of 10^{-11} to -10^{-10} cm^2 s^{-1}. Typically this result is of the same order of magnitude as hydrogen diffusion coefficients measured in titanium and Ti-based alloys by nuclear resonance[59]: about 5×10^{-11} cm^2 s^{-1} at 450 K.

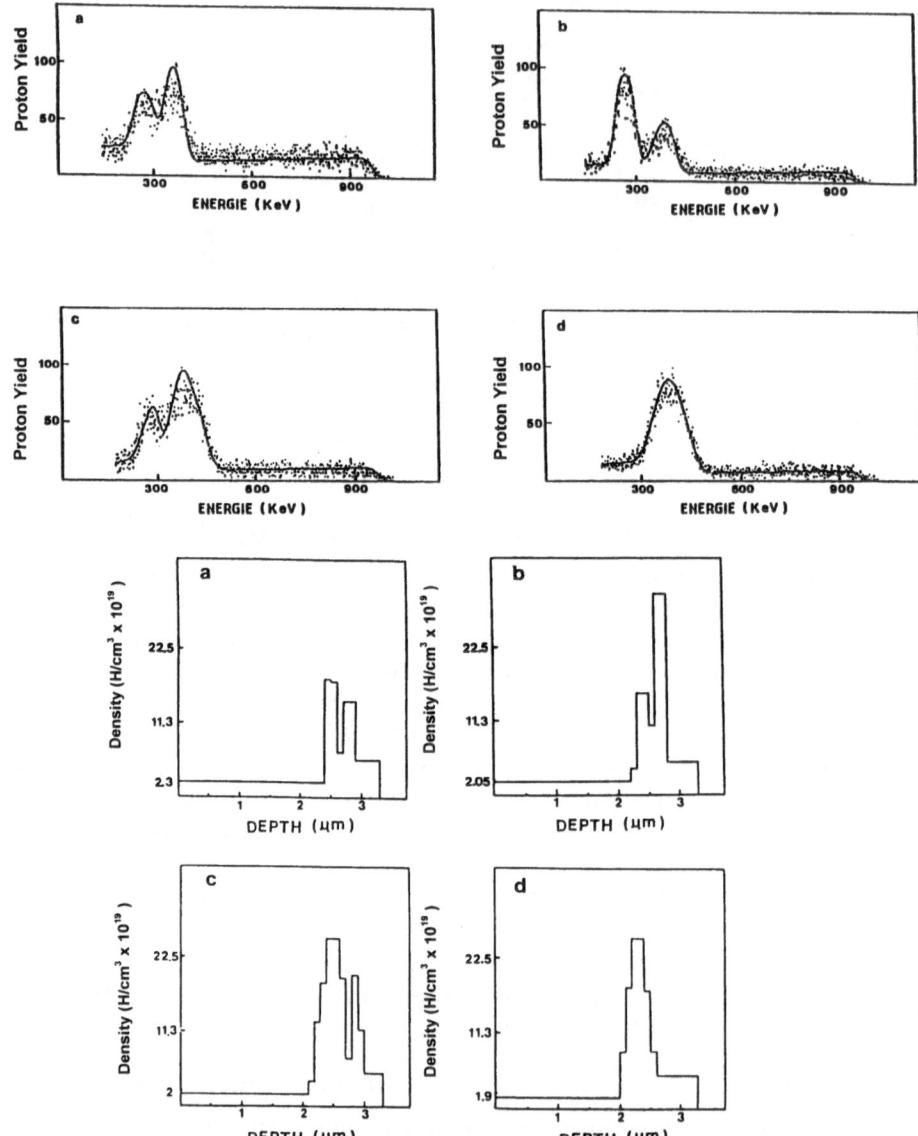

Figure 14.5. Successive ERDA energy spectra obtained in transmission geometry for a 12.5-μm-thick polyimide film preirradiated by 230-MeV krypton ions (fluence = 1.4 10^{19} cm^{-2}s^{-1}): ^{4}He$^+$ energy = 2.05 MeV, t = 500 s, beam diameter = 30 μm, and corresponding depth profiles. (Data from Ref. 58)

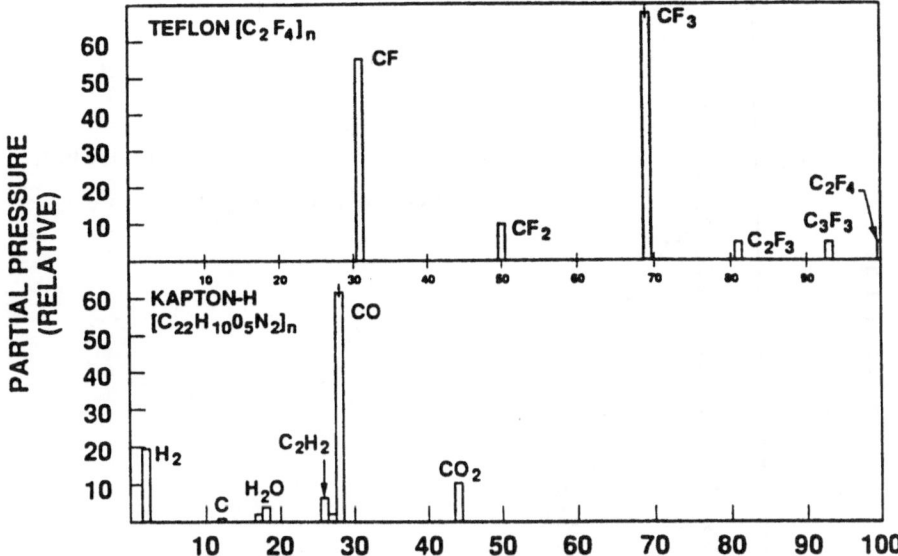

Figure 14.6. Residual gas analysis of Teflon and Kapton during 200-keV silicon irradiation. Beam flux was approximately 60 nA/cm^2, spectra taken during the first minute of irradiation, analyzer range was 100 amu. (Data from Ref. 3)

Measuring hydrogen depth profiles by means of the 6.385-MeV $^1H(^{15}N, \alpha \gamma)^{12}C$ resonance in leached glasses, Trocellier has showed that hydrogen tends to move to the glass surface when both current density or total charge increase above threshold values,[60] as shown on Figure 14.7. Hydrogen migration is clearly assisted by the surface electrical field induced by ion bombardment.

A similar effect was observed in archeological ceramics (sixteenth century) in which weathering alteration produces a thick hydrated layer. Initial hydrogen loss rate can be reduced by a factor of 2.5 (–10% instead of –25%) by decreasing the $^{15}N^{3+}$ current density by a factor of 4 (5 nA in 4 mm^2 instead of 20 nA). Total hydrogen loss can be reduced by a factor of 3 (–20% instead of –60%) by decreasing the total dose by a factor of 4 (4 μC instead of 16 μC with 5 nA in 4 mm^2) without affecting the counting statistics.[61]

Gallien and Cachoir[62] recently studied the alteration products of uranium dioxide corroded by granitic groundwater and particularly the growth of schoepite crystals (UO_3, $2H_2O$). Using a 3.5-MeV proton microbeam (10 μm^2, 100 pA), they found that microbeam bombardment induced dehydration in crystals (measured by microERDA) as well as partial reduction and a certain degree of fracturing in the crystals (see Figure 14.8).

14.3.3. Irradiation-Enhanced Hydrogen Desorption

Hydrogen desorption induced by ion beam irradiation can severely affect IBA quantitative data for several types of materials. In the case of metallic alloys, semicon-

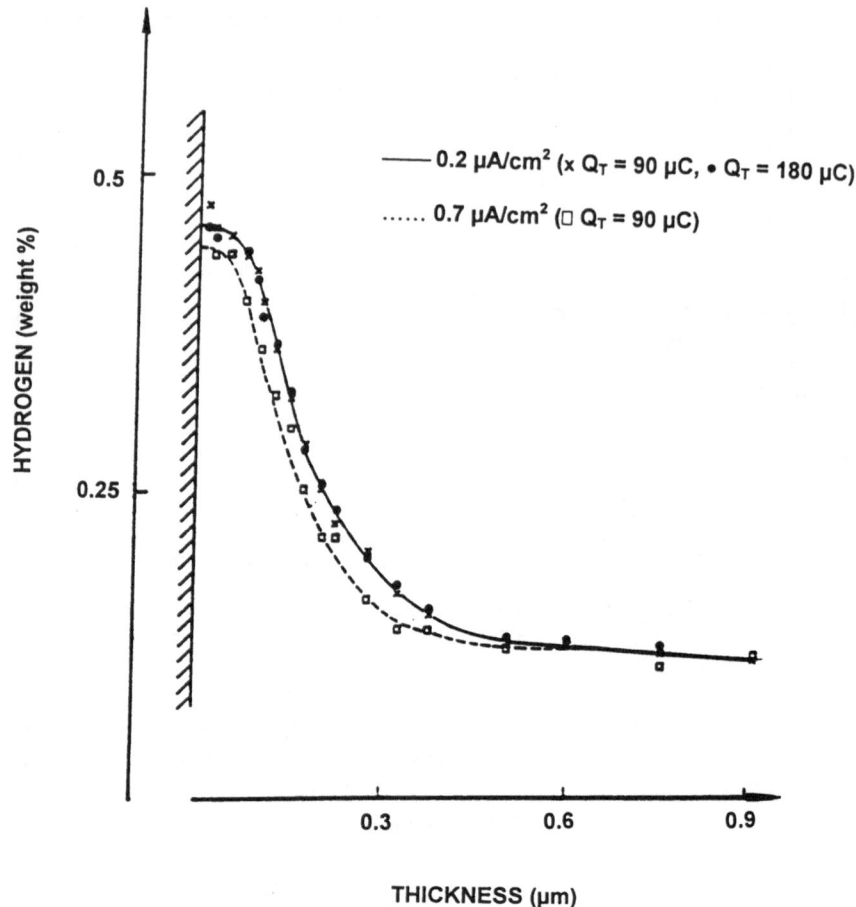

Figure 14.7. The $^{15}N^{3+}$ current density and dose dependence on hydrogen depth profile in the near-surface region of a borosilicate glass leached during 25 days at 20 °C in deionized water. (Data from Ref. 60)

ductors, or ceramics, such hydrogen desorption corresponds to surface hydrogen outgasing, although in the case of hydrated minerals, it corresponds to water desorption or water loss. Microbeam-induced water loss was shown by Mosbah and coworkers,[63] who measured the water content of melt inclusions trapped in volcanic glasses from Pantelleria Island. Nevertheless surface hydrogen desorption can sometimes be an excellent way of reducing surface contamination.

Using a 350-keV $^4He^+$ beam and ion fluences from 1.2×10^{14} to 1.0×10^{17} $^4He^+/cm^2$, Ross and Richard observed depletion of hydrogen implanted or included in several materials, such as amorphous carbon–boron deposits, beryllium, pyrolitic carbon, vitreous carbon, silicon nitride, titanium carbide, or stainless steel.[64] Desorption is practically negligible in scandium, silicon, and silicon carbide. The desorption degree largely depends on initial hydrogen concentration. Figure 14.9 shows examples of H

Figure 14.8. The SEM image of a schoepite crystal grown on the surface of a uranium dioxide sample leached in ground water. (a) At the end of the leaching test, (b) after its analysis by a 3.5-MeV proton microbeam (100 pA, 10 μm^2, 0.1 μC in 45 mn). (Data from Ref. 62)

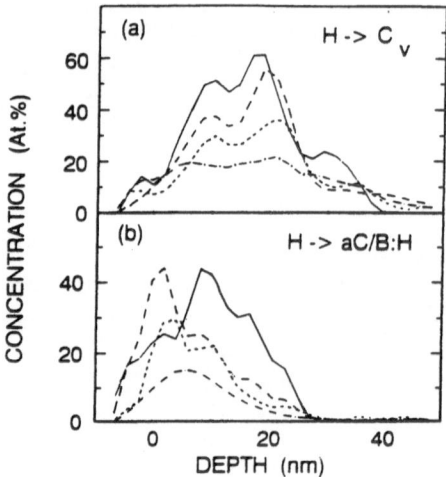

Figure 14.9. (a) Depth profiles of H implanted to saturation in vitreous carbon and analyzed with fluences 3.05×10^{14}(—), 9.13×10^{14}(---), 3.15×10^{15}(...), and 2.32×10^{16}(- -) $^4He^+/cm^2$. (b) Depth profiles of H included in aC/B : H and analyzed with fluences 2.88×10^{14}(—), 7.19×10^{14}(---), 2.89×10^{15}(. . .), and 2.88×10^{16}(-.-.)$^4He^+/cm^2$. (Data from Ref. 64)

depth profiles in vitreous carbon or in C compounds varying with $^4He^+$ fluence. Generally a large H depletion occurs at the beginning of the analysis and lasts until beam fluence reaches a saturation value around 2×10^{16} per cm^2. Moreover a displacement of implanted hydrogen to the surface before desorption was observed.

Nagata and coworkers showed that the ion-induced release of H and D implanted in Be, C, Si, and SiC at the energy of 5–8 keV is induced both by low-energy (5–20 keV) and medium-energy (2-MeV) $^4He^+$ ions.[65] The release extent for medium-energy ions is an order of magnitude larger than for low-energy ions, as shown on Figure 14.10.

Recently, Berthier and coworkers, using the nuclear microprobe facility at Laboratoire Pierre Süe in Saclay to characterize the hydrogen distribution within Zircaloy nuclear fuel claddings submitted to water vapor alteration at high temperature, found local ZrH_2 precipitates beneath the zirconia layer grown at the cladding surface.[66] They have observed that these hydrides may decompose under 2.4-MeV $^4He^+$ microbeam irradiation. Hydrogen loss can be more than one-half the total initial hydrogen content with a $2 \times 5 \ \mu m^2$ beam spot and a beam intensity around 100 pA, constituting a total ion dose of about 0.2 μC.

14.3.4. Behavior of Other Elements

The behavior of alkali cations, and especially of sodium in natural glasses, was studied by Mosbah and coworkers.[67] These researchers showed that microbeam irradiation (2-MeV protons) induces damaging in SiO_2-Na_2O glasses all along the path

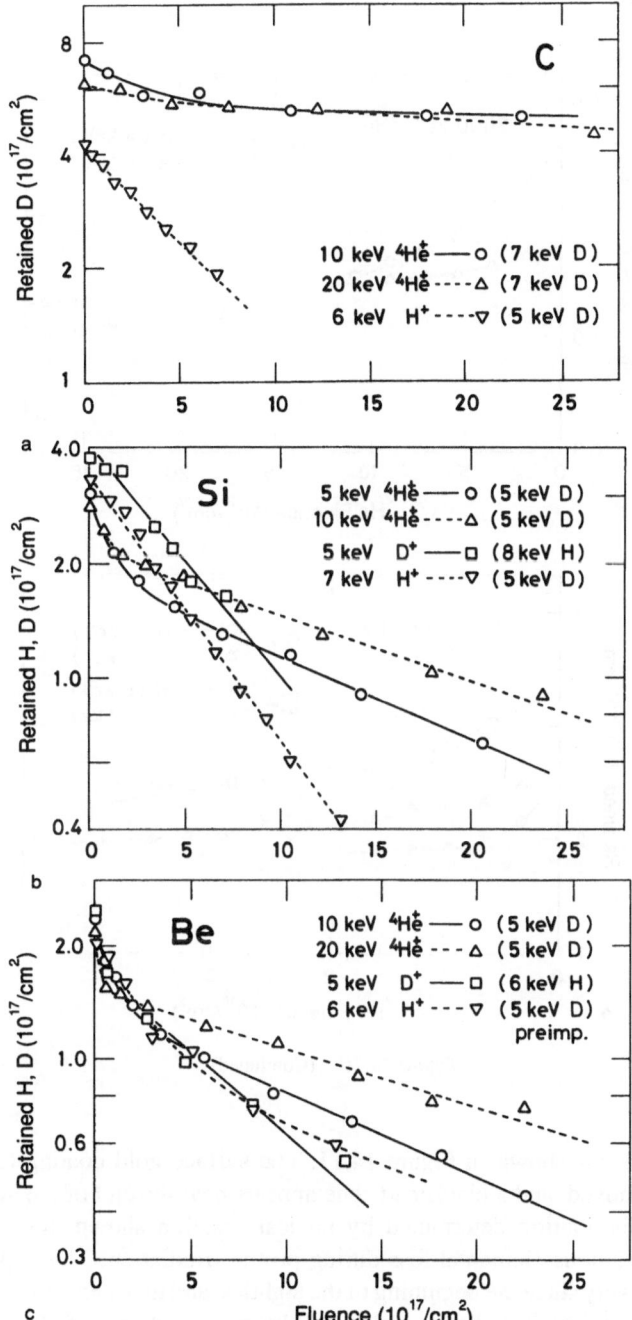

Figure 14.10. Retained amounts of H and D implanted in C, Si, and Be as a function of fluence during (a,b,c) low-energy (5–20 keV) hydrogen and helium bombardment, (d,e) high-energy (2000 keV) helium bombardment. (Data from Ref. 65)

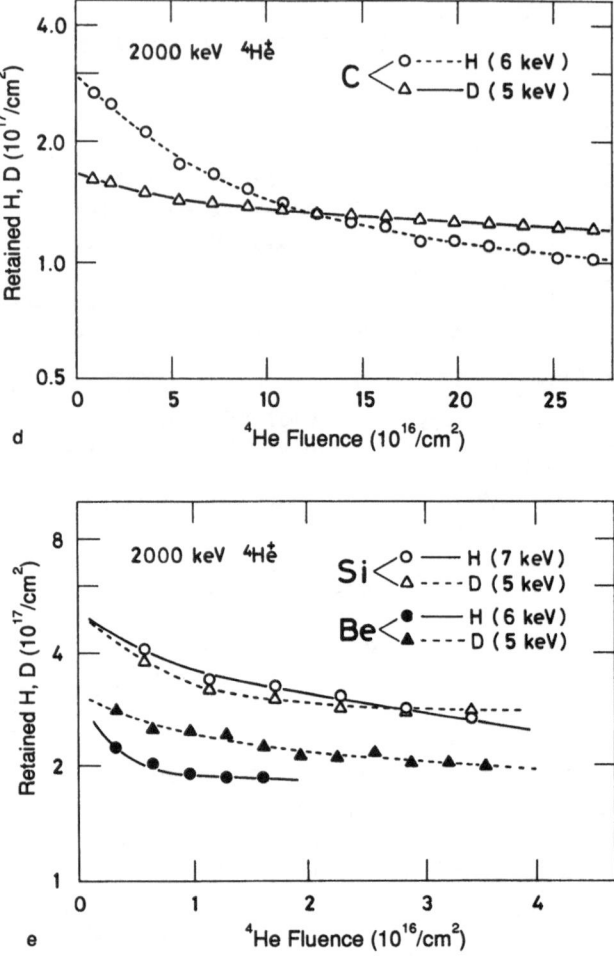

Figure 14.10. (Continued)

of incident ions, as shown in Figure 14.11. The surface gold coating (25 nm) was completely removed, and a blackened zone appears near the end of the proton range. The sodium distribution determined by nuclear reaction shown in Figure 14.11c exhibits a continuous elemental loss during proton irradiation. This topic has been studied extensively since the beginning of the eighties, and many excellent papers have been published.[10,15,23,68] Radiation-enhanced diffusion and loss of alkalis in glasses were discussed in Section 14.2.4.

Biological samples are particularly sensitive to ion beam irradiation. Massiot and coworkers[69] investigated the stability of nitrogen content in soybean leaves under

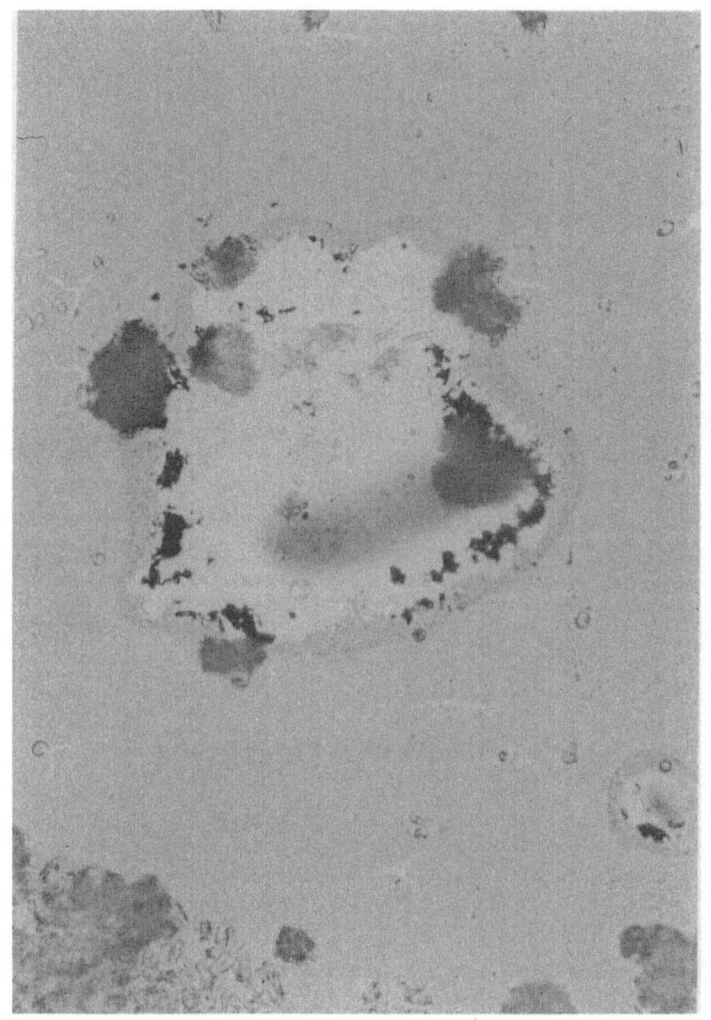

Figure 14.11. Optical images of a SiO$_2$-Na$_2$O glass sample after a 2-MeV proton microbeam irradiation (beam density 90 pA/μm^2). (a) Focusing is done at the glass surface, (b) Focusing is done about 50 μm beneath the glass surface, (c) sodium distribution measured through the ^{23}Na(p,p'γ)^{23}Na reaction, showing continuous sodium loss during irradiation [distributions from the 511-keV line and the 1.634-MeV line of the ^{23}Na(p,αγ)^{20}Ne reaction are also visible]. (Data from Ref. 67)

a

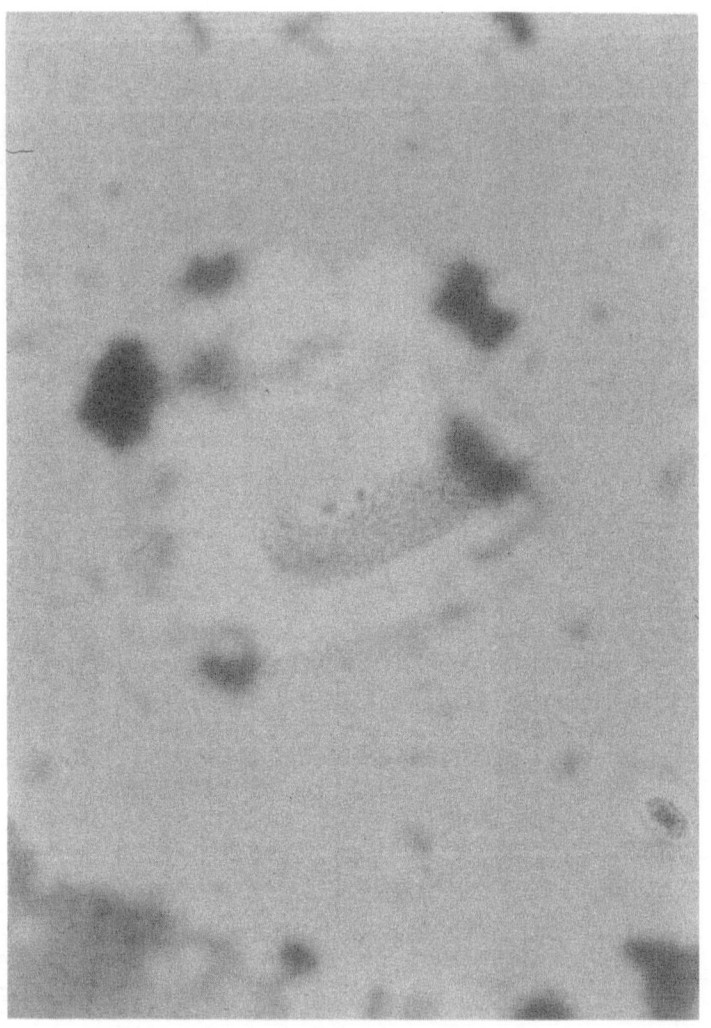

Figure 14.11. Continued

b

Glass sample : high density current

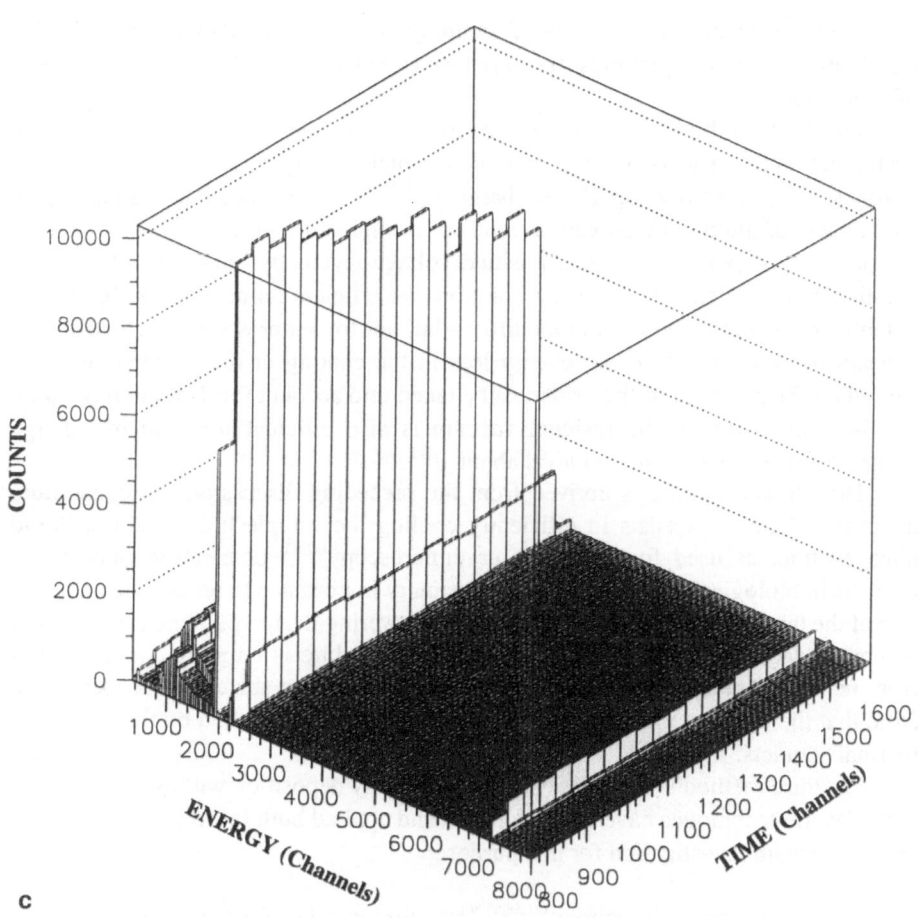

c

Figure 14.11. Continued.

3-MeV protons (15 pA/μm^2) to be able to use the local $^{14}N/^{15}N$ ratio to study a metabolic mechanism. They observed a 50–60% nitrogen loss, with a saturation for a fluence of about 0.3–0.4 μC, which allowed them to adopt this depletion level as a reference. For a more detailed discussion of radiation damage induced by ion irradiation in solids, see Refs. 70–74.

14.4. REDUCTION OF RADIATION DAMAGE

Considering the impossibility of avoiding beam-induced damaging in investigated targets, the physicist must try to reduce as much as possible secondary effects of irradiation.

The first method involves avoiding or limiting the consequences of charge accumulation in the near-surface region of the irradiated target. Several techniques are available for preventing superficial charging.[68,75–81] A thin metallic coating (a few nanometers of aluminum, copper, or gold) or a conducting grid can be used and the sample surface polarized with a positive voltage (typically from 45–300 V). As mentioned in Section 4.3.5, all these solutions are efficient for decreasing the surface potential or making the field more uniform, but not for suppressing the electric field beneath the surface. Moreover energy loss in the coating or the occultation effect caused by the presence of the grid must be taken into account for data interpretation. A slight ionization of the residual vacuum is also efficient for reducing charge accumulation even at a vacuum level about 10^{-8}.

The second method is derived from the preceding discussion on temperature increase effects. It consists in efficiently cooling the sample holder with a liquid nitrogen trap, as used for electronic cryomicroscopy[82,83] or electron microprobe analysis in biological materials.[84] It is of primary importance to reduce the temperature of the target surface to room temperature. Lowering the temperature further is not recommendable because the drawback of this procedure is trapping contamination from residual atmosphere in the vacuum chamber onto the surface to be examined. Note that the efficiency of sample cooling is essentially controlled by the quality of thermal contacts.

The third method is to limit severely total energy deposition within the irradiated area. Several techniques have been proposed and applied both for millibeam analysis and microbeam investigation for this purpose:

- To investigate thin samples[21,58,85,86] to eliminate the effects of beam damaging in regions beyond the analyzed zone; power per unit volume dissipated in the investigated zone is unchanged.
- To minimize beam density in the investigated zone, lowering beam current or maximizing spot size, or to reduce the total dose applied.[23,87,88]
- To use beam scanning with a relatively high-frequency rate as proposed by Llabador,[89] Cholewa,[28] and Themner[57] or to use a sample scanning stage.[83]

Each option has advantages and drawbacks, the question to be solved is *how to guarantee that the detected signal really represents the initial target composition.* Studying thin layers or preparing self-supported targets is not always possible; metallic coatings, polymer films, and geologic samples are generally well-adapted for this kind of measurement.[82] Reducing both beam density and total dose must be limited to obtain good counting statistics; this depends essentially on the cross sections of the processes involved. A compromise must always be found between these contradictory requirements. Nevertheless fast elemental release occurring during the first few seconds of irradiation cannot be prevented. Beam scanning or sample scanning are compatible only with microbeam investigations. These two options reduce only volumic energy deposition for a constant total fluence.

14.5. CHOICE, PREPARATION, AND STABILITY OF STANDARD SAMPLES

14.5.1. Introduction

Before developing this section we recall that using any analytic standard sample consists of three steps:

1. Prepare a reference sample avoiding any contamination effect.
2. Completely characterize the sample (stoichiometry, elemental contents).
3. Assess the stability of the standard composition under experimental conditions adopted for samples being investigated.

14.5.2. Hydrogen Standard Samples

In any analytic problem, using a standard sample (also called a reference sample) proves to be very efficient and at the same time very critical, particularly in terms of representing the investigated sample. In the case of hydrogen determination, a problem arises due to the high mobility of the hydrogen atom in condensed matter. In addition there scarcely exists an analytic method allowing reproducible and accurate determination of hydrogen stoichiometry.

For ^1H several types of standard samples can be considered. We can imagine using the media containing the largest content of hydrogen, which means polymers, for example polyester (Mylar) and polyimide (Kapton).[58] The first advantage of such materials is that they can be found as thin films (from 3–20 μm) or as thick samples (from 50 μm up to several hundred micrometers). The second advantage is that the hydrogen composition can be assumed to be homogeneous in depth or laterally. Nevertheless there are two limitations:

- The thickness heterogeneity for thin polymer films, especially below 20 μm
- The sensitivity of polymer structure under ion beam bombardment leading to elemental losses

Thus polymers cannot be considered as stable standards but only as a possible compromise, taking into consideration the possibility of relying on the reproducibility of losses.

Another kind of reference material can be obtained from hydrogen-implanted monocrystals, such as Si <111>[20] (Figure 14.12). The main interest of H-implanted samples lies in the versatility of the available depth profile (mean depth, full width at half-maximum, maximum content), but it is difficult to predict accurately implantation profiles and assess the conservation of the hydrogen profile for long periods of time. Another solution is to use hydrated minerals of perfectly known composition, like mica-muscovite $(K_2O, 3Al_2O_3, 6SiO_2, 2H_2O)$,[70] but the primary limitation comes from their behavior under ion beam irradiation because dehydration may probably occur. Hjörvarsson and coworkers proposed using hydrogenated tantalum as a calibration matrix.[90] They have shown the long-term stability of the average atomic H/Ta ratio when profiling hydrogen by using the 6.385-MeV resonance of the nuclear reaction $^1H(^{15}N,\alpha\gamma)^{12}C$. These calibration samples are found to be ultrahigh vacuum compatible up to 150 °C. The last solution consists in using metal hydrides with a known stoichiometry as TiH_2.[58] The limiting parameter does not lie in the stability of the crystalline structure of the hydride, but it is certainly possible to control its stoichiometry exactly.

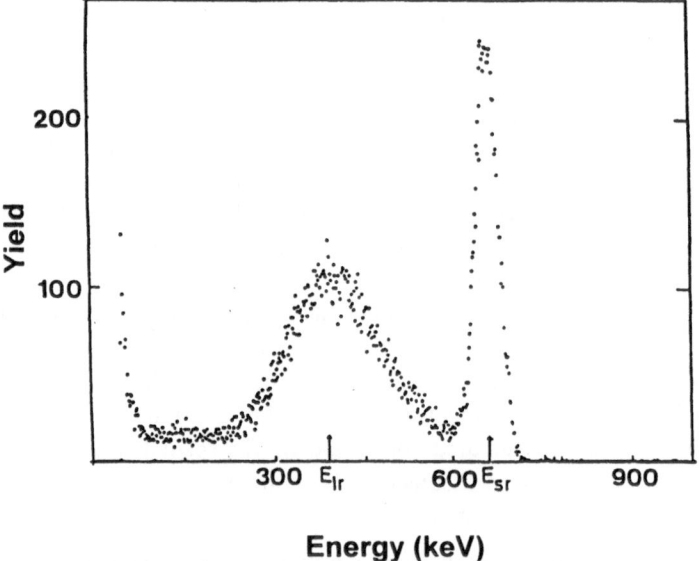

Figure 14.12. Typical recoil spectrum from a <111> silicon monocrystal implanted with 10^{16} H/cm^2 at 10 keV. (Data from Ref. 20)

For ^2H only three of the four solutions previously considered for ^1H are usable. Deuterated polymers[91] (Figure 14.13) can be considered as well as deuterium-implanted monocrystals[92,93] (Figure 14.14). Metal deuterides, like titanium and nickel deuterides,[94,95] have also been considered. Nevertheless, once again the primary limiting parameter is the irradiation sensitivity of the sample structure.

Thus no ideal solution exists and whatever the final choice, the quantitative interpretation of ERDA data first requires studying the irradiation behavior of the reference sample under conditions selected to optimize the experimental configuration and minimize beam-damaging effects. A possible approach involves determining the rate of hydrogen loss in a standard sample as similar as possible to the investigated material, possibly using event-by-event ERDA (cf. Section 4.3.4), and then extrapolating its initial content and deducing the initial content of the unknown sample. This is approximately the procedure adopted by Massiot and coworkers[69] for ^{14}N-^{15}N determination in soybean leaves.

14.5.3. Other Standards

Many papers discuss using standard samples for calibration procedures. We quote SiO_2/Si standards prepared by wet oxidation of (111)Si that show a quite stable

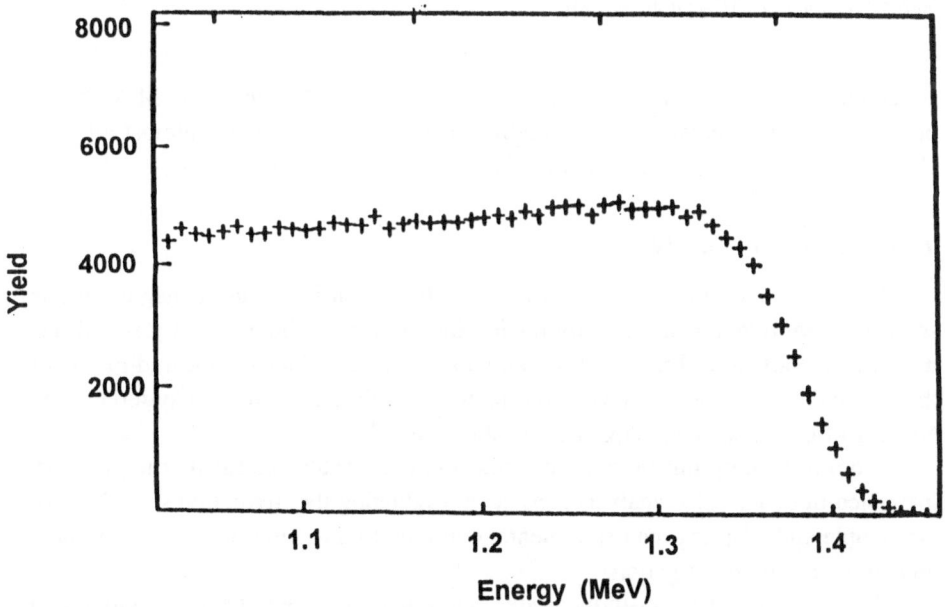

Figure 14.13. A 2.8-MeV helium ERDA recoil yield versus detected energy profile of ^2H in an ^1H-PS (40%)/^2H-PS (60%) mixture. (Data from Ref. 91)

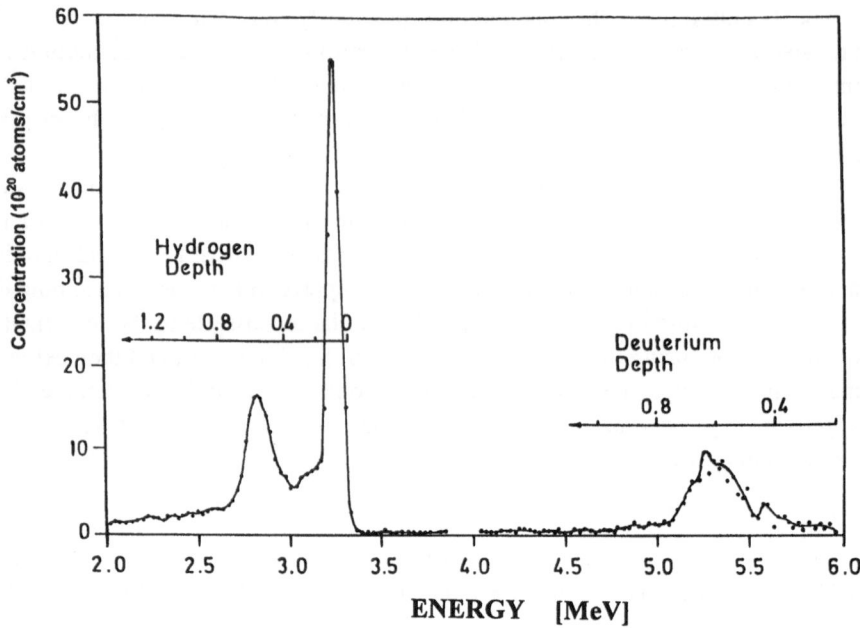

Figure 14.14. Depth profile of a silicon wafer both implanted with hydrogen and deuterium using a 50-MeV S^{6+} ion beam. (Data from Ref. 93)

stoichiometry and oxygen loss less than 5% under 2.10^{14} $^{81}Br^{8+}/cm^2$ of 48 MeV.[96] The preparation, characterization, and stability of reference samples for almost all light elements from Li to Si are abundantly discussed in Refs. 23, 69, 87, 97–102.

14.6. CONCLUSION

It is relatively simple to summarize this chapter on ion beam damaging in one sentence, "**Sample degradation under ion beam bombardment is easy to induce, difficult to control and impossible to avoid,**"[103] as concluded at the end of one of the special workshops organized during the Fourth International Conference on Nuclear Microprobe Technology and Applications.[104]

From a more optimistic point of view, the main challenge facing the physicist interested in quantitative analysis consists in evaluating the irradiation sensitivity of the sample and adapting the experimental configuration to limit unavoidable sample damaging during investigation.

Another type of beam-induced degradation must be emphasized, which can be called differed damage. It corresponds to the unexpected and delayed apparition of colored stains on the sample surface due to the relaxation of mechanical and physicochemical constraints. This type of damage is extremely difficult to avoid, and its effects are particularly serious for precious archeological artifacts or gems.[105]

REFERENCES

1. Calcagno, L., Compagnini, G., and Foti, G., Structural modification of polymer films by ion irradiation, *Nucl. Instrum. Methods Phys. Res. Sect. B* **65**, 413 (1992).
2. Rickards, J., and Zironi, E. P., Chlorine loss from polyvinyl chloride under proton bombardment, *Nucl. Instrum. Methods Phys. Res. Sect. B* **56/57**, 687 (1991).
3. Lewis, M. B., and Lee, E. H., Residual gas and ion beam analysis of ion-irradiated polymers, *Nucl. Instrum. Methods Phys. Res. Sect. B* **61**, 457 (1991); see also G-values for gas production from ion-irradiated polystyrene, *J. Nucl. Mater.* **203**, 224 (1993).
4. Venkatesan, T., Edelson, T., and Brown, W. L., Ionization-induced decomposition and diffusion in thin polymer films, *Nucl. Instrum. Methods Phys. Res. Sect. B* **1**, 286 (1984); see also Venkatesan, T., Wolf, T., Allara, D., Wilkens, B. J., and Taylor, G. N., Synthesis of novel organic films by ion beam irradiation of polymer films, *Appl. Phys. Lett.* **43** (10), 934 (1983).
5. Primak, W., Ion bombardment of insulators, *J. Nucl. Mater.* **53**, 238 (1974).
6. Stoneham, A. M., Radiation effects in insulators, *Nucl. Instrum. Methods Phys. Res. Sect. A* **91**, 1 (1994).
7. Davenas, J., and Thévenard, P., Multiaspects of ion beam modification of insulators, *Nucl. Instrum. Methods Phys. Res. Sect. B* **80/81**, 1021 (1993).
8. Devine, R. A. B., Macroscopic and microscopic effects of radiation in amorphous SiO_2, *Nucl. Instrum. Methods Phys. Res. Sect. B* **91**, 378 (1994).
9. Dooley, S. P., Jamieson, D. N., and Prawer, S., He^+ and H^+ microbeam damage, swelling, and annealing in diamond, *Nucl. Instrum. Methods Phys. Res. Sect. B* **77**, 484 (1993).
10. Mazzoldi, P., and Miotello, A., Radiation effects in glasses, *Radiation Effects* **98**, 39 (1986).
11. Dreschhoff, G., and Zeller, E. J., Effect of space charge on F centers near the stopping region of monoenergetic protons, *J. Appl. Phys.* **48**, 4544 (1977).
12. Trocellier, P., Sputtering yields of thin films for MeV microbeam irradiation: Preliminary results, *Radiation Effects* **132**, 305 (1994).
13. Trocellier, P., Behaviour of thin films and bulk oxides under microbeam irradiation, *Nucl. Instrum. Methods Phys. Res. Sect. B* **104**, 571 (1995).
14. Boutard, D., and Berthier, B., Irradiation-induced modifications in metal–silicon interfaces under MeV helium ion irradiation: influence of the beam current density, *Nucl. Instrum. Methods Phys. Res. Sect. B* **106**, 1106 (1995).
15. Auciello, O., and Kelly, R., *Ion Bombardment Modification of Surfaces: Fundamentals and Applications*, Beam Modification of Materials, 1 (Elsevier, Amsterdam, (1984); see also Mazzoldi, P., and Arnold, G. W., *Ion Beam Modification of Insulators*, Beam Modification of Materials, 2 (Elsevier, Amsterdam, 1987).
16. Robinson, M. T., Basic physics of radiation damage production, *J. Nucl. Mater.* **216**, 1 (1994).
17. Hobbs, L. W., Clinard, F. W., Jr., Zinkle, S. J., and Ewing, R. C., Radiation effects in ceramics, *J. Nucl. Mater.* **216**, 291 (1994).
18. Tombrello, T. A., Damage in metals from MeV heavy ions, *Nucl. Instrum. Methods Phys. Res. Sect. B* **95**, 501 (1995).
19. Trocellier, P., Les réactions nucléaires, Compte-rendu des Journées LPS de Caen, 22–24 mars 1995, unpublished data.
20. Tirira, J., Frontier, J. P., Trocellier, P., and Trouslard, P., Development of a simulation algorithm for energy spectra of elastic recoil spectrometry, *Nucl. Instrum. Methods Phys. Res. Sect. B* **54**, 328 (1991).

21. Trocellier, P., Tirira, J., Massiot, Ph., Gosset, J., and Costantini, J. M., Nuclear microprobe study of the composition degradation induced in polyimides by irradiation with high-energy heavy ions, *Nucl. Instrum. Methods Phys. Res. Sect. B* **54**, 118 (1991).

22. Tirira, J., Trocellier, P., Mosbah, M., and Metrich, N., Study of hydrogen content in solids by ERDA and radiation-induced damage, *Nucl. Instrum. Methods Phys. Res. Sect. B* **56/57**, 839 (1991).

23. Mercier, F., Toulhoat, N., Trocellier, P., Durand, C., and Bisiaux, M., Characterization of organic matter in oil-related rocks: Experimental approach and illustration, *Nucl. Instrum. Methods Phys. Res. Sect. B* **77**, 492 (1993).

24. Cahill, T. A., McColm, D. W., and Kusko, B. H., Control of temperature in thin samples during ion beam analysis, *Nucl. Instrum. Methods Phys. Res. Sect. B* **14**, 38 (1986).

25. Gloystein, F., and Richter, F. W., A radiation thermometer for temperature control of thin samples during PIXE analysis, *Nucl. Instrum. Methods Phys. Res. Sect. B* **22**, 45 (1987).

26. Cookson, J. A., Specimen damage by nuclear microbeams and its avoidance, *Nucl. Instrum. Methods Phys. Res. Sect. B* **30**, 324 (1988).

27. Talmon, Y., and Thomas, E., Beam heating of a moderately thick cold-stage specimen in the SEM/STEM, *J. Microsc.* **111** (2), 151 (1977).

28. Cholewa, M., and Legge, G. J. F., Temperature estimation of organic foil for particle beams, *Nucl. Instrum. Methods Phys. Res. Sect. B* **40/41**, 651 (1989); see also Cholewa, M., Bench, G., Kirby, B. J., and Legge, G. J. F., Changes in organic materials with scanning particle microbeams, *Nucl. Instrum. Methods Phys. Res. Sect. B* **54**, 101 (1991).

29. McColm, D. W., and Cahill, T. A., Central temperature of convectively cooled thin targets during ion beam analysis, *Nucl. Instrum. Methods Phys. Res. Sect. B* **51**, 196 (1990).

30. McColm, D. W., and Cahill, T. A., Temperature profile of a microprobe, *Nucl. Instrum. Methods Phys. Res. Sect. B* **54**, 91 (1990).

31. CRC *Handbook of Chemistry and Physics*, 75th ed. (D. R. Lide, ed.) (CRC 1995, Boca Raton), pp. 9-86–9-98.

32. Trouslard, P., Pyrole: un logiciel au service des analyses par Faisceau d'ions, *Rapport CEA-R-5703*, (1995).

33. Marin, N., and Serruys Y., Contribution a l'étude de la diffusion d'impurités métalliques dans un film polymère (Kapton) sous et hors irradiation Rapport CEA-R. 5713, 1995.

34. Biron, I., and Barbu, A., Radiation effects on phase separation and viscosity in a B_2O_3-PbO glass, *Appl. Phys. Lett.* **48**, 1645 (1986).

35. Desgranges, C., Serruys, Y., and Gorse, D., unpublished results, (1993).

36. Doan, N. V., and Tietze, H., Molecular dynamics simulations of displacement cascades in metallic systems, *Nucl. Instrum. Methods Phys. Res. Sect. B* **102**, 58 (1995).

37. Sigmund, P., Theory of sputtering I: Sputtering yield of amorphous and polycrystalline targets, *Phys Rev.* **184**, 383 (1969).

38. Goldman, D. T., and Simon, A., Theory of sputtering by high-speed ions, *Phys. Rev.* **111**, 383 (1958).

39. Pease, R. S., Sputtering of solids by penetrating ions, Rendiconti della Scuola Internazionale di Fisica Enrico Fermi, Corso XIII, 158 (1959).

40. Pleshitsev, N. V., Sputtering of copper by hydrogen ions with energies up to 50 keV, *Soviet Phys. JETP* **37** (10), 878 (1960).

41. Kaminsky, M., Sputtering experiments in the Rutherford collision region, *Phys. Rev.* **126**, 1267 (1962).

42. Sigmund, P., in *Sputtering by Particle Bombardment I*, Topics in Applied Physics, vol. 47 (R. Behrisch, ed.) (Springer, Berlin, 1981), pp. 9–71.

43. Cazaux, J., The role of the Auger mechanism in the radiation damage of insulators, *Microsc. Microanal. Microstr*, in press.

44. Battaglin, G., Della Mea, G., De Marchi, G., Mazzoldi, P., and Puglisi, O., Modification of sodium concentration profiles after proton irradiation of glasses, *Radiation Effects* **64**, 99 (1982).

45. Battaglin, G., Della Mea, G., De Marchi, G., Mazzoldi, P., Miotello, A., and Guglielmi, M., Field-assisted sodium migration in glasses during medium-energy proton irradiation, *J. Phys. C: Solid State Phys.* **15**, 5623 (1982).
46. Mazzoldi, P., Properties of ion-implanted glasses, *Nucl. Instrum. Methods Phys. Res.* **209/210**, 1089 (1983).
47. Arnold, G. W., Peercy, P. S., and Doyle, B. L., Ion-implantation-induced phase separation and crystallization in lithia-silica glasses, *Nucl. Instrum. Methods Phys. Res.* **182/183**, 733 (1981).
48. Arnold, G. W., Alkali depletion and ion beam mixing in glasses, *Nucl. Instrum. Methods Phys. Res. Sect. B* **1**, 516 (1984).
49. Battaglin, G., Della Mea, G., De Marchi, G., Mazzoldi, P., and Miotello, A., Alkali migration in ion-irradiated glasses, *Nucl. Instrum. Methods Phys. Res. Sect. B* **1**, 511 (1984).
50. Battaglin, G., Boscoletto, A., Della Mea, G., De Marchi, G., Mazzoldi, P., Miotello, A., and Tiveron, B., Heavy-ion irradiation in glasses: Enhanced diffusion and preferential sputtering of alkali elements, *Radiation Effects* **98**, 101 (1986).
51. Battaglin, G., Della Mea, G., De Marchi, G., Mazzoldi, P., Miotello, A., Boscolo-Boscoletto, A., and Tiveron, B., Characteristics of glass composition modification during heavy-ion irradiation, *Nucl. Instrum. Methods Phys. Res. Sect. B* **19/20**, 948 (1987).
52. Arnold, G., Battaglin, G., Della Mea, G., De Marchi, G., Mazzoldi, P. and Miotello, A., Enhanced diffusion processes during heavy-ion irradiation of glasses, *Nucl. Instrum. Methods Phys. Res. Sect. B* **32**, 315 (1988).
53. Miotello, A., and Mazzoldi, P., Sputtering process during ion implantation in glasses: Mathematical and physical analysis, *J. Phys. C: Solid State Phys.* **16**, 221 (1983).
54. Calmon, P., and Serruys, Y., in *Modifications Induced by Irradiation in Glasses*, E-MRS Symp. Proc., vol. 29 (P. Mazzoldi, ed.) (Elsevier, Amsterdam, 1992), pp. 33–38.
55. Matzke, Hj., Toscano, E., and Linker, J., Alkali diffusion and radiation effects in the waste glass VG 98/12, *Nucl. Instrum. Methods Phys. Res. Sect. B* **32**, 508 (1988).
56. Régnier, P., Serruys, Y., and Zemskoff, A., Electric-field-stimulated sodium depletion of glass and related penetration of environmental atomic species, *Phys. Chem. Glasses* **27**, 185 (1986).
57. Themner, K., Elemental losses from organic material caused by proton irradiation, *Nucl. Instrum. Methods Phys. Res. Sect. B* **54**, 115 (1991).
58. Tirira, J., Trocellier, P., Frontier, J. P., Massiot, P., Costantini, J. M., and Mori, V., 3D hydrogen profiling by elastic recoil detection analysis in transmission, *Nucl. Instrum. Methods Phys. Res. Sect. B* **50**, 135 (1990); see also Tirira, J., Contribution à l'étude de la collision hélion-4/proton et à la spectrométrie de recul elastique, Rapport CEA R-5529.
59. Brauer, E., Doerr, R., Gruner, R., and Rauch, F., A nuclear physics method for the determination of hydrogen diffusion coefficients, *Corros. Sci.* **21**, 449 (1981).
60. Trocellier, P., Nens, B., and Engelmann, C., Measurements of the hydrogen, sodium, and aluminum concentration versus depth in the near-surface region of glasses by resonant nuclear reactions, *Nucl. Instrum. Methods Phys. Res.* **197**, 15 (1982).
61. Germain-Bonne, D., and Biron, I., private communication.
62. Cachoir, C., Guittet, M. J., Gallien, J. P., and Trocellier P., Uranium-dioxide-leaching mechanism in a synthetic granitic groundwater, communication to the International Conference Migration 95, Saint-Malo (France), 10–15 September 1995.
63. Mosbah, M., Clocchiatti, R., Tirira, J., Gosset, J., Massiot, P., and Trocellier, P., Study of hydrogen in melt inclusions trapped in quartz with a nuclear microprobe, *Nucl. Instrum. Methods Phys. Res. Sect. B* **54**, 298 (1991).
64. Ross, G. G., and Richard, I., Influence of the ion-beam-induced desorption on the quantitative depth profiling of hydrogen in a variety of materials, *Nucl. Instrum. Methods Phys. Res. Sect. B* **64**, 603 (1992).
65. Nagata, S., Yamaguchi, S., Bergsäker, H., and Emmoth, B., Ion-induced release of H and D implanted in Be, C, Si, and SiC, *Nucl. Instrum. Methods Phys. Res. Sect. B* **33**, 739 (1988).

66. Berthier, B., Tessier, C., Auffore, L., Perrot, M., and Couvreur, F., Emploi de la microanalyse nucléaire pour mesurer des profils élémentaires de répartition dans des gaines de zircaloy, unpublished data, (1995).

67. Mosbah, M., personal communication, 1995.

68. Engelmann, Ch., Loeuillet, M., Nens, B., and Trocellier P., Profils de concentration du sodium dans la région superficielle d'un verrre lixivié à 20, 60, et 100 °C, *Silicates Industriels* **2**, 47 (1981).

69. Massiot, P., Sommer, F., Thellier, M., and Ripoll, C., Simultaneous determination of ^{14}N and ^{15}N by elastic backscattering and nuclear reaction: Application to biology, *Nucl. Instrum. Methods Phys. Res. Sect. B* **66**, 250 (1992); see also Massiot, P., Michaud, V., Sommer, F., Grignon, N., Gojon, A., and Ripoll, C., Nuclear microprobe analysis of ^{14}N and ^{15}N in soybean leaves, *Nucl. Instrum. Methods Phys. Res. Sect. B* **82**, 465 (1993).

70. Proc. of the Sixth International Conference on Ion Beam Modification of Materials, Tokyo, Japan, 12–17 June 1988, *Nucl. Instr. Meth. Phys. Res. B* **39** 1 (1989).

71. Proc. of the Fifth International Conference on Radiation Effects in Insulators, Hamilton, Canada, 19–23 June 1989, *Nucl. Instr. Meth. Phys. Res. B* **46** 1 (1990).

72. Proc. of the Seventh International Conference on Ion Beam Modification of Materials, Knoxville, TN, USA, 9–14 Sept. 1990, *Nucl. Instr. Meth. Phys. Res. B* **59/60** 1 (1991).

73. Proc. of the Eighth International Conference on Ion Beam Modification of Materials, Heidelberg, Germany, 7–11 Sept. 1992, *Nucl. Instr. Meth. Phys. Res. B* **80/81** 1 (1993).

74. Proc. of the Ninth International Conference on Ion Beam Modification of Materials, Canberra, Australia, 5–10 Feb. 1995, to be published in a special issue of *Nucl. Instr. Meth. Phys. Res.*

75. Blanchard, B., Carriere, P., Hilleret, N., Marguerite, J. L., and Rocco, J. C., Utilization des canons à ions pour l'étude des isolants à l'analyseur ionique, *Analusis* **4**, 180 (1976).

76. Calmon, P., Contribution de l'analyse RBS à l'étude des effets d'irradiation sur la diffusion dans les verres d'oxydes, Rapport CEA R-5560, (1991).

77. Mercier, F., Caractérisation par différentes techniques de surface des associations organo-minérales dans des milieux modèles de roches-réservoirs de pétrole, thesis, Université Paris-IX-Orsay, (1994).

78. Cazaux, J., Some considerations on the electric field induced in insulators by electron bombardment, *J. Appl. Phys.* **59**, 1418 (1986).

79. Cazaux, J., Electrostatics of insulators charged by incident electron beams, *J. Microsc. Electron.* **11**, 293 (1986).

80. Cazaux, J., and Lehuede, P., Some physical descriptions of the charging effects of insulators under incident particle bombardment, *J. Electron Spectrosc. Relat. Phenom.* **59**, 49 (1992).

81. Cazaux, J., in *Ionization of Solids by Heavy Particles*, vol. 306, (R. A. Baragiola, ed.) (NATO ASI Series B, 1993), (Plenum, New York), pp. 325–50.

82. Hall, T. A., and Gupta, B. L., Beam-induced loss of organic mass under electron microprobe conditions, *J. Microsc.* **100**, 177 (1974).

83. Talmon, Y., Frozen hydrated specimens, *Electron Microscopy* **1**, 25 (1982).

84. Vis, R. D., *Proton Microprobe Applications in the Biomedical Field.* (CRC, Boca Raton, Fl, 1985), pp. 128–33.

85. Tirira, J., Trocellier, P., and Frontier, J. P., Analytical capabilities of ERDA in transmission geometry, *Nucl. Instrum. Methods Phys. Res. Sect. B* **45**, 147 (1990).

86. Mosbah, M., Tirira, J., Clocchiatti, R., Gosset, J., and Massiot, P., Hydrogen determination in geological materials using elastic recoil detection analysis (ERDA), *Nucl. Instrum. Methods Phys. Res. Sect. B* **49**, 340 (1990).

87. Toulhoat, N., Courel, P., Trocellier, P., and Gosset, J., Stability and distribution of lithium and boron in minerals, *Nucl. Instrum. Methods Phys. Res. Sect. B* **77**, 436 (1993).

88. Toulhoat, N., Trocellier, P., Akram, N., and Gosset, J., Determination of carbon and oxygen in fluid inclusions, *Nucl. Instrum. Methods Phys. Res. Sect. B* **77**, 472 (1993).

89. Llabador, Y., Bertault, D., Gouillaud, J. C., and Moretto, Ph., Advantages of high-speed scanning for microprobe analysis of biological samples, *Nucl. Instrum. Methods Phys. Res. Sect. B* **49**, 435 (1990).

90. Hjörvarsson, B., Ryden, J., Ericsson, T., and Karlsson, E., Hydrogenated tantalum: A convenient calibration substance for hydrogen profile analysis using nuclear resonance reactions, *Nucl. Instrum. Methods Phys. Res. Sect. B* **42**, 257 (1989).
91. Green, P. F., and Doyle, B. L., Hydrogen and deuterium profiling in polymers using light and heavy ion elastic recoil detection, in Proc. of the High-Energy and Heavy-Ion Beams in Materials Analysis Workshop, Albuquerque, 14–16 June 1989 (J. R. Tesmer, C. J. Maggiore, M. Nastasi, J. C. Barbour, and J. W. Mayer, eds.) (Materials Research Society, Pittsburgh, 1990), pp. 87–101.
92. Shull, K. R., Dai, K. H., Kramer, E. J., Fetters, L. J., Antonietti, M., and Silescu, H., Diffusion by constraint release in branched macromolecular matrices, *Macromolecules* **24**, 505 (1991).
93. Alurralde, M., Garcia, E., Abriola, D., Filevich, A., Garcia Bermudez, G., Aucouturier, M., and Siejka, J., Identification and depth profiling of light elements elastically recoiled by heavy ion beams, in Proc. of the High Energy and Heavy Ion Beams in Materials Analysis Workshop, Albuquerque, 14–16 June 1989 (J. R. Tesmer, C. J. Maggiore, M. Nastasi, J. C. Barbour, and J. W. Mayer, eds.) (Materials Research Society, 1990), pp. 119–27.
94. Petitpierre, O., Möller, W., and Scherzer, B. M. U., Low-temperature dynamic depth profile measurements of implanted deuterium in nickel : Precipitation of nickel deuteride, *J. Appl. Phys.* **68**, 3169 (1990).
95. Trocellier, P., and Ducroux, R., in Proc. of the 11th ICXOM, London, Ontario, 4–8 Aug. 1986 (J. D. Brown and R. H. Packwood, eds.), pp. 159–62.
96. Whitlow, H. J., Andersson, A. B. C., and Petersson, C. S., Thermally grown SiO_2 standards for elastic recoil detection analysis, *Nucl. Instrum. Methods Phys. Res. Sect. B* **36**, 53 (1989).
97. Mosbah, M., Métrich, N., and Massiot, P., PIGME fluorine determination using a nuclear microprobe with application to glass inclusions, *Nucl. Instrum. Methods Phys. Res. Sect. B* **58**, 227 (1991).
98. Rio, S., Métrich, N., Mosbah, M., and Massiot, P., Lithium, boron, and beryllium in volcanic glasses and minerals studied by nuclear microprobe, *Nucl. Instrum. Methods Phys. Res. Sect. B* **100**, 141 (1995).
99. Courel, P., Trocellier, P., Mosbah, M., Toulhoat, N., Gosset, J., Massiot, P., and Piccot, D., Nuclear reaction microanalysis and electron microanalysis of light elements in minerals and glasses, *Nucl. Instrum. Methods Phys. Res. Sect. B* **54**, 429 (1991).
100. Michaud, V., Toulhoat, N., Trocellier, P., and Courel, P., Elemental and isotopic distributions of boron and lithium in tourmalines using nuclear microprobe, *Nucl. Instrum. Methods Phys. Res. Sect. B* **85**, 881 (1994).
101. Mosbah, M., Bastoul, A., Cuney, M., and Pironon, J., Nuclear microprobe analysis of ^{14}N and its application to the study of ammonium bearing minerals, *Nucl. Instrum. Methods Phys. Res. Sect. B* **77**, 450 (1993).
102. Mercier, F., Toulhoat, N., Trocellier, P., and Durand, C., Characterization of organic matter/minerals associations in oilfield rocks using the nuclear microprobe, *Nucl. Instrum. Methods Phys. Res. Sect. B* **85**, 874 (1994).
103. Revel, G., personal communication.
104. Zhu, J., Proc. of the Fourth International Conference on Nuclear Microprobe Technology and Applications, Shanghai, China, 10–14 Oct., 1994, to be published in a special issue of *Nucl. Instrum. Methods Phys. Res. Sect. B* in 1995.
105. Bouquillon, A., and Queré, G., unpublished data.

BIBLIOGRAPHY

Besenbacher, F., Bøttiger, J., Graversen, O., Hansen, J. L., and Sørensen, H., Erosion of frozen argon by swift helium ions, *Nucl. Instrum. Methods Phys. Res.* **191**, 221 (1981).
Boutard, D., Scherzer, B. M. U., and Möller, W., Ion-induced depletion of hydrogen from a soft carbonized layer, *J. Appl. Phys.* **65**, 3833 (1989).

Calcagno, L., and Foti, G., Hydrogen profile in ion implanted polyethylene, *Appl. Phys. Lett.* **47**, 363 (1985).

Calcagno, L., Musumeci, P., Percolla, R., and Foti, G., Calorimetric measurements of MeV-ion-irradiated polyvinylidene fluoride, *Nucl. Instrum. Methods Phys. Res. Sect. B* **91**, 461 (1994).

Carter, G., and Nobes, M. J., Ion-beam-induced topography and compositional changes in depth profiling, *Surface Interface Analysis* **19**, 39 (1992).

Costantini, J. M., Flament, J. L., Mori, V., Sinopoli, L., Trochon, J., Uzureau, J. L., Zuppiroli, L., Forro, L., Ardonceau, J., and Lesueur, D., Conductivity of irradiated kapton in relation to energy loss of ions and electrons, *Rad. Eff. Def. Sol.* **115**, 83 (1990).

Dran, J. C., Radiation effects in radioactive waste storage materials, *Solid State Phenomena* **30/31**, 367 (1993).

Dran, J. C., Petit, J. C., Prot, T., Lameille, J. M., and Montagne, M., Radiation-enhanced solubility of calcite: Implications for actinide retention, *Nucl. Instrum. Methods Phys. Res. Sect. B* **65**, 330 (1992).

Egerton, R. F., Measurement of radiation damage by electron energy loss spectroscopy, *J. Microsc.* **118**, 389 (1980).

Egerton, R. F., Organic mass loss at 100 K and 300 K, *J. Microsc.* **126**, 95 (1981).

Endisch, D., Rauch, F., Götzelmann, A., Reiter, G., and Stamm, M., Application of the ^{15}N nuclear reaction technique for hydrogen analysis in polymer thin films, *Nucl. Instrum. Methods Phys. Res. Sect. B* **62**, 513 (1992).

Fenyö, D., Hakansson, P., and Sundqvist, B. U. R., On the ejection of hydrogen ions from organic solids impacted by MeV ions, *Nucl. Instrum. Methods Phys. Res. Sect. B* **84**, 31 (1994).

Ingarfield, S. A., McKenzie, C. D., Short, K. T., and Williams, J. S., Semiconductor analysis with a channeled helium microbeam, *Nucl. Instrum. Methods Phys. Res.* **191**, 521 (1981).

Kohno, H., Yamanaka, C., Ikeya, M., Ikeda, S., and Horino, Y., An ESR study of radicals in $CaCO_3$ produced by 1.6-MeV He$^+$ and γ-irradiation, *Nucl. Instrum. Methods Phys. Res. Sect. B* **91**, 366 (1994).

Matzke, Hj., Turos, A., and Linker, G., Polygonization of single crystals of the fluorite-type oxide UO_2 due to high-dose ion implantation, *Nucl. Instrum. Methods Phys. Res. Sect. B* **91**, 294 (1994).

Pivin, J. C., Della Mea, G., Rigato, V., Tonidandel, M., and Carturan, S., Correlation between the rheological properties and damaging of ion-implanted kapton, *Nucl. Instrum. Methods Phys. Res. Sect. B* **91**, 455 (1994).

Sofield, C. J., Sugden, S., Bedell, C. J., Graves, P. R., and Bridwell, L. B., Ion beam modification of polymers, *Nucl. Instrum. Methods Phys. Res. Sect. B* **67**, 432 (1992).

Tombrello, T. A., Ubiquity of C–H bond breaking by MeV ion irradiation, *Nucl. Instrum. Methods Phys. Res. Sect. B* **27**, 517 (1987).

Vieu, C., Ben Assayag, G., and Gierak, J., Observation and simulation of focused ion-beam-induced damage, *Nucl. Instrum. Methods Phys. Res. Sect. B* **93**, 439 (1994).

Wang, T., Wang, W., and Chen, B., Irradiation effect of MeV protons on diamondlike carbon films, *Nucl. Instrum. Methods Phys. Res. Sect. B* **71**, 186 (1992).

15

Hydrogen Determination by Nuclear Resonance

15.1. INTRODUCTION

Hydrogen determination in solids can be carried out by a large number of analytic techniques from nuclear magnetic resonance to infrared absorption spectroscopy, from electron paramagnetic resonance to neutron scattering, from ion-induced photon spectroscopy or secondary ion mass spectrometry to MeV IBA. Most of these techniques are not considered further, because in Chapter 15 we focus on hydrogen depth profiling in the framework of ion beam techniques. Chapter 15 discusses and compares the respective capabilities and limitations of both nuclear resonant reaction analysis and ERDA for hydrogen determination. Thus we compare hydrogen (^1H) depth-profiling approaches using either ERDA or nuclear resonance.

Due to its low atomic number and mass and its irradiation sensitivity, hydrogen is the most difficult chemical element to be determined in solids. Powerful analytic techniques based on electron spectroscopy or X-ray induced emission are inoperant, and Rutherford backscattering spectrometry cannot be used. If we exclude ERDA, physicists have only four other ways of performing hydrogen analysis in the near-surface region of solids: SIMS, sputter-induced photon spectroscopy (SIPS), neutron elastic recoil detection (NERD), and NRA.

Despite the high sensitivity and the high resolution of SIMS, its basic principle—erosion of the solid surface by primary ions—makes a quantitative analysis difficult especially for hydrogen determination,[1] but this disadvantage tends to be slowly decreasing due to present technical progress. Nevertheless the ion-mixing process induced by the primary beam remains a strong disadvantage.

SIPS was first proposed by Tsong.[2] It involves bombarding a solid surface with low-energy heavy ions, such as 10-keV Ar^+, to detect induced photon emission. Combining this detection with progressive erosion of the surface by incident ions, depth profiles can be extracted. In fact SIPS like SIMS presents three major limitations for hydrogen determination: the difficulty of obtaining quantitative data, surface

Table 15.1. Characteristics of Nonresonant Nuclear Reactions
Available to Determine Hydrogen Isotopes[a]

Reaction	Incident Energy (MeV)	Product Energy (MeV)[b]	Cross Section (mb/sr)
D (d, p) T	1	2.3	5.2
D (^{3}He, p) ^{4}He	0.7	13	61
T (d, n) ^{4}He	0.3	2.4	1

[a]From Ref. 4.
[b]For a laboratory emission angle of 150°.

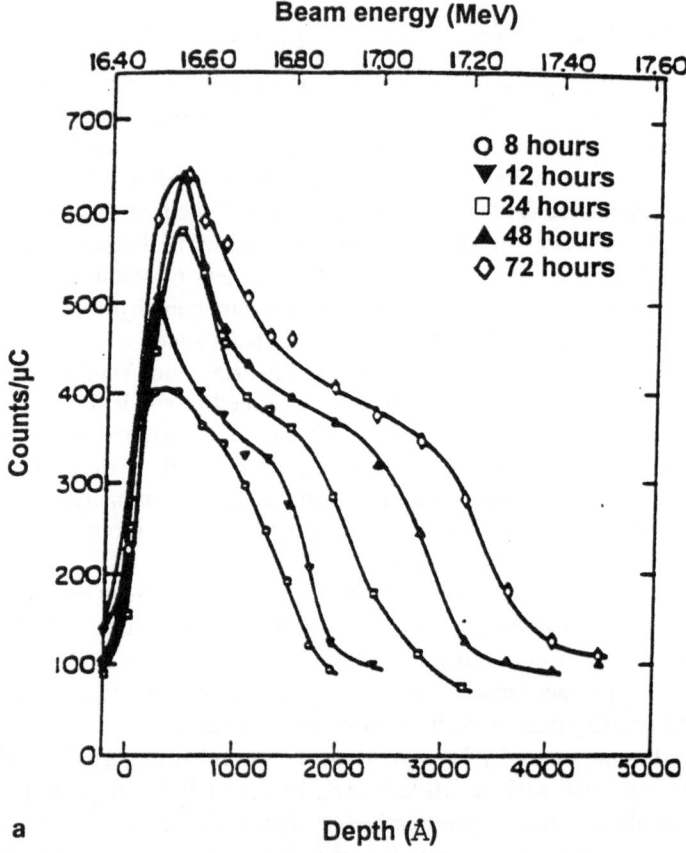

a

Depth (Å)

Figure 15.1. Comparison of hydrogen profiles in various samples obtained by means of SIMS or resonant NRA: (a) ^{19}F hydrogen profiling in obsidians hydrated at 90 °C, (b) SIMS ^{1}H$^+$ and ^{23}Na$^+$ profiling in the obsidian hydrated during 24 hours. (Data from Ref. 25)

roughness induced by the primary beam, and the problem of finding a reference matrix for making relative measurements.

Based on nuclear reactions as (n, p), (n, d) or (n, t) induced by energetic neutrons (2.5–14 MeV), depth profiling hydrogen and its isotopes is possible at a probing depth higher than 10 μm. We do not give more details on NERD, since a serious review of its analytic capabilities is presented by Skorodumov and coworkers.[3]

NRA enables us to determine most of the isotopes of the first 17 elements in the periodic classification.[4] Many reactions are available to measure ^{1}H, [2]D, and [3]T distributions[4]; for example, deuterium and tritium can currently be analyzed by using nonresonant nuclear reactions leading to the emission of energetic charged particles, – [2]D (d, p) [3]T,[5] – [2]D ([3]He, p) [4]He,[6,7] and – [3]T (d,n) α,[8,9] as shown in Table 15.1.

On the other hand, ^{1}H can be determined only by using resonant nuclear reactions, corresponding to inverse reactions of those allowing the analysis of some isotopes of light elements with a proton beam:

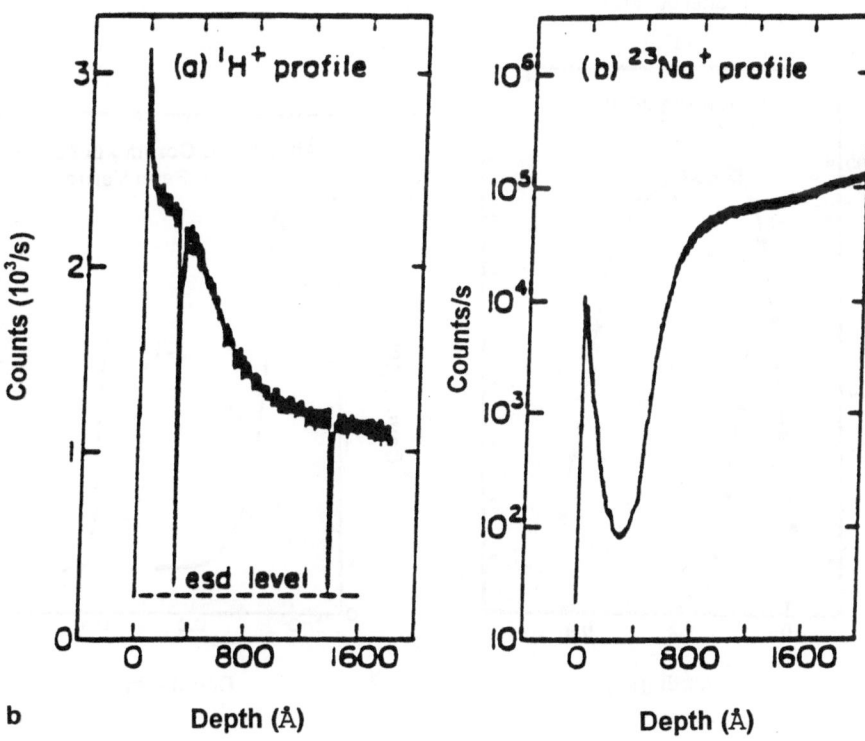

b

Figure 15.1. Continued.

Proton Beam Analysis	Resonant Nuclear Reactions
^{7}Li (p,γ) $^{8}Be^{(10)}$	^{1}H (^{7}Li, γ) $^{8}Be^{(18)}$
^{11}B (p, α) $^{8}Be^{(11)}$	^{1}H (^{11}B, α) $^{8}Be^{(19)}$
^{13}C (p, γ) $^{14}N^{(12)}$	^{1}H (^{13}C, γ) $^{14}N^{(20)}$
^{15}N (p, α γ) $^{12}C^{(13)}$	^{1}H (^{15}N, α γ) $^{12}C^{(21,22)}$
^{18}O (p, α) $^{15}N^{(14,15)}$	^{1}H (^{18}O, α) $^{15}N^{(23)}$
^{19}F (p, α γ) $^{16}O^{(16,17)}$	^{1}H (^{19}F, α γ) $^{16}O^{(24)}$

Some authors have compared the respective analytic capabilities of SIMS, SIPS, and resonant NRA to determine the hydrogen profile induced by hydration in minerals and glasses[25,26] or by implantation in semiconductors[27] (Figures 15.1–15.3).

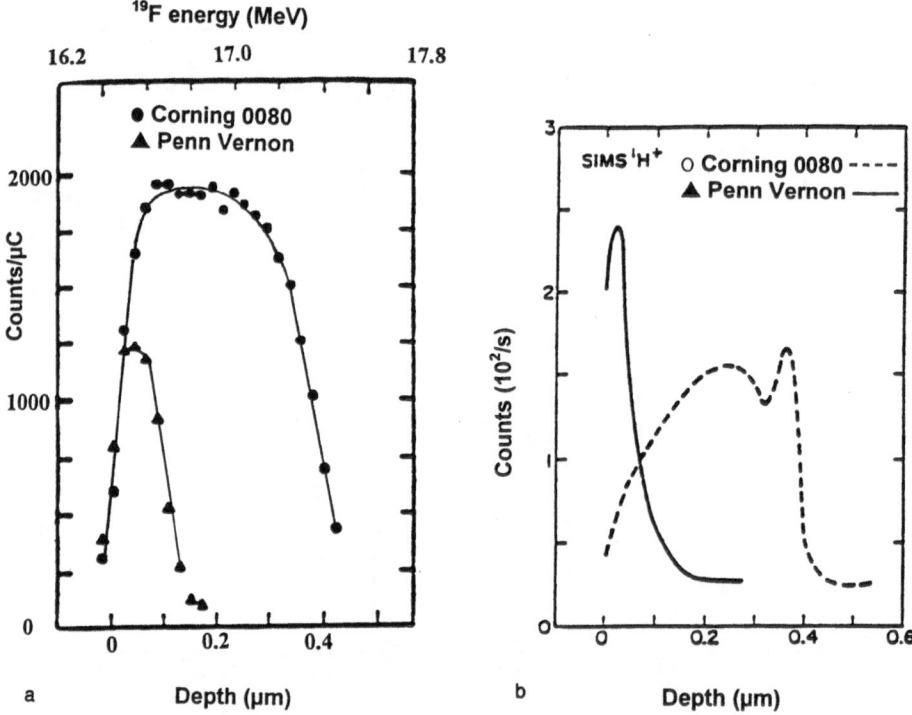

Figure 15.2. (a) The ^{19}F hydrogen profiling in two leached commercial glasses, (b) SIMS hydrogen depth profiles, (c) SIPS depth profiles. (Data from Ref. 26)

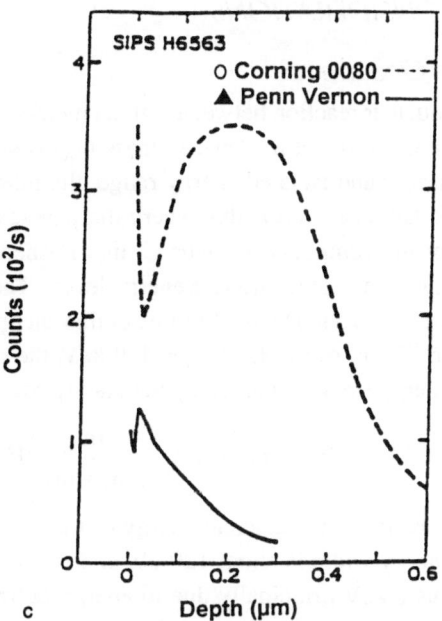

Figure 15.2. Continued

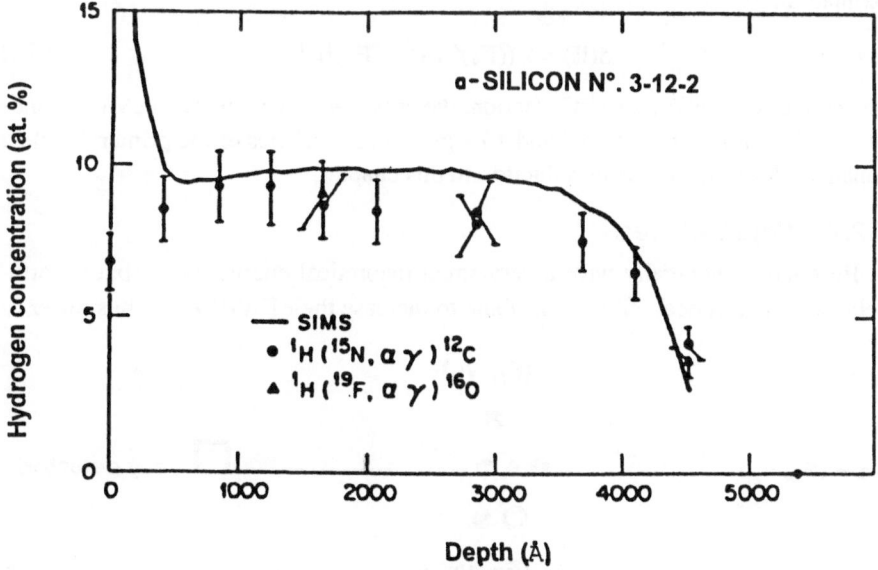

Depth (Å)

Figure 15.3. Hydrogen depth profile in an amorphous-Si film obtained by ^{15}N, ^{19}F resonance, and SIMS analysis. (Data from Ref. 27)

15.2. GENERAL CONSIDERATIONS

15.2.1. Resonance Occurrence

The principle of a nuclear reaction between a given incident ion a (mass m_a) with a given energy E_a and a given target nucleus A (mass m_A) is shown in Figure 15.4. Sometimes in a very narrow and isolated energy range, the interaction cross-section (i.e., the probability of Reaction a + A) exhibits a very sharp increase. Indeed excitation energy E^*_C available for the compound nucleus C (m_C) formed with the interaction has exactly the same value as one of its discrete energy levels. The energy width of the resonance Γ_R is thus partly controlled by the lifetime of this energy level. For example, considering the reaction ^{15}N (p, α, γ) ^{12}C at E_R = 429 keV, the compound nucleus is ^{16}O, and its excitation energy (in MeV) given by the classic equation:

$$E_c^* = 931.48\,(m_a + m_A - m_c) + \left(\frac{m_A}{m_a + m_A}\right)E_a \qquad (15.1)$$

is 12.528 MeV. It corresponds to the discrete energy of the seventeenth energy level of ^{16}O. Its theoretical energy width is about 120 eV, and the measured experimental resonance width is about 1 keV principally due to energy definition in the incident beam. Table 15.2 gives the corresponding excited states attained during nuclear resonances described in the introduction to this chapter.

Resonance cross-section $\sigma(E)$ can be expressed by using the Breit-Wigner approximation:

$$\sigma(E) = k\,((\Gamma_R)^2 + (E - E_R)^2)^{-1} \qquad (15.2)$$

In the case of the ^{15}N (p, α γ) ^{12}C reaction, the cross section at the 429-keV resonance is about 1650 mb.[29] Tables 15.3 and 15.4 give characteristics of the primary nuclear resonances described in the introduction to this chapter.

15.2.2. Physical Effects

Resonances generally have a very small theoretical energy width, but unfortunately external physical effects contribute to increase their FWHM and thus lower

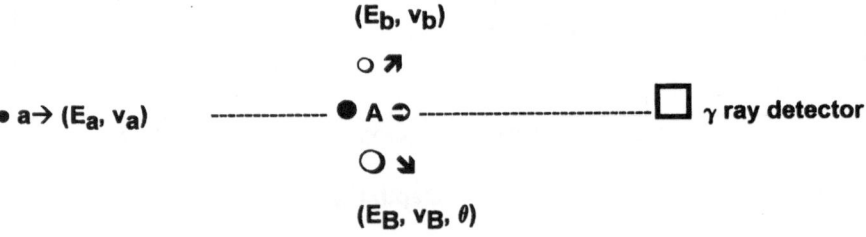

Figure 15.4. Schematic description of the nuclear reaction A(a, b γ)B.

depth resolution: energy straggling, recoil of the emitting nucleus, Doppler broadening, and Lewis effect. We discuss each of these phenomena in detail.

15.2.2.1. Energy Straggling

The energy straggling, in more simple words the difference of energy loss between two identical particles traversing the same target material in the same conditions was already treated in Section 2.5.1. Straggling is basically due to the statistical fluctuations of the number of electron collisions along the particle trajectory. The standard deviation of the energy loss distribution has been first expressed by Bohr[30] (Eq. 2.18) that can be simplified as:

$$s = 0.395 \, z_a \left(Z_A \frac{x}{A_A} \right)^{1/2} \tag{15.3}$$

with x the particle path length (g/cm^2).

$$x_{limit} \leq 3 \cdot 10^{-4} \frac{A_A}{Z_A} \left(\frac{E_a}{m_a} \right)^2 \tag{15.4}$$

Table 15.2. Corresponding Excited States Reached during Nuclear Resonance Occurrence

Reaction (energy)	Compound Nucleus	Energy Level (MeV)	Theoretical Energy Width (keV)
$^7Li + p$ (441 keV)	^8Be	17.642	12.2
$^{11}B + p$ (163 keV)	^{12}C	16.107	5.5
$^{13}C + p$ (1748 keV)	^{14}N	9.172	0.075
$^{15}N + p$ (429 keV)	^{16}O	12.528	0.120
$^{18}O + p$ (629 keV)	^{19}F	8.590	2
$^{19}F + p$ (340 keV)	^{20}Ne	13.168	2.3

[a]From Ref. 28.

Table 15.3. Current Reactions Used to Determine Isotopes of Light Elements by Proton Bombardment

Nuclear Reaction	E_R (keV)	$\sigma(E_R)$ (mb)
^7Li (p, γ) ^8Be	441	5–6
^{11}B (p, α) ^8Be	163	120
^{13}C (p, γ) ^{14}N	1748	340
^{15}N (p, $\alpha\gamma$) ^{12}C	429	1650
	897	1050
^{18}O (p, α) ^{15}N	152	
	629	75
	1165	
^{19}F (p, $\alpha\gamma$) ^{16}O	340	88
	872	440

Table 15.4. Main Resonant Reactions Used to
Determine ^{1}H in Solids[a]

Nuclear Reaction	E_R (keV)[b]	Γ_R (keV)
^{1}H(^{7}Li, γ) ^{8}Be	3.075	80
^{1}H(^{11}B, α) ^{8}Be	1.793	60
^{1}H(^{13}C, γ) ^{14}N	22.553	1
^{1}H(^{15}N, $\alpha\,\gamma$) ^{12}C	6.385	1.8[c]
	3.351	19
^{1}H(^{18}O, α) ^{15}N	11.250	43
^{1}H(^{19}F, $\alpha\,\gamma$) ^{16}O	6.418	44
	16.440	86

[a]From Ref. 29.

[b]Energy values of Column 2 can be obtained from Table 15.3 by
multiplying the values in Column 2 by (m_a / m_A).

[c]The last value published by Ref. 28 gives 0.6 keV; see also Ref. 32
and Figure 15.5.

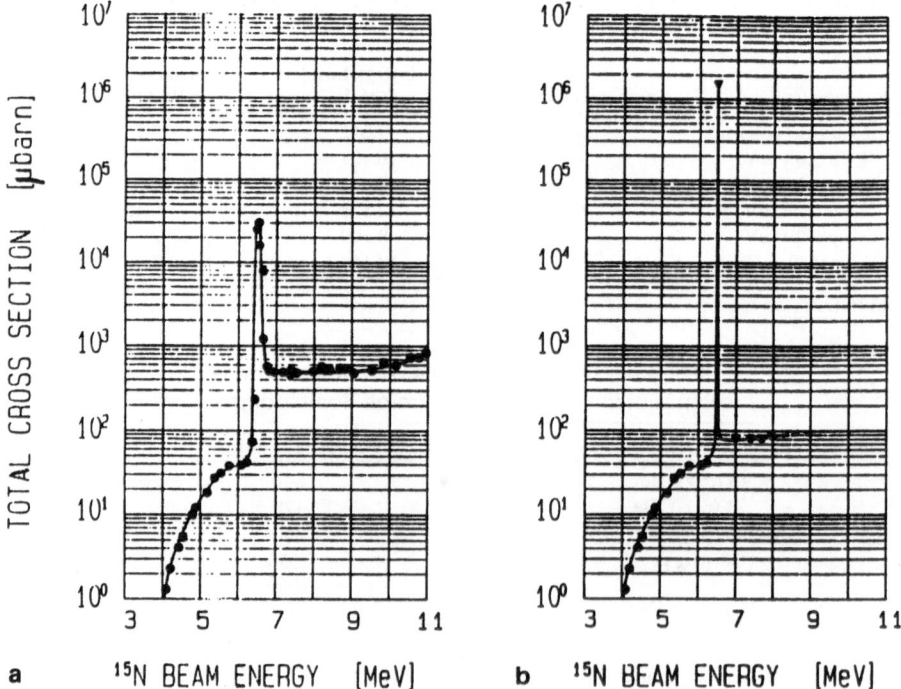

Figure 15.5. Old (a) and new (b) data for the ^{1}H(^{15}N, $\alpha\,\gamma$)^{12}C cross section. (Data from Ref. 32)

Vavilov has calculated the energy distribution under the approximation of small energy loss assuming the stopping power to be constant along the particle trajectory. It is the well-known thin-target approximation with the usual limitation expression:[31]

For energy range higher than this limit, the Bohr model is suitable for calculating the standard deviation of the energy loss distribution.

The main consequence of straggling is to degrade depth resolution. In the case of $^{15}N^{2+}$ ions in Si or SiO^2, the depth resolution for hydrogen profiling becomes poorer and poorer with increasing depth. It decreases by a factor of 5 at a 1-μm depth and by a factor of 11 at a 3-μm depth to the surface, where depth resolution is controlled only by resonance width and beam energy definition.

15.2.2.2. Recoil

A second physical effect has to be considered in the case of an A (a, b γ) B reaction: The recoil of the residual nucleus B, which tends to shift the energy of the emitted gamma-ray to a lower value. The total momentum conservation gives

$$\frac{E_\gamma}{c} = m_B V_B \qquad (15.5)$$

with $V_B = 2E_B/m_B$. This energy shift can be expressed by:

$$\Delta E_\gamma = 5.37 \times 10^{-4} \frac{E_\gamma^2}{A_B} \qquad (15.6)$$

It is for example less than 1 keV for the 4.439-MeV emission induced by the ^1H (^{15}N, $\alpha \gamma$) ^{12}C reaction.

15.2.2.3. Doppler Broadening

The third broadening cause is due to the Doppler effect corresponding to the de-excitation of a recoiling nucleus. Much effort was devoted to discussing and quantifying this phenomenon.[32-35] Doppler effect generates two different results. First the experimental resonance width (Γ_R^{exp}) appears to be larger than the theoretical value (Γ_R^{th}):

$$\Gamma_R^{exp} = [(\Gamma_R^{th})^2 + (\Gamma_{beam})^2 + (\Gamma_D)^2]^{1/2} \qquad (15.7)$$

with Γ_{beam} and Γ_D beam energy definition and Doppler broadening, respectively.

Typically Doppler broadening for the reaction ^1H (^{15}N, $\alpha \gamma$) ^{12}C can be about 5–10 keV. Thus it is the primary reason why experimental width of the 6.385-MeV resonance was given as 6–14 keV instead of 1.8 keV, which corresponds to the theoretical value.[20,21,29,32]

Secondly energy emitted by the gamma-ray may subsequently be broadened. This last effect obeys:

$$E\gamma = E_{th}\left(1 + \frac{V_B}{c}\cos\theta\right) \qquad (15.8)$$

where θ is the angle between the direction of the recoil nucleus B and the gamma-ray emission direction. This energy spreading is less than 1 keV in the case of the ^{1}H (^{15}N, $\alpha\,\gamma$) ^{12}C reaction at 6.4 MeV.

Since both recoil and Doppler effects are present, the observed gamma ray energy is given by:

$$E'_\gamma = \left(E\gamma - \frac{E_\gamma^2}{2M_Bc^2}\right)\left(1 + \frac{V_B}{c}\cos\theta\right) \qquad (15.9)$$

15.2.2.4. Lewis Effect

The discreteness of the energy loss process for a charged particle due to elastic collisions with the electrons of the atomic constituents of the target induces oscillation effects in the thick-target yield curve for a sharply resonant reaction. The energy extension of these oscillations is typically around 1 keV, and criterion for its occurrence is that the combined energy resolution of the experimental system (including the natural and Doppler widths of the resonance together and the energy definition of the incident beam) should be much less than maximum energy loss for the particle. The near-surface resonance yield is expressed by:

$$Y_R = \frac{k\sigma(E_R)\Gamma_{Rexp}N_A}{(dE/dx)} \qquad (15.10)$$

where dE/dx, the energy loss of incident ions near the resonance energy, is not continuous but occurs in discrete jumps with a maximum value of $4m_e/M_a\,E$. This maximum energy loss value is about 0.94 keV for a collision between a ^{15}N$^+$ ion with an electron that is much less than the theoretical resonance width.

A detailed study of the Lewis effect was conducted by Maurel[36] and Vickridge.[37] This effect is strongly influenced by an instrumental factor term, including surface roughness, surface contamination, beam energy definition, etc.

15.3. HYDROGEN PROFILING BY NUCLEAR RESONANCE

15.3.1. Basic Principle

To use a resonant nuclear reaction to profile hydrogen, a beam of well-chosen ions is directed onto the investigated target, then the yield of characteristic gamma-rays is measured. Because the reaction is resonant, the cross section is large at E_R and small off resonance. Hence when the sample is bombarded at E_R, the gamma-ray yield is proportional to the surface hydrogen content. If beam energy is raised above E_R, there are no longer reactions with surface hydrogen due to small cross sections, but as incident ions loose their energy while penetrating the sample, they reach the resonance energy in a deeper layer. The gamma-ray yield is now proportional to the hydrogen content present at this depth. Thus by raising the incident beam energy and measuring induced gamma-ray yield, the hydrogen depth profile is determined (Figure 15.6).

In the case of a reaction leading to the emission of a charged particle, such as ^1H (^{18}O, α) ^{15}N, the basic principle of depth profiling is exactly the same as in the case of gamma-ray emission. Nevertheless interpretation of the yield/incident energy curve in terms of depth profile is a little bit more complicated due to calculations required by the energy loss process occurring with the detected particle.

The quantitative aspect of the resonance method is based on two main relations. The first one connects the incident energy E_a with the reaction depth x and the second,

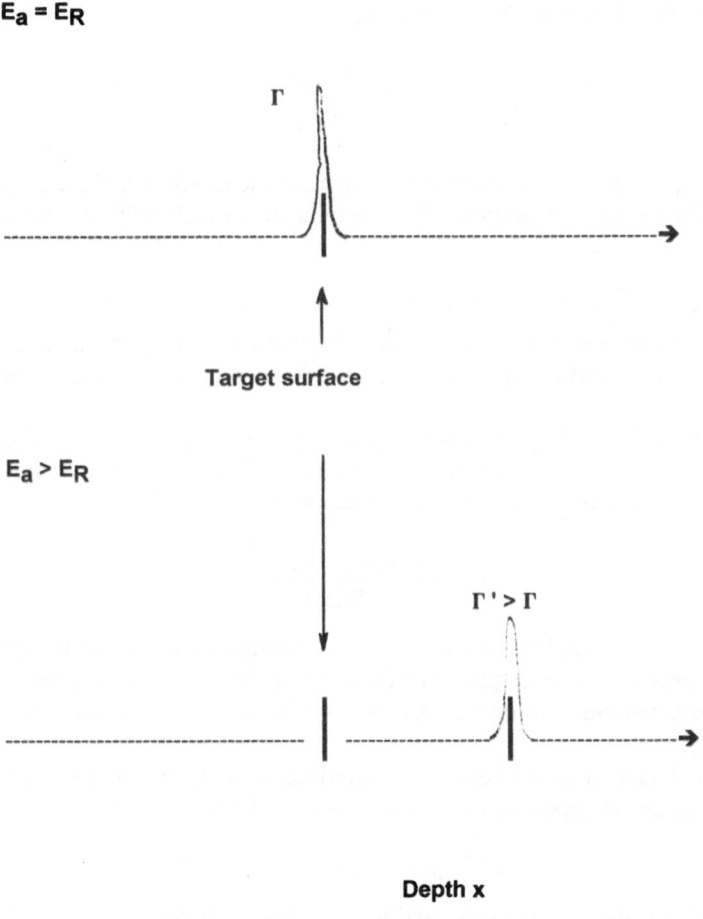

Figure 15.6. Illustration of depth profile measurement using nuclear resonance.

the reaction yield Y_b or Y_γ with the hydrogen content at the reaction depth $C(x)$. In first approximation we can write

$$x = \frac{(E_a - E_R)}{S(E_R)}$$ (15.11)

$$Y_B(E_a) \quad \text{or} \quad Y_\gamma(E_a) = kC(x)$$ (15.12)

where the proportionality factor k depends on the number of incident ions N_a, the efficiency of the detection device ε, the detection solid angle $\Delta\Omega$, and the differential cross-section of the considered reaction $d\sigma/d\Omega(E_a)$. By analyzing a standard sample containing a known amount of hydrogen under the same conditions, this factor can easily be simplified, so Eq. 15.12 becomes

$$C_{sam}(x) = \frac{C_{stan}(x)Y_{sam}S_{sam}(E_a)}{Y_{stan}S_{stan}(E_a)}$$ (15.13)

where subscripts *sam* and *stan* correspond to the sample and the standard, respectively. A more rigorous treatment can be used by applying numerical methods similar to those presented in Chapter 11.

15.3.2. Analytic Characteristics

Three analytic parameters are sufficient for characterizing performances allowed in depth profiling: surface resolution, analyzable depth, and minimum detectable concentration.

Surface resolution (R_{surf}) depends exclusively on resonance width, but in the case of a very narrow resonance (less than 100 eV), the definition of beam energy (ΔE_{beam}) may become the controlling parameter. We can write

$$R_{surf} = \frac{(\Delta E_{beam}^2 + \Gamma_R^2)^{1/2}}{S(E_R)}$$ (15.14)

where Γ_R includes Doppler broadening, recoil contribution, and Lewis effect; and $S(E_R)$ represents the stopping power of the target at the resonance energy value. For example depth resolution using the ^1H (^{15}N, $\alpha\,\gamma$) ^{12}C reaction at 6.4 MeV is as low as 1 nm in SiO_2.

It is evident that depth resolution deteriorates with increasing depth due to straggling effects. At depth x the upper term in Eq. 15.14 becomes:

$$\Delta E = (\Delta E_{beam}^2 + \Delta E_S^2 + \Delta E_D^2 + \Gamma_R^2)^{1/2}$$ (15.15)

with ΔE_S and ΔE_D the respective dispersion terms due to straggling effect and Doppler broadening.

Table 15.5. Performances Allowed for Hydrogen Depth Profiling in Si[a]

Reaction	$S(E_R)^b$ $(MeV \, cm^2 \, mg^{-1})$	Surface Resolution (nm)	Analyzable Depth (μm)	MDL (wt. ppm)
$^1H\,(^7Li, \gamma)\,^8Be$	1.952	160	5	10–100
$^1H\,(^{11}B, \alpha)\,^8Be$	3.583	62	0.4	100–1000
$^1H\,(^{13}C, \gamma)\,^{14}N^c$	4.047	1		
$^1H\,(^{15}N, \alpha \gamma)\,^{12}C$				
6.385 MeV	6.126	2–5	4.8	1–100
13.351 MeV		24	3.3	20
$^1H\,(^{18}O, \alpha \gamma)\,^{15}N$	7.255	18	3	10
$^1H\,(^{19}F, \alpha \gamma)\,^{16}O$				
6.418 MeV	8.156	25	1.3	150
16.440 MeV		46	0.6	10

[a]From Refs. 29, 38, 40.

[b]These stopping powers were calculated in Si according to Refs. 42 and 43 for each resonance energy value.

[c]This reaction has never been used for hydrogen profiling.

The analyzable depth (X_{max}) generally depends on the energy difference between two successive resonances in the excitation function of the considered reaction provided the cross section does not vary rapidly versus energy; in such a case, the combined influence of the off-resonance region and the tails of other resonances can be rather damaging.[38] If the difference energy between two successive resonances is not too high, we can easily write

$$X_{max} = \frac{(E_{R2} - E_{R1})}{S((E_{R1} + E_{R2})/2)} \tag{15.16}$$

For example, the analyzable depth using the $^1H\,(^{15}N, \alpha \gamma)\,^{12}C$ reaction is in the range of 3–4 μm in Si or SiO_2. In the case of a reaction leading to the emission of a charged particle, the outgoing range of this particle may become the controlling parameter

The minimum detectable content (MDC) essentially depends on two opposing factors: cross-section resonance and background yield. Using gamma-ray spectrometry, the background radiation from natural and cosmic sources limits the MDL, especially when the detection system is not shielded.[29,39,40] Typical sensitivity encountered in hydrogen determination using nuclear resonance is in the range of 10–1000 at. ppm.[41]

Table 15.5 gives typical performances allowed in hydrogen profiling using the resonant reactions induced by heavy incident ions.

15.4. COMPARISON WITH ELASTIC RECOIL SPECTROMETRY

15.4.1. Introduction

Comparing analytic techniques is always a very complex exercise; indeed it is extremely difficult for the writers to maintain objectivity because they are generally involved with one of the methods. Moreover truth is never the intrinsic property of one of the techniques and the question can often be cleared by adjusting the capabilities of the technique to the analytic problem to be solved.

Concerning hydrogen depth profiling, several authors have tried to compare the respective advantages and limitations of nuclear resonance and ERDA from a synthetic point of view: Umezawa in 1988,[44] Khabibullaev in 1989,[38] Lanford in 1992,[29] and Arnoldbik in 1993.[45] Many authors have also simultaneously used resonant nuclear reaction analysis and ERDA to determine hydrogen content and depth profile in solids.[46–51]

15.4.2. Technical Comparison

Regarding the basic physical principles of each technique, the characteristics, and the performances allowed by NRA and ERDA in performing hydrogen profiling, several general comments can be made. In terms of minimum detectable content, NRA and ERDA offer almost the same limits. For instance Endisch in a recent paper[52] showed that the $^1H\,(^{15}N, \alpha\,\gamma)\,^{12}C$ reaction may be as sensitive as 10 at. ppm for hydrogen analysis, using a new shielding concept. At the Fourth International Conference on Nuclear Microprobe Technology and Applications Sie and coworkers have presented hydrogen microERDA results in volcanic minerals and glasses at a 10 at. ppm level. Sometimes it is possible to improve the detection limit for recoil spectrometry, as demonstrated by Wielunski and coworkers. They obtained a limit of 1 at. ppm for both H and D isotopes in 25 μm or less thin foils by using 4–6 MeV $^4He^+$ ions.[54]

It is clear that the best surface resolution is obtained by using nuclear resonance. Particularly with the 6.385-MeV ^{15}N resonance, it may reach 1 nm in high silica content media. Nevertheless degradation of depth resolution versus incident ion penetration is much more rapid in the case of $^{15}N^{2+}$ or $^{15}N^{3+}$ than $^4He^+$.

The largest analyzable depth is provided by ERDA in transmission geometry (about 7 μm in glass for 3-MeV $^4He^+$), but this detection configuration is not systematically applicable to any kind of materials. Consequently NRA generally allows greater analyzable depths (several μm) than ERDA in reflexion geometry (≤ 1 μm). Moreover for routine analysis, typical irradiation times are about a few tenths of minutes in ERDA, while NRA hydrogen profiling requires around 90 minutes. However two very important remarks must be made in this case: First the mathematical treatment of an ERDA signal to obtain the hydrogen profile is much more complex than for NRA because it requires successive spectrum simulations with several iterations, for example using the GABY code proposed by Tirira.[55] The second point is that sample damaging by ^{15}N is undoubtedly much more important than by ^4He for a comparable dose. For example initial energy transferred by a 6.4-MeV ^{15}N ion is about 1450 eV/nm in silicon, while for a 3-MeV ^4He, it is only 20 eV/nm.[55]

Another parameter that must be considered is the surface roughness of the sample under investigation. For instance in the case of ERDA in reflection geometry, the tilt of the target surface induces broadening in the beam spot (a factor 4 for a tilt angle of 85°); thus surface irregularities can play a perturbating role. In the case of a resonant nuclear reaction the beam is generally normal to the target surface and roughness effects are minimized. As concerns emitted gamma-rays absorption differences can be neglected and only the inward path is perturbated.

When hydrogen must be determined in solid phases having small dimensions, typically a few square micrometers, a microbeam is strictly required. The ERDA is the only method that can be applied routinely: ^{15}N microbeams are not currently produced and used for microanalytical purposes; moreover sample damaging would be a strongly restrictive parameter. Another case in which ERDA is preferable to NRA corresponds to simultaneously determining hydrogen isotopes ^1H and ^2H.

15.4.3. Arguments for a Practical Choice

Which parameters are going to determine the analyst's choice? We can easily list the following points, divided in two groups: selective criteria or destructive (forbidding one of the options) criteria. The main destructive criteria are

- Characteristics of the available Van de Graaff accelerator (single-ended or tandem type, heavy-ion source, maximum accelerating voltage, etc.)
- Available data processing set-up
- Availability of special detection devices, such as TOF spectrometers, telescopes, *ExB* filters
- Necessity of simultaneously determining hydrogen and deuterium
- Feasibility of preparing the sample to be investigated (bulk sample or thin layer, surface roughness, etc.)

The main selective criteria are

- Sample sensitivity to ion beam damaging (organic-based materials, crystals, ceramics, etc.)
- Hydrogen concentration level in the target (at. % or a few tenths of at. ppm)
- Analytic requirement for a specific performance (optimum depth resolution, small or large probing depth)
- Risk of hydrogen surface contamination of the sample capable of affecting measurement data.

15.5. CONCLUSION

It would not be reasonable to say that NRA or ERDA offers the best analytic compromise for carrying on quantitative hydrogen measurement in the near-surface region of solids. In fact the two methods present their own specificities in terms of analytic performances and practical limitations.

From a practical point of view, it appears that nuclear resonance is preferable for hydrogen millibeam measurement, and ERDA is preferable for hydrogen microbeam determination and simultaneous determination of ^{1}H and ^{2}H isotopes or H and other light element distribution. The final choice largely depends on both the target material and the aim of the analytic investigation.

REFERENCES

1. Brundle, C. R., Evans, C. A., Jr., and Wilson, S., *Encyclopedia of Materials Characterization: Surfaces, Interfaces, and Thin Films* (Butterworth Heinemann, Boston, 1992).
2. Tsong, I. S. T., Photon emission from sputtered particles during ion bombardment, *Phys. Status Solidi A* 7, 451 (1971).
3. Skorodumov, B. G., Ulanov, V. G., Zhukovska, E. V., Zhukovsky, O. A., Coad, J. P., and Wu, C., Simultaneous measurement of hydrogen isotopes in carbon surface layers by the NERD technique and Monte Carlo simulation of recoil energy spectra for hydrogen isotopes concentration depth profiling by NERD method, Proc. of the Twelfth International Conference on Ion Beam Analysis, Tempe, AZ, 22–26 May 1995, to be published in a special issue of *Nucl. Instrum. Methods Phys. Res. Sect. B*.
4. Mayer, J. W., and Rimini, E., *Ion Beam Handbook for Materials Analysis* (Academic, New York 1977); see also Tesmer, J. R., and Nastasi, M., *Handbook of Modern Ion Beam Materials Analysis* (Materials Research Society, Pittsburgh, 1995).
5. Moller, W., Hufschmidt, H., and Kamke, D., Large depth profile measurements of D, ^{3}He, and ^{6}Li by deuteron-induced nuclear reactions, *Nucl. Instrum. Methods Phys. Res.* 140, 157 (1977).
6. Pronko, P. P., and Pronko, J. G., Depth profiling of ^{3}He and ^{2}H in solids using the ^{3}He(d, p)^{4}He resonance, *Phys. Rev. B: Condens. Matter* 9, 2870 (1974).
7. Alstetter, C. J., Behrisch, R., Bottiger, J., Pohl, F., and Scherzer, B. M. U., Depth profiling of deuterium implanted into stainless steel at room temperature, *Nucl. Instrum. Methods Phys. Res.* 149, 59 (1978).
8. Conner, J. P., Bonner, T. W., and Smith, J. R., A study of the ^{3}H(d, n)^{4}He reaction, *Phys. Rev.* 88, 468 (1952).
9. Sawicki, J. A., Depth profiling of tritium in materials for fusion technology, *Fusion Technol.* 14, 884 (1988).
10. Lappalainen, R., Räisänen, J., and Anttila, A., Analysis of boron using the (p, α) reaction, *Nucl. Instrum. Methods Phys. Res. Sect. B* 9, 55 (1985).
11. Ligeon, E., and Bontemps, A., Nuclear reaction analysis of boron and oxygen in silicon, *J. Radioanal. Chem.* 12, 335 (1972).
12. Pierce, T. B., Peck, P. F., and Henry, W. M., Determination of carbon in steels by measurement of the prompt γ-radiation emitted during proton bombardment, *Nature* 204, 571 (1964).
13. Maurel, B., and Amsel, G., A new measurement of the 429-keV ^{15}N(p, α γ)^{12}C resonance. Applications of the very narrow width found to ^{15}N and ^{1}H depth location, *Nucl. Instrum. Methods Phys. Res.* 218, 159 (1983).
14. Amsel, G., and Samuel, D., Microanalysis of the stable isotopes of oxygen by means of nuclear reactions, *Anal. Chem.* 39, 1689 (1967).
15. Lorenz-Wirzba, H., Schmalbrock, P., Trautvetter, H. P., Wiescher, M., Rolfs, C., and Rodney, W. S., The ^{18}O(p, α)^{15}N reaction at stellar energies, *Nucl. Phys. A* 313, 346 (1979).
16. Dieumegard, D., Maurel, B., and Amsel, G., Microanalysis of fluorine by nuclear reactions : I. ^{19}F(p, α_0)^{16}O and ^{19}F(p, α γ)^{16}O reactions, *Nucl. Instrum. Methods Phys. Res.* 168, 93 (1980).
17. Dababneh, S. O. F., Toukan, K., and Khubeis, I., Excitation function of the nuclear reaction ^{19}F(p, α γ)^{16}O in the proton range 0.3–3.0 MeV, *Nucl. Instrum. Methods Phys. Res. Sect. B* 83, 319 (1993).
18. Padawer, G. M., and Schneid, E. J., A lithium microprobe technique for measuring the concentration profile of hydrogen at surfaces, *Trans. Am. Nucl. Soc.* 12, 493 (1969).

19. Ligeon, E., and Guivarc'h, A., A new utilization of ^{11}B ion beams: Hydrogen analysis by ^{1}H(^{11}B, α) α α reaction, *Rad. Eff.* **22**, 101 (1974).

20. Trocellier, P., and Engelmann, Ch., Hydrogen depth profile measurement using resonant nuclear reaction: An overview. *J. Radioanal. Nucl. Chem.* **100**, 117 (1986).

21. Lanford, W. A., Trautvetter, H. P., Ziegler, J. F., and Keller, J., New precision technique for measuring the concentration versus depth of hydrogen in solids, *Appl. Phys. Lett.* **28**, 566 (1976).

22. Thomas, J. P., Pijolat, C., and Fallavier, M., Emploi de la réaction résonnante ^{1}H(^{15}N, α γ) pour l'obtention de profils de concentration d'hydrogène dans les matériaux. *Rev. Phys. Appl.* **13**, 433 (1978).

23. Leich, D. A., and Tombrello, T. A., A technique for measuring hydrogen concentration versus depth in solid samples, *Nucl. Instrum. Methods Phys. Res.* **108**, 67 (1973).

24. Barnes, C. A., Overley, J. C., Zwitkowski, Z. E., and Tombrello, T. A., Measurement of hydrogen depth distribution by resonant nuclear reactions, *Appl. Phys. Lett.* **31**, 239 (1977).

25. Tsong, I. S. T., Smith, G. A., Michels, J. W., Winterberg, A. L., Miller, P. D., and Moak, C. D., Dating of obsidian artifacts by depth profiling of artificially hydrated surface layers, *Nucl. Instrum. Methods Phys. Res.* **191**, 403 (1981).

26. Houser, C. A., Tsong, I. S. T., White, W. B., Winterberg, A. L., Miller, P. D., and Moak, C. D., Ion beam depth profiling studies of leached glasses, *Rad. Eff.* **64**, 103 (1982).

27. Clark, G. J., White, C. W., Allred, D. D., Appleton, B. R., Magee, C. W., and Carlson, D. E., The use of nuclear reactions and SIMS for quantitative depth profiling of hydrogen in amorphous silicon, *Appl. Phys. Lett.* **31**, 582 (1977).

28. Ajzenberg-Selove, F., Energy levels of light nuclei : A = 5–10, A = 11–12, A = 13–15, A = 16–17, A = 18–20, *Nucl. Phys A* **152** (1970), *A* **166** (1971), *A* **190** (1972), *A* **227** (1974) , and *A* **248** (1975).

29. Lanford, W. A., Analysis for hydrogen by nuclear reaction and energy recoil detection, *Nucl. Instrum. Methods Phys. Res. Sect. B* **66**, 65 (1992).

30. Bohr, N., On the decrease of velocity of swiftly moving electrified particles in passing through matter, *Philos. Mag.* **30**, 581 (1915).

31. Vavilov, P. V., Ionization losses of high-energy heavy particles, *Soviet Phys. JETP* **5**, 749 (1957).

32. Lanford, W. A., Guo, X. S., and Rodbell, K., High-energy ion beam analysis and application of the difference method, Proc. of the High-Energy and Heavy-Ion Beams in Materials Analysis Workshop, Albuquerque, NM, 14–16 June 1989 (J. R. Tesmer, C. J. Maggiore, M. Nastasi, J. C. Barbour, and J. W. Mayer, eds.) (Materials Research Society, Pittsburgh, 1990), pp. 203–19.

33. Amsel, G., and Maurel, B., High-resolution techniques for nuclear reaction narrow resonance width measurements and for shallow depth profiling, *Nucl. Instrum. Methods Phys. Res.* **218**, 183 (1983).

34. Horn, K. M., and Lanford, W. A., Observation of the bond-dependent Doppler broadening of the p(^{15}N, α γ)^{12}C nuclear reaction, *Nucl. Instrum. Methods Phys. Res. Sect. B* **29**, 609 (1988).

35. Battistig, G., Amsel, G., d'Artemare, E., and Vickridge, I., A very narrow resonance in ^{18}O(p, α)^{15}N near 150 keV: Application to isotopic tracing, *Nucl. Instrum. Methods Phys. Res. Sect. B* **61**, 369 (1991); *ibid.* **66**, 1 (1991).

36. Maurel, B., Théorie stochastique de la perte d'énergie dans la matière des particules chargées rapides. Application aux courbes d'excitation autour de résonances nucléaires étroites et à la mesure de profils de concentration près de la surface des solides, thesis, Université Paris VII, 1980.

37. Vickridge, I., Théorie stochastique de la perte d'énergie des ions rapides et sa mise en oeuvre pour le profilage à haute résolution en profondeur par résonances nucléaires étroites. Applications à des expériences de traçage par isotopes stables en Sciences des Matériaux, thèse, Université Paris VII, 1990.

38. Khabibullaev, P. K., and Skorodumov, B. G., *Determination of Hydrogen in Materials: Nuclear Physics Methods*, Springer Tracts in Modern Physics, vol. 117 (Springer Verlag, Berlin 1989).

39. Damjantschitsch, H., Weiser, M., Heusser, G., Kalbitzer, S., and Mannsperger, H., An in-beam-line low-level system for nuclear reaction γ-rays, *Nucl. Instrum. Methods Phys. Res.* **218**, 129 (1983).

40. Clark, G. J., White, C. W., Allred, D. D., Appleton, B. R., Koch, F. B., and Magee, C. W., The application of nuclear reactions for quantitative hydrogen analysis in a variety of different materials problems, *Nucl. Instrum. Methods Phys. Res.* **149**, 9 (1978).

41. Torri, P., Keikonen, J., and Nordlund, K., A low-level detection system for hydrogen analysis with the reaction $^1H(^{15}N, \alpha\gamma)^{12}C$, *Nucl. Instrum. Methods Phys. Res. Sect. B* **84**, 105 (1994).

42. Northcliffe, L. C., and Schilling, R. F., Range and stopping power tables for heavy ions, *Nucl. Data Tables A* **7**, 233 (1970).

43. Ziegler, J. F., and Biersack, J. P., in *Treatise on Heavy-Ion Science*, vol. 6 (D. Allan Bromley, ed.) (Plenum, New York, 1985), pp. 95–129.

44. Umezawa, K., Yamane, J., Kurol, T., Shoji, F., Oura, K., and Hanawa, T., Nuclear reaction analysis and elastic recoil detection analysis of the retention of deuterium and hydrogen implanted into Si and GaAs crystals, *Nucl. Instrum. Methods Phys. Res. Sect. B* **33**, 638 (1988).

45. Arnoldbik, W. M., and Habraken, F. H. P. M., Elastic recoil detection, *Rep. Prog. Phys.* **56**, 859 (1993).

46. Ascheron, C., Lehmann, D., Neelmeijer, C., Schindler, A., and Bigl, F., Hydrogen depth profiling in bevelled proton-implanted semiconductors by ^4He ERDA and ^{19}F NRA, *Nucl. Instrum. Methods Phys. Res. Sect. B* **63**, 412 (1992).

47. Bottiger, J., A review on depth profiling of hydrogen and helium isotopes within the near-surface region of solids by use of ion beams, *J. Nucl. Mater.* **78**, 161 (1978).

48. Rauch, F., Kuhn, D., and Wagner, W., Hydrogen profiling using MeV ion beams, *Nucl. Tracks Radiat. Meas.* **19**, 939 (1991).

49. Stensgaard, I., Surface studies with high-energy ion beams, *Rep. Prog. Phys.* **55**, 989 (1992).

50. Kramer, E. J., Depth-profiling methods that provide information complementary to neutron reflectivity, *Physica B* **173**, 189 (1991).

51. Knapp, J. A., Barbour, J. C., and Doyle, B. L., Ion beam analysis for depth profiling, *J. Vac. Sci. Technol. A* **10**, 2685 (1992).

52. Endisch, D., Sturm, H., and Rauch, F., Nuclear reaction analysis of hydrogen at levels below 10 at. ppm, *Nucl. Instr. Meth. Phys Res. Sect. B* **84**, 380 (1994).

53. Sie, S., Chekhmir, A., and Green, T. H., Microbeam recoil detection for hydration of minerals studies, Proc. of the Fourth International Conference on Nuclear Microprobe Technology and Applications, Shanghai, China, 10–14 October 1994, to be published in a special issue of *Nucl. Instrum. Methods Phys. Res. Sect. B*.

54. Wielunski, L. S., Benenson, R. E., Horn, K., and Lanford, W. A., High-sensitivity hydrogen analysis using elastic recoil, *Nucl. Instrum. Methods Phys. Res. Sect. B* **15**, 469 (1986).

55. Tirira, J., Frontier, J. P., Trocellier, P., and Trouslard, P., Development of a simulation algorithm for energy spectra of elastic recoil spectrometry, *Nucl. Instr. Meth. Phys Res. Sect. B* **54**, 328 (1991).

56. Ziegler, J. F., Biersack, J. P., and Littmark, U., *Transport of Ions in Matter*, TRIM 92.11 manual 1992; see also *Stopping and Range of Ions in Matter* (Pergamon, New York, 1985).

BIBLIOGRAPHY

Adler, P. N., Kamykowski, E. A., and Padawer, G. M., in *American Society for Metals*, Washington, DC (1974), pp. 623–30.

Baumann, H., Lenz, Th., and Ruch, F., Gettering of hydrogen in iron and nickel caused by the implantation of titanium ions, *Mater. Sci. Eng.* **69**, 421 (1985).

Bøttiger, J., Picraux, S. T., and Rud, N., in *Ion Beam Surface Layer Analysis*, vol. 2 (O. Meyer, G. Linker, and F. Kappeler, eds.) (Plenum, New York, 1976), pp. 811–19.

Boutard, D., Hydrogène et microfaisceau, *Ann. Chim. Fr.* **19**, 269 (1994).

Brauer, E., Doerr, R., Gruner, R., and Rauch, F., A nuclear physics method for the determination of hydrogen diffusion coefficients, *Corrosion Sci.* **21**, 449 (1981).

Bugeat, J. P., and Ligeon, E., Influence of ion beam bombardment in hydrogen surface layer analysis, *Nucl. Instrum. Methods Phys. Res.* **159**, 117 (1979).

Fallavier, M., Chartoire, M. Y., and Thomas, J. P., UHV equipment for in situ studies of hydrogen interaction with materials based on NRA, *Nucl. Instrum. Methods Phys. Res. Sect.* B **15**, 712 (1986).

Fallavier, M., Thomas, J. P., Frigerio, J. M., and Rivory, J., Hydrogen depth profiling in sputtered Cu_xZr_{1-x} amorphous alloys studied by nuclear reaction analysis: Evolution as a function of annealing temperature, *Sol. State Comm.* **57**, 59 (1986).

Hjörvarsson, B., Ryden, J., Ericsson, T., and Karlsson, E., Hydrogenated tantalum: A convenient calibration substance for hydrogen profile analysis using nuclear resonance reactions, *Nucl. Instrum. Methods Phys. Res. Sect.* B **42**, 257 (1989).

Hoffmann, B., Baumann, H., and Rauch, F., Hydrogen uptake by palladium-implanted titanium studied by NRA and RBS, *Nucl. Instrum. Methods Phys. Res. Sect.* B **15**, 361 (1986).

Hoffmann, B., Baumann, H., Rauch, F., and Bethge, K., Ion implantation of Ti into Fe in the presence of residual and backfilled gases, *Nucl. Instrum. Methods Phys. Res. Sect.* B **28**, 336 (1987).

Horn, K. M., Lanford, W. A., Rodbell, K., and Ficalora, P., Nuclear reaction analysis for hydrogen in nonvacuum environments, *Nucl. Instr. Meth. Phys. Res. Sect.* B **26**, 559 (1987).

Kamykowski, E. A., Kuehne, F. J., Schneid, E. J., and Schulte, R. L., Application of bulk hydrogen standards to the calibration of ion beam surface analysis, *Nucl Instrum. Methods. Phys. Res.* **165**, 573 (1979).

Lanford, W. A., Glass hydration: A method of dating glass objects, *Science* **196**, 975 (1977).

Lanford, W. A., 1 ^{15}N hydrogen profiling: scientific applications, *Nucl. Instrum. Methods Phys. Res.* **149**, 1 (1978).

Lanford, W. A., and Burman, C., Technique for the quantitative measurement of the three-dimensional distribution of hydrogen in solids, *Appl. Phys. Lett.* **41**, 473 (1982).

Laursen, T., and Lanford, W. A., Hydration of obsidian, *Nature* **276**, 153 (1978).

Lee, R. R., Leich, D. A., Tombrello, T. A., Ericson, J. E., and Friedman, I., Obsidian hydration profile measurements using a nuclear reaction technique, *Nature* **250**, 44 (1974).

Ligeon, E., and Guivarc'h, A., Hydrogen implantation in silicon between 1.5 and 60 keV, *Rad. Eff.* **27**, 129 (1976).

Markwitz, A., Bachmann, M., Baumann, H., Bethge, K., and Krimmel, E., Hydrogen profiles of thin PVD silicon nitride films using elastic recoil detection analysis, *Nucl. Instrum. Methods Phys. Res. Sect.* B **68**, 218 (1992).

Martinsson, B. G., and Kristiansson, P., A high-sensitivity method for hydrogen analysis in thin targets, *Nucl. Instrum. Methods Phys. Res. Sect.* B **82**, 589 (1993).

Marwick, A. D., and Young, D. R., Measurements of hydrogen in metal-oxider-semiconductor structures using nuclear reaction profiling, *J. Appl. Phys.* **63**, 2291 (1988).

Marwick, A. D., Liu, J. C., and Rodbell, K. P., Hydrogen redistribution and gettering in AlCu/Ti thin films, *J. Appl. Phys.* **69**, 7921 (1991).

Ogawa, M., Saneyoshi, K., Harada, T., and Imai, K., Hydrogen depth profiling using the $^1H(^{15}N, \alpha \gamma)^{12}C$ reaction in stainless steel pre-irradiated with helium ions, *Fusion Technol.* **14**, 719 (1988).

Olafsson, S., Hjörvarsson, B., and Stilleesjö, F., On the near-surface yield of the $^1H(^{15}N, \alpha \gamma)^{12}C$ nuclear reaction and initial charge state equilibration, *Nucl. Instrum. Methods Phys. Res. Sect.* B **94**, 345 (1994).

Padawer, G. M., Larson, D. J., Jr., and Adler, P. N., A comparison of bulk and surface hydrogen concentration in NBS hydrogen-in-titanium standards, *Met. Trans.* **2**, 2287 (1971).

Peercy, P. S., Stein, H. J., Doyle, B. L., and Picraux, S. T., Hydrogen concentration profiles and chemical bonding in silicon nitride, Report SAND78-0404C, CONF-780684-1, (1978).

Persson, L., Whitlow, H. J., Keinonen, J., Torri, P., Maximov, I., Samuelsson, L., Knox, J., and Malmqvist, K. G., $^1H(^{15}N, \alpha \gamma)^{12}C$ nuclear resonance broadening measurements of hydrogen incorporation during plasma etching of GaAs and $Ga_xIn_{(1-x)}As$ quantum wells, *Nucl. Instrum. Methods Phys. Res.* B **89**, 346 (1994).

Rauch, F., Wagner, W., and Bange, K., Nuclear reaction analysis of hydrogen in optically active coatings on glass, *Nucl. Instrum. Methods Phys. Res. Sect.* B **42**, 264 (1989).

Sellshop, J. P. F., Connell, S. H., Madiba, C. C. P., Sideras-Haddad, E., Stemmet, M. C., Bharut-Ram, K., Appel, H., Kundig, W., Patterson, B., and Holzschuh, E., Hydrogen in and on natural and synthetic diamond, *Nucl. Instrum. Methods Phys. Res. Sect. B* **68**, 133 (1992).

Sofield, C. J., Singleton, J. F., Hotston, E. S., and McCracken, G. M., Measurement of the hydrogen concentration in the walls and limiters of the DITE Tokomak, *J. Nucl. Mater.* **76/77**, 348 (1978).

Stern, A., Khatamian, D., Laursen, T., Weatherly, G. C., and Perz, J. M., Hydrogen and deuterium profiling at the surface of zirconium alloys II: Effects of oxidation, *J. Nucl. Mater.* **148**, 257 (1987).

Switkowski, Z. E., and Dayras, R. A., Target thickness determinations from a study of contaminant hydrogen distributions, *Nucl. Instrum. Methods Phys. Res.* **128**, 9 (1975).

Wesser, M. Kalbitzer, S., Zinke-Allmang, M., Damjantschitsch, H., and Freck, G., Low-temperature implantation and analysis techniques for hydrogen in metals, *Mater. Sci. Eng.* **69**, 411 (1985).

Xiong, F., Rauch, F., Shi, C., Zhou, Z., Liviand, R. P., and Tombrello, T. A., Hydrogen depth profiling in solids: A comparison of several resonant nuclear reaction techniques, *Nucl. Instrum. Methods Phys. Res. Sect. B* **27**, 432 (1987).

Ziegler, J. F., Wu, C. P., Williams, P., White, C. W., Terreault, B., Scherzer, B. M. U., Schulte, R. L., Scheid, E. J., Magee, C. W., Ligeon, E., L'Ecuyer, J., Lanford, W. A., Kuehne, F. J., Kamykowsky, E. A., Hofer, W. O., Guivarc'h, A., Filleux, C. H., Deline, V. R., Evans, C. A., Jr., Behrisch, R., Appleton, B. R., and Allred, D. D., Profiling hydrogen in materials using ion beams, *Nucl. Instrum. Methods Phys. Res.* **149**, 19 (1978).

General Conclusion

Although ERDA is a relatively novel technique for profiling light elements, it is also the oldest if dated from its first description. It is now one of the most promising ion beam techniques, and it is widely used in the energy range from 50 keV–300 MeV. Since the special session of the IBA-12 Conference, a general agreement in favor of the acronym ERDA has replaced a multitude of names and acronyms.

The basic concepts of ion beam interaction with matter are common to various methods in modern elastic recoil spectrometry. Indeed kinematics, scattering cross-section, energy loss, energy straggling, and multiple scattering were described in the past for the purposes of such well-known techniques as RBS, PIXE, or NRA. However many subtle differences justify a special development for ERDA. We tried to give an extensive review of theoretical and practical aspects of elastic recoil analysis with particular attention to the peculiarities proper to this technique. Recent progress in theory and determining physical parameters and experimental equipment have been attentively examined.

Geometrical configurations for practical experiments are presented on the basis of kinematic concepts. The mass ratio is treated as a determining factor in energy transferred during collision and in the relation between scattering and recoil angles. The intrinsic difficulties of ERDA—mass–depth and recoil–projectile ambiguities—require the choice of an adequate experimental configuration. To resolve these ambiguities, several methods have been developed, from the simple absorber foil allowing analysis of a single element when it is lighter than the projectile to such mass-selective devices as telescopes, TOF spectrometers, or ExB filters. These techniques are described in detail and discussed in relation with various possible experimental conditions and analytic aims.

The interpretation of elastic recoil spectrometry relies on calculating ion stopping powers and ranges. A large number of papers on stopping powers has been published for the past 40 years, but the important problem of multielemental targets that do not obey Bragg's rule has never found a general solution. It is well-known that deviations exist between experimental data and this simple rule, but these deviations are quite variable, from less than 5% for metallic alloys to about 40% for a hydrogenated carbon target. To resolve these discrepancies, alternative rules exist to take into account chemical bond effects in stopping powers and to give fair agreement with experimental data. The cores and bonds theory is for the moment the most promising approach for calculating stopping powers in compounds. Nevertheless this theory has been checked for only a limited set of compounds, and progress is still expected in light of new experimental data.

Energy straggling is discussed as a function of the ratio of the mean energy loss to the average energy of the particle along its path. The well-known Landau–Vavilov formulation for thin films and Bohr's model for medium-path lengths can be applied. However special attention must be given to thick targets, since the energy straggling distribution cannot be interpreted using the simple model of Bohr. Although models exist describing energy straggling in thick targets, the problem is not yet completely solved, and measurements may be needed for a particular projectile–target system. Nevertheless we think that theoretical approaches presented herein are quite reasonable and sufficient for typical applications in ERDA experiments.

The well-known Rutherford formula applies only to purely coulombic interactions, which is not the case for the most usual ERDA investigations. Evaluating non-Rutherford scattering cross-sections is intricate, and it can be discussed only as far as data exist in the energy region of interest. There may be several possible analytic expressions that fit accurately enough the available set of data, depending on the energy range under consideration. We tried to privilege the most general formulae, but for each projectile target couple, we must consider the energy range to which a particular expression can be applied. In other words different analyses may accurately represent data over small parts of the entire energy range. In addition, there are more theoretical treatments for high energies than measured values for energies considered in ERDA.

Although it is not possible to give a complete analysis for a large energy interval, we present some key expressions that can be used for typical problems in ERDA. Note that semiempirical expressions result from both experimental data and semitheoretical analyses for each energy interval. Polynomial fits presented herein allow us to evaluate the scattering cross-sections within less than 10% with reference to the whole body of each data base. These expressions cannot be extrapolated outside the energy region in which they were deduced. Furthermore for some collision systems we give expressions of the scattering cross-section for particular angles. Their applicability is thus limited, so new measurements for intermediate angles should be preferred to questionable interpolations.

Among phenomena involved in IBA, MS is certainly one of the most difficult to be properly taken into account. When total energy loss of particles over their path is negligible, several approaches have been discussed. Now when energy loss is not neglected, theoretical treatment relies on less general hypotheses. Although the influence of energy loss on the variation of physical parameters related to collisions is well-known, there is no really universal treatment concerning the MS. Moreover it is difficult to check the validity of models that can be applied to practical problems because experimental data are not abundant.

In our brief outline we give the main expressions needed for calculating the half-width of the angular MS distribution. The lateral half-width can be deduced from the angular one, since they are correlated parameters. However once these distributions are determined, there still remains the problem of combining them correctly to find the final width of the resulting energy distribution. A Gaussian function is often used to represent angular and lateral distributions. Then the final half-width results from the summation in quadrature of both terms. This is much too rough an approximation, particularly when path length is not large. Indeed the MS distribution is never Gaussian, even when the thickness is very large. Nevertheless a Gaussian description is acceptable *in first approximation* when the reduced thickness is very large.

Moreover for ERDA in reflection geometry, we must use the projected MS distribution instead of the spatial MS distribution. In general the different formalisms we report are complex, so it is not easy to carry out a practical evaluation in a simple way. However several tabulated values have already been published that can be used in problems related to ERDA.

Concerning the recoil yield, we try to give a general development of analytic expressions applicable to ERDA. In particular these relationships allow us to explore the sensitivity of the elastic recoil spectrum to stopping powers, scattering cross-section, and energy spreading. Yet the usual theoretical treatment of the recoil yield assumes implicitly both an *ideal* experimental setup and an *ideal* target sample. We complete this discussion by stressing problems arising from the usual imperfections of *real* setups and targets and by presenting some possibilities to overcome them, at least approximatively. In particular we briefly treat the complex problem of lateral nonuniformity in the target structure.

We also discuss fundamental parameters that usually characterize an analytic technique: total energy resolution, depth resolution, and mass resolution. The concept of depth resolution is discussed not only in terms of the classic theoretical development related to evaluating total energy spreading but also from a more practical point of view, taking into account the possibilities of data processing. This latter approach leads us to introduce the concept of *final depth resolution*, describing the ability of the recoil technique to resolve depth intervals when available knowledge about energy spreading is included in data processing. By the way, it highlights the role of statistical counting fluctuations in the practical limits of depth profiling. Concerning mass resolution, it essentially depends on kinematic conditions; however, for detection systems that record individual signals for each mass, it also includes parameters proper to the detection system, or it must even be replaced by the concept of mass selectivity.

Note that some questions related to basic parameters—scattering cross-sections, energy straggling, MS, Bragg's rule, or even depth resolution and mass resolution—represent universal problems in IBA. Although these have already been studied for a long time, they remain hitherto open to discussion and require altogether further theoretical analysis, more complete data bases, and progress in experimental equipment. In addition data processing has made significant advances recently thanks to modern computer hardware, but it is probably far from reaching the limits of its possibilities.

Another general if not universal problem of analysis techniques is the choice of suitable standards. This is particularly important for ERDA, which generally does not have the self-calibration possibilities of some other techniques. Moreover reliable, stable, and universal standards for hydrogen, one of the main elements of interest, are extremely difficult to find. Further research on standards appears to be a priority.

Radiation damage is also a problem of paramount importance in IBA. Materials analyzed by ERDA, particularly hydrogen-containing ones, are often highly sensitive to irradiation. Despite the growing interest of ERDA using microbeams, irradiation effects are most certainly its main limitation. There are many types of ion beam damage and no radical nor universal means of avoiding them. We can only stress the necessity of checking that irradiation effects have a weak enough influence on analysis results.

We have preferred to bypass the deceptive question of the optimal experimental setup. This choice is too dependent on the particular target under consideration and on the inherent goals of each experiment. For the same reason, we cited some elements of comparison between ERDA and nuclear reaction analysis without attempting to establish their respective fields of application.

Our main goal in writing this book was to give any ion beam user a practical and helpful guide concerning ERDA. We thus had to take a position on some questions, in particular the choice of fitting formulae, trying to take carefully into account the state of present knowledge and the whole body of existing data. In other cases where theory or practical methods are still under progress, we felt that we could not draw a clear conclusion. These points remain open to discussion in light of future research. We hope to have presented ERDA as attractive enough for many young scientists to try to advance this technique.

Acknowledgments

First of all we are very indebted to Nick Dytlewski, Hans Hofsäss, and Guy Ross for having kindly accepted to contribute the major part of quite specialized chapters concerning TOF-ERDA, *ExB* filters, and coincidence ERDA. We have particularly appreciated the collaboration of such eminent specialists and the cordiality of our relationship with them.

We wish to acknowledge the following institutions for their kind authorization to reproduce some figures extracted from their published materials:

American Chemical Society	Gordon and Breach Science Publishers
American Institute of Physics	Il Nuovo Cimento
American Physical Society	Institute of Physics Publishing
Elsevier Science BV	Materials Research Society

The authors wish to thank the following authors for their kind permission to use figures extracted from their publications to illustrate various aspects of ERDA spectrometry:

M. Alurralde	N. J. Chou	P. Goppelt-Langer
R. Anne	W. K. Chu	R. G. Gordon
E. Arai	G. J. Clark	P. F. Green
G. W. Arnold	B. L. Cohen	K. O. Groeneveld
W. A. Arnoldbik	G. Compagnini	S. C. Gujrathi
V. Antoni	J. J. Cuomo	F. H. P. M. Habraken
W. Assmann	G. Dollinger	H. C. Hofsäss
J. C. Barbour	B. L. Doyle	C. A. Houser
D. Boutard	N. Dytlewski	M. Hult
C. Cachoir	R. D. Edge	S. S. Klein
J. Cazaux	J. S. Forster	O. Kruse
M. Cholewa	B. Gebauer	W. A. Lanford

J. L'Ecuyer	F. Pászti	R. Siegele
M. B. Lewis	D. Peebles	G. Spahn
N. Marin	R. Pretorius	E. Szylágyi
J. W. Martin	V. M. Prozesky	G. Terwagne
Hj. Matzke	V. Quillet	C. Tschälär
M. K. Mehta	H. Rijken	I. S. T. Tsong
J. A. Moore	G. G. Ross	U. Wahl
M. Mosbah	B. Roux	M. Wang
S. Nagata	J.R. Sabin	Y. Z. Wang
C. Neelmeijer	F. Schiettekatte	J. H. J. Whitlow

Special thanks are due to Bernard Berthier, Isabelle Biron, Dominique Boutard, Christelle Cachoir, Jean Paul Gallien, Dominique Germain-Bonne, Nicolas Marin, Michelle Mosbah, Guy Terwagne, and Philippe Trouslard for permitting us to use unpublished data.

The authors are very indebted to Wim Arnoldbik, Walter Assmann, Bernard Berthier, William Lanford, Peter Sigmund, and Georges Amsel for their efficient collaboration and useful comments.

Thanks are also due to Georges Amsel for providing us with the abstracts booklet of the Twelfth International Ion Beam Analysis Conference and with important papers to be published.

Appendix A

Basic Data References

References for $^1H(^4He, {}^4He)^1H$ differential cross-section data in the range of $E_{He} = 1{-}4$ MeV.

A1. Ingram, D. C., McCormick, A. W., Pronko, P. P., Carlson, J. D., and Woollan, J. A. Hydrogen analysis as a function of depth for hydrogenous films and polymers by proton recoil detection. *Nucl. Instrum. Methods Phys. Res. Sect. B* **6**, 430 (1985). $\phi = 30°$; $E_{He} = 1.5{-}4$ MeV (assumed uncertainty < 10%).

A2. Nagata, S., Yamagushi, S., Fujino, Y., Hori, Y., Sugiyama, N., and Kamada, K. Depth resolution and recoil cross section for analyzing hydrogen in solids using elastic recoil Detection with ^4He beam. *Nucl. Instrum. Methods Phys. Res. Sect. B* **6**, 533 (1985). $\phi = 20°$, $21.5°$, $31.3°$, $41.5°$; $E_{He} = 1.5{-}3$ MeV (assumed uncertainty < 20%).

A3. Pászti, F., Kótai, E., Mezey, G., Manuaba, A., Pócs, L., Hildebrandt, D., and Strusny, H. Hydrogen and deuterium measurements by elastic recoil detection using alpha particles. *Nucl. Instrum. Methods Phys. Res. Sect. B* **15**, 486 (1986). $\phi = 23°$, $E_{He} = 1.5{-}3.3$ MeV (uncertainty not listed).

A4. Wang Hong and Zhou, G. Q. Measurement of the ^4He-p recoil cross sections. *Nucl. Instrum. Methods Phys. Res. Sect. B* **34**, 145 (1988). $\phi = 20°$, $30°$, $E_{He} = 1.3$, 1.4, 1.6, 1.8, 1.9, 2.1 MeV (assumed uncertainty < 12%).

A5. Szilágyi, E., Pászti, F., Manuaba, A., Hajdu, C., and Kótai, E. Cross section measurements of the $^1H(^4He, {}^4He)^1H$ elastic recoil reaction for ERD analysis. *Nucl. Instrum. Methods Phys. Res. Sect. B* **43**, 502 (1989). $\phi = 16°$, $18°$, $21°$, $26°$, 31, $36°$, $41°$, $E_{He} = 1{-}3.5$ MeV (systematic error < 10–20%).

A6. Tirira, J., Trocellier, P., Frontier, J. P., and Trousland, P. Theoretical and experimental study of low-energy ^4He-induced ^1H elastic recoil reaction for ERD analysis. *Nucl. Instrum. Methods Phys. Res. Sect. B* **45**, 203 (1990).

$\phi = 20°, 22°, 28°, 34°, 38°$ $E_{He} = 1.5, 1.65, 1.8$ MeV (assumed uncertainty < 9%).

A7. Baglin, J. E., Kellock, A. J., Crockett, N. A., and Shih, A. H. Absolute cross section for hydrogen forward scattering. *Nucl. Instrum. Methods Phys. Res. Sect. B* **64**, 469 (1992). $\phi = 20°, 25°, 30°, 35°$; $E_{He} = 1-3$ MeV (assumed uncertainty < 5%).

A8. Quillet, V., Abel, F., and Schott, M. Absolute cross section measurements for H and D elastic recoil using 1 to 2.5 MeV ^{4}He ions, and for the ^{12}C(d, p)^{13}C and ^{16}O(d, p$_1$)^{17}O nuclear reactions. *Nucl. Instrum. Methods Phys. Res. Sect. B* **83**, 47 (1993). $\phi = 10°, 20°, 30°$, $E_{He} = 1-2.5$ MeV (assumed uncertainty < 4-7%).

A9. Kido, Y., Miyauchi, S., Takeda, O., Nakayama, Y., Sato, M., and Kusao, K. Precise determination of H recoil recoil cross sections for 1.5–3.0 MeV He ions. *Nucl. Instrum. Methods Phys. Res. Sect. B* **82**, 474 (1993). $\phi = 20°, 30°$, 33°, 40°, $E_{He} = 1.5-3.5$ MeV (assumed uncertainty > 10%).

References for ^{2}H(^{4}He, ^{4}He)^{2}H differential cross-section data in the range of $E_{He} = 1-3.5$ MeV.

A10. Galonsky, A., Douglas, R. A., Haerberli, W., McEllistrem, M. T., and Richards, H. T. Deuteron-helium differential scattering cross sections. *Phys. Rev.* **98**, 586 $\phi = 0°-35°$, $E_{He} = 0.9-2.3$ MeV, and $\phi = 24.4°, 27.7°$, $E_{He} = 1.8-2.5$ MeV (assumed uncertainty 3%).

A11. Nagata, S., Yamagushi, S., Fujino, Y., Hori, Y., Sugiyama, N., and Kamada, K. Depth resolution and recoil cross section for analyzing hydrogen in solids using elastic recoil detection with ^{4}He beam. *Nucl. Instrum. Methods Phys. Res. Sect. B* **6**, 533 (1985). $\phi = 21.5°, 31.3°, 41.5°$; $E_{He} = 1.9-3$ MeV (assumed uncertainty < 20%). $\phi = 10°-41°$, $E_{He} = 1.5-3.5$ MeV (assumed uncertainty > 10%).

A12. Pászti, F., Kótai, E., Mezey, G., Manuaba, A., Pócs, L., Hildebrandt, D., and Strusny, H. Hydrogen and deuterium measurements by elastic recoil detection using alpha particles. *Nucl. Instrum. Methods Phys. Res. Sect. B* **15**, 486 (1986). $\phi = 30°$, $E_{He} = 1.6-3$ MeV (uncertainty not listed).

A13. Besenbacher, F., Stensgaard, I., and Vase, P. Absolute cross sections for recoil detection of deuterium. *Nucl. Instrum. Methods Phys. Res. Sect. B* **15**, 459 (1986). $\phi = 0°-35°$, $E_{He} = 0.9-2.3$ MeV (assumed uncertainty 5%).

A14. Quillet, V., Abel, F., and Schott, M. Absolute cross section measurements for H and D elastic recoil using 1 to 2.5 MeV ^{4}He ions, and for the ^{12}C(d, p)^{13}C and ^{16}O(d, p$_1$)^{17}O nuclear reactions. *Nucl. Instrum. Methods Phys. Res. Sect. B* **83**, 47 (1993). $\phi = 10°, 20°, 30°$, $E_{He} = 1-2.5$ MeV (assumed uncertainty < 6-10%).

References for ^{3}H(^{4}He, ^{4}He)^{3}H differential cross-section data in the range of E_{He} = 1–3 MeV.

A15. Chuang, L. S. The elastic scattering reactions ^{4}He(t,t)^{4}He and ^{4}He(τ, τ)^{4}He near 2 MeV. *Nucl. Phys. A* **174**, 399 (1971). ϕ = 31.6°, 25.3°, 21.5° E_{He} = 2.84 MeV, and ϕ = 30° (assumed uncertainty 7–10%).

A16. Sawicki, J. A. Measurements of the differential cross sections for recoil tritons in ^{4}He-^{3}T scattering at energies between 0.5 and 2.5 MeV. *Nucl. Instrum. Methods Phys. Res. Sect. B* **30**, 123 (1988). ϕ = 10°–40°, E_{He}= 0.8, 1.4, 2 MeV, and ϕ = 30°, E_{He}= 0.5–2.4 MeV (assumed uncertainty < 10%).

References for ^{1}H(^{3}He, ^{3}He)^{1}H differential cross-section data.

A17. Terwagne, G. Cross-section measurements of the ^{1}H(^{3}He,^{1}H)^{3}He elastic reaction between 1.9 and 3 MeV. *Nucl. Instrum. Methods Phys. Res.* To be published.

A18. Benenson, R. E., Wielunski, L. S., and Lanford, W. A. Computer simulation of helium-induced forward recoil proton spectra for hydrogen concentration determinations. *Nucl. Instrum. Methods Phys. Res. Sect. B* **15**, 4 43 (1986).

A19. Tombrello, T. A., Miller Jones, C., Phillips, C. G., and Weil, J. L. The scattering of protons from ^{3}He. *Nucl. Phys.* **39**, 541 (1962).

References for non-Rutherford scattering cross-section data.

A20. Martin, J. A., Nastasi, M., Tesmer, J. R., and Maggiore, C. J. High-energy elastic backscattering of helium ions for compositional analysis of high-temperature superconductor thin films. *Appl. Phys. Lett.* **52**, 2177 (1988).

A21. Goppelt, P., Gebauer, B., Fink, D., Wilpert, M., Wilpert, Th., and Bohne, W. High-energy ERDA with very heavy ions using mass- and energy-dispersive spectrometry. *Nucl. Instrum. Methods Phys. Res. Sect. B* **68**, 235 (1992).

A22. Bozoian, M., Hubbard, K. M., and Nastasi, M. Deviations from Rutherford scattering cross section. *Nucl. Instrum. Methods Phys. Res. Sect. B* **51**, 311 (1990).

A23. Bozoian, M. A useful formula for departures from Rutherford backscattering. *Nucl. Instrum. Methods Phys. Res. Sect. B* **82**, 602 (1993).

A24. Bozoian, M. Threshold of non-Rutherford nuclear cross section for ion beam analysis. *Nucl. Instrum. Methods Phys. Res. Sect. B* **56/57**, 740 (1991).

A25. Räisänen, J., Rauhala, E., Hnox, J., and Harmon, J. F. Non-Rutherford cross section in heavy-ion elastic recoil spectrometry: 40–70 MeV 32S ions on carbon, nitrogen, and oxygen. *J. Appl. Phys.* **75**, 3273 (1994).

A26. Räisänen, J., and Rauhala, E. J. Angular distribution of ^{12}C, ^{14}N, and ^{16}O ion elastic scattering by sulfur near the Coulomb barrier and the high-energy limits of heavy-ion Rutherford scattering. *J. Appl. Phys.* **77**, 1761 (1995).

A27. Spiger, R. J., and Tombrello, T. A. Scattering of ^{4}He by ^{3}He and of ^{3}He by tritium. *Phys. Rev.* **163**, 964 (1967).

A28. Stammbach, Th., and Walter, R. R-matrix formulation and phase shifts for n-^{4}He scattering for energies up to 20 MeV. *Nucl. Phys. A* **180**, 225 (1972).

A29. Satchler, G. R., Owen, W., Elwyn, A. J., Morgan, G. L., and Walte, R. An optical model for the scattering of nucleons from ^{4}He at energies below 20 MeV. *Nucl. Phys. A* **112**, 1 (1968).

References for energy-straggling data.

A30. Tschalär C., and Maccabee, H. D. Energy-straggling measurements of heavy charged particles in thick absorbers. *Phys. Rev. B: Condens. Matter* **1**, 2863 (1970).

A31. Landau, L. On the energy loss of fast particles by ionization. *J. Phys. JETP* **8** (4), 201 (1944).

A32. Vavilov, P. V. Ionization losses of high-energy heavy particles. *S. Phys. JETP* **5** (4), 749 (1957).

A33. Deconninck, G., and Fouille, Y. Energy-spreading calculations and consequences. In *Ion Beam Surface Layer Analysis* (O. Meyer, G. Linker, and F. Käppeler, eds.) (Plenum, New York, 1976), pp. 87.

A34. Bonderup, E., and Hvelplund, P. Stopping power and energy straggling for swift protons. *Phys. Rev. A: Gen. Phys.* **4**, 562 (1971).

A35. Chu, W. K. Calculation of energy straggling for protons and helium ions. *Phys. Rev. A: Gen. Phys.* **13**, 2057 (1976).

A36. Tukahashi, T., Awaya, Y., Tonuma, T., Kumagai, H., Izumo, K., Hashizume, A., Uchiyama, S., and Hitachi, A. Energy straggling of C and He ions in metals foils. *Nucl. Instrum. Methods Phys. Res.* **166**, 587 (1979).

A37. Belery, P., Delbar, T., and Gregoire, G. Multiple scattering and energy straggling of heavy ions in solid targets. *Nucl. Instrum. Methods Phys. Res.* **179**, 1 (1981).

A38. Harris, J. M., Chu, W. K., and Nicolet, M.-A. Energy straggling of ^{4}He below 2 MeV in Pt. *Thin Solid Films* **19**, 259 (1973).

A39. Kido, Y. Energy and Z_2 dependences of energy straggling for fast proton beams passing through solids. *Phys. Rev. B: Condens. Matter* **34**, 73 (1986).

A40. Livingston, M. S., and Bethe, H. A. Nuclear dinamics, Experimental. *Rev. Mod. Phys.* **9** (3), 245 (1937).

A41. Sofield, C. J., Cowern, N. E. B., and Freeman, J. M. The role of charge exchange in energy loss sraggling. *Nucl. Instrum. Methods Phys. Res.* **191**, 462 (1981).

417

A42. Briere, M. A., and Biersack, J. P. Energy loss straggling of MeV ions in thin solid films. *Nucl. Instrum. Methods Phys. Res. Sect. B* **64**, 693 (1992).

A43. Vollmer, O. Der Einfluss von Ladungsfluktuationen auf die Energiever- lustverteilung geladener Teilchen. *Nucl. Instrum. Methods Phys. Res.* **121**, 373 (1974).

A44. Besenbacher, E., Andersen, J. U., and Bonderup, E. Straggling in energy loss of energetic hydrogen and helium ions. *Nucl. Instrum. Methods Phys. Res.* **168**, 1 (1980).

A45. Friedland, E., and Kotze, C. P. Energy loss straggling of protons, deuterons and α-particles in copper. *Nucl. Instrum. Methods Phys. Res. Sect. B* **191**, 490 (1981).

A46. Friedland, E., and Lombaard, J. M. Energy loss straggling of alpha particles in Al, Ni, and Au. *Nucl. Instrum. Methods Phys. Res.* **168**, 25 (1981).

A47. Symon, K. R. Ph.D. diss., Harvard University, 1948. See Symon's theory discussion in M. G. Payne. Energy straggling of heavy particles in thick absorbers. *Phys. Rev* **185**, 611 (1969).

A48. Tschälär, C. Straggling distribution of large energy losses. *Nucl. Instrum. Methods Phys. Res. Sect. B* **61**, 141 (1968).

A49. L'Hoir, A. L., and Schmaus, D. Stopping power and energy straggling for small and large energy losses of MeV protons transmitted through polyester films. *Nucl. Instrum. Methods Phys. Res. Sect. B* **4**, 1 (1984).

A50. Schwab, Th., Geissel, H., Armbruster, P., Gillibert, A., Mitting, W., Olson, R. E., Witerborn, K. B., Wollnik, H., and Münzenberg, G. Energy and angular distribution for Ar ions penetrating solids. *Nucl. Instrum. Methods Phys. Res. Sect. B* **48**, 69 (1990).

A51. Wilken, B., and Fritz, T. A. Energy distribution functions of low-energy ions in silicon absorbers measured for large relative energy losses. *Nucl. Instrum. Meth. Phys. Res.* **138**, 331 (1976).

A52. Antolak, A. J., Handy, B. N., Morse, D. H., and Pontau, A. E. Energy loss and straggling measurements of ions in solid absorbers. *Nucl. Instrum. Methods Phys. Res. Sect. B* **59/60**, 13 (1991).

A53. Cohen, D. D., and Rose, E. K. Large energy loss straggling of protons and He ions in Mylar foils. *Nucl. Instrum. Methods Phys. Res. Sect. B* **64**, 672 (1992).

A54. Andersen, H. H., Besenbacher, F., and Goddiksen, P. Stopping power and straggling of 80–500 keV lithium ions in C, Al, Ni, Cu, Se, Ag, and Te. *Nucl. Instrum. Methods Phys. Res.* **168**, 75 (1980).

A55. Yang, Q., and O'Connor, D. J. Empiral formulae for energy loss straggling of ions in matter. *Nucl. Instrum. Methods Phys. Res. Sect. B* **61**, 49 (1991).

References for stopping power data.

A56. Bohr, N. On the theory of the decrease of velocity of moving electrified particles on passing through matter. *Philos. Mag.* **25**, 10 (1913).

A57. Bohr, N. On the decrease of velocity of swiftly moving electrified particles in passing through matter. *Philos. Mag.* **30**, 581 (1915).

A58. Bethe, H. A. Zur theorie des Durchgangs schneller Korpuskularstrahlen durch Materia. *Ann. Phys.* **5**, 325 (1930).

A59. Bethe, H. A. Bremsformel für elektronen relativistischer Geschwindigkeit. *Z. Phys.* **76**, 293 (1932).

A60. Bethe, H. A., and Heitler, W. On the stopping of fast particles and on the creation of positive electrons. *Proc. R. Soc. London* A **146**, 83 (1934).

A61. Bloch, F. Zur Bremsung rasch bewegter Teilchen beim Durchgang durch Materia. *Ann. Phys.* **16**, 285; Bremsvermögen von Atomen mit mehreren Elektronen. *Z. Phys.* **81**, 363 (1933).

A62. Lindhard, J., and Scharff, M. Energy loss in matter by fast particles of low charge. *Mat. Fys. Medd. Dan. Vid. Selsk.* **27** (15), 1 (1953).

A63. Lindhard, J. On the properties of a gas of charged particles. *Mat. Fys. Medd. Dan. Vid. Selsk.* **28** (8), 1 (1954).

A64. Lindhard, J., and Scharff, M. Energy dissipation by ions in the keV region. *Phys. Rev.* **124**, 128 (1961).

A65. Lindhard, J., Scharff, M., and Shiott, H. E. Range concepts and heavy-ion ranges. (Notes on atomic collisions, II) *Mat. Fys. Medd. Dan. Vid. Selsk.* **33** (14), 1 (1963).

A66. Bohr, N. Penetration of atomic particles through matter. *Mat. Fys. Medd. Dan. Vid. Selsk.* **18** (8) (1948).

A67. Firsov, O. B. A qualitative interpretation of the mean electron excitation energy in atomic collisions. *Sov. Phys. JETP.* **36**, 1076 (1959).

A68. Fano, U. Penetration of protons, alpha particles, and mesons. *Ann. Rev. Nucl. Sci.* **13** (1), 1 (1963).

A69. Jackson, J. D. C. *Classic Electrodynamics* (Wiley, New York, 1975), chap. 13.

A70. Sigmund, P. *Radiation Damage Process in Materials* (Noordhoff, Leyden, 1975), chap. 1.

A71. Andersen, H. H., and Ziegler, J. F. *Hydrogen Stopping Powers and Ranges in All Elements*, V.3 (Pergamon, New York, 1977).

A72. Ziegler, J. F., and Andersen, H. H. *Helium Stopping Powers and Ranges in All Elements*, V.4 (Pergamon, New York, 1977).

A73. Ziegler, J. F., Biersack, J. P., and Littmark, U. *Stopping Powers and Range of Ions in Solids* (Pergamon, New York, 1985).

A74. Littmark, U., and Ziegler, J. F. *Handbook of Range Distribution for Energetic Ions in All Elements* (Pergamon, New York, 1980).

A75. Montenegro, E. C., Cruz, S. A., and Vargas-Aburto, C. A universal equation for the electronic stopping of ions in solids. *Phys. Lett.* A **92**, 195 (1982).

A76. Brice, D. K. *Ion Implantation Range and Energy Deposition Distribution* (Plenum, New York, 1975).

A77. Gibbons, J. F., Johnson, W. S., and Mylroie, S. W. *Projected Range Statistics* (Dowden, Hutchinson, and Ross, Stroundsburg, PA, 1975).

A78. Eisen, F. H., Welch, B., Westmoreland, J., and Mayer, J. W. In *Atomic Collision Processes in Solids* (D. W. Palmer, M. W. Thompson, and P. D. Townsend, eds.) (Pergamon, London, 1970).

A79. Biersack, J. P., and Haggmark, L. G. A Monte Carlo computer program for the transport of energetic ions in amorphous targets. *Nucl. Instrum. Methods Phys. Res.* **174**, 257 (1980).

A80. Molière, G. Theorie der Streuung schneller geladener Teilchen I. *Z. Naturforsch. A* **2**, 133 (1947).

A81. Northcliffe, L. C., and Schilling, R. F. Range and stopping-power tables for heavy ions. *Nucl. Data Tables A* **7**, 233 (1970).

A82. Bauer, P. How to measure absolute stopping cross section by backscattering and transmission methods. Part I. Backscattering. *Nucl. Instrum. Methods Phys. Res. Sect. B* **27**, 301 (1987).

A83. Mertnes, P. How to measure absolute stopping cross section by backscattering and transmission methods. Part II. Transmission. *Nucl. Instrum. Methods Phys. Res. Sect. B* **27**, 315 (1987).

A84. Trouslard, P., Pyrole, a program for ion beam analysis. Rapport CEA-R-5703 (1995).

References for multiple-scattering data.

A85. Scott, W. T. The theory of small-angle multiple scattering of fast charged particles. *Reports Mod. Phys.* **35**, 231 (1963).

A86. Sigmund, P., and Winterbon, K. B. Small-angle multiple scattering of ions in the screened Coulomb region. I. *Nucl. Instrum. Methods Phys. Res.* **119**, 541 (1974).

A87. Marwick, A. D., and Sigmund, P. Small-angle multiple scattering of ions in the screened Coulomb region. II. *Nucl. Instrum. Methods Phys. Res.* **126**, 317 (1975).

A88. Bothe, W. Das allgemeine Fehlergesetz, die Schwankungen der Feldestärke in einem Dielektrikum und die Zerstreuung der α-Strahlen. *Z. Phys.* **5**, 63 (1921).

A89. Wentzel, G. Zur Theorie der Streuung von β-Strahlen. *Ann. Phys.* (Leipzig) **69**, 335 (1922).

A90. Williams, E. J. The straggling of β particles. *Proc. Roy. Soc. A* **125**, 420 (1929).

A91. Williams, E. J. Multiple scattering of fast electrons and alpha particles, and curvature of cloud tracks due to scattering. *Rev. Mod Phys.* **17**, 292 (1940).

A92. Williams, E. J. Application of ordinary space-time concepts in collision problems and relation of classical theory to Born's approximation. *Phys. Rev.* **58**, 217 (1945).

A93. Goudsmit, S., and Saunderson, J. L. Multiple scattering of electrons. *Phys. Rev.* **57**, 24 (1940).

A94. Molière, G. Theorie der Streuung schneller geladener Teilchen. *Z. Naturforsch.* **2a**, 133 (1947) and *Z. Naturforsch.* **3a**, 78 (1948).

A95. Marion, J. B., and Young, F. C. *Nuclear Reaction Analysis* (North-Holland, Amsterdam, 1968).

A96. Meyer, L. Plural and multiple scattering of low-energy heavy particles in solids. *Phys. Status Solidi B* **44**, 253 (1971).

A97. Sigmund, P., Heinemeier, J., Besenbacher, F., Hvelplund, P., and Knudsen, H. Small-angle multiple scattering of ions in the screened Coulomb region. III. Combined angular and lateral spread. *Nucl. Instrum. Methods Phys. Res.* **150**, 221 (1978).

A98. Sigmund, P. *Interaction of Charged Particles with Solids and Surfaces* (A. Gras-Marti, H. Urbassek, and N. Arista, eds.) (Plenum, New York, 1991), pp. 73.

A99. Winterbon, K. B. Finite-angle multiple scattering. *Nucl. Instrum. Methods Phys. Res. Sect. B* **21**, 1 (1987).

A100. Winterbon, K. B. Finite-angle multiple scattering: Revised. *Nucl. Instrum. Methods Phys. Res. Sect. B* **43**, 146 (1989).

A101. Fermi, E. Referred to by W. T. Scott. *Phys. Rev.* **97**, 12 (1949).

A102. Valdés, J. E., and Arista, N. R. Energy loss effects in multiple-scattering angular distribution of ions in matter. *Phys. Rev.* **49**, 2690 (1994).

A103. Amsel, G., Battistig, G., Pászti, F., and Szilágyi, E. Projected small-angle multiple scattering and lateral spread distribution and their combination. II. Analytical approximation. *Nucl. Instrum. Methods Phys. Res.* To be published.

A104. Battistig, G., Amsel, G., d'Artemare, E., and L'Hoir, A. Multiple-scattering-induced resolution limits in grazing incidence resonance depth profiling. *Nucl. Instrum. Methods Phys. Res. Sect. B* **85**, 572 (1994).

A105. Szilágyi, E., Pászti, F., and Amsel, G. Theoretical approximation for depth resolution calculation in IBA methods. *Nucl. Instrum. Methods Phys. Res. Sect. B* **100**, 103 (1995).

A106. Schmaus, D., and L'Hoir, A. Lateral and angular spread up to large energy losses for MeV protons transmitted through polyester. *Nucl. Instrum. Methods Phys. Res. Sect. B* **4**, 317 (1984).

A107. Högberg, G., Nordén, H., and Berry, H. G. Angular distribution of ions scattered in thin carbon foils. *Nucl. Instrum. Methods Phys. Res.* **90**, 283 (1970).

A108. Andersen, H. H., and Bøttiger, J. Multiple scattering of heavy ions of keV energies transmitted through thin carbon films. *Phys. Rev. B: Condens. Matter* **4**, 2105 (1971).

A109. Meyer, L., and Krygel, P. The determination of screening parameters from measurements of multiple scattering of low-energy heavy particles. *Nucl. Instrum. Methods Phys. Res.* **98**, 381 (1972).

A110. Andersen, H. H., Bøttiger, J., Knudsen, H., and Møller Petersen, P. Multiple scattering of heavy ions of keV energies transmitted through thin films. *Phys. Rev. A: Gen. Phys.* **10**, 1568 (1974).

A111. Knudsen, H., and Andersen, H. H. Multiple scattering of MeV gold and carbon ions in carbon and gold targets. *Nucl. Instrum. Methods Phys. Res.* **136**, 199 (1976).

A112. Hooton, B. W., Freeman, J. M., and Kane, P. P. Small-angle multiple scattering of 12–40 MeV heavy ions from thin foils. *Nucl. Instrum. Methods Phys. Res.* **124**, 29 (1975).

A113. Spahn, G., and Groeneveld, K. O. Angular straggling of heavy and light ions in thin solid foils. *Nucl. Instrum. Methods Phys. Res* **123**, 425 (1975).

A114. Schmaus, D., Abel, F., Bruneaux, M., Cohen, C., L'Hoir, A., Della Mea, G., Drigo, A. V., Lo Russo, S., and Bentini, G. G. Multiple scattering of MeV light ions through thin amorphous anodic SiO_2 layers formed on silicon single crystals. *Phys. Rev. B: Condens. Matter* **19**, 558 (1979).

A115. Schmaus, D., and L'Hoir, A. Multiple scattering of MeV light ions transmitted through thin Al_2O_3 films: Detailed analysis of angular distributions. *Nucl. Instrum. Methods Phys. Res.* **194**, 75 (1982).

A116. Schmaus, D., L'Hoir, A., and Cohen, C. Ion multiple scattering: A tool for studying the thickness topography of self-supporting targets. *Nucl. Instrum. Methods Phys. Res.* **194**, 81 (1982).

A117. Schmaus, D., and L'Hoir, A. Multiple-scattering angular distribution of MeV ^4He ions transmitted through Ta_2O_5 targets. *Nucl. Instrum. Methods Phys. Res. Sect. B* **2**, 187 (1984).

A118. Anne, R., Herault, J., Bimbot, R., Gauvin, H., Bastin, G., and Hubert, F. Multiple angular scattering of heavy ions ($^{16,17}O$, ^{36}Kr, ^{40}Ar, ^{86}Kr, and ^{100}Mo) at intermediate energies (20–90 MeV/u). *Nucl. Instrum. Methods Phys. Res. Sect. B* **34**, 295 (1988).

A119. Lindhard, J., Nielsen, V., and Scharff, M. Approximation method in classical scattering by screened Coulomb fields. *Mat. Fys. Medd.* **36** (10) (1968).

A120. Dieumegard, D., Dubreuil, D., and Amsel, G. Analysis and depth profiling of deuterium with the D(3He,p)4He reaction by detecting the protons at backward angles. *Nucl. Instrum. Methods Phys. Res.* **166**, 431 (1979).

A121. Hooton, B. W., Freeman, J. M., and Kane, P. P. Small-angle multiple scattering of 12–40 MeV heavy ions from thin solids. *Nucl. Instrum. Methods Phys. Res.* **124**, 29 (1975).

A122. Bird, J. R., and Williams, J. S. *Ion Beams for Materials Analysis* (Academic, Sydney, 1989), pp. 620.

A123. Amsel, G., Battistig, G., and L'Hoir, A. Projected small-angle multiple scattering and lateral spread distribution and their combination. I. Basic formulae and numerical results. *Nucl. Instrum. Methods Phys. Res.* To be published.

Appendix B

Calculation Of The Detection Solid Angle

Determining precisely the detection solid angle $\Delta\Omega$ for an ERDA experiment is necessary to convert yields into absolute concentrations, yet it is not an easy task. In fact the definition of a solid angle usually refers to a point source P from which a surface S is observed; then:

$$\Delta\Omega = \int_S \frac{\vec{n} \cdot \vec{r}}{|\vec{r}|^3} \, dS \qquad (\text{B.1})$$

where $\vec{r}$ is a position vector by reference to origin P and $\vec{n}$ the unit normal vector of surface S. The problem is thus trivial for an infinitely thin beam, but in practice the beam impact is a strongly eccentric ellipse (except in some transmission configurations) in the plane of the target surface, and the latter is not parallel to the detector window. Contours of the beam impact and the detector window do not really define a solid angle even when the window is circular. Moreover as stated in Section 10.5.1.3, the detector window may be a rectangle or curvilinear quadrilateral.

Thus we must consider more accurately the definition of the so-called detection solid angle before discussing methods for its evaluation. The recoil yield emitted between depths x and x + dx around a recoil angle ϕ is usually written in the simple form (cf. Eq. 4.20):

$$\frac{d\text{N}(x)}{dx} \delta L_1 = N \, Q \delta\sigma[\phi, E_0 (x)] \Delta\Omega \delta L_1 \qquad (\text{B.2})$$

This yield can also be written as an integral of the number of recoils emitted from different points of the interaction volume. Since the analyzable depth and lateral spread due to MS are small in comparison with the beam section, the detector area and the target-to-detector distance, this volume can be assimilated to a planar source located at the target surface. The yield is thus expressed as an integral over the beam impact area S_0:

$$\frac{d\mathrm{N}(x)}{dx}\delta\mathrm{L}_1 = \int\limits_{S_0} \int \mathrm{N_P q_P}\delta\sigma[\phi_\mathrm{P}, \mathrm{E}_0(x)]\Delta\Omega_\mathrm{P}\,d\mathrm{S}\,\delta\mathrm{L}_1 \tag{B.3}$$

where P refers to the current point in the beam impact and q_P is the areal fluence of impinging particles. Implicit assumptions of such equations as Eq. 4.20 are lateral uniformity of the target, homogenity of the beam intensity, and a negligible variation of the recoil angle inside the interaction volume. Accordingly the energy dispersion of recoils reaching the detector is not taken into consideration here. Then N_P, q_P, and ϕ_P are constants, respectively, N, $q = Q/S_0$ and ϕ. Equation B.3 can then be rewritten as:

$$\frac{d\mathrm{N}(x)}{dx}\delta\mathrm{L}_1 = \mathrm{N}\,q\delta\sigma[\phi, \mathrm{E}_0(x)]\int\limits_{S_0}\int \Delta\Omega_\mathrm{P}\,d\mathrm{S}\delta\mathrm{L}_1$$

$$= \mathrm{N}Q\delta\sigma[\phi, \mathrm{E}_0(x)]\int\limits_{S_0}\int \frac{\Delta\Omega_\mathrm{P}}{S_0}\,d\mathrm{S}\,\delta\mathrm{L}_1$$

$$= \mathrm{N}Q\,\delta\sigma[\phi, \mathrm{E}_0(x)]\,\Delta\Omega\delta\mathrm{L}_1 \tag{B.4}$$

which makes $\Delta\Omega$ appear to be an average value of the solid angle $\Delta\Omega_P$ associated to each point P of the beam impact.

The difficulty resides in computing $\Delta\Omega_P$ as a function of coordinates inside S_0 and in integrating Eq. B.4. In the general case an analytic resolution appears quite intricate if not untractable. In principle it is possible to determine $\Delta\Omega$ experimentally by measuring the yield from a pure, perfectly stable standard sample containing a precisely known amount of an element of well-known cross section. Beyond these conditions—purity, stability, stoichiometry, well-defined cross section—such a measurement relies on the accuracy of beam current integration, and it must be repeated as soon as the geometry has to be changed to optimize conditions for a new sample.

Nevertheless, several variants of numerical computation are possible. These consist in averaging over the source area S_0 the solid angle $\Delta\Omega_P$ subtended by the detector area at different points P. There are two options for this step:

- Points P can be chosen as a regular array over the beam impact, with a constant associated differential area. For a cylindrical beam of radius r, $\delta\beta = 2\pi/n$, and $\delta r_i = ((i/n)^{1/2} - ((i-1)/n)^{1/2})r$, $i = 1,2 \ldots n$ define a subdivision in elements of equal area that can be projected onto the target surface.

- Another possibility[1-3] is to select points P at random in the elliptical beam impact area. The condition of homogeneous areal density of probability can be fulfilled by using two random values X and Y uniformly distributed between zero and unity, taking polar coordinates in the circular section of the beam as $\beta = 2\pi X$ and $\rho = r(Y)^{1/2}$ and projecting them onto the impact ellipse. For a rapid calculation, polar coordinates in the elliptical impact area are taken in Refs. 1–3 as $\beta = 2\pi X$ and $\rho = r_\beta (Y)^{1/2}$, where r_β is the radius vector describing the ellipse.

The second step consists in determining the solid angle $\Delta\Omega_p$ subtended by the detector at each point P, or equivalently the probability $\Delta\Omega_p / 4\pi$ for a straight line issued from P to reach the detector. We first present this calculation in the case of a circular detector, then turn to other shapes. A direction from point P can be defined by two polar angles α $(0 \le \alpha \le 2\pi)$ and θ $(0 \le \theta \le \pi)$. The joint probability density for an isotropic emission is

$$p(\alpha, \theta)d\alpha \, d\theta = \frac{d\Omega}{4\pi} = \frac{\sin \theta \, d\theta \, d\alpha}{4\pi}$$

with

$$p(\theta)d\theta = \int_0^{2\pi} d\alpha \, \frac{\sin\theta}{4\pi} \, d\theta = \frac{1}{2} \sin \theta \, d\theta \qquad p(\alpha)d\alpha = \int_0^{\pi} \sin \theta \, d\theta \, \frac{d\alpha}{4\pi} = \frac{d\alpha}{2\pi}$$

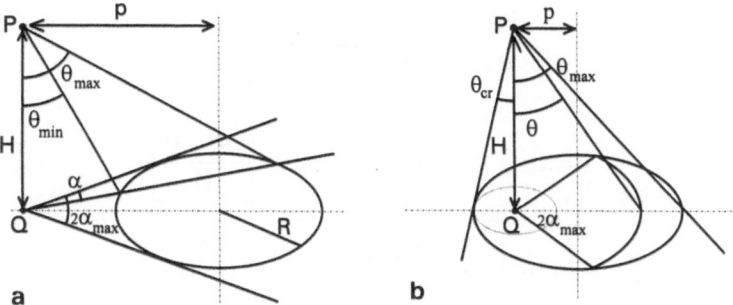

Figure B.1. Definition of geometrical parameters for a circular dectector depending on the position of the current point of source relative to the detector.

Two cases have to be considered.[4] The projection Q of point P onto the detector plane normal to this plane is outside the detector area. The detector can be reached (see Figure B.1a) only if $-\alpha_{max} \le \alpha \le \alpha_{max}$ and if for such a value of α, $\theta_{min} \le \theta \le \theta_{max}$ with:

$$\theta_{min} = \arctan\left\{\frac{[p\cos\alpha - (R^2 - p^2\sin^2\alpha)^{1/2}]}{H}\right\}$$

$$\theta_{max} = \arctan\left\{\frac{[p\cos\alpha + (R^2 - p^2\sin^2\alpha)^{1/2}]}{H}\right\} \qquad (B.5)$$

Here, R is the radius of the detector, p the distance from Q to the center of the detector, and H the distance from P to the detector plane. The probability of reaching the detector is then:

$$\frac{\Delta\Omega_P}{4\pi} = \frac{\displaystyle\int_{-\alpha_{max}}^{\alpha_{max}}\int_{\theta_{min}}^{\theta_{max}} \frac{1}{2}\sin\theta\,d\theta\,(d\alpha/2\pi)}{\displaystyle\int_{\alpha=0}^{2\pi}\int_{\theta=0}^{\pi} \frac{1}{2}\sin\theta\,d\theta\,(d\alpha/2\pi)} = \int_{-\alpha_{max}}^{\alpha_{max}} \frac{1}{2}(\cos\theta_{min} - \cos\theta_{max})\frac{d\alpha}{2\pi} \qquad (B.6)$$

Equation B.6 can be evaluated by taking regularly spaced values of α or uniformly distributed random values according to a Monte-Carlo method and using Eq. B.5. For a rapid calculation,[1-4] only one random value ($-\alpha_{max} \le \alpha \le \alpha_{max}$) can be taken, relying on the large number of points P to provide a complete enough exploration of the whole probability density distribution. This is acceptable if the distribution is almost identical for any point P, i.e., if the target-to-detector distance is much larger than the impact and detector dimensions.

When Q is inside the detector area, the detector is reached (see Figure B.1b) if $0 \le \theta \le \theta_{max}$ and $-\alpha_{max} \le \alpha \le \alpha_{max}$ with:

$$\theta_{max} = \arctan\left[\frac{(R+p)}{H}\right]$$

$$\alpha_{max} = \pi \quad \text{if} \quad \theta \le \theta_{cr} = \arctan\left[\frac{(R-p)}{H}\right]$$

$$\alpha_{max} = \arccos\left[\frac{(p^2 + H^2\tan^2\theta - R^2)}{2Hp\tan\theta}\right] \quad \text{if} \quad \theta > \theta_{cr} \qquad (B.7)$$

Equation B.6 becomes

$$\frac{\Delta\Omega_P}{4\pi} = \frac{\displaystyle\int_0^{\theta_{max}} \int_{-\alpha_{max}}^{\alpha_{max}} \frac{1}{2} \sin\theta \, d\theta \, (d\alpha/2\pi)}{\displaystyle\int_{\alpha=0}^{2\pi} \int_{\theta=0}^{\pi} \frac{1}{2} \sin\theta \, d\theta \, (d\alpha/2\pi)} = \int_0^{\theta_{max}} \frac{\alpha_{max}}{\pi} \frac{1}{2} \sin\theta \, d\theta \tag{B.8}$$

It can be solved in the same way as Eq. B.6, using Eq. B.7.

Since at least a thousand sample points are necessary for a low enough variance of a Monte-Carlo calculation, this option or the choice of an array of points with an uniform areal density are almost equivalent. Yet the advantage of a Monte-Carlo method becomes evident in the case of more complex impact or detector shapes.

Note that the calculation is considerably simplified if the marginal distributions of angles α and θ were independent or in other words if θ_{min} and θ_{max} (respectively, α_{max}) did not depend on α (respectively, θ). This is precisely the case in the approximation of a point source for curved slits, as recommended by Brice and Doyle[5] for minimizing the kinematic spread of recoils (cf. Section 10.5.1.3). For an infinitely small impact area and a curved slit, Eq. B.6 reduces to[5]:

$$\Delta\Omega_P = 2\alpha_{max} (\cos\theta_{min} - \cos\theta_{max}) \tag{B.9}$$

For a rectangular slit[5]:

$$\Delta\Omega_P = 2 \int_{\theta_{min}}^{\theta_{max}} \alpha_m (\theta) \sin\theta \, d\theta \tag{B.10}$$

where $\alpha_m (\theta)$, θ_{min}, and θ_{max} must be determined from slit dimensions.

When the impact area is too large for Eqs. B.9 and B.10 to be a suitable approximation, $\Delta\Omega_P$ is still given for any point P by Eq. B.10, with expressions $\alpha_m (\theta)$, θ_{min}, and θ_{max} depending on the coordinates of point P in the case of a rectangular slit. On the other hand, in the case of curved slits, assimiliting the slit to a section of a cone with vertex P is only approximate when P is not the center of the ellipse, and a similar modification of Eq. B.9 is less reliable. In such cases a Monte-Carlo method may be more appropriate.

REFERENCES

1. Tirira, J., Contribution à l'étude de la collision hélion-4 proton et à la spectrométrie de recul élastique. Rapport CEA R-5529 (1990).
2. Tirira, J., Frontier, J.-P., Trocellier, P., and Trouslard, P., Development of a simulation algorithm for energy spectra of elastic recoil spectrometry, *Nucl. Instrum. Methods Phys. Res. Sect. B* **54**, 328 (1991).
3. Trouslard, Ph., Pyrole: A program for ion beam analysis, Rapport CEA R-5703 (1995).

4. Wielopolski, L., The Monte-Carlo calculation of the average solid angle subtended by a right circular cylinder from distributed sources, *Nucl. Instrum. Methods Phys. Res.* **143**, 577 (1977).
5. Brice, D. K., and Doyle, B. L., A curved detection slit to improve ERD energy and depth resolution, *Nucl. Instrum. Methods. Phys. Res. Sect. B* **45**, 265 (1990).

Appendix C

Specific Units, Physical Constants, and Conversion Factors

Specific units

Absorbed dose: Gray, 1 Gy = 1 J·kg^{-1} = 100 rad
Absorbed dose rate: 1 Gy·s^{-1} = 100 rad·s^{-1} = 3.6 × 10^5 rad·h^{-1}
Activity: Becquerel, 1 Bq = 1 disintegration·s^{-1}, Curie, 1 Ci = 3.7 × 10^{10} Bq
Atomic mass unit: 1 amu = 1.6605402 × 10^{-27} kg = 931.4943 MeV·c^{-2}
Beam current density: nA·mm^{-2} or pA·mm^{-2}
Chemical bond energy: 1 kcal·mol^{-1} = 4185.5 J·mol^{-1} = 2.612 × 10^{22} eV·mol^{-1}
Cross section: barn, 1 b = 10^{-24} cm^2
Differential cross section: mb·sr^{-1} = 10^{-3} b·sr^{-1}
Diffusion coefficient: m^2·s^{-1} or cm^2·s^{-1}
Ion fluence: ion·cm^{-2}
Nuclear radius: Fermi, 1fm = 10^{-15} m
Pressure: 1 Torr = 1 mmHg = 133.3224 Pa
Specific heat: J·mol^{-1}·K^{-1}
Sputtering yield: atom/ion
Stopping power: eV·cm^2·atom^{-1} or eV·cm^2·g^{-1} or eV·Å^{-1} or keV·µm^{-1}
Thermal conductivity: W·m^{-1}·K^{-1}

Physical constants

Avogadro number: N_0 = 6.0221367 × 10^{23} mol^{-1}
Bohr radius: a_0 = 0.529177 × 10^{-10} m
Bohr velocity: v_0 = 2.18769 × 10^6 m s^{-1}
Boltzmann constant: k = 1.3807 × 10^{-23} J·K^{-1} = 0.8617 × 10^{-4} eV·K^{-1}
Fine structure constant: α = 7.29735 × 10^{-3}

Light velocity in vacuum: $c = 2.99792458 \times 10^8$ m s^{-1}
Planck constant: $h = 6.62618 \times 10^{-34}$ J·s; $\hbar = h/2\pi = 1.05459 \times 10^{-34}$ J·s
Stefan-Boltzmann constant: $C = 5.67051 \times 10^{-8}$ J m^{-2} K^{-4} s^{-1}

Specific mass and charge values

Alpha particle mass: $m_\alpha = 6.64476 \times 10^{-27}$ kg $= 4.0015062$ amu; $m_\alpha c^2 = 3.72741$ MeV
Deuteron mass: $m_d = 3.34364 \times 10^{-27}$ kg $= 2.013553$ amu; $m_d c^2 = 1875.3$ MeV
Electron charge: $e = 1.60219 \times 10^{-19}$ C; $e^2 = 1.439965 \times 10^{-13}$ MeV·cm
Electron mass: $m_e = 9.1095 \times 10^{-31}$ kg $= 5.4858 \times 10^{-4}$ amu
Neutron mass: $m_n = 1.67495 \times 10^{-27}$ kg $= 1.008665$ amu; $m_n c^2 = 939.57$ MeV
Proton mass: $m_p = 1.67265 \times 10^{-27}$ kg $= 1.007276$ amu; $m_p c^2 = 938.28$ MeV

Appendix D

Recent References

Bergmaier, A., Dollinger, G., and Frey, C. M. Quantitative elastic recoil detection. *Nucl. Instr. Meth. Phys. Res. Sect. B* **99**, 488 (1995).

Boutard-Gabillet, D. Hydrogène et microfaisceau. *Ann. Chim. Fr.* **19**, 269 (1994).

Burkhart, J. H. and Barbour, J. C. Material analysis using combined elastic recoil detection and Rutherford/enhanced Rutherford backscattering spectrometry. *Nucl. Instr. Meth. Phys. Res. Sect. B* **99**, 484 (1995).

Chambers, G. P. and Hubler, G. K. A particle telescope system for nuclear reaction identification. *Nucl. Instr. Meth. Phys. Res. Sect. B* **99**, 669 (1995).

Kitamura, A., Saitoh, T., and Itoh, H., In situ elastic recoil detection analysis of hydrogen isotopes during deuterium implantation into materials, *Fusion Technol.* **29**, 372 (1996).

Mosbah, M., Clocchiatti, R., Métrich, N., Piccot, D., Rio, S., and Tirira, J., The characterization of glass inclusions through nuclear microprobe, *Nucl. Instrum. Methods Phys. Res. Sect. B* **104**, 271 (1995).

Martin, J. W., Cohen, D. D., Russell, G. J., Dytlewski, N., and Evans, P. J., MEVVA ion implantation of high T_c superconductors, *Nucl. Instrum. Methods Phys. Res. Sect. B* **106**, 624 (1995).

Ruckman, M. W., Strongin, M., Lanford, W. A., and Turner, W. C., ^{15}N hydrogen depth profiling measurements of candidate superconducting supercollider beam pipe materials, *J. Vac. Sci. Technol.* **A13** (4), 1994 (1995).

Stannard, W. B., Johnston, P. N., Walker, S. R., Bubb, I. F., Scott, J. F., Cohen, D. D., Dytlewski, N., and Martin, J. W. Heavy-ion recoil spectrometry of barium strontium titanate films. *Nucl. Instr. Meth. Phys. Res. Sect. B* **99**, 447 (1995).

Wätjen, U., Maier-Komor, P., Pengo, R., Zaika, N.I., Budnar, M., and Valkovic, V. Intercalibration standards for accelerator based analytical techniques. *Nucl. Instr. Meth. Phys. Res. Sect. B* **99**, 376 (1995).

Appendix E

Acronyms

ADC: analogic digital converter
AES: Auger electron spectroscopy
amu: atomic mass unit
ANSTO: Australian nuclear scientific and technical organization
ANTARES: Australian national tandem for research
CERDA: coincident elastic recoil detection analysis
CERN: centre européen de recherche nucléaire
CSIRO: Commonwealth scientific and industrial research organization
DLC: diamondlike carbon
EDXS: energy-dispersive X-ray spectrometry
ERCS: elastic recoil coincidence spectrometry
ERD(A): elastic recoil detection (analysis), *also the name of an analysis program*
FRS: forward recoil spectrometry
FWHM: full-width at half-maximum
GABY: *simulation program for elastic recoil spectrum*
GISA: *program for ion-backscattering analysis from Jaakko Saarilahti (VTT, Espoo, Finland)*
Hi-ERD(A): heavy-ion elastic recoil detection (analysis)
HP: high purity
IBA: ion beam analysis
KFK: Kernforschungzentrum Karlsruhe
LARN: laboratoire d'analyse par réactions nucléaires, Namur, Belgium
LORI: *Ion beam analysis program developed at the University of Aarhus (Denmark)*
LSS: Lindhardt–Scharff–Schiott theory
MCA: multichannel analyser
MCP: multichannel plate
MDC: minimum detection content
MS: multiple scattering

NERD: neutron elastic recoil detection
NRA: nuclear reaction analysis
PAW: physics analysis workstation
PERM: *petit équipement du Rutherford-maniaque or primary equipment for Rutherfordmania*
PIGE: proton-induced gamma-ray emission
PIPS: passivated implanted planar silicon (detectors)
PIXE: proton-induced X-ray emission
PMMA: poly(methyl methacrylate)
PRAL: projected range algorithm
PS: polystyrene
PSD: position-sensitive detector or pulse shape discrimination
PYROLE: toolbox for IBA parameter calculations
RBS: Rutherford-backscattering analysis
RBX: *Ion beam analysis program developed in the Research Institute for Particle and Nuclear Physics, Budapest (Hungary)*
RF: radio frequency
RUMP: *Rutherford universal manipulation program from Cornell University*
SBD: surface barrier detector
SEM: scanning electron microscopy
SENRAS: simulation and evaluation of nuclear analysis spectra
SIMOX: silicon metal oxide
SIMS: secondary ion mass spectrometry
SIPS: sputter-induced photon spectroscopy
SRCS: scattering recoil coincidence spectroscopy
SSBD: silicon surface barrier detector
STEM: scanning transmission electron microscopy
STIMS: sputtering thermoionization mass spectrometry
TAC: time analogic converter
TOF: time of flight
TRIM: transport of ion in matter
XPS: X-ray photoelectron spectroscopy
ZGOUBI: *ion trajectory tracing code from CERN, Geneva*

Index